CW00703786

Springer
Tokyo
Berlin
Heidelberg
New York
Barcelona
Hong Kong
London
Milan
Paris

K. Omasa, H. Saji,
S. Youssefian, N. Kondo (Eds.)

Air Pollution and Plant Biotechnology

Prospects for Phytomonitoring and Phytoremediation

With 103 Figures

Kenji Omasa, Ph.D.
Professor, The University of Tokyo
Yayoi 1-1-1, Bunkyo-ku, Tokyo 113-8657, Japan

Hikaru Saji, Ph.D.
Head of Molecular Ecotoxicology Section
Environmental Biology Division
National Institute for Environmental Studies
Onogawa 16-2, Tsukuba, Ibaraki 305-8506, Japan

Shohab Youssefian, Ph.D.
Associate Professor, Akita Prefectural University
Minami 2-2, Ohgata-mura, Akita 010-0444, Japan

Noriaki Kondo, Ph.D.
Professor, The University of Tokyo
Hongo 7-3-1, Bunkyo-ku, Tokyo 113-0033, Japan

ISBN 4-431-70216-4 Springer-Verlag Tokyo Berlin Heidelberg New York

Printed on acid-free paper

© Springer-Verlag Tokyo 2002
Printed in Japan
This work is subject to copyright. All rights are reserved whether the whole or part of the material is concerned, specifically the rights of translation, reprinting, reuse of illustrations, recitation, broadcasting, reproduction on microfilms or in other ways, and storage in data banks.
The use of registered names, trademarks, etc. in this publication does not imply, even in the absence of a specific statement, that such names are exempt from the relevant protective laws and regulations and therefore free for general use.

Typesetting: Camera-ready by the editors and authors
Printing and binding: Hicom, Japan
SPIN: 10656138

Foreword

Organisms have encountered numerous environmental changes since their first appearance on earth some 4 billion years ago. Such changes generally occurred gradually at slow rates, but sometimes they occurred transiently, violently, and at rapid rates. In most cases, changes in environmental factors were induced by geological variations or by cosmic events, such as bombardment by giant meteorites. Pollutants produced by human activities would be classified under rapid change in the global environment. Organisms themselves also brought about changes in the environment, as typically observed with the increasing concentrations of atmospheric dioxygen produced by oxygenic photosynthetic organisms that originated from cyanobacteria. When faced with such environmental changes, organisms that possessed systems enabling them to tolerate the new conditions or factor could survive, whereas organisms lacking these tolerant systems either disappeared from the new biosphere or escaped into the limited biosphere of the previous environment. Only those organisms that were tolerant to the new environmental factor could subsequently develop the appropriate mechanisms to utilize this factor and to occupy a dominant position in the new biosphere.

A typical environmental factor, to which organisms have shown an evolutionary acquisition of tolerance and which they have been able to effectively utilize, is atmospheric dioxygen. Prior to the appearance of cyanobacteria, the amount of dioxygen in the atmosphere, produced by UV-photolysis of water, was minute, with estimated concentrations of 0.002%, around 10^{-4} that of present atmospheric levels. Even these extremely low dioxygen concentrations appear to have been lethal to anaerobes, the sole organisms 3 billion years ago. Many anaerobes, including anaerobic photosynthetic bacteria, contain superoxide dismutase, an essential enzyme that scavenges superoxide radicals and protects against dioxygen damage. Hence, prior to the accumulation of dioxygen in the atmosphere resulting from the activity of oxygenic photosynthetic organisms, some anaerobes must already have possessed systems that enabled them to tolerate the dioxygen-induced damage that resulted mainly from oxidative breakdown of target molecules by reactive species of oxygen. Clearly, therefore, systems that scavenge such reactive molecules are absolutely indispensable for the survival of organisms in atmospheres with even extremely low dioxygen concentrations.

Cyanobacteria, the first biological donors of dioxygen, are thought to have arisen from the fusion of anaerobic purple bacteria and green sulfur photosynthetic bacteria, through which the two photochemical reaction centers became associated and the additional water oxidation system in the thylakoids was acquired. Fusion with superoxide dismutase-containing photosynthetic bacteria presumably provided the cyanobacteria with protection against the dioxygen that they themselves produced. However, this dioxygen from cyanobacteria was a dangerous pollutant gas, as are present pollutants, for neighboring anaerobes, and caused lethal oxidative damage to cells that had limited potential to scavenge reactive species of oxygen. Some such anaerobes were able to escape to anaerobic environments, which are still present today. On the other hand, those anaerobes that acquired the scavenging systems of reactive species of oxygen were able to survive even in the atmosphere containing dioxygen at higher levels, produced by prokaryotic algae and then by the eukaryotic algae and terrestrial plants. The oxygen-tolerant organisms thus evolved and subsequently acquired the aerobic respiratory systems that enabled them to produce 19-fold higher amounts of bioenergy, in the form of adenosine 5'-triphosphate (ATP), than that generated by fermentation.

The atmospheric oxygen was, at first, a toxic pollutant to anaerobes, but it finally came to be used as an effective electron acceptor for respiration and as a substrate for the biosynthesis of various essential metabolites via oxidase and oxygenase reactions. However, dioxygen is still toxic to all organisms including aerobes, and no organisms on earth survive without effective systems and mechanisms of scavenging the reactive species of oxygen, of rapidly repairing and synthesizing de novo the oxidized target molecules, and of fine-tuning the protecting systems into an orchestrated response to oxidative stress.

Among all organisms, plants are exposed to the severest environments with respect to oxidative damage. Cellular concentrations of dioxygen in leaf tissues (more than 2.5×10^{-4} M) are the highest of all organisms; for comparison, the concentration in the vicinity of mitochondria in mammalian hepatic cells is 10^{-7} to 10^{-8} M. Furthermore, plants are always exposed to strong sunlight, with a maximal 2×10^{-3}mol m^{-2} s^{-1} in the visible range, which induces photosensitized oxidations. To maintain photosynthetic activity, plants are equipped with the most effective mechanisms of photooxidative stress protection. Even so, photosynthesis is appreciably photoinhibited in nature either by defects in the protection system or by overproduction of reactive species of oxygen generated by other environmental stresses. Atmospheric pollutants may also enhance oxidative damage, especially under sunlight, indicating that the reactive species of oxygen participate in such pollutant damage, as also presented in this monograph. The effect of environmental stresses, such as excess light, UV, drought, salt, temperature, and CO_2 on photosynthesis is generally to stimulate the photoproduction of reactive species of oxygen, by which the pollutants synergistically accelerate photodamage.

No one could possibly doubt the importance of maintaining plant photosynthesis levels for protection of the global environment. Annual CO_2 emissions by fossil fuel combustion and cement production (6.4×10^9 tons as carbon) comprise only 6% of the annual global fixation of CO_2 by photosynthetic organisms (1.1×10^{11} tons, in net production as carbon). Thus, even a minor decrease in the rate of the global CO_2 fixation, as affected by pollutants and other environmental stresses, would greatly accelerate the increase in atmospheric CO_2 and result in further amplification of the numerous stress factors that affect photosynthesis. Conversely, even a minor increase in the rate of CO_2 fixation could negate the increase in atmospheric CO_2 resulting from fossil fuel combustion.

The molecular breeding of pollutant-tolerant plants, as described in this monograph, is indispensable for the sustainable protection of the global environment. To achieve this objective, the mechanisms of tolerance, not only to pollutants but also to environmental stresses need to be considered in the light of breeding strategies. An in-depth understanding of the cross talk between environmental and pollutant-induced stresses would most certainly allow the breeding of plants suited to specific environments. Furthermore, the incorporation of the capacity of organisms living in extreme environmental conditions into plants would also be an effective strategy of developing novel tolerant plants. In these respects I believe that the present monograph, *Air Pollution and Plant Biotechnology*, will serve as a milestone in mankind's future progress in identifying ways of maintaining and improving our global environments.

Kozi Asada
Department of Biotechnology
Faculty of Life Engineering
Fukuyama University
Fukuyama, Hiroshima, Japan

Preface

Air pollution is ubiquitous in industrialized and densely populated districts. It is not only toxic per se to organisms, but is also responsible for a wide range of environmental crises, including acidic deposition, stratospheric ozone-layer destruction, and global warming. It is therefore critical that we both monitor and reduce the levels of atmospheric pollutants to preserve our natural environment.

Plants can serve as useful tools for such purposes, and some species have already been exploited as detectors (for phytomonitoring) or as scavengers (for phytoremediation) of air pollutants. However, accelerating progress in biotechnology, especially in the fields of plant tissue culture and genetic manipulation, now makes it possible to improve various plant characteristics rapidly so as to extend their utility in these areas. An essential prerequisite to the development of such novel plants is a systematic accumulation of up-to-date information on the physiological and biochemical responses of plants, on the regulation of plant gene expression, and on the effects and mechanisms of pollutant metabolism in plant cells.

To address the need for such information, we present in this volume current topics that deal with various aspects of phytomonitoring and phytoremediation. The book is thus divided into four main sections: section I concerns plant responses and phytomonitoring; section II deals with resistant plants and phytoremediation; section III examines systems for imaging diagnosis of plant responses and gas exchange; and section IV focuses on the generation and use of novel transgenic plants. We review the basic physiological and biochemical properties of plants as a necessary background to several of these topics, especially where attempts have been made to apply modern methods of biotechnology to air pollution control. We also evaluate current concepts and techniques for the early detection of plant stress and for the screening of tissues and plants with characteristics relevant to phytoremediation and phytomonitoring. This book will therefore be of considerable value to researchers and students who are interested in these new technologies and who are considering areas in which to utilize their knowledge of and expand their skills in plant biotechnology.

Kenji Omasa
Hikaru Saji
Shohab Youssefian
Noriaki Kondo

Contents

IV. Generation of Transgenic Plants

Contributors

Aono, Mitsuko, Environmental Biology Division, National Institute for Environmental Studies, Onogawa 16-2, Tsukuba, Ibaraki 305-8506, Japan

Arai, Kazushi, Environmental Ecology Division, Tokyo Metropolitan Forestry Experiment Station, Hirai 2753-1, Hinode, Nishitama-gun, Tokyo 190-0182, Japan

Arimura, Gen-Ichiro, Department of Mathematical and Life Sciences, Graduate School of Science, Hiroshima University, Kagamiyama 1-3-1, Higashi-Hiroshima, Hiroshima 739-8526, Japan

Barnes, Jeremy, Department of Agricultural and Environmental Science, Ridley Building, Newcastle University, Newcastle upon Tyne, NE1 7RU, UK

Burchett, Margaret, Centre for Ecotoxicology, Faculty of Science, University of Technology, Sydney (UTS), Westbourne St, Gore Hill, NSW 2065, Australia

Chappelle, Emmett W., Biospheric Sciences Branch, Laboratory for Terrestrial Physics, NASA/GSFC, Greenbelt, MD 20771, USA

Daughtry, Craig S. T., Hydrology and Remote Sensing Laboratory, USDA Agricultural Research Service, Beltsville, MD 20705, USA

De Kok, Luit J., Laboratory of Plant Physiology, University of Groningen, P.O. Box 14, 9750 AA Haren, The Netherlands

Ebinuma, Hiroyasu, Pulp and Paper Research Laboratory, Nippon Paper Industries Co., LTD, Oji 5-21-1, Kita-ku, Tokyo 114-0002, Japan

Endo, Saori, Pulp and Paper Research Laboratory, Nippon Paper Industries Co., LTD, Oji 5-21-1, Kita-ku, Tokyo 114-0002, Japan

Kim, Hak Y., Institute of Agricultural Science and Technology, Kyungpook National University, 1370 Sankyuk-dong, Puk-ku, Taegu 702-701, Korea

Kim, Moon S., Instrumentation and Sensing Laboratory, USDA Agricultural Research Service, Beltsville, MD 20705, USA

Kondo, Noriaki, Department of Biological Sciences, Graduate School of Science, The University of Tokyo, Hongo 7-3-1, Bunkyo-ku, Tokyo 113-0033, Japan

Kondo, Takayuki, Air Quality Section, Toyama Prefectural Environmental Science Research Center, Nakataikouyama 17-1, Kosugi, Toyama 939-0363, Japan

Kubo, Akihiro, Environmental Biology Division, National Institute for Environmental Studies, Onogawa 16-2, Tsukuba, Ibaraki 305-8506, Japan

Kuno, Haruko, Environmental Ecology Division, Tokyo Metropolitan Forestry Experiment Station, Hirai 2753-1, Hinode, Nishitama-gun, Tokyo 190-0182, Japan

Lyons, Tom, Department of Agricultural and Environmental Science, Ridley Building, Newcastle University, Newcastle upon Tyne, NE1 7RU, UK

Matsunaga, Etsuko, Pulp and Paper Research Laboratory, Nippon Paper Industries Co., LTD, Oji 5-21-1, Kita-ku, Tokyo 114-0002, Japan

McMurtrey, James E., Hydrology and Remote Sensing Laboratory, USDA Agricultural Research Service, Beltsville, MD 20705, USA

Morikawa, Hiromichi, Department of Mathematical and Life Sciences, Graduate School of Science, Hiroshima University, Kagamiyama 1-3-1, Higashi-Hiroshima, Hiroshima 739-8526, Japan

Morita, Shigeto, Faculty of Agriculture, Kyoto Prefectural University, Shimogamo-Hangicho 1-5, Sakyo-ku, Kyoto 606-8522, Japan

Mousine, Rachid, Centre for Ecotoxicology, Faculty of Science, University of Technology, Sydney (UTS), Westbourne St, Gore Hill, NSW 2065, Australia

Mulchi, Charles L., Department of Natural Resource Sciences, University of Maryland, College Park, MD 20742, USA

Nakajima, Nobuyoshi, Biodiversity Conservation Research Project, National Institute for Environmental Studies, Onogawa 16-2, Tsukuba, Ibaraki 305-8506, Japan

Nouchi, Isamu, Agro-Meteorology Group, National Institute for Agro-Environmental Sciences, Kannondai 3-1-3, Tsukuba, Ibaraki 305-8604, Japan

Omasa, Kenji, Department of Biological and Environmental Engineering, Graduate School of Agricultural and Life Sciences, The University of Tokyo, Yayoi 1-1-1, Bunkyo-ku, Tokyo 113-8657, Japan

Osmond, Barry, Biosphere 2 Center, Columbia University, 32540 S Biosphere Road, PO Box 689, Oracle AZ 85623, USA

Park, Yong-Mok, Department of Life Science, College of Natural Science and Engineering, Chongju University, Chongju, 360-764, Korea

Saji, Hikaru, Environmental Biology Division, National Institute for Environmental Studies, Onogawa 16-2, Tsukuba, Ibaraki 305-8506, Japan

Sakaki, Takeshi, Department of Bioscience and Technology, School of Engineering, Hokkaido Tokai University, Minami-sawa 5-1-1-1, Minami-ku, Sapporo 005-8601, Japan

Srivastava, Hari S., Department of Plant Science, Rohilkhand University, Bareilly 243006, India

Stuiver, C. Elisabeth E., Laboratory of Plant Physiology, University of Groningen, P.O. Box 14, 9750 AA Haren, The Netherlands

Stulen, Ineke, Laboratory of Plant Physiology, University of Groningen, P.O. Box 14, 9750 AA Haren, The Netherlands

Tabei, Yutaka, Plant Biotechnology Department, National Institute of Agrobiological Sciences, Kan-nondai 2-1-2, Tsukuba, Ibaraki 305-8602, Japan

Takahashi, Misa, Department of Mathematical and Life Sciences, Graduate School of Science, Hiroshima University, Kagamiyama 1-3-1, Higashi-Hiroshima, Hiroshima 739-8526, Japan

Takayama, Kotaro, Department of Biological and Environmental Engineering, Graduate School of Agricultural and Life Sciences, The University of Tokyo, Yayoi 1-1-1, Bunkyo-ku, Tokyo 113-8657, Japan

Tanaka, Kunisuke, Faculty of Agriculture, Kyoto Prefectural University, Shimogamo-Hangicho 1-5, Sakyo-ku, Kyoto 606-8522, Japan

Tarran, Jane, Centre for Ecotoxicology, Faculty of Science, University of Technology, Sydney (UTS), Westbourne St, Gore Hill, NSW 2065, Australia

Tobe, Kazuo, Laboratory of Intellectual Fundamentals for Environmental Studies, National Institute for Environmental Studies, Onogawa 16-2, Tsukuba, Ibaraki 305-8506, Japan

Westerman, Sue, Laboratory of Plant Physiology, University of Groningen, P.O. Box 14, 9750 AA Haren, The Netherlands

Yamada-Watanabe, Keiko, Pulp and Paper Research Laboratory, Nippon Paper Industries Co., LTD, Oji 5-21-1, Kita-ku, Tokyo 114-0002, Japan

Yoneyama, Tadakatsu, Department of Applied Biological Chemistry, Graduate School of Agricultural and Life Sciences, The University of Tokyo, Yayoi 1-1-1, Bunkyo-ku, Tokyo 113-8657, Japan

Youssefian, Shohab, Biotechnology Institute, Faculty of Bioresource Sciences, Akita Prefectural University, Minami 2-2, Ohgata-mura, Akita 010-0444, Japan

Zheng, Youbin, Department of Plant Agriculture, Bovey Building, Gordon Street, University of Guelph, Guelph, Ontario, Canada

I. Plant Responses and Phytomonitoring

1 Responses of Whole Plants to Air Pollutants

Isamu Nouchi

Agro-Meteorology Group, National Institute for Agro-Environmental Sciences, Kannondai 3-1-3, Tsukuba, Ibaraki 305-8604, Japan

1. Introduction

Air pollution refers to the condition in which the existence of toxic substances in the atmosphere, generated by various human activities and natural phenomena such as volcanic eruptions, results in damaging effects on the welfare of human beings and the living environment. Air pollution in advanced nations has treaded the following path of historical changes. Air pollution in urbanized cities first appeared as smoke (SO_x, fly ash, or fumes), produced by the burning of coal by industrialized societies after the industrial revolution (i.e., "London-smog type" pollution). When the major fuel use switched from coal to petroleum and natural gas, the extent of smoke pollution decreased rapidly. However, rapid increases in population and transportation, in addition to industrial growth, resulted in a new form of pollution caused by auto exhaust and photochemical smog (i.e., "Los Angeles smog type" pollution). Photochemical smog is produced in the atmosphere by complex photochemical reactions involving nitrogen oxides and hydrocarbons from sources such as auto exhaust gases and electric power plants. A similarly serious air pollution problem has now emerged in large urban cities in developing countries. More recently, there has been growing concern about various global changes, notably planetary warming due to elevated concentrations of greenhouse gases, such as carbon dioxide, methane, nitrous oxide, and chlorofluorocarbons in the atmosphere, increases in ultraviolet-B radiation on the earth's surface caused by stratospheric ozone depletion, and higher levels of acid

Air Pollution and Plant Biotechnology
–Prospects for Phytomonitoring and Phytoremediation–
Edited by K. Omasa, H. Saji, S. Youssefian, and N. Kondo
© Springer -Verlag Tokyo 2002

rain resulting from the incorporation of sulfur and nitrogen oxides into precipitation by conversion to sulfuric acid or nitric acid during long-distance transport.

Plants cannot select or move their living conditions as animals can, and thus must live responding to their surrounding environment. Whenever environmental components, such as temperature, water in the soil, nutrients, or air pollutants, either exceeds the range to which the plants can adapt or become limiting, the plant develops abnormal symptoms or growth. Several forms of air pollutants, such as sulfur dioxide, nitrogen oxides, ozone, peroxyacetyl nitrate, halogens, and acid rain can damage plants, and I attempt to summarize here the responses of plants (injury symptoms, physiological and metabolical changes, and growth or yield reductions) to these air contaminants. For more detailed accounts of the effects of air pollutants on plants, the reader is referred to the books by Mudd and Kozlowski (1975), Treshow (1984), Guderian (1985a), Heck et al. (1988), Treshow and Anderson (1991), Wellburn (1994a), Alscher and Wellburn (1994), Sandermann et al. (1997) and De Kok and Stulen (1998).

2. Sulfur Oxides

2.1 Injury Symptoms

Sulfur dioxide (SO_2) was one of the first air pollutants recognized as harmful to humans and ecosystems. Sulfur dioxide, produced during the refining of metal ores, was found to cause dieback of forest trees and growth or yield reduction of crops near the metal refinery. For example, the National Research Council of Canada (1939) described a particularly detailed and prolonged study of the effects of smelter emissions on vegetation around the Trail smelter. Visible injuries caused by SO_2 in higher plants include necrosis (Fig. 1), discoloration, chlorosis of the foliage, early abscission of needles or leaves, inhibition of growth, a decrease in yield, and premature aging. Two types of visible symptoms on foliage are commonly observed, acute and chronic. Acute injury of leaves, which results from the absorption of high pollutant concentrations during short-term exposure, is characterized by bleached or brown bifacial necrosis. In contrast, chronic injury, caused by repeated exposure to sublethal concentrations of a pollutant, most commonly results in chlorosis. During the last few decades, a dramatic reduction in the industrial emission of SO_2 in advanced nations has been achieved as a consequence of regulatory legislation. However, atmospheric SO_2 is still occurring in large areas of the world.

2.2 Damage (Symptoms and Growth) Occurrence and SO_2 Concentration

Plant leaves absorb SO_2 and other atmospheric gases, such as ozone (O_3), nitrogen dioxide (NO_2), peroxyacetyl nitrate (PAN), and hydrogen fluoride (HF), predominantly through stomata. The high rate of SO_2 uptake is caused by its high solubility and rapid hydration in the aqueous phase of the plant. Acute foliar injury on sensitive plants, such as spinach, cucumber, and oat, may occur when the plants are exposed to 50-500 ppb SO_2 for 8 h (Mudd 1975a). Daily 4-h exposures of soybeans in open-top field chambers (101 days) demonstrated that 26 ppb SO_2 did not depress yields (Heagle et al. 1983). Bell (1982) summarized the results of various studies on the effects of low SO_2 concentrations on growth and productivity of forage grasses, and suggested that growth suppression and effects on senescence were observed after fumigation with concentrations below 38 ppb. A decrease in yield, a reduction in dry weight, and inhibition of growth were observed in several tree species in response to low SO_2 concentrations (less than 75 ppb). Recently, Shaw et al. (1993), using long-term open air fumigation

Fig. 1. Brown necrosis on *Viburnum Awabuki* K. Koch caused by SO_2 (0.50 ppm, 24 h)

experiments, reported the threshold mean SO_2 concentration for visible injury on Scots pine (*Pinus sylvestris* L.) in the UK to be in the range 6-8 ppb (measured during the critical period of needle expansion). However, a wide interspecies and intraspecies variation in susceptibility has been observed (Taylor 1978; Alscher et al. 1987). Prolonged exposure of plants to SO_2 can result in a decrease in biomass production, changes in plant morphology, and premature abscission of leaves (Malhorta and Khan, 1984). Phytotoxicity strongly depends on environmental conditions, SO_2 concentration, and duration of the exposure, and is influenced by the sulfur status of the plant (Bell 1980).

2.3 Physiological and Biochemical Mechanisms

Although several of the mechanisms by which SO_2 could possibly interfere with normal plant functions have been resolved, the essential mechanism of SO_2 phytotoxicity is still unclear. Sulfite (SO_3^{2-}) and bisulfite (HSO_3^-), which have been shown to be phytotoxic to many biochemical and physiological processes, are formed when SO_2 dissolves in cellular fluid (Fig. 2). In such a reaction, the cellular pH is also influenced by the generation of protons (H^+). Oxidation of SO_3^{2-} and HSO_3^- in plant cells can occur by both enzymatic and nonenzymatic processes, with the subsequent accumulation of sulfate (SO_4^{2-}). Because SO_3^{2-} is

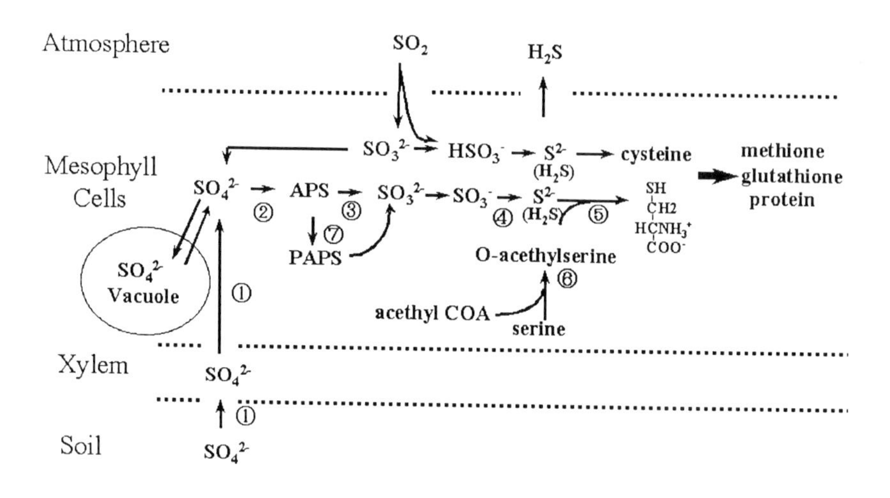

Fig. 2. Uptake and metabolic pathway of sulfur dioxide into plant tissues from the atmosphere (modified Kondo and Saji, 1992, with permission). APS, adenosine 5'-phosphosulfate; PAPS, 3'-phoshphoadenosine 5'-phosphosulfate; ①, sulfate transporter; ②, ATP sulfurylase; ③, APS reductase; ④, sulfite reductase; ⑤, cysteine synthase; ⑥, serine acetyltransferase; ⑦, APS kinase

regarded to be 30 fold more toxic than SO_4^{2-}, the oxidation of SO_3^{2-} to SO_4^{2-} might represent one of several detoxification pathways of cells. Sulfite is not only oxidized but can be reduced to produce cysteine (Fig. 2). During these processes, H_2S appears to be released (Rennenberg 1984; Ghisi et al. 1990); and such release in response to excess sulfur has been well documented for vascular plants (Rennenberg 1984).

From the concomitant observation of photosynthesis, as measured by CO_2 absorption, and of stomatal closure, as determined by measuring transpiration, a rapid decrease in net photosynthesis and a slight decrease in stomatal closure was found on exposure of sunflower leaves to 1.5 ppm SO_2 (Furukawa et al. 1980), suggesting that chloroplast activities were the primary site of SO_2 attack. Inhibition of photosynthesis in chloroplasts, which consists of light and dark reactions, is generally considered to be one of the first effects of SO_2 on plants. Shimazaki and Sugahara (1979) showed that SO_2 fumigation in the light inhibited the activity of photosystem (PS) II but not of PS I. Inhibition of PS II activity by SO_2 exposure was accompanied by a similar inhibition of noncyclic photophosphorylation, but not of cyclic photophosphorylation, which is driven by PS I. However, this inhibition occurred just before or when visible injury appeared. In contrast, ribulose-1, 5-bisphosphate carboxylase/oxygenase (Rubisco) and phosphoenol pyruvate carboxylase (PEP), which effect the first step of CO_2 fixation in the C_3 and C_4 pathways of photosynthesis, respectively, are enzymes that may be affected by SO_2. Ziegler (1972) reported that the inhibition of photosynthesis in isolated spinach chloroplasts was due to competition between bicarbonate (HCO_3^-) and SO_3^{2-} for the active site of Rubisco. However, Gezeilus and Hallgren (1980) showed that the SO_3^{2-} inhibition of Rubisco was not competitive with respect to HCO_3^-. As Rubisco requires a pH change in the stroma from 7.5 to 9.0 to become active, a lowering of the pH due to SO_2 is a likely cause of Rubisco inhibition.

In the presence of SO_3^{2-} and HSO_3^-, more superoxide (O_2^-) is formed by free radical chain oxidation than otherwise (Fig. 3), such as in chloroplasts (Asada 1980; Scandalios 1994). Metabolism of O_2^- by superoxide dismutase (SOD) results in the production of hydrogen peroxide (H_2O_2) and O_2 (Asada 1980). H_2O_2 can then inactivate the sulfhydryl (SH) group of enzymes (Tanaka et al. 1982; Hossain and Asada 1984) or react nonenzymatically with O_2^- to form an OH radical, which can bring about lipid peroxidation (Thompson et al. 1987). The accumulation of such active oxygen species (AOS) in living cells leads to the breakdown of proteins, peroxidation of membrane lipids, and DNA cleavage (Thompson et al. 1987). Plants have developed natural scavengers that protect them from these injurious effects and, in general, no apparent damage is observed on leaves (Asada 1980). Many researchers point to the existence of a correlation between SOD activity and sensitivity to SO_2 (Asada and Kiso 1973; Tanaka and Sugahara 1980). Tanaka et al. (1982) showed that H_2O_2 accumulation in spinach leaves and their chloroplasts, immediately after SO_2 fumigation, inactivated thiol enzymes of the reductive pentose phosphate cycle and resulted in depression of

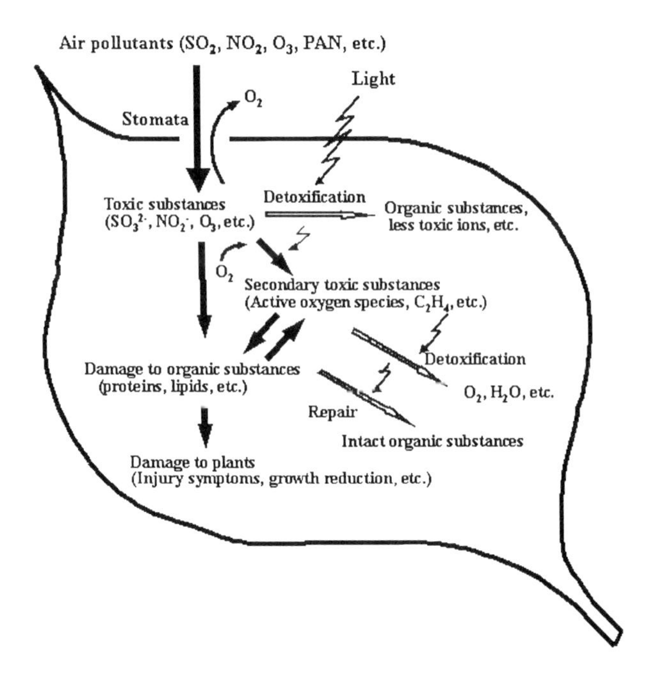

Fig. 3. Responses to air pollutants in plant leaves (Kondo and Saji, 1992, with permission). Solid arrows show response that causes damage; open arrows indicate response avoiding damage

photosynthetic CO_2 uptake. While it appears likely that AOS and free radicals, which are involved in sulfite oxidation, are responsible for this phenomenon, their occurrence and significance at realistic SO_2 concentrations have been questioned (De Kok and Stulen 1993).

3. Ozone

Ozone (O_3) is generally conceived as occurring only in the form of beneficial stratospheric ozone, a natural screen from the harmful effects of ultraviolet radiation. Although stratospheric ozone is decreasing because of destruction by anthropogenic chlorofluorocarbons and NO_x, the concentration of surface ozone in the troposphere is stable or increasing (Oltmans et al. 1998). Increases in the surface ozone levels in many industrialized countries are largely the result of photochemical oxidant pollution, produced in the atmosphere by complex photochemical reactions involving NO_x and hydrocarbons. The surface ozone in the troposphere may reduce the growth and productivity of both agricultural crops

and forest trees (Guderian 1985b; Lefohn 1991; Sandermann et al. 1997). Thus, while the beneficial role of stratospheric ozone as an attenuator of UV-B radiation decreases, the harmful effects of surface ozone as widespread phytotoxic air pollutants are increasing (Lefohn 1991).

3.1 Injury Symptoms

Soon after unusually high concentrations of photochemical oxidants (mainly O_3) were recorded, plants such as tobacco, spinach, Chinese rape, Chinese cabbage, turnip, radish, taro, Welsh onion, and morning glory showed injury symptoms on the upper surface of their leaves. Visible symptoms of O_3 damage on the leaves of many herbaceous plants, such as morning glory, spinach, radish, and tobacco, include numerous bleached spots, relatively large bleached areas, and large bifacial necrotic bleached areas on interveinal areas (Fig. 4). In contrast, foliar symptoms of O_3 damage in the gramineous plants, leguminous plants, and broadleaf woody plants, such as Japanese zelkova (*Zelkova serrata*) and plantanus (*Plantanus acerifolia*), are numerous reddish-brown stipples. The palisade parenchyma cells of the leaves are the most sensitive to ozone, and their collapse in the leaves of morning glory, spinach, and radish is associated with transformation of the cell walls, leaving larger intercellular spaces. The reddish-brown stipples that develop on the leaves of gramineous plants, leguminous plants, and broadleaf woody plants are the result of the accumulation of black or red pigment in dead cells in the palisade tissues.

Fig. 4. White, bleached areas on Chinese rape var. Komatsuna (*Brassica rapa* L.) leaves caused by ambient O_3

3.2 Damage Occurrence and O_3 Concentration

In field surveys in the Tokyo area of Japan, acute foliar injury on sensitive plants such as spinach, taro, and morning glory were often observed when exposed to 60-90 ppb of daily maximum oxidant concentration. Foliar injury of spinach and radish also occurred when exposed to 70-90 ppb O_3 for 3 h in controlled environment chambers (Nouchi 1979). Foliar injury is a principal plant response, although not always associated with growth reduction. Similarly, growth reduction may occur with no visible foliar injury symptoms.

The effect of O_3 on crop production has been quantitatively assessed on both regional and national scales in the USA (Heck et al. 1988) and Japan (Kobayashi 1992). Studies conducted by the National Crop Loss Assessment Network (NCLAN) in the USA have clearly shown that ambient O_3 levels over the 40-75 ppb range (7 h/day [900 to 1600] seasonal mean) significantly reduce the growth and yield of many crops, as predicted by the Weibull model, using results from open-top chambers (Heck et al. 1988). The results indicate that O_3 affects on a wide range of plant species with the particular response differing between crops; some species, such as soybean and spinach, show a rapid reduction in yield with increasing O_3, whereas others, such as wheat and corn, do not. The threshold concentrations for a 10% yield reduction ranged from 40 ppb for sensitive crops (as in soybean and turnips) to 75 ppb and more for the most resistant crops. Furthermore, NCLAN reported that the decreased yield due to current seasonal O_3 concentrations costs the USA approximately 3 billion dollars annually in crop productivity (Adams et al. 1988).

Kobayashi (1992) assessed the yield loss of rice by O_3 in the Kanto district of Japan from 1981 to 1985 using a growth model that incorporated the effects of O_3 on light-use efficiency (LUE) and weather variation on crop growth processes. LUE is defined as the ratio of dry matter production to cumulative light absorption and is reduced by O_3, especially in the reproduction stage (Kobayashi and Okada 1995). The total production loss by ambient O_3 in the Kanto district, which includes Tokyo and the surrounding six prefectures, ranged from 16,000 metric tons in 1981 to 78,500 metric tons in 1985, corresponding to 1.1% and 4.6%, respectively, of the total productivity of the district.

The European Community has also investigated the effects of air pollution on agricultural crops and forest trees using open-top chambers (Unsworth and Geisser 1993). In Europe, the concepts of "critical loads of pollutants" and "threshold of injury" are given more attention than the assessment of crop loss. The current threshold concentration adopted by the United Nations Economic Commission for Europe (UN-ECE), above which adverse effects begin to occur, is the AOT 40 (the accumulated number of daytime hours of exposure above 40 ppb) (UN-ECE 1994, 1996). The 40-ppb threshold is defined just above the range of background O_3 concentrations in Europe.

3.3 Mechanisms of O_3 Action

The growth reduction caused by O_3 may result from a direct or indirect inhibition of photosynthesis and the extra cost of repairing any cellular and metabolic damage (Miller 1987; Runeckles and Chevone 1991). Yield loss is caused by the cumulative effects of chronic daily O_3 exposures over the growth period. Early senescence decreases photosynthetic activity in plant leaves, thereby decreasing the plant growth rate, total photosynthate production, and hence final yield.

Ozone enters the plant through the stomata, dissolves in the water of the moist cellular surface, and reacts with apoplastic structures and plasma membranes forming AOS, such as O_2^-, H_2O_2, and OH radical (Fig. 3). Formation of such AOS in plant tissues after O_3 exposure has been demonstrated by electron spin resonance (ESR) spectroscopy (Mehlhorn et al. 1990). Ozone or AOS can oxidize various cellular components, such as sulfhydryl (SH) groups, amino acids, and unsaturated fatty acids (Fig. 2) (Heath 1980; Kangasjärvi et al. 1994). It is believed that the initial site of O_3 injury is the plasma membrane, and that this is due to lipid peroxidation and/or ozonolysis (Wellburn 1994b). Membrane functions, such as membrane fluidity, permeability, K^+-exchange via ATPase reactions, and Ca^{2+} exclusion, are rapidly lost on exposure (Heath and Taylor 1997).

A number of studies have documented the destruction of unsaturated fatty acids when suspensions of unicellular algae or isolated membranes of chloroplasts and microsomes were exposed to O_3, and also when whole plants were exposed to extremely high concentrations of O_3 (more than 500 ppb) (Frederick and Heath 1975; Pauls and Thompson 1981). However, there are few data showing that lipids in whole plant leaves are attacked by ambient O_3 levels. For example, during exposures lasting up to 8 h, although obvious water-soaked injury symptoms appeared on leaf surfaces of morning glory at 6 h (typical injury symptoms occurred after 1 day of exposure), 150 ppb O_3 had no effect on the unsaturated fatty acids or glycolipids of either bean or morning glory (Nouchi and Toyama 1988). In this work, phospholipids initially increased during exposure, and glycolipids significantly decreased after 24 h, suggesting that in the initial stages ozone may alter the lipid content by affecting lipid metabolic processes. Sakaki et al. (1985, 1990a, 1990b, 1990c) demonstrated that glycolipids were metabolically decomposed from the onset of O_3 exposure, followed by a rapid increase in the amount of triacylglycerol (TG), synthesized from acyl moieties of monogalactosylglycerol (MGDG), and increased the amount of free fatty acid produced by decomposition of MGDG.

Recently, a few critical sulfhydryl or ring amino acids of proteins have been recognized as sites of O_3 attack, primarily because of their high reactivity in a water environment (Mudd 1996). Heath and coworkers demonstrated that sulfhydryl groups on K^+-stimulated ATPase on the plasma membrane were sensitive to O_3 (Dominy and Heath 1985), and that plasma membrane Ca^{2+}-transport systems were altered by O_3 exposure (Castillo and Heath 1990). Heath and Taylor (1997) recently proposed that ionic changes in cells increase the loss of a wide range of metabolites, induce enzyme activation, and alter normal gene transcription. Thus, the membrane is altered principally via protein changes (i.e., proteins being more sensitive than lipids to O_3).

Plant growth and yield are inherently linked to photosynthetic carbon assimilation and the subsequent partitioning of this photoassimilate. Although the adverse effects of O_3 on the overall photosynthetic process, as revealed by gas exchange measurements, are well known, there are few reports of its effects on the processes of electron transport on thylakoid membranes and CO_2 fixation in stroma. Sugahara et al. (1984) found that 500 ppb O_3 inhibited electron transport in both PS I and PS II of isolated chloroplasts from several species, whereas a 4-h exposure to 100 ppb O_3 did not suppress electron transport in either photosystem. No changes in electron transport of photosystem and NADPH levels in poplar leaves exposed to 180 ppb O_3 for 3 h were also reported (Sen Gupta et al. 1991). On the other hand, Schreiber et al. (1978) and Shimazaki (1988), using chlorophyll fluorescent transients, observed that the initial damage was on the donor site of PS II (H_2O splitting enzyme system) and not directly on the reaction center of PS II, thus leading to a suppression of the production of ATP and NADPH required for energy and reducing power. The effects on chloroplastic ATP levels would then be expected to influence the dark reactions of the Calvin cycle in the stroma.

Fumigation with O_3 was found to reduce the activity of Rubisco, a key enzyme in carbon dioxide fixation during photosynthesis, in rice (Nakamura and Saka 1978), potato (Dann and Pell 1989), alfalfa (Pell and Pearson 1983), and radish and poplar (Pell et al. 1992). Reduction of Rubisco activity in O_3-stressed plants is usually associated with a decrease in amount of Rubisco. A loss in Rubisco may also play an important role in the acceleration of premature senescence as Rubisco is a major leaf protein (Pell et al. 1994). Ozone was also found to reduce mRNA levels of both the small (rbcS) and large (rbcL) subunits of Rubisco which, in turn, lowered Rubisco levels within the chloroplast (Pell et al. 1994), and thus reduced the rate of CO_2 fixation and productivity. The reductions in mRNA levels of both rbcS and rbcL mean loss in transcription. The activities of other enzymes of the Calvin cycle are also affected (i.e., inhibited or activated), leading to a disruption of normal carbohydrate metabolism in chloroplasts.

Ozone differentially alters photosynthate partitioning as a function of both O_3 concentration and the plant developmental stage (Cooley and Manning 1987). During the vegetative growth stage, relatively low concentrations of O_3 (less than 150 ppb) generally reduce root growth more than shoot growth in a wide range of plant species. The mature lower leaves, which act as the main source of photosynthates for root growth, are the most damaged by O_3, offering an explanation for decreases in partitioning to roots (Cooley and Manning 1987; Okano et al. 1984). At flowering, and as seeds or fruit develop, these reproductive organs generate a high demand for photosynthates under normal conditions, and divert photosynthates from leaves and roots. Ozone reduces the number of flowers, fruits, and seeds, but the remaining reproductive organs are often able to attain normal or larger sizes (Cooley and Manning 1987).

It is well known that stress-induced evolution of ethylene (C_2H_4) or ethane (C_2H_6) occurs in plants exposed to air pollutants such as O_3 and SO_2 (Fig. 3) (Craker 1971; Tingey et al. 1976; Bressan et al. 1979; Peiser and Yang 1979; Langebartels et al. 1991), which was thought to be a typical wounding and injury response. Ethane appeared to be evolved from visibly damaged leaves, although enhanced evolution of C_2H_4 was observed within a short time after the start of O_3 fumigation. Plants produce C_2H_4 from l-methionine via S-adenosyl-l-methionine (SAM) and 1-aminocyclopropane-1-carboxylic acid (ACC) (Yang and Hoffman 1984), whereas C_2H_6 might be released as a result of peroxidation of lipids by free radicals (John and Curtis 1977). ACC synthase, an enzyme of conversion of SAM to ACC, is considered to be a key enzyme in the regulation of the production of C_2H_4. Mehlhorn and Wellburn (1987) showed that the rate of O_3-stimulated evolution of C_2H_4 was decreased by pretreatment with aminoethoxyvinylglycine (AVG), a specific inhibitor of ACC synthase, and also almost abolished the visible foliar injury normally caused by O_3 exposure. These results indicated that cession of C_2H_4 release prevented the formation of visible injury. On the other hand, Bae et al. (1996) observed that evolution of C_2H_4 in ozone-treated tomato plants resulted from not only increases in activities of ACC synthase but also of ACC oxidase, an enzyme of conversion ACC to C_2H_4.

One possibility that reaction products between O_3 and C_2H_4 formed via wounding induced by O_3 produce O_3 injury is proposed. Elstner et al. (1985) and Elstner (1987) suggested that forming hydrogen peroxide (H_2O_2) and formaldehyde (HCHO) in nonenzymatic reaction between C_2H_4 and O_3 might damage the waxy layer of leaves. Furthermore, Mehlhorn and Wellburn (1987) and Mehlhorn et al. (1991) proposed that injury due to O_3 is attributed to a free radical-forming reaction between the O_3 coming from the atmosphere and stress C_2H_4 emerging from the plant cells in guard cells or substomatal cavities. However, no conclusive evidence for these proposals has yet been reported.

Recently, organic peroxides such as hydroxymethyl hydroperoxide (HMHP, $HOCH_2OOH$) were reported to be produced by reactions of ozone with volatile hydrocarbons emitted by plants, which were detected in the forest air and in the plant leaves (Hewitt et al. 1990). Low environmentally relevant HMHP concentrations inhibited peroxidases located in acidic aqueous phase of cell walls (Polle and Junkermann 1994). Therefore, HMHP is thought to play a part in forest decline (Hewitt et al. 1990). Further studies should be done to clarify relationships between ambient concentrations of organic peroxides and phytotoxicity.

3.4 Detoxification of Active Oxygen Species

The toxicity of AOS, such as O_2^-, H_2O_2, OH radical, and $^1O_2$ (singlet oxygen), associated with O_3 was previously reported (Runeckles and Chevone 1991). These species are formed either directly, by O_3 decomposition in the aqueous phase of cells (Heath 1987), or indirectly, by oxygen reduction in chloroplasts (Asada 1980; Scandalios 1994). The accumulation of these AOS in living organisms leads to the breakdown of proteins, peroxidation of membrane lipids, and cleavage of DNA (Thompson et al. 1987). Moreover, Sakaki et al. (1983) proposed that pigment bleaching and lipid decomposition in leaf disks of O_3-fumigated plants result from accumulation of AOS in leaves due to the breakdown of physiological defense processes against oxygen toxicity.

Reduced ascorbate and glutathione act as antioxidants, not only by reacting directly with O_3 with high rate constants (7.0×10^8 $M^{-1}s^{-1}$ for glutathione and 6×10^7 $M^{-1}s^{-1}$ for ascorbate at pH 6-7; Kanofsky and Sima 1995), but also by participating in enzymatic and nonenzymatic O_2^- and H_2O_2 degradation (Runeckles and Chevone 1991). By using a mathematical analysis of mass transfer accompanying the chemical reactions, Chamcides (1989) suggested that ascorbate within the cell wall prevents the penetration of O_3 into plasma membranes. However, the extent of protection afforded by apoplastic ascorbate has recently been questioned (Jacob and Heber 1998). Apoplastic ascorbate is generated by reduction of its oxidation products presumably inside the cytoplasm (Rautenkranz et al. 1994). Comparison of the rates between ozone uptake by leaves and loss of apoplastic ascorbate suggests that only about 5% to 10% of entering ozone is detoxified by ascorbate (Luwe et al. 1993). These facts suggested that transport of oxidized ascorbate across the plasmalemma into the cytoplasm of leaf cells and import of regenerated ascorbate back into the apoplast is slow. Therefore, the majority of ozone molecules would escape from being scavenged by ascorbate and have injurious effects on plasma membranes and intracellular components (Sakaki 1998).

An enzymatic system, including SOD, ascorbate peroxidase (AP), dehydroascorbate reductase (DHAR), and glutathione reductase (GR), termed the ascorbate/glutathione cycle, in which AP decomposes H_2O_2 and GR recycles GSSG to GSH to maintain the reducing power, is an effective protection system in chloroplasts against the toxic AOS. It has been reported that antioxidant levels and related enzyme activities in higher plants increase in response to oxidant stress. For example, Mehlhorn et al. (1986) reported that the contents of ascorbate, glutathione, and α-tocopherol increased in fir and spruce needles after long-term exposure to 12 ppb (winter) and 4 ppb (summer) of SO_2, 37 ppb O_3, and a combination of both. Tanaka et al. (1985, 1988) also reported the increased activity of AP and GR in spinach leaves after fumigation with 70 or 100 ppb O_3. Changes in the levels of antioxidants, such as ascorbate and glutathione, and the activities of related enzymes in rice leaves exposed to high (500 ppb) or low (50 and 100 ppb) O_3 concentrations were also investigated by Nouchi (1993). Exposure to high concentrations for 8 h resulted in the oxidation of ascorbate to form dehydroascorbate, an increase in the total glutathione level, and the rapid decline of enzyme activities, such as AP and SOD. The inherent protection system against O_3 stress is considered to be partially damaged by exposure to high concentrations of O_3. On the other hand, exposure to low (50 ppb) or relatively low (100 ppb) O_3 concentrations for 5 weeks was found to increase both ascorbate levels and the activities of enzymes such as AP, GR, and SOD, possibly as an adaptation mechanism to maintain growth under mild O_3 stress. However, Sen Gupta et al. (1991) found no changes in SOD level of poplar seedlings exposed to 180 ppb O_3, and Benes et al. (1995) found little change in SOD, AP, and guaicol-peroxidase when ponderosa pine trees were exposed to low, long-term concentrations of O_3. Therefore, as in the case of SO_2 exposure, the role of antioxidants in O_3 exposure is still unclear, and more extensive research is required to clarify their exact role.

4. Peroxyacetyl Nitrate (PAN)

4.1 Injury Symptoms

PAN is a minor component of photochemical oxidants that result from photochemical reactions but is highly toxic to plants even at fairly low concentrations. Characteristic symptoms, such as glazing, bronzing, and silvering, were first observed on the lower surface of younger leaves of several vegetables (Fig. 5), such as lettuce, garden beet, and spinach, in the Los Angeles area in 1944 (Middleton et al. 1950). The injury symptoms are clearly distinct from those produced by SO_2 or O_3. For many plants, PAN injury usually consists of the collapse and pigmentation of the spongy parenchyma cells. Although the global distribution and effects of PAN are poorly understood, PAN concentrations in southern California appear to be five- to ten-fold higher than those reported in eastern North America, western Europe, or Japan (Temple and Taylor 1983). In Riverside, California, PAN averaged 8-9 ppb in the summer to autumn of 1980 (Temple and Taylor 1983). PAN-induced injury symptoms on plants in the USA, Netherlands, and Japan have been reported (Temple and Taylor 1983). In the Los Angeles area, the threshold for foliar injury of susceptible plants such as lettuce and leaf beet in the field is about 15 ppb PAN for 4 h; severe injury was observed at 25-30 ppb for 4 h (Temple and Taylor 1983). In a field survey in Japan, PAN-induced injury on the leaves of a most sensitive petunia variety was observed after a few hours of exposure to 5 ppb PAN (Nouchi et al. 1984b). In PAN exposure experiments, Nouchi (1979) found that the threshold PAN concentrations that induced leaf injury in petunia (white flower variety, a most sensitive species to PAN) were 32, 14, and 7 ppb for exposure durations of 1, 3, and 8 h, respectively.

4.2 Mechanism of PAN Action

The physiological and biochemical mechanisms of PAN action were vigorously studied in the 1960s. However, since the early 1970s, less research has been conducted, largely because PAN causes less severe injury in the field than O_3. Therefore, while not much is known about the mechanisms by which PAN induces damage, the physiological and biochemical effects of PAN on photosynthesis, lipids, and enzymes are known to be very different to those induced by O_3, even though both are oxidants.

Of all the plant reactions, photosynthesis must be regarded as one of the most critical. Chloroplasts isolated from plants exposed to 600 ppb PAN for 30 min showed inhibited oxygen evolution whereas photophosphorylation was unaffected (Dugger et al. 1965). In addition, Coulson and Heath (1975) demonstrated that exposure of isolated spinach chloroplasts to PAN inhibited electron transport in both PS I and PS II. On the other hand, photosynthesis studies using the chamber method (net absorption of CO_2) produced somewhat different results. Nouchi (1988) observed that although exposure to PAN concentrations as high as 95 ppb for 4 h resulted in visible foliar damage to kidney bean plants, photosynthetic and transpiration rates remained at almost normal levels during exposure, and only decreased abruptly and considerably after the development of water-soaked symptoms on the leaves. These results indicate that PAN does not affect stomatal

Fig. 5. Glazing on lower surface of leaf beet (*Beta vurgaris* L. var. flavescens DC.) caused by ambient PAN

closure until appearance of the initial visible symptoms, and that PAN may strongly attack stroma and membrane structures of chloroplasts even after cessation of exposure.

PAN also reacts with the double bond of olefine to form epoxides (Mudd 1975b), and also affects lipid biosynthesis. Such reactions could affect both the proteins and lipids of membranes. In addition, PAN can oxidize NADPH and prevent the incorporation of acetate into long-chain fatty acids (Mudd and Dugger 1963). Therefore, exposure to PAN seems to alter the membrane lipids in leaf tissue. Nouchi and Toyama (1988) studied the effect of PAN on fatty acids and lipids. When kidney bean plants were exposed to 100 ppb PAN for 8 h, the phospholipid, glycolipid, and total fatty acid content in leaves remained unchanged for the first 4 h of the exposure, during which no visible foliar symptoms were observed. However, the phospholipid, glycolipid, and total fatty acid content decreased considerably and the MDA content increased sharply as the water-soaked and wilting symptoms began to appear on the leaves 6 h after the start of exposure. It was assumed that PAN directly and strongly attacks lipids of the chloroplast thylakoid membranes, leading to the abrupt collapse of membrane structures and ultimate death of the cells. In addition, Nouchi (1988) examined the participation of active oxygen, which may be formed either directly by decomposition of PAN in solution or indirectly by biochemical processes in cells during the process of PAN injury. In this study, various active oxygen scavengers were exogenously applied to leaf disks that were previously exposed to PAN. The scavenger experiments indicated that the decomposition of chlorophyll and the formation of MDA were apparently caused by a type of active oxygen, O_2^-. However, as O_2^- has no relation to the decomposition of polar lipids in leaves caused by PAN, the results suggested that the oxidative action of PAN proceeds in at least two ways; that is, an early stage, in which PAN by itself acts as an oxidant on polar lipids, and at a late stage, in which O_2^- acts as an oxidant on chlorophyll and fatty acids.

PAN reacts strongly with the SH groups of enzymes and with low molecular weight sulfur-containing compounds, such as amino acids, to form either disulfide or S-acetyl groups. There is considerable evidence that PAN damages the SH groups of several enzymes (Taylor 1969; Mudd 1975b). The conditions that protect enzymes from PAN were equally effective in protecting from SH reagents; the enzymes that had no SH groups were not affected by PAN (Mudd 1963). The close correlation of inhibition by PAN and SH reagents is taken as strong evidence that one point of inhibition by PAN is the enzyme SH group (Mudd 1963). It is well known that illumination is required before, during, and after exposure to PAN for the appearance of visible injury (Mudd 1975b). There is also an absolute requirement for light during and after exposure if any inhibition of photosynthesis is to be seen (Koukol et al. 1967). These facts indicate that PAN may cause damage to plants through reactions between PAN and some components of the plant photochemical pathway (Dugger et al. 1963) and that the requirement for light is most likely related to the oxidation of SH groups of a photoreducible protein (Ziegler 1973). Furthermore, Wellburn (1994b) has proposed that light initiates the formation of additional free radicals, which then overwhelm the antioxidant mechanisms.

5. Nitrogen Oxides

Of the several oxides of nitrogen that may be found in the atmosphere, including nitrogen dioxide (NO_2), nitric oxide (NO), nitrous oxide (N_2O), nitrogen trioxide (N_2O_3), and nitrogen pentoxide (N_2O_5), the most serious air pollutants are NO_2 and NO. Although recent studies have focused on N_2O, because it is one of the greenhouse gases and has also been implicated in stratospheric ozone depletion, it has little effect on plants. Nitrogen oxides, like SO_x, are formed mainly as a result of the burning of fossil fuels. Nitric oxide is mainly formed in the heat of combustion from the combination of atmospheric nitrogen and oxygen ($N_2 + O_2 \rightarrow 2NO$). There is then a spontaneous, but not necessarily rapid, reaction between NO and oxygen to form nitrogen dioxide ($2NO + O_2 \rightarrow 2NO_2$). Plants absorb gaseous NO_2 more rapidly than NO (Bennett and Hill 1975), because NO_2 reacts rapidly with water while NO is almost insoluble (Malhotra and Khan 1984). The uptake of NO_2 per unit leaf area was almost three times that of NO when the two gases were present at the same concentration (Law and Mansfield 1982). Therefore, NO_2 is more toxic than NO (Bennett and Hill 1973; Ormrod 1978).

5.1 Injury Symptoms and Effects of Growth

The phytotoxicity of NO_2 is relatively less than SO_2 and O_3, and acute leaf injury on sensitive plants, such as radish and alfalfa, occurs after exposure to 3 ppm NO_2 for 4-8 h. Visible symptoms resulting from NO_2 exposure are relatively large white or brown necrotic areas. Visible injury due to NO_x is very rare in the field, although injury symptoms accidentally occur on eggplant, cucumber, and tomato in greenhouses from internally generated NO_2 (approximately 20 ppm) from fertilizer by the activity of soil bacteria (Hashida 1965). In addition, visible injury or invisible injury on pepper plants was observed in greenhouses equipped with propane or kerosene burners to provide an enriched atmospheric CO_2 (Law and Mansfield 1982). The burners generate a relatively high level of NO_x (400-2000 ppb NO with 50-500 ppb NO_2) in the greenhouse as well as provide about 1000 ppm CO_2 (Capron and Mansfield 1975; Law and Mansfield 1982). The effects of NO_2 on growth are not clear because growth is likely to be stimulated by low concentrations (less than 100 ppb) but to be reduced by high concentrations (more than 1 ppm) of NO_2. Long-term exposures of plant species to realistic concentrations of NO_x (under 100 ppb) are even less well understood. Wright (1987) found both stimulation and reduction in growth of different clones of birch exposed to a weekly mean concentration of 62 ppb NO_2. Ashenden et al. (1990)

also showed similar contrasting responses of different fern species to long-term fumigation with 60 ppb NO_2.

5.2 Mechanism of NO_2 Action

When NO and NO_2 dissolve in the extracellular water within a leaf, they form nitrate (NO_3^-) and nitrite (NO_2^-) in equal amounts, and protons (H^+). The plants absorb NO_3^- from the root as a nitrogen source and have a nitrogen metabolic pathway that synthesizes amino acids and proteins from NO_3^- (Fig. 6). Because NO_3^- and NO_2^- formed atmospheric NO_2 are also used as substrates of the nitrogen metabolic pathway, these are utilized for the synthesis of amino acids and proteins (Yoneyama and Sasakawa 1979). Therefore, a low concentration of NO_2 has an effect as a fertilizer for plants, and growth may be promoted. Especially, the growth-promoting effect is often found in plants grown under nitrogen deficiency in soil. On the other hand, plants exposed to a high concentration of NO_2 will cause visible symptoms and growth reduction.

NO_3^- from atmospheric NO_2 is reduced, first to NO_2^- by nitrate reductase (NaR) located in cytosol, and then to the ammonium ion (NH_4^+) by nitrite reductase (NiR) located in chloroplasts (Fig. 5). Furthermore, NH_4^+ is incorporated into glutamate by the combined action of glutamine synthetase (GS) and glutamate synthase (GOGAT) in chloroplasts, and then converted to other amino acids (Yoneyama et al. 1979; Lea et al. 1990; Wellburn 1990). The activity

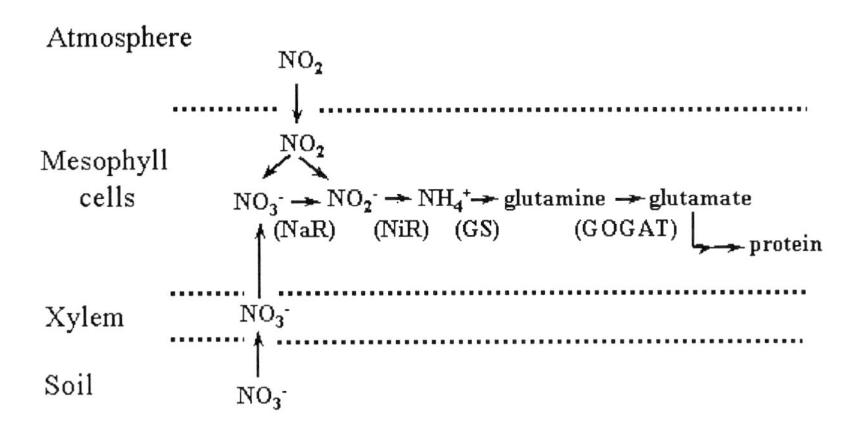

Fig. 6. Uptake and metabolic pathways of nitrogen dioxide into plant tissues from the atmosphere (from Kondo and Saji, 1992, with permission). NaR, nitrate reductase; Nir, nitrite reductase; GS, glutamine synthetase; GOGAT, glutamate synthase; glutamine, $COOHCHNH_2(CH_2)_2COOH$; glutamate, $CONH_2(CH_2)_2CHNH_3^+COO^-$

of these enzymes is higher in the light than in the dark. Especially, the reduction of NO_2^- to NH_4^+ by NiR requires energy of reduced ferredoxin and ATP provided by photosynthesis (Malhotra and Khan 1984). It is well known that NO_2^- is more toxic than NO_3^- (Mudd 1973). Plants generally do not accumulate NO_2^- in leaves, because the activity of NiR is about two to ten times higher than that of NaR in the light. However, a large NO_2^- accumulates in the leaves resulting from lowered NiR activity, when plants were exposed to NO_2 in the dark or grown under nitrogen deficiency conditions. In general, when plants are exposed to ambient levels of NO_2, NO_2^- levels rarely rise (Srivastava and Ormrod1984). Zeevaart (1976) reported large increases of NO_2^- rather than NO_3^- when peas were exposed to high concentration of NO_2 (8.4 ppm) for 1-2 h.

What materials do cause cell damage and death in leaves by exposure to NO_2? As the usual concentration of NO_3^- in leaf cells is approximately 10 mM, and this concentration of NO_3^- remained approximately unchanged by exposure to NO_2, NO_3^- seems unlikely to be a contributor to injury. On the other hand, NO_2^- is usually kept to a low concentration in cells, 1 mM, but the concentration of NO_2^- rapidly increases after the start of NO_2 exposure (Takeuchi et al. 1985). There is a close relationship between the accumulation of NO_2^- and visible injury. For example, kidney bean plants, a species sensitive to NO_2, accumulated large NO_2^- and showed foliar injury, while corn plants, a species tolerant to NO_2, accumulated less NO_2^- and visible injury could not recognized (Yoneyama et al. 1979). In addition, when kidney bean and spinach plants were exposed to NO_2 (3.5 ppm for 1.5 h) in the dark, NO_2^- accumulated in both plants (more than 100 nmol cm^{-2}). Kidney bean plants retained considerable NO_2^- accumulation (approximately 75 nmol cm^{-2}) in the leaves, however, and severe injury symptoms appeared when plants were transferred to the light, while in spinach, a species tolerant to NO_2, NO_2^- disappeared almost completely within 2 h when plants were illuminated after fumigation (Shimazaki et al. 1992). The balance between production and consumption of NO_2^- might regulate the concentration of NO_2^- in leaves. Shimazaki et al. (1992) concluded that plant species that have a high capacity to reduce NO_2^- in leaf cells may have a higher tolerance to NO_2, as is the case of spinach plants, in which the concentration of NO_2^- in leaves fell more rapidly than in the leaves of kidney bean plants.

NaR activity could be induced by NO_3^- as a nutrient source of nitrogen (Solomoson and Barber 1990), as well as by light. A higher level of NO_2^- accumulated in the first trifoliate leaves of kidney bean plants that had been supplied with NO_3^- (abbreviated NO_3^--plants) than in those of plants that had been supplied with NH_4^+ (abbreviated NH_4^+-plants) after fumigation of NO_2 (2.2 ppm in the dark; Shimazaki et al. 1992). In this case, visible injuries were more marked in NO_3^--plants than in NH_4^+-plants. These results indicate that the NO_3^--plants might be more susceptible to NO_2 fumigation because the higher NaR activity might cause enhanced accumulation of NO_2^- in the leaves. Thus, it is believed that the acute injury of NO_2 to plants is determined by the extent of the accumulation of NO_2^- in leaves (Zeevaart 1976; Shimazaki et al. 1992). However, the mechanism of NO_2 phytotoxicity is not perfectly clarified yet.

SO_2 and O_3 inhibit photosynthetic rate by closing stomata as well as affecting the photosystem in chloroplasts, while NO_2 lowered the photosynthetic rate without closing the stomata (Furukawa 1984). In addition, the lowering of the photosynthetic rate becomes clear only when plants were exposed to considerable high concentrations of NO_2 (500-700 ppb and above) in short-term fumigations (<8 h) and 250 ppb during a 20-h period (Darrall 1989). Therefore, NO_2 toxicity is much less than that of SO_2 and O_3. Furukawa (1984) found that NO_2^- accumulated in the leaf when sunflower was exposed to NO_2 (2-4 ppm), and that the accumulated NO_2^- in the leaf rapidly decreased after the exposure was stopped. This change in NO_2^- agreed with the inhibition and recovery of the photosynthetic rate, indicating that the lowering of the photosynthetic rate is attributed to accumulation of NO_2^- in the leaf from atmospheric NO_2. Since it is known that NO_2^- inhibited carbonic anhydratase activity (Bamberger and Avron 1975), this seems to be one of mechanisms of photosynthesis inhibition by NO_2; when this enzyme is inhibited by NO_2^-, the supply of CO_2 to Rubisco is suppressed in the carbon fixation system. On the other hand, Shimazaki et al. (1992) found that light has two distinct roles in the development of visible injuries after fumigation with NO_2. Light suppressed the destruction of chlorophyll by reducing the concentration of NO_2^- in leaves while, conversely, it stimulated the destruction of chlorophyll by generating active oxygen species (AOS). They proposed that generation of AOS may increase in the presence of NO_2^- in the light, because NO_2^- may inhibit the photosynthetic fixation and, consequently, accelerate the one-electron reduction of O_2 on the reducing side of PS I as a result of the absence of a physiological acceptor of $NADP^+$. This result indicates the possibility that AOS is also concerned in visible injury caused by NO_2.

6. Fluoride

Fluoride, mostly in the gaseous HF form, is released into the atmosphere whenever the clays, rocks, and ores in which it is present are heated. Plant damage from fluorides frequently occurs near aluminum refineries using cryolite (Na_3AlF_6) in the electrolysis of aluminum oxides, phosphorous fertilizer plants using phosphor rock, and ceramic factories using fluorine compounds for coloring tiles, bricks, and enamel.

Injury symptoms from fluorides occur mostly on the leaf tips and leaf margins. A characteristic symptom of necrosis is the presence of a sharply separating reddish-brown band between the necrotic and healthy tissue. Injury symptoms on sensitive plants, such as gladiolus, apricot, and grape, may appear when leaf concentrations of fluoride exceed 20 ppm (Treshow and Anderson 1991). As enzymes such as phosphoglucomutase and glucose-6-phosphate dehydrogenase are inhibited by fluoride, a number of metabolic processes may be affected (Treshow and Anderson 1991).

7. Acid Rain

It has long been recognized that the emissions of sulfur and nitrogen oxides from metal refineries, electric power plants, and auto exhaust gases to the atmosphere cause local air pollution problems. However, it is now being recognized that increases in these emissions could affect areas several thousand kilometers distant from the main emission source, and result in acidification of the atmosphere with various adverse ecological effects in many regions of the world. Therefore, acid deposition (including both wet and dry deposition) has recently become a subject of widespread concern in relation to global changes. This deposition occurs mainly in highly industrialized regions, most notably Europe and the eastern part of North America, as evidenced by reports of changes in the chemistry and biology of lakes and marshes in the 1970s and of terrestrial ecosystems in the 1980s.

The SO_x and NO_x emitted into the atmosphere may be transformed into particulates that can be transported over long distances. Particulate sulfur and nitrous compounds and SO_x or NO_x may be incorporated into precipitation by conversion of atmospheric chemical reactions to H_2SO_4 or HNO_3 and result in a decrease in the pH of the precipitation. Under nonpolluted conditions, the natural acidity of precipitation is around pH 5.6, a value that is in equilibrium with the gas-liquid phase reactions between the atmospheric CO_2 concentration (about 350 ppm) and cloud water. Therefore, acid precipitation may be defined as rain, snow, fog, or other forms of precipitation with pH values less than 5.6 and a hydrogen ion concentration above 2.5×10^{-6} mol/l.

7.1 Injury Symptoms

Acid rain or mist induces bleaching or brown necrotic lesions on the upper surface of leaves and bleached spots on petals. In simulated acid rain experiments, visible foliar injury of most herbaceous plants and some sensitive woody plants occurred at pH values below 3.5 and 3.0, respectively (Lee at al. 1981). On the other hand, damage to petals of sensitive horticultural species, such as morning glory, has been observed when the petals were exposed to natural or simulated acid rain at pH 4.3 or lower (Nouchi 1992).

7.2 Growth or Yield Effects on Crops

There have been very few reports to date on visible injury or crop loss under field conditions caused by naturally occurring acid precipitation, except for the case of volcanic eruptions. Therefore, most studies on the effects of acid rain on the growth or yield of agricultural crops have been carried out using simulated acid rain.

7.2.1 Controlled Environment Experiments

Plant growth may be stimulated, inhibited, or not affected by exposure to simulated acid rain. Lee et al. (1981) conducted experiments using 28 crops grown in pots in field chambers. Simulated sulfuric acid rain was applied at pH levels of 5.6, 4.0, 3.5, and 3.0. Marketable yield was inhibited in 5 crops (radish, beet, carrot, mustard green, and broccoli) and stimulated in 6 crops (tomato, green pepper, strawberry, alfalfa, orchard grass, and timothy). They observed no consistent effects on the other 16 crops, and noted that foliar injury was not generally related to the effects on yield.

Hosono and Nouchi (1992, 1994) exposed eight crop species (radish, spinach, bush bean, turnip, packchoi (*Brassica campestris* L.), lettuce, carrot and rice) to simulated acid rain (SO_4^{2-}: $NO3^-$: Cl^- = 2:1:2, equivalent ratio) at pH levels of 5.6, 4.0, 3.0, and 2.7 (or 2.5) throughout the growth period in four greenhouses in Japan. Treatment at pH 3.0 or below produced visible foliar injury on leaves of all the test plants. In many cases, visible foliar injury was more severe in the early stages of growth than at the later stages. Treatment at pH 3.0 or higher did not significantly affect the whole plant dry weight or yield of any of the test species. When the plants were exposed to simulated acid rain at pH 2.7 or 2.5, the dry weight of the radish hypocotyl and whole plant dry weights of spinach, pakchoi, turnip, carrot, lettuce, and bush bean decreased significantly compared to plants exposed to rain at pH 5.6 (Fig. 7). On the other hand, although the dry weight of rice plants exposed to simulated acid rain below pH 3.0 was reduced compared

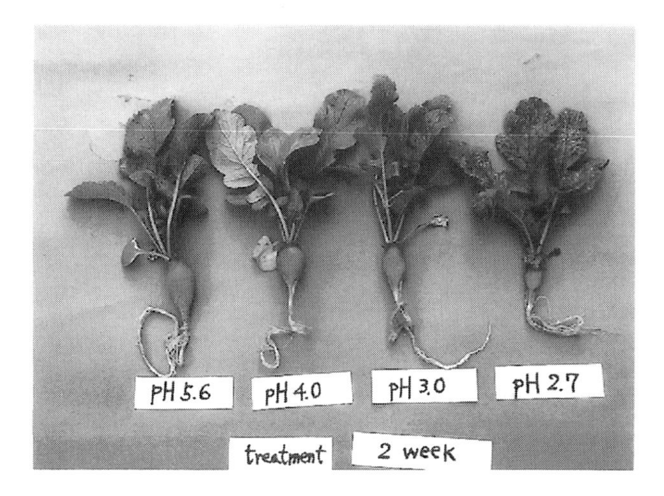

Fig. 7. Growth reduction of radish caused by simulated acid rain. When radish plants were exposed to simulated acid rain at pH 2.7, the whole plant was reduced compared with plants exposed to rain at pH 5.6

with that at pH 5.6 during the early growth stage, the difference was not significant during the middle growth stage or after. Moreover, the yield of rice was not reduced even at pH 2.5.

Ashenden and Bell (1987, 1989) found that normal ambient pH range (3.5-4.5) of simulated acid rain caused growth or yield reductions of some crops. Their study suggested 9%-17% yield reductions in winter barley grown on a range of British soils in response to the critical pH range of rainfall of 3.5-4.5 (Ashenden and Bell 1987). In addition, one of the legume species, *Vicia faba* L., was found to be sensitive to the pH 4.5 treatment, with an 18% reduction in total plant weights compared to the plants grown in the pH 5.6 treatment (Ashenden and Bell 1989).

One of the most comprehensive dose-response studies illustrating the possible impact of acidic precipitation on crop yield was conducted on greenhouse-grown radish by Irving (1985). Radish plants were exposed to simulated acid rain throughout a growing season at pH values ranging from 2.6 to 5.6. The results indicated the presence of a threshold for significant yield loss from rain acidity for pH values between 3.0 and 3.4 throughout the growing season. Jacobson et al. (1988), using Irving's radish data described above and a Mitcherlich function, calculated the threshold pH value associated with a 10% reduction in hypocotyl yield to be 3.3 ± 0.3.

7.2.2 Field Experiments

Hypocotyl (root) yields of radish plants were reduced after exposure to simulated acidic rainfall under controlled environmental conditions. In contrast to these results, Troiano et al. (1982) recorded higher radish yields under high-acidity rainfall in field conditions compared with controls, while Evans et al. (1982) did not observe a significant yield loss in radish. Generally, the growth or yield of plants cultivated under growth chamber or greenhouse conditions appears, for unknown reasons, to be more adversely affected by acid rain compared with plants grown under field conditions (Irving 1983).

One of the most important objectives of research related to the effects of acidic rain on agricultural crops is to determine the impact of acidic rain on the growth and yield of crops cultivated under field conditions. In such experiments in the USA, for example, automatically mobile rainfall-exclusion shelters have been employed for acidic deposition trials conducted in the field (Banwart 1988). These shelters, which automatically move in response to rain, cover the vegetation and exclude ambient precipitation when natural precipitation occurs and when artificial treatments are applied, allowing experimental plants to grow in a microclimate comparable to the normal agricultural field except during natural or simulated precipitation. Open-top field chambers, which have been used to expose plants to air pollutants such as O_3, have also been used with automatic rain exclusion shelters to expose plants to acid rain.

Irving (1987) summarized the results of a field experiment on eight crops (corn, soybean, wheat, timothy/clover, tobacco, potato, oat, and snap bean) using simulated acid rain and ambient rain exclusion conditions in the National Acid Precipitation Assessment Program (NAPAP). None of the crops tested exhibited consistent yield reductions from simulated rain within the average ambient acidity range (pH 4.1-5.1) during the growing season, or even from higher acidity levels found in occasional rain events in the eastern part of the USA (pH 3.0-4.0), when compared with the yields of plants that were exposed to rain without strong acidity (pH 5.6). However, variable results were obtained for some cultivars of soybean. Evans et al. (1984, 1986) reported that simulated acid rain at pH values from 3 to 4 reduced the yields of field-grown crops, such as soybean, when compared to exposures at pH 5.6, apparently due to a decrease in the number of pods per plant.

7.3 Forest Decline

Symptoms of a new type of damage to forest trees first appeared in West Germany in the late 1970s, initially on silver fir, *Abies alba*, and then on Norway spruce, *Picea abies*. For example, Norway spruce trees showed symptoms such as yellowing and early loss of needles, loss of fine roots, and decreased growth, leading to premature death. This phenomenon is currently widespread in Europe and North America, but the causes are complex and controversial.

The complexity of plant-environment interactions, including the acid precipitation phenomena, in the forest decline problem was reviewed by Schütt and Cowling (1985) from both a historical perspective as well as on the basis of common symptoms, and five hypotheses on its causes were presented:

1. The acidification-aluminum toxicity hypothesis: natural acidification of soil increases as a direct or indirect result of deposition of acidic or acidifying substances from the atmosphere. Increased acidity in the soil leads to an increase in concentration of soluble aluminum ions. The resulting aluminum toxicity causes the necrosis of fine roots, leading to subsequent water and nutrient stress and eventual death of the trees.
2. The ozone hypothesis: ozone induces physiological disturbances, which may lead to adverse growth and productivity and consequent dieback. The primary cause of forest decline in North America is considered to be O_3.
3. The Mg-deficiency hypothesis: this hypothesis is based primarily on field observations of yellowing symptoms and low concentrations of magnesium in both soil and leaves at high elevations. Element leaching from foliage is presumably accelerated by O_3 or frost damage of cuticles and cell membranes.
4. The excess nitrogen hypothesis: nitrogen is an essential nutrient generally in limited supply in a forest. Most of the nitrogen is provided by root uptake, but some can be absorbed through the foliage as nitrate (NO_3^-), ammonium (NH_4^+), or ammonia (NH_3) from the atmosphere. Nitrogen inputs from the atmosphere to forests have increased with industrialization, thus promoting growth and increasing the demand for all other essential nutrients, and leading

to deficiencies in these elements. Furthermore, excess nitrogen inhibits the activity of mycorrhizae, increases the susceptibility to frost, root disease fungi, changes the root-shoot ratio, and alters the patterns of nitrification, denitrification, and possibly N fixation.

5. The multiple-stress hypothesis: air pollutants lead to a decrease in net photosynthesis and modify the carbohydrate status. The decrease in the amount of energy to the roots leads to poor development of fine roots and of mycorrhizae, and foliar decline symptoms. The reduced energy status also increases the susceptibility of the trees to other stress factors, such as drought, nutrient deficiency, and biotic pathogens. Thus, the multiple-stress hypothesis, proposed to explain forest decline, suggests that a complex of biotic and abiotic factors interact to influence tree responses to air pollutants.

The cause of forest decline in Europe and North America has not yet been unequivocally resolved. At first, the forest decline was thought to be caused by soil acidification and aluminum toxicity resulting from acid rain (Ulrich et al. 1983). However, from more recent investigations, using controlled and field O_3 exposures and field observations, the cause of these forest declines in central Europe and North America is considered to be primarily O_3 (Miller et al. 1997; Skelly et al. 1997; Rennenberg et al. 1997). On the other hand, another report suggests that O_3 is not currently causing significant declines in Swiss and Austrian forests (Matyssek et al. 1997). However, the cause of forest decline appears to be the result of complex interactions between natural ecosystems and environmental stress factors, including cold injury, drought stress, soil acidity, aluminum toxicity, insect attacks, high concentrations of O_3, and acid precipitation. Forest decline may be caused by a combination of factors, even when one factor is dominant.

8. Combination of Air Pollutants

Plants in the natural environment often are exposed simultaneously to a combination of many air pollutants. The effect of the combination of two or three air pollutants has been examined in phytotrons or open-top field chambers. Most reports discuss a mixture of two air pollutants, such as SO_2 + O_3, SO_2 + NO_2, and NO_2 +O_3, while a combination of three or more pollutants has rarely been reported. From the research that examined such combinations of air pollutants, it was proven that the combined effects may be the same as when each pollutant is applied alone (additive), greater than the sum of each applied alone (synergistic), or less than if each were applied alone (antagonistic). Because responses of the plants differ greatly in combination of gases, pollution concentration, exposure duration, species and cultivars of tested plant, and environmental conditions under exposure experiments, it is not possible to deduce an explicit conclusion. The effects of combinations of air pollutants on plants have been reviewed by Reinert et al. (1975), Ormrod (1982), Runeckles (1984), Guderian (1985b), and Darrall (1989). Over the past 10 years, however, the direction of research has shifted from the combination of phytotoxic air pollutants to the combination of air pollutants and global environmental problems, such as acid rain, ultraviolet-B radiation, and doubling of ambient CO_2 concentration.

8.1 $SO_2 + O_3$

As SO_2 and O_3 are phytotoxic air pollutants and are commonly encountered in the atmosphere, studies on the effects of the mixture of these pollutants on a range of herbaceous and tree species were carried out. Menser and Heggestad (1966) firstly clearly reported that tobacco leaves were injured more than additively by a mixture of SO_2 and O_3. Jacobson and Colavito (1976) also confirmed this phenomenon on tobacco, but found that an antagonistic effect occurred in bean. The extent of foliar injury by the mixture tended to be greater than that from the two pollutants alone at concentrations near the threshold for the individual gases. In other words, synergistic foliar responses may occur at a low SO_2 or O_3 level that induced little injury by itself, and antagonistic foliar effects at a high SO_2 or O_3 level that induced severe injury.

Exposures to mixture of O_3 (50 ppb) and SO_2 (50 ppb) for 8 h/day, 5 day/week for 3-5 weeks caused greater than additive inhibition of early root growth of soybean, additive inhibition of tobacco growth, and less than additive inhibition of alfalfa growth (Tingey and Reinert 1975). Growth reduction in soybean exposed to O_3 (100 ppb) or SO_2 (100 ppb) for 6 h/day throughout the growing season in the field exposure chambers were approximately additive (Heagle et al. 1974). Further, Heagle et al. (1983) exposed to soybean plants O_3 (25 -125 ppb, 7 h seasonal mean for 111 days) and SO_2 (0-367 ppb, 4-h seasonal mean for 101 days) and found growth reduction. They concluded that the resultant growth reductions were the consequence of the independent action of the two gases. Thus, the mixture of SO_2 and O_3 tended to have an additive effect on growth or yield but not synergistic.

8.2 $SO_2 + NO_2$

This combination of pollution occurs in various industrial areas with burning facilities. Exposure to a combination of SO_2 and NO_2 is often more toxic to grasses and woody plants than exposures to the pollutants applied alone. For example, long-term exposures of four grasses to a combination of SO_2 (68 ppb) and NO_2 (68 ppb), the weekly average concentrations, synergistically reduced growth parameters such as leaf area, number of tillers, root weight, and total dry weight (Ashenden and Mansfield 1978). Although the mechanism of this interaction has not yet been elucidated, Wellburn et al. (1981) proposed a mechanism of additive effects of this mixture. NO_2 induced an increase in activity of nitrite reductase (NiR) in chloroplasts; meanwhile, it is known that SO_2 alone had almost no effect on it. NiR leads to NO_2^-, one of the phytotoxic substances, to ammonia and amino acid synthesis, and then to become harmless. Wellburn et al. (1981) found that NiR activity induced by NO_2 caused perfect inhibition when SO_2 was present. As a result, plants may accumulate NO_2^- in cells and cause foliar injury under the combination of SO_2 and NO_2. However, this mechanism fails to explain greater-than-additive effects. Thereafter, from the fact that symptoms of injury caused by the mixtures of SO_2 and NO_2 often resemble those due to O_3 alone (Reinert et al. 1975), Wellburn (1990) proposed that damage caused by the mixture is induced by free radicals generated by either SO_2 and NO_2, similar to that caused by O_3 alone.

8.3 O_3 + NO_2

Both O_3 and NO_2 are constituents of photochemical oxidant pollution, but few studies have been made of their combined effects on plants. Exposure of combination of NO_2 (100 ppb) and O_3 (100 ppb) to Virginia and loblolly pines for 6 h/d for 28 consecutive days considerably reduced plant height (Kress and Skelly, 1982). In addition, Shimizu et al. (1984) reported that the mixture of NO_2 (100 ppb) and O_3 (100 ppb) for a 12-day exposure increased the extent of foliar injury and reduced the growth of sunflower plants. On the other hand, the interaction effects of NO_2 and O_3 on the growth of radish and azalea plants were not significant (Reinert and Gray 1981; Sanders and Reinert 1982). The combination of O_3 and NO_2 frequently produced complex response.

8.4 O_3 + PAN

O_3 and PAN exist in photochemical air pollution, but little is known of how such a combination might affect plants. Using a high PAN concentration (180 ppb), Kohut et al. (1976) showed that a PAN-tolerant clone of hybrid poplar generally developed more injury when exposed to O_3 (180 ppb) and PAN (180 ppb) simultaneously than when exposed to O_3 alone. Using a lower PAN concentration (50 ppb), a combination of O_3 and PAN induced less injury on ponderosa pine needles than did O_3 alone (Davis 1977). Simultaneous exposures to O_3 (100-400 ppb) and PAN (10-40 ppb for petunia and 30-100 ppb for 6- to 7-day-old kidney bean) induced significantly less injury to petunia and kidney bean than either pollutant alone (Nouchi et al. 1984a). Generally, the antagonistic response may occur at the low PAN concentration in the combination of PAN and O_3.

8.5 O_3 + SO_2 + NO_2

A Mixture of O_3 + SO_2 + NO_2 is more realistic pollution than the combination of SO_2 and NO_2. Turf grasses exposed to these gases at a concentration of 150 ppb each showed more foliar injury and greater reduction in leaf area compared with exposures to the pollutants applied singly (Elkiey and Ormrod 1980). In addition, white clover was exposed to combinations of SO_2 (40 ppb) + NO_2 (40 ppb) + O_3 (40 ppb + additional peaks of 2×3 h at 80 ppb and 1×1 h at 110 ppb per week) for 15 weeks. Gas pollution treatments caused substantial reductions in total dry weight (17% for NO_2 + SO_2, 44% for O_3 and 59% for O_3 + SO_2 + NO_2) in comparison with control plants (charcoal-filtered air) (Ashenden et al. 1995). The mixture of three air pollutants tended to show additive or synergistic actions.

8.6 O_3 + Acid Rain

Large-scale forest decline is currently widespread in Europe and North America, but the causes are complex and controversial. Not only O_3 or acid rain alone but also the combination of O_3 and acid rain has been noted as the cause of the forest decline. In reports to date, physiological and growth responses of crops and trees caused by the combination tended to show single effects of each O_3 and acid rain. For example, in the soybean of a greenhouse experiment there was no interaction in growth parameters (Norby and Luxmoor 1983) or physiological processes (Norby et al. 1985). Similarly, no interactive effect between acid rain and O_3 was found on growth parameters of radish and alfalfa (Johnson and Shriner 1986; Rebbek and Brennan 1984).

As for tree species, Matsumura et al. (1998) investigated dry matter production and gas exchange rates of Japanese cedar, Nikko fir, Japanese white birch, and Japanese zelkova seedlings exposed to O_3 and simulated acid rain for 20 weeks. There were no significant combined effects of O_3 (18, 37, 67, and 98 ppb) and simulated acid rain (pH 3.0) the whole plant dry mass, leaf area, net photosynthetic rate, dark respiration rate, gas and liquid phases diffusive conductance to CO_2, carboxylation efficiency of photosynthesis, and CO_2-saturated net photosynthetic rate of Japanese cedar, Nikko fir, Japanese white birch and Japanese zekova seedlings for 20 weeks, except for the dry mass ratio of top and root (T/R ratio) in all species. However, there are some reports that growth and photosynthesis were decreased by the combination of acid rain and ozone in some tree species (Chappelka et al. 1985; Chappelka and Chevone 1986). Further studies on the long-term effects of O_3 and acid rain on forest decline are needed.

8.7 O_3 + CO_2

The tropospheric concentrations of CO_2 and O_3 have increased. Elevated CO_2 potentially enhances photoassimilation and biomass production, whereas O_3 causes reductions in photoassimilation and significant yield losses in agricultural and tree plants. Reports of the effects of such combinations are beginning to appear. For example, Mulchi et al. (1992) have reported an open-top chamber study on soybean conducted in 1989. In growth and physiological characteristics, such as shoot weight, photosynthetic gas exchange, and stomatal resistance, increased CO_2

was found to ameliorate the harmful effects of O_3. However, in combination, the effects of elevated CO_2 and O_3 on plant growth and physiology are not simply additive (Mulchi et al. 1992; McKee et al. 1997). Heagle et al. (1998) conducted an experiment on the yield of soybean in open-top field chambers using a combination of O_3 and enhanced CO_2. They found a strong interaction in that CO_2-induced stimulation was greater for plants stressed by O_3 than for nonstressed plants. For example, CO_2 at 2.0 times ambient increased 2-year mean seed yield by 16%, 24%, and 81% at O_3 levels of 0.4, 0.9, and 1.5 times ambient, respectively. In other words, although the increase in the O_3 concentration decreased the yield of soybean in the present ambient CO_2 concentration, influence of O_3 was almost lacking in the later ambient doubling CO_2 concentration. As CO_2 enrichment decreased stomatal conductance and transpiration of most species, these effects of CO_2 would be expected to decrease the entry of pollutant gases into leaves and thereby decrease adverse pollutant effects (Fiscus et al. 1997).

8.8 O_3 + UV-B

The amount of stratospheric ozone is being reduced due to anthropogenic emission of chlorofluorocarbon and N_2O. This reduction of stratospheric ozone results in increased solar ultraviolet-B radiation (280-320 nm) reaching the surface of the earth. Growth of plants will generally be reduced under enhanced UV-B radiation (Caldwell and Flint 1994; Runeckles and Krupa 1994). Miller et al. (1994) conducted exposure tests to a combination of UV-B radiation and O_3 on soybean in the field. O_3 remarkably lowered growth and yield, while enhanced UV-B radiation did not show any change in growth and yield singly. When enhanced UV-B radiation was added simultaneously to O_3 exposure, the influence of the O_3 did not change the experimental result on their soybean by enhanced UV-B. Since it was considered, by many experiments, that enhanced UV-B by itself has little effect on growth and yield under field conditions (Fiscus and Booker 1995; Kim et al. 1996), enhanced UV-B will not affect any interaction with ozone.

References

Adams RM, Glyer JD, McCarl RA (1988) The NCLAN economic assessment: approach, findings and implications. In: Heck WW, Taylor OC, Tingey DT (eds) Assessment of crop loss from air pollutants. Elsevier, London, pp 473-504

Alscher RG, Bower JI, Zipfel W (1987) The basis for different sensitivities of photosynthesis to SO_2 in two cultivars of pea. J Exp Bot 38:99-108

Alscher RG, Wellburn AR (1994) Plant responses to the gaseous environment. Chapman & Hall, London

Asada K (1980) Formation and scavenging of superoxide in chloroplasts with relation to injury by sulfur oxides. In: National Institute of Environmental Studies (ed) Studies on the effects of air pollutants on plants and mechanisms of phytotoxicity. Res Report No.

11, pp 165-179
Asada K, Kiso K (1973) Initiation of aerobic oxidation of sulphite by illuminated spinach chloroplasts. Eur J Biochem 33:253-257
Ashenden TW, Bell SA (1987) Yield reductions in winter barley grown on a range of soils and exposed to simulated acid rain. Plant Soil 98:433-437
Ashenden TW, Bell SA (1989) Growth responses of three legume species exposed to simulated acid rain. Environ Pollut 62:21-29
Ashenden TW, Mansfield TA (1978) Extreme pollution sensitivity of grasses when SO_2 and NO_2 are present in the atmosphere together. Nature 273:142-143
Ashenden TW, Bell SA, Rafarel CR (1990) Effects of nitrogen dioxide pollution on the growth of three fern species. Environ Pollut 66:301-308
Ashenden TW, Bell SA, Rafarel CR (1995) Responses of white clover to gaseous pollutants and acid mist: implications for setting critical levels and loads. New Phytol 130:89-96
Bae GY, Nakajima N, Ishizuka K, Kondo N (1996) The role in ozone phytotoxicity of the evolution of ethylene upon induction of 1-aminocyclopropane-1-carboxylic acid synthase by ozone fumigation in tomato plants. Plant Cell Physiol 37:129-134
Bamberger ES, Avron M (1975) Site of action of inhibitors of carbon dioxide assimilation by whole lettuce chloroplasts. Plant Physiol 56:481-485
Banwart WL (1988) Field evaluation of an acid rain-drought stress interaction. Environ Pollut 53:123-133
Bell JNB (1980) Response of plants to sulphur dioxide. Nature 284:399-400
Bell JNB (1982) Sulphur dioxide and growth of grasses. In: Unsworth MH, Ormrod DP (eds) Effects of gaseous air pollution in agriculture and horticulture. Butterworths, London, pp 225-246
Benes SE, Murphy TM, Anderson PD, Houpis JLJ (1995) Relationship of antioxidants enzymes to ozone tolerance in branches of mature ponderosa pine (*Pinus ponderosa*) trees exposed to long-term, low concentration, ozone fumigation and acid precipitation. Physiol Plant 94:123-134
Bennett JH, Hill AC (1973) Inhibition of apparent photosynthesis by air pollutants. J Environ Qual 2:526-530
Bennett JH, Hill AC (1975) Interaction of air pollutants with canopies of vegetation. In: Mudd KB, Kozlowski TT (eds) Responses of plants to air pollution. Academic Press, New York, pp 273-306
Bressan RA, LeCureux L, Wilson LG, Filner P (1979) Emission of ethylene and ethane by leaf tissue exposed to injurious concentration of sulfur dioxide or bisulfite ion. Plant Physiol 63:924-930
Caldwell MM, Flint SD (1994) Stratospheric ozone reduction, solar UV-B radiation and terrestrial ecosystems. Clim Change 28:375-394
Capron TM, Mansfield TA (1975) Generation of nitrogen oxide pollutions during CO_2 enrichment of glasshouse atmospheres. J Hortic Sci 50:233-238
Castillo FJ, Heath RL (1990) Ca^{2+} transport in membrane vesicles from pinto bean leaves and its alteration after ozone exposure. Plant Physiol 94:788-795
Chameides WL (1989) The chemistry of ozone deposition to plant leaves: role of ascorbic acid. Environ Sci Technol 19:1206-1213
Chappelka AH, Chevone BI (1986) White ash seedling growth response to ozone and simulated acid rain. Can J For Res 16:786-790
Chappelka AH, Chevone BI, Burk TE (1985) Growth response of yellow-poplar

(*Liriodendron tulipifera* L.) seedlings to ozone, sulfur dioxide, and simulated acidic precipitation, alone and in combination. Environ Exp Bot 25:233-244

Cooley DR, Manning WJ (1987) The impact of ozone on assimilate partitioning in plants: a review. Environ Pollut 47:95-113

Coulson CL, Heath RL (1975) The interaction of peroxyacetyl nitrate (PAN) with the electron flow of isolated chloroplasts. Atmos Environ 9:231-238

Craker LE (1971) Ethylene production from ozone injured plants. Environ Pollut 1:299-304

Dann MS, Pell EJ (1989) Decline of activity and quantity of ribulose bisphosphate carboxylase/oxygenase in ozone-treated potato foliage. Plant Physiol 91:427-432

Darrall NM (1989) The effect of air pollutants on physiological processes in plants. Plant Cell Environ 12:1-30

Davis DD (1977) Response of ponderosa pine primary needle to separate and simultaneous ozone and PAN exposures. Plant Dis Rep 61:640-644

De Kok LJ, Stulen I (1993) Role of glutathione in plants under oxidative stress. In: De Kok LJ, Stulen I, Rennenberg H, Brunold C, Rauser WE (eds) Sulfur nutrition and sulfur assimilation in higher plants. SPB Academic, The Hague, pp 295-313

De Kok LJ, Stulen I (eds) (1998) Responses of plant metabolism to air pollution and global change. Backhuys, Leiden.

Dorminy PJ, Heath RL (1985) Inhibition of the K^+-stimulated ATPase of the plasmalemma of pinto bean leaves by ozone. Plant Physiol 77:43-45

Dugger WM Jr, Mudd JB, Koukol J (1965) Effect of PAN on certain photosynthetic reactions. Arch Environ Health 10:195-200

Dugger WM Jr, Taylor OC, Klein WM, Shropshire W (1963) Action spectrum of peroxyacetyl nitrate damage to bean plants. Nature 198:75-76

Elkiey T, Ormrod PP (1980) Response of turf grass cultivars to ozone, sulfur dioxide, nitrogen dioxide, or their mixtures. J Am Soc Hortic Sci 105:664-668

Elstner EF (1987) Ozone and ethylene stress. Nature 328:482

Elstner EF, Osswald W, Youngman RJ (1985) Basic mechanisms of pigment bleaching and loss of structural resistance in spruce (*Picea abies*) needles: advances in phytomedical diagnostics. Experientia 41:591-597

Evans KS, Lewin KF, Cunningham EA, Patti MJ (1982) Effects of simulated acid rain on yields of field-grown crops. New Phytol 91:429-441

Evans LS, Lewin KF, Patti MJ (1984) Effects of simulated acid rain on yields of field-grown soybeans. New Phytol 96:207-213

Evans LS, Lewin KF, Owen EL, Santucci KA (1986) Comparison of yields of several cultivars of field-grown soybeans exposed to simulated acidic rainfalls. New Phytol 102:409-417

Fiscus EL, Booker FL (1995) Is increased UV-B a threat to crop photosynthesis and productivity? Photosynth Res 43:81-92

Fiscus EL, Reid CD, Miller JE, Heagle AS (1997) Elevated CO_2 reduces O_3 flux and O_3-induced yield losses in soybeans: possible implications for elevated CO_2 studies. J Exp Bot 48:307-313

Fredrick P, Heath RL (1975) Ozone-induced fatty acid and viability changes in *Chlorella*. Plant Physiol 55:15-19

Furukawa A (1984) Photosynthesis inhibition of higher plants by various air pollutants. Res Rep Natl Inst Environ Studies Jpn 64:131-139 (in Japanese)

Furukawa A, Natori T, Totsuka T (1980) The effect of SO_2 on net photosynthesis in

sunflower leaf. Res Rep Natl Inst Environ Studies Jpn 11:1-8
Gezeilus K, Hallgren JE (1980) Effect of SO_3^{2-} on the activity of ribulose biphosphate carboxylase from seedlings of *Pinus sylvestris*. Physiol Plant 49:354-358
Ghisi R, Dittrich APM, Herber U (1990) Oxidation versus reductive detoxification of SO_2 by chloroplasts. Plant Physiol 92:842-849
Guderian R (ed) (1985a) Air pollution by photochemical oxidants. Springer-Verlag, Berlin
Guderian R (1985b) Effects of pollutant combination. In: Guderian R (ed) Air pollution by photochemical oxidants. Springer-Verlag, Berlin, pp 246-275
Hashida S (1965) Soil scientific and nutritional problems of cultivation in plastic greenhouses (in Japanese). Dojou Hiryou Gaku Zasshi (Jpn J Soil Sci Plant Nutr) 36:274-284
Heagle AS, Body BE, Nealy GE (1974) Injury and yield responses of soybean to chronic doses of ozone and sulfur dioxide in the field. Phytopathology 64:132-136
Heagle AS, Heck WW, Rawlings JO, Philbeck RB (1983) Effects of chronic doses of ozone and sulfur dioxide on injury and yield of soybeans in open-top chambers. Crop Sci 23:1184-1191
Heagle AS, Miller JE, Pursley WA (1998) Influence of ozone stress on soybean response to carbon dioxide enrichment. III. Yield and seed quality. Crop Sci 38:128-134
Heath RL (1980) Initial events in injury to plants by air pollutants. Annu Rev Plant Physiol 31:395-431
Heath Rl (1987) The biochemistry of ozone attack on the plasma membrane of plant cells. Rec Adv Photochem 21:29-54
Heath RL, Taylor GE (1997) Physiological processes and plant responses to ozone exposure. In: Sandermann H, Wellburn AR, Heath RL (eds) Forest decline and ozone, Springer-Verlag, Berlin, pp 317-368
Heck WW, Taylor OC, Tingey DT (eds) (1988) Assessment of crop loss from air pollutants. Elsevier, London
Hewitt CN, Kok GL, Fall R (1990) Hydroperoxides in plants exposed to ozone mediate air pollution damage to alkene emitter. Nature 344:56-58
Hosono T, Nouchi I (1992) Effects of simulated acid rain on the growth of radish, spinach and bush bean plants. Taiki Osen Gakkaishi (J Jpn Soc Air Pollut) 27:111-121 (in Japanese with English summary)
Hosono T, Nouchi I (1994) Effects of simulated acid rain on growth, yield and net-photosynthesis of several agricultural crops. Nougyo Kisho (J Agric Meteorol) 50:121-127 (in Japanese with English summary)
Hossain MA, Asada K (1984) Inactivation of ascorbate peroxidase in spinach chloroplasts on dark addition of hydrogen peroxide: its protection by ascorbate. Plant Cell Physiol 25:1285-1295
Irving PM (1983) Acidic precipitation effects on crops: a review and analysis of research. J Environ Qual 12:442-453
Irving PM (1985) Modeling the response of greenhouse-grown radish plants to acid rain. Environ Exp Bot 25:327-338
Irving PM (1987) Effects on agricultural crops. In: National Acid Precipitation Assessment Program (NAPAP). Interim assessment: the cause and effects of acidic deposition, vol IV. NAPAP, Washington, DC, pp 6.1-6.50
Jacob B, Heber U (1998) Apoplastic ascorbate does not prevent the oxidation of fluorescent amphiphilic dyes by ambient and elevated concentrations of ozone in leaves. Plant Cell Physiol 39:313-322

Jacobson JS, Colavito LJ (1976) The combined effect of sulfur dioxide and ozone on bean and tobacco plants. Environ Exp Bot 16:277-285

Jacobson JS, Irving PM, Kuja A, Lee J, Sjriner DS, Troiano J, Perrigan S, Cullinan V (1988) A collaborative effort to model plant response to acidic rain. J Air Pollut Control Assoc 38:777-783

John WW, Curtis RW (1977) Isolation and identification of the precursor of ethane in *Phaseolus vulgaris* L. Plant Physiol 59:521-522

Johnson JW, Shriner DS (1986) Yield responses of Davis soybean to simulated acid rain and gaseous pollutants. New Phytol 103:695-707

Kangasjärvi J, Talvien J, Utriainen M, Karjalainen R (1994) Plant defense system induced by ozone. Plant Cell Environ 17:783-794

Kanofsky JR, Sima PD (1995) Reactive absorption of ozone by aqueous biomolecule solutions: implications for the role of sulfhydryl compounds as targets for ozone. Arch Biochem Biophys 316:52-62

Kim HY, Kobayashi K, Nouchi I, Yoneyama T (1996) Enhanced UV-B radiation has little effect on growth ^{13}C values and pigments of pot-grown rice (*Oryza sativa*) in the field. Physiol Plant 96:1-5

Kobayashi K (1992) Modeling and assessing the impact of ozone on rice growth and yield. In: Bergland RL (ed) Tropospheric ozone and the environment. II: Effects, modeling and control. Air and Waste Management Association, Pittsburgh, pp 537-551

Kobayashi K, Okada M (1995) Effects of ozone on the light use of rice (*Oryza sativa* L.) plants. Agric Ecosyst Environ 53:1-12

Kohut RJ, Davis DD, Merrill W (1976) Response of hybrid poplar to simultaneous exposure to ozone and PAN. Plant Dis Rep 60:777-780

Kondo N, Saji H (1992) Tolerance of plants to air pollutants (in Japanese with English summary). Taiki Osen Gakkaishi (J Jpn Soc Air Pollut) 27:273-288

Koukol J, Dugger WM Jr, Palmer RL (1967) Inhibitory effect of peroxyacetyl nitrate on cyclic photophosphorylation by chloroplasts from black valentine bean leaves. Plant Physiol 42:1419-1422

Kress LW, Skelly JM (1982) Response of several eastern forest tree species to chronic doses of ozone and nitrogen dioxide. Plant Dis 66:1149-1152

Langebartels C, Kerner K, Leonardi S, Schraudner M, Trost M, Heller W, Sandermann H (1991) Biochemical plant responses to ozone. I. Differential induction of polyamine and ethylene biosynthesis in tobacco. Plant Physiol 95:882-889

Law RM, Mansfield TA (1982) Oxides of nitrogen and the greenhouse atmosphere. In: Unsworth MH, Ormrod DP (eds) Effects of gaseous air pollution in agriculture and horticulture. Butterworths, London, pp 93-112

Lea PJ, Robinson SA, Stewart GR (1990) The enzymology and metabolism of glutamine, glutamate and asparagines. In: Miflin BJ, Lea PJ (eds) The biochemistry of plants, vol, 16. Academic Press, New York, pp 121-159

Lee JJ, Neely GE, Perrjiean SC, Grothaus LC (1981) Effects of simulated sulfuric acid rain on yield, growth and foliar injury of several crops. Environ Exp Bot 21:171-185

Lefohn AS (ed) (1991) Surface-level ozone exposures and their effects on vegetation. Lewis, Chelsea

Luwe M, Takahama U, Heber U (1993) Role of ascorbate in detoxifying ozone in the apoplast of spinach (*Spinacia oleracea* L.) leaves. Plant Physiol 101:969-976

Malhotra SS, Khan AA (1984) Biochemical and physiological impact of major pollutants. In: Treshow M (ed) Air pollution and plant life. Wiley, Chichester, pp 113-157

Matsumura H, Kobayashi T, Kohno Y (1998) Effects of ozone and/or simulated acid rain on dry weight and gas exchange rates of Japanese cedar, Nikko fir, Japanese white birch and Japanese zelkova seedlings (in Japanese with English summary). Taiki Kankyo Gakkaishi (J Jpn Soc Atmos Environ) 33:16-35

Matyssek R, Havranek WM, Wieser G, Innes JL (1997) Ozone and forests in Austria and Switzerland. In: Sanderman H, Wellburn AR, Heath RL (eds) Forest decline and ozone. Springer-Verlag, Berlin, pp 95-134

McKee IF, Bullimore JF, Long SP (1997) Will elevated CO_2 concentrations protect the yield of wheat from O_3 damage ? Plant Cell Environ 18:215-225

Mehlhorn H, Wellburn AR (1987) Stress ethylene formation determines plant sensitivity to ozone. Nature 327:417-418

Mehlhorn H, Tabner B, Wellburn AR (1990) Electron spin resonance evidence for the formation of free radicals in plants exposed to ozone. Physiol Plant 79:377-383

Mehlhorn H, O'Shea JM, Wellburn AR (1991) Atmospheric ozone interacts with stress ethylene formation by plants to cause visible plant injury. J Exp Bot 42:17-24

Mehlhorn H, Seufert G, Schmidt A, Kunert KJ (1986) Effects of SO_2 and O_3 on production of antioxidants in conifers. Plant Physiol 82:336-338

Menser HA, Heggestad HE (1966) Ozone and sulfur dioxide synergism: injury to tobacco plants. Science 153:424-425

Middleton JR, Kendrick JB Jr, Schwalm HW (1950) Injury to herbaceous plants by smog or air pollution. Plant Dis Rep 34:245-252

Miller JE (1987) Effects of ozone and sulfur dioxide stress on growth and carbon allocation in plants. Rec Adv Phytochem 21:55-100

Miller PR, Arbaugh MJ, Temple PJ (1997) Ozone and its known and potential effects on forests in western United States. In: Sanderman H, Wellburn AR, Heath RL (eds) Forest decline and ozone. Springer, Berlin, pp 39-67

Miller JE, Booker FL, Fiscus EL, Heagle AS, Pursley WA, Vozzo S, Heck WW (1994) Ultraviolet-B radiation and ozone effects on growth, yield and photosynthesis of soybean. J Environ Qual 23:83-91

Mudd JB (1963) Enzyme inactivation by peroxyacetyl nitrate. Arch Biochem Biophys 102:59-65

Mudd JB (1996) Biochemical basis for the toxicity of ozone. In: Iqbal M, Yunus M (eds) Plant responses to air pollution. Willy, Chichester, pp 267-283

Mudd JB (1973) Biochemical effects of some air pollutants on plants. Adv Chem Ser 122:31-47

Mudd JB (1975a) Sulfur dioxide. In: Mudd JB, Kozlowski TT (eds) Responses of plants to air pollution. Academic Press, New York, pp 9-22

Mudd JB (1975b) Peroxyacyl nitrates. In: Mudd JB, Kozlowski TT (eds) Responses of plants to air pollution. Academic Press, New York, pp 97-119

Mudd JB, Dugger WM Jr (1963) The oxidation of pyridine nucleotides by peroxyacyl nitrates. Arch Biochem Biophys 102:52-58

Mudd JB, Kozlowski TT (eds) (1975) Responses of plants to air pollution, Academic Press, New York

Mulchi CL, Slaughter L, Saleem M, Lee EH, Pausch R, Rowland R (1992) Growth and physiological characteristics of soybean in open-top chambers in response to ozone and increased atmospheric CO_2. Agric Ecosyst Environ 38:107-118

Nakamura H, Saka H (1978) Photochemical oxidants injury in rice plants, III: Effect of ozone on physiological activities in rice plants (in Japanese with English summary).

Nippon Sakumotsu Gakkai Kiji (Jpn J Crop Sci) 47:707-714
Norby RJ, Luxmoore RJ (1983) Growth analysis of soybean exposed to simulated acid rain and gaseous air pollutants. New Phytol 95:277-287
Norby RJ, Richer DD, Luxmoore RJ (1985) Physiological process in soybean inhibited by gaseous pollutants but not by acid rain. New Phytol 100:79-85
Nouchi I (1979) Effects of ozone and PAN concentrations and exposure duration on plant injury (in Japanese with English summary). Taiki Osen Gakkaishi (J Jpn Soc Air Pollut) 14:489-496
Nouchi I (1988) Leaf injury of plants and mechanism of injury by photochemical oxidants (ozone and peroxyacetyl nitrate) (in Japanese with English summary). Bull Natl Inst Agro-Environ Sci 5:1-121
Nouchi I (1992) Acid precipitation in Japan and its impact on plants. 1. Acid precipitation and foliar injury. JARQ 26:171-177
Nouchi I (1993) Changes in antioxidant levels and activities of related enzymes in rice leaves exposed to ozone. Soil Sci Plant Nutr 39:309-320
Nouchi I, Mayumi H, Yamazoe F (1984a) Foliar injury response of petunia and kidney bean to simultaneous and alternate exposures to ozone and PAN. Atmos Environ 18:453-460
Nouchi I, Ohashi T, Sofuku M (1984b) Atmospheric PAN concentrations and foliar injury to petunia indicator plants in Tokyo (in Japanese with English summary). Taiki Osen Gakkaishi (J Jpn Soc Air Pollut) 19:392-402
Nouchi I, Toyama S (1988) Effects of ozone and peroxyacetyl nitrate on polar lipids and fatty acids in leaves of morning glory and kidney bean. Plant Physiol 87:638-646
Okano K, Ito O, Takeba G, Shimizu A, Totsuka T (1984) Alteration of ^{13}C-acculimate partitioning in plants of *Phaseolus vurgaris* exposed to ozone. New Phytol 97:155-163
Oltmans DJ, Lefohn AS, Scheel HE, Harris JM, Levy H II, Galbally IE, Brunke EG, Meyer CP, Lathrop JA, Johnson BJ, Shadwick DS, Cuevas E, Schmidlin FJ, Tarasick DW, Claude H, Kerr JB, Uchino O, Mohnen V (1998) Trends of ozone in the troposphere. Geophys Res Lett 25:139-142
Ormrod DP (ed) (1978) Pollution in horticulture. Elsevier, New York
Ormrod DP (1982) Air pollutant interactions in mixtures. In: Unsworth MH, Ormrod DP (eds) Effects of gaseous air pollution in agriculture and horticulture. Butterworths, London, pp 307-331
Pauls KP, Thompson JE (1981) Effects of in vitro treatment with ozone on physical and chemical properties of membranes. Physiol Plant 53:255-262
Peiser GD, Yang SF (1979) Ethylene and ethane production from sulfur dioxide-injured plants. Plant Physiol 63:142-145
Pell EJ, Pearson NS (1983) Ozone-induced reduction in quantity of ribulose-1,5-bisphosphate carboxylase in alfalfa foliage. Plant Physiol 73:185-187
Pell EJ, Eckardt NA, Enyedi AJ (1992) Timing of ozone stress and resulting status of ribulose bisphosphate acrboxylase/oxygenase and associated net photosynthesis. New Phytol 120:387-405
Pell EJ, Landry LG, Eckardt NA, Glick RE (1994) Air pollution and Rubisco: effects and implications. In: Alscher RG, Wellburn AR (eds) Plant responses to the gaseous environment. Chapman & Hall, London, pp 239-253
Polle A, Junkermann W (1994) Inhibition of apoplastic and symplastic peroxidase activity from Norway spruce by the photooxidant hydroxymethyl hydroperoxide. Plant Physiol 104:617-621

Rautenkranz AAF, Li L, Machler F, Martinoia E, Oertli JJ (1994) Transport of ascorbic and dehydroascorbic acids across protoplast and vacuole membranes isolated from barley (*Hordeum vulgare* L. cv Gerbel) leaves. Plant Physiol 106:187-193

Rebbeck J, Brennan E (1984) The effect of simulated acid rain and ozone on the yield and quality of glasshouse-grown alfalfa. Environ Pollut Ser A 36:7-16

Reinert RA, Gray TN (1981) The response of radish to nitrogen dioxide, sulfur dioxide, and ozone, alone and in combination. J Environ Qual 10:240-243

Reinert RA, Heagle AS, Heck WW (1975) Plant response to pollutant combination. In: Mudd JB, Kozlowski TT (eds) Responses of plants to air pollution. Academic Press, New York, pp 159-177

Rennenberg H (1984) The fate of excess sulfur in higher plants. Annu Rev Plant Physiol 35:121-153.

Rennenberg H, Polle A, Reuther M (1997) Role of ozone in forest decline on Wank mountain (Alps). In: Sanderman H, Wellburn AR, Heath RL (eds) Forest decline and ozone. Springer-Verlag, Berlin, pp 135-162

Runeckles VC (1984) Impact of air pollutant combinations on plants. In: Treshow M (ed) Air pollution and life. Wiley, Chichester, pp 239-258

Runeckles VC, Chevone BI (1991) Crop responses to ozone. In: Lefohn AS (ed) Surface level ozone exposures and their effects on vegetation. Lewis, Chelsea, pp 157-270

Runeckles VC, Krupa SV (1994) The impact of UV-B radiation and ozone on terrestrial vegetation. Environ Pollut 83:191-213

Sakaki T (1998) Photochemical oxidants: toxicity. In: De Kok LJ, Stulen I (eds) Responses of plant metabolism to air pollution and global change. Backhuys, Leiden, pp 117-129

Sakaki T, Kondo N, Sugahara K (1983) Breakdown of photosynthetic pigments and lipids in spinach leaves with ozone fumigation: role of active oxygens. Physiol Plant 59:28-34

Sakaki T, Ohnishi J, Kondo N, Yamada M (1985) Polar and neutral lipid changes in spinach leaves with ozone fumigation: triacylglycerol synthesis from polar lipids. Plant Cell Physiol 26:253-262

Sakaki T, Saito K, Kawaguchi A, Kondo N, Yamada M (1990a) Conversion of monogalacosyldiacylglycerols to triacylglycerols in ozone-fumigated spinach leaves. Plant Physiol 94:766-772

Sakaki T, Kondo N, Yamada M (1990b) Pathway for the synthesis of triacylglycerols from monogalactosyldiacylglycerols in ozone-fumigated spinach leaves. Plant Physiol 94:773-780

Sakaki T, Kondo N, Yamada M (1990c) Free fatty acids regulate two galactosyltransferases in chloroplast envelope membranes isolated from spinach leaves. Plant Physiol 94:781-787

Sanders JS, Reinert RA (1982) Screening azalea cultivars for sensitivity to nitrogen dioxide, sulfur dioxide, and ozone alone and in mixtures. J Am Soc Hortic Sci 107:87-90

Sandermann H, Wellburn AR, Heath RL (ed) (1997) Forest decline and ozone. Springer-Verlag, Berlin

Scandalios JG (1994) Molecular biology of superoxide dismutase. In: Alscher RG, Wellburn AR (eds) Plant responses to the gaseous environment. Chapman & Hall, London, pp 147-164

Schreiber U, Vidaver W, Runeckles VC, Rosen P (1978) Chlorophyll fluorescence assay for ozone injury in intact plants. Plant Physiol 61:80-84

Schütt P, Cowling EB (1985) Waldsterben, a general decline of forest in central Europe:

symptoms, development and possible causes. Plant Dis 69:548-558
Sen Gupta A, Alsher RG, McCune D (1991) Response of photosynthesis and cellular antioxidants to ozone in Populus leaves. Plant Physiol 96:650-655
Shaw PJA, Holland MR, Darrall NM, McLead AR (1993) The occurrence of SO_2-related foliar symptoms on Scots pine (*Pinus sylvestris* L.) in an open-air forest fumigation experiment. New Phytol 123:143-152
Shimazaki K (1988) Thylakoid membrane reactions to air pollutants. In: Schulte-Hostede S, Darrnall NM, Blank LW, Wellburn AR (eds) Air pollution and plant metabolism. Elsevier, London, pp 116-133
Shimazaki K, Sugahara K (1979) Specific inhibition of photosystem II activity in chloroplasts by fumigation of spinach leaves with SO_2. Plant Cell Physiol 20:26-35
Shimazaki K, Yu SW, Sakaki T, Tanaka K (1992) Differences between spinach and kidney bean plants in terms of sensitivity to fumigation with NO_2. Plant Cell Physiol 33:267-273
Shimizu H, Oikawa T, Totsuka T (1984) Effects of low concentration of NO_2 and O_3 alone and in mixture on growth of sunflower plants. Res Rep Natl Inst Environ Studies Jpn 65:121-136
Skelly JM, Chappelka AH, Laurence JA, Frsdericksen TS (1997) Ozone and its known and potential effects on forests in Eastern United States. In: Sanderman H, Wellburn AR, Heath RL (eds) Forest decline and ozone. Springer-Verlag, Berlin, pp 69-93
Solomonson LP, Barber MJ (1990) Assimilatory nitrate reductase: functional properties and regulation. Annu Rev Plant Physiol 41:225-253
Srivastava HS, Ormrod DP (1984) Effects of nitrogen dioxide and nitrate nutrition on growth and nitrite assimilation in bean leaves. Plant Physiol 76:418-423
Sugahara K, Ogura K, Takimoto M, Kondo N (1984) Effects of air pollutant mixtures on photosynthetic electron transport systems. Res Rep Natl Inst Environ Studies Jpn 65:155-165
Takeuchi Y, Nihira J, Kondo N, Tezuka T (1985) Change in nitrate-reducing activity in spinach seedlings with NO_2 fumigation. Plant Cell Physiol 26:1027-1035
Tanaka K, Sugahara K (1980) Role of superoxide dismutase in defense against SO_2 toxicity and an increase in superoxide dismutase activity with SO_2 fumigation. Plant Cell Physiol 21:601-611
Tanaka K, Kondo N, Sugahara K (1982) Accumulation of hydrogen peroxide in chloroplasts of SO_2^- fumigated spinach leaves. Plant Cell Physiol 23:999-1007
Tanaka K, Suda Y, Kondo N, Sugahara K (1985) O_3 tolerance and the ascorbate-dependent H_2O_2 decomposing system in chloroplasts. Plant Cell Physiol 26:1425-1431
Tanaka K, Saji H, Kondo N (1988) Immunological properties of spinach glutathione reductase and inductive biosynthesis of enzyme with ozone. Plant Cell Physiol 29:637-642
Taylor OC (1969) Importance of peroxyacetyl nitrate (PAN) as a phytotoxic air pollutant. J Air Pollut Control Assoc 19:347-351
Taylor GE (1978) Genetic analysis of ecotypic differentiation within annual plant species, *Geranium carolinianum* L., in response to sulfur dioxide. Bot Gaz 139:362-368
Temple PJ, Taylor OC (1983) World-wide ambient measurements of peroxuacetyl nitrate (PAN) and implications for plant injury. Atmos Environ 17:1583-1587
The National Research Council of Canada (1939) Effect of sulphur dioxide on vegetation. National Research Council of Canada Publication 815
Thompson JE, Legge RL, Barber RF (1987) The role of free radicals in senescence and

wounding. New Phytol 105:317-344
Tingey DT, Reinert RA (1975) The effect of ozone and sulfur dioxide singly and in combination on plant growth. Environ Pollut 9:117-125
Tingey DT, Standley C, Field RW (1976) Stress ethylene evolution: a measure of ozone effects on plants. Atmos Environ 10:969-974
Treshow M (ed) (1984) Air pollution and plant life. Wiley, Chichester
Treshow M, Anderson FK (eds) (1991) Plant Stress from Air Pollution. Wiley, Chischester, pp 61-76
Troiano J, Heller L, Jacobson JS (1982) Effects of added water and acidity of simulated rain on growth of field-grown radish. Environ Pollut Ser A 29:1-11
Ulrich B, Mayer R, Khanna PK (1983) Chemical changes due to acid precipitation in a loess-derived soil in central Europe. Soil Sci 130:193-199
Unsworth MH, Geissler P (1993) Results and achievements of European Open-top Chamber Network. In: Jäger HJ, Unsworth MH, De Temmerman L, Mathy P (eds) Effects of air pollution on agricultural crops in Europe. Commission of the European Communities, Brussels, pp 5-22
UN-ECE (1994) Workshop report 16. ECE Critical Levels Workshop (Fuher J, ed), March 1988. Bad Harzburg, Germany
UN-ECE (1996) Critical levels for ozone in Europe: testing and finalisting the concepts. Department of Ecology and Environment Science, University of Kuopio, Finland
Wellburn AR (1990) Why are atmospheric oxides of nitrogen usually phytotoxic and not alternative fertilizers? New Phytol 115:395-429
Wellburn AR (ed) (1994a) Air pollution and climate change, 2nd edn. Longman, Essex
Wellburn AR (1994b) Ozone, PAN and photochemical smog. In: Wellburn AR (ed) Air pollution and climate change, 2nd edn. Longman, Essex, pp 123-144
Wellburn AR, Higginson C, Robinson D, Walmsley C (1981) Biochemical explanations of more than additive inhibitory low atmospheric levels of SO_2 + NO_2 upon plants. New Phytol 88:223-237
Wright EA (1987) Effects of SO_2 and NO_2, singly or in mixture, on the macroscopic growth of three birch clones. Environ Pollut 46:209-221
Yang SF, Hoffman NE (1984) Ethylene biosynthesis and its regulation in higher plants. Annu Rev Plant Physiol 35:155-189
Yoneyama T, Sasakawa H (1979) Transformation of atmospheric NO_2 absorbed in spinach leaves. Plant Cell Physiol 20:263-266
Yoneyama T, Sasakawa H, Ishizuka S (1979) Absorption of atmospheric NO_2 by plants and soils. II: Nitrite accumulation, nitrite reductase activity, and diurnal change, of NO_2 absorption in leaves. Soil Sci Plant Nutr 25:267-275
Zeevaart AJ (1976) Some effects of fumigating plants for short periods with NO_2. Environ Pollut 11:97-108
Ziegler I (1972) The effects of SO_3^{2-} on the activity of ribulose-1,5-diphosphate carboxylase in isolated spinach chloroplasts. Planta 103:155-163
Ziegler I (1973) The effect of air polluting gases on plant metabolism. Environ Qual Safe 2:182-208

2
Plants as Bioindicators of Air Pollutants

Isamu Nouchi

Agro-Meteorology Group, National Institute for Agro-Environmental Sciences, Kannondai 3-1-3, Tsukuba, Ibaraki 305-8604, Japan

1. Introduction

Plants cannot select and move their living places as can animals, and thus live their lives responding to their surrounding environment. Whenever environmental components, such as temperature, soil water content, nutrients, and air pollutants, exceed the range to which the plants can adapt or become limiting, the plant develops abnormal symptoms or growth. The appearance of such abnormal symptoms or growth is a good indicator of the dangers of environmental pollution to human beings. A number of air pollutants, such as sulfur dioxide, nitrogen oxides, ozone, peroxyacetyl nitrate, halogens, and acid rain, can damage plants. Therefore, plants offer an excellent alarm system for detecting the presence of excessive concentrations of these air pollutants and often provide the very first evidence that the air is polluted. Plant responses, especially characteristic foliar symptoms, have long been used as indicators of air pollutants. In addition, the amount of metal accumulation has also been used as a bioindicator.

At present, the most widespread phytotoxic air pollutant that exerts the greatest effect on the terrestrial ecosystem, is ozone (O_3). Therefore, there is considerable research into the identification of bioindicators for O_3. Hence, an outline of indicator plants, with O_3 as the central figure, is presented in this chapter.

Air Pollution and Plant Biotechnology
–Prospects for Phytomonitoring and Phytoremediation–
Edited by K. Omasa, H. Saji, S. Youssefian, and N. Kondo
© Springer -Verlag Tokyo 2002

2. Bioindicators for Sulfur Dioxide

Many plants are known to be injured by sulfur dioxide (SO_2) under natural and experimental exposure conditions. Several hours of exposure to 100-300 ppb SO_2 can cause visible symptoms in sensitive plants such as alfalfa, buckwheat, and hybrid poplar. The indicator and accumlator plants used for monitoring of effects of SO_2 in the National Monitoring Network for Air Pollution in the Netherlands were alfalfa (*Medicago sativa* L.) and buckwheat (*Fagopyrum esculentum* Mönch) (Posthumus 1984).

Lower plants, such as lichens and mosses, are also sensitive to SO_2, and some make excellent bioindicators of atmospheric SO_2 (LeBlanc and DeSloover 1970; Taoda 1972). A survey of the distribution and abundance of lichens on tree trunks around industrial and urban areas showed that SO_2 was the prime cause of the disappearance of lichens and mosses. It was observed that an annual mean SO_2 concentration of more than 50-60 ppb is sufficient to kill lichens (LeBlanc and Rao 1975). LeBlanc et al. (1972) had reported that the normal lichen distribution zone (corresponding to the area of minimum or relatively no pollution) would occur when the annual SO_2 concentration reached less than 5 ppb. In addition, DeSloover and LeBlanc (1968) developed an Index of Atmospheric Purity (IAP), based on a mathematical formula that correlates the lichen and bryophyte vegetation of an area with the air quality around urban areas or point sources of SO_2. Although the IAP is a useful air quality evaluation method, its ecological index depends on the method by which the area investigated is selected, and thus the IAP values of different areas cannot be compared. Several modified IAP methods for correlating SO_2 pollution in different areas have been developed (for example, Nakagawa and Kobayashi 1990).

Furthermore, because SO_2 accumulates in the leaves as sulfurous compounds, SO_2 injury can be identified by chemical analysis of leaf tissues. Although sulfur (S) is a natural constituent of plant tissues, and a healthy plant without SO_2 impact contains S, SO_2 injury can be diagnosed by comparing the S content of the same species from a nonpolluted region. In general, the S content in a healthy plant grown in a nonpolluted area is about 1500-3000 ppm on a dry weight basis.

3. Bioindicators for Hydrogen Fluoride

Hydrogen fluoride (HF) is one of the most phytotoxic air pollutants. Monocotyledonous horticultural plants, such as tulips and gladioli, and fruit species, such as plums and apricots, are specially sensitive to HF. The F ion accumulates in the leaf tips and leaf margins, causing necrosis of these sites. Because the F ion accumulates in the leaves, as does sulfur from SO_2, fluoride injury can be diagnosed by chemical analysis. In general, the F content in the leaves of healthy plants is less than 20 ppm.

Gladiolus (*Gladiolus hortulanus* Bailey) is the most widely used plant for biomonitoring fluoride (Manning and Feder 1980). In a field polluted by atmospheric fluoride, the accumulation fluoride in gladiolus leaves was correlated with fluoride concentrations in ambient air (Yamamoto et al. 1975). Furthermore, the percent leaf injury could be correlated with fluoride concentrations in leaves (Vasiloff and Smith 1974).

4. Bioindicators for Ethene (Ethylene)

Although ethene (C_2H_4) is a plant hormone, it is a known and important phytotoxic air pollutant. External sources of C_2H_4, such as from auto exhaust, may induce the same effects as endogenous C_2H_4, including inhibition of stem elongation and leaf expansion, promotion of senescence and abscission, induction of epinasty, and a reduction of flowering and fruit set. A reduction of flowering and fruitset was found in petunia, cucumber, soybean, kidney bean, and wheat when exposed to 25 ppb of C_2H_4 (Abeles and Heggestad 1973). Epinasty in tomato was proposed as a useful biomonitoring symptom for atmospheric C_2H_4 (Abeles and Heggestad 1973). Posthumus (1983) suggested the use of petunia (*Petunia axilliaris hybrida* multiflora, cv. White Joy) as an indicator plant for C_2H_4 in the Netherlands, as its flower size decreased and the flower-bud abortion rate increased in response to C_2H_4. Pleijel et al. (1994) used potted petunia (*Petunia hybrida*), placed at distances of 10, 20, 40, 80, and 120 m from a motorway with approximately 30,000 vehicles/day, as an indicator plant for C_2H_4 in Sweden in 1989. The petunia flowers were significantly smaller on plants closer to the motorway than those more distant. In addition, the abortion of flower buds was more frequent and the rate of fruit ripening was higher close to the motorway. The authors concluded from the survey that the C_2H_4 concentrations close to the motorway were high enough to influence the petunia reproductive structures.

5. Bioindicators for Ozone

Ozone (O_3) is a widespread phytotoxic air pollutant, and some excellent bioindicator plants have been used for many years to evaluate its presence. The Bel W_3 tobacco (*Nicotiana tabacum* L.) has been widely used as a bioindicator for O_3 all over the world (Heggestad 1991). Morning glory (*Pharbitis nil* L. or *Ipomea nil* L.) in Japan (Matsunaka 1977; Nouchi and Aoki 1979; Matsumaru 1994) and clover in Sweden (Karlsson et al. 1995) have also been used as indicator plants for O_3. In addition, a growth reduction in radish (*Raphanus sativus* L.) has been used as an indicator for O_3 in Japan (Izuta et al. 1991, 1993) and Egypt (Hassan et al. 1995).

Fig. 1. Bleached stipples on older leaves of tobacco ver. Wel-W_3 caused by ambient O_3

5.1 Tobacco

"Weather fleck" symptoms on tobacco (Fig. 1) were first reported to be caused by ambient O_3 in 1959 (Heggestad and Middleton 1959). The cultivar Bel-W_3 was selected as an extremely sensitive cultivar for detecting low concentrations of O_3 (Heggestad and Menser 1962), with injury symptoms appearing after exposure to 100 ppb O_3 for 2 h (Menser et al. 1963). Larsen and Heck (1976), using an empirical dose-response model based on ozone exposure experiments, reported that the threshold for foliar injury to Bel-W_3 plants was 30 ppb O_3 for 8 h, the lowest for all species tested. In contrast, cultivar Bel-B was found to be tolerant to O_3, and injury occurrence required exposure to 220 ppb O_3 for 2 h (Menser et al. 1963). Therefore, a set including both Bel-W_3 and Bel-B, as sensitive and tolerant cultivars, respectively, was recommended as a bioindicator for ozone (Manning and Feder 1980). Heggestad (1991) summarized the results from 14 different countries (Australia, Denmark, Germany, Holland, India, Israel, Italy, Japan, Mexico, Sweden, Taiwan, United Kingdom, and USA). A pilot study for detecting O_3 and determining its phytotoxity in a forest area of Poland was also initiated in 1991 using two Bel-W_3 and Bel-B cultivars (Bytnerowicz et al. 1993). Two major surveys, reported in the United Kingdom and Holland, are presented below.

5.1.1 Nationwide Surveys of O_3 in the British Isles

Nationwide surveys of O_3 using Bel-W_3 were carried out at 60 sites throughout the

UK in the summer of 1977 (Ashmore et al. 1978, 1980). Eight-week-old tobacco plants were transplanted and exposed to ambient O_3 for 4 weeks. The plants were replaced by a fresh batch of the same 8-week-old plants. The amount of leaf injury was recorded at weekly intervals and the increase in leaf injury during each week was calculated. The amount of leaf injury in certain urban areas was relatively low, while it was high in some rural areas, such as North Wales. Only in northern Scotland was no leaf damage found. These facts revealed that ozone concentrations are not restricted to urban areas, but can occur far from sources of pollution. The geographic distribution of foliar injury appeared to be related to sunshine levels. Because sunlight is required to drive the reactions leading to O_3 formation, O_3 concentrations tend to be high when sunshine levels are high.

5.1.2 National Monitoring Network for Air Pollution Within the Netherlands

A nationwide network for monitoring air pollution effects on indicator plants, such as gladiolus and tulip for HF, clover for SO_2, tobacco for ozone, *Urtica urens* L. for peroxyacetyl nitrate (PAN), and petunia for C_2H_4, existed in the Netherlands from 1973 to 1988 (Posthumus 1983; Tonneijck and Posthumus 1987; Tonneijck and Bugter 1991). This biological network was integrated into the nationwide automated network for the measurement of air pollution concentrations. In this network, the acute and chronic effects of several air pollutants on different indicator and accumulator plants were monitored at about 40 experimental fields, more or less regularly distributed over the Netherlands, from 1976 onward. The concentrations of SO_2, NO_x, O_3, and CO were continuously measured in ambient air with specific analyzers. Biological effect monitoring was performed only during the vegetation period of the indicator and accumulator plants (from April until November). The plants were grown in standard soil in plastic containers with ceramic filter candles in the soil for automated watering. The extent of foliar injury due to acute exposures was assessed weekly on sensitive indicator plants.

From the results of the ozone-induced injury on tobacco Bel-W_3, it was found that foliar injury was more severe in the western half of the country than in the eastern half. Field data for 1983 were in good agreement with the experimentally derived exposure-response function (Tonneijck and Posthumus 1987). Ozone concentrations exceeding the 24-h-averaged 31-36 ppb always resulted in injury to tobacco plants (Tonneijck and Posthummus 1987).

As for the validity of using tobacco Bel-W_3 as a representative of other indicator plants, the relationships between ozone injury to tobacco Bel-W_3 and subterranean clover and the levels of ambient ozone were analyzed using the data set of 1988 (Tonneijck and Bugter 1991). Although the relationships between injury intensity on tobacco Bel-W_3 and various O_3 exposure indices were all significant, foliar injury was poorly related to the level of ambient O_3. Furthermore, the foliar injury response of tobacco Bel-W_3 to ambient ozone did not relate to the responses of subterranean clover to O_3. The authors concluded that O_3 injury on tobacco Bel-W_3 is not an adequate indicator of the ambient ozone concentrations, nor is it a good indicator of the risk of O_3 damage to other plant species or to vegetation as a whole. Nevertheless, they stated that indicator plants are very useful for demonstrating the occurrence of ozone-induced injury and for studying the spatial and temporal distribution of these effects.

5.2 Morning Glory

Plant damage by photochemical smog has been reported on numerous occasions in Japan since 1970. Tobacco has been extensively used in the USA for studying plant damage caused by photochemical ozone but, in Japan, its free cultivation was illegal until 1985. Morning glory (*Pharbitis nil* or *Ipomoea nil*), a traditional horticultural plant in Japan, has therefore been used as a bioindicator for photochemical oxidants (mainly O_3), especially in view of its characteristics: (1) high sensitivity to photochemical oxidants; (2) ease of cultivation; (3) its contribution production of new leaves throughout the growing season, allowing for long-term observations; (4) its cultivation period, which almost coincides with the season of photochemical air pollution; and (5) the injury symptoms on its leaves, which are specifically characteristic and identifiable.

5.2.1 Injury Symptoms and Relationship Between Leaf Injury and Ozone Concentration

Injury symptoms observed on morning glory leaves subjected to ambient oxidants consist of bleached spots and occasional necrosis on the interveinal areas of leaves (Fig. 2). Nouchi and Aoki (1979) proposed an empirical dose-response model based on ozone exposure experiments. The extent of leaf damage of morning glory (cv. Scarlet O'Hara) due to O_3, on the basis of its concentration and exposure duration, was approximated by

$$S = (0.278\, ln\, (C^{2.2} \times t) + 0.999) \times 100$$

where S is the percentage of damaged leaf area to the whole leaf area, C is the ozone concentration (ppm), and t is the exposure duration (h). This equation suggests that even under the same dose, that is, the product of the ozone concentration (C) and exposure duration (t), the percentage of damaged leaf area is at higher concentrations. The calculated threshold ozone concentrations for the occurrence of leaf injury on morning glory were 196 ppb for 1 h, 119 ppb for 3 h, and 76 ppb for 8 h.

However, foliar injury rarely occurred in the field in the Tokyo area even when plants were exposed to a daily maximum concentration of 30-50 ppb.

5.2.2 National Japan Survey

A 3-year project of monitoring photochemical oxidants using morning glory (cv.

Fig. 2. Small white stipples and brown necrosis on morning glory leaves caused by ambient O_3

Scarlet O'Hara) throughout Japan from 1974 to 1976 was conducted under the sponsorship of all 47 prefectural governments and one of the largest newspaper companies (Yomiuri Shinbun) in Japan (Matsunaka 1977). In this program, Scarlet O'Hara, a morning glory variety with high sensitivity to oxidants, was observed by teachers and students of biology or science clubs in more than 120 junior high schools. At least two schools were selected in each prefecture. Observations were made every day in July of 1974, and every Tuesday and Friday in July and weekly in August of 1975 and 1976, with respect to the injuries (if any) and the percentage area of each leaf damaged.

The distribution of injury occurrence on morning glory, shown in Fig. 3, demonstrates nationwide injury except for 10 prefectures, such as Hokkaido and Okinawa. In 1975, only 2 prefectures announced warnings against photochemical oxidant pollution but could not observe foliar injury symptoms, whereas foliar damage was noted in 15 prefectures that did not announce warnings. As for the relationship between the occurrence of foliar injury and oxidant level, foliar injury often occurred on exposure to 72 ppb or at a dose above 400-480 ppb-h near the air pollution monitoring stations (different locations from the observing schools and monitoring stations). This study revealed, for the first time, that photochemical oxidant pollution extended nationwide and, as such, represented an epoch-making investigation.

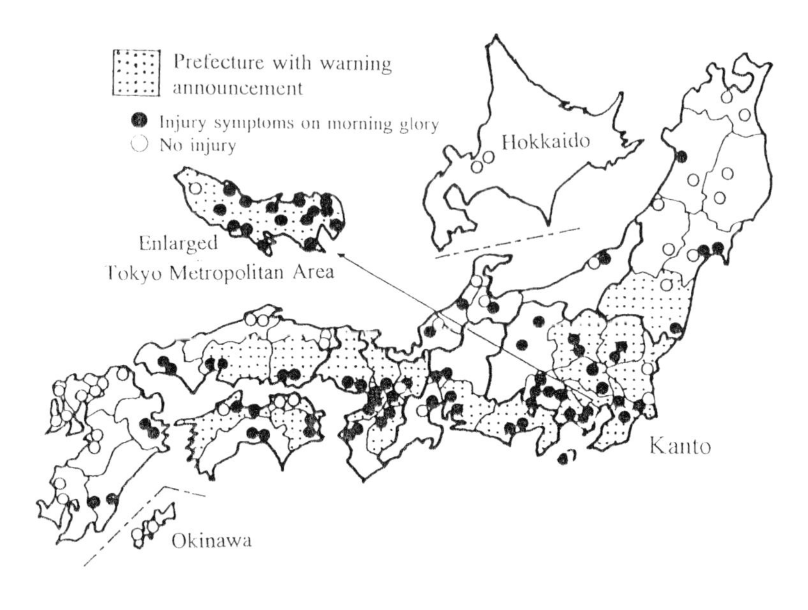

Fig. 3. Distribution of Japanese prefectures that announced photochemical smog warnings, and of junior high schools which observed injury symptoms on morning glory leaves in 1975 (from Matsunaka 1977). Black circles represent the appearance of injury symptoms at the school location in 1975; dotted areas show the distribution of prefectures where official warnings of photochemical smog occurrence, when the average concentration of the oxidants exceeded 120 ppb (corrected values according to the present), were issued

5.2.3 Long-Term Field Surveys of Morning Glory in the Kanto District and Its Surrounding Prefectures of Japan

To evaluate the distribution of plant damage by photochemical oxidants, a project committee, specializing in the study of plant damage by photochemical oxidants, was established in 1974 by seven prefectures of the Kanto district of Japan; Ibaraki, Tochigi, Gunma, Saitama, Chiba, Tokyo, and Kanagawa. The committee was composed of researchers of agricultural experiment stations and research institutes for environmental protection. Since then, annual field surveys of plant damage by photochemical oxidants have been carried out at the end of July at about ten observation points per prefecture throughout the Kanto region. The field surveys expanded further into Nagano Prefecture in 1987, Yamanashi Prefecture in 1988, and Shizuoka Prefecture in 1989. Currently, the field surveys using morning glory and taro are conducted annually by numerous researchers of the Tokyo metropolis and nine prefectures of the Kanto district and its surrounding prefectures, at about 100 different observation points, with the aim of investigating long-term changes in the distribution of plant damage.

Every year, seeds of morning glory (var. Scarlet O'Hara) are sown in mid-May, grown in a greenhouse equipped with activated charcoal filters to remove ambient pollutants, and then five seedlings are transplanted to the cultural grounds in late June. At the end of July, the damaged areas of three plants showing normal growth are visually assessed to record the percentage leaf area injured using 10% increments. Figure 4 shows the extent of morning glory leaf injury (percentage of injured leaves to the total number of leaves in each plant) in the Kanto district and its surrounding prefectures in 1996 (69 observation sites) (Department of Air Pollution, Environmental Protection Measures Promotion Headquarters, Kanto District Governors Association 1997). Plant damage occurred throughout almost all the Kanto and surrounding districts, and the distribution of cumulative foliar damage coincided, to some extent, with the distribution of photochemical oxidants. However, in some areas, no correlation between the severity of foliar injury and oxidant concentrations was observed. Annual changes in the ratio of the sum of

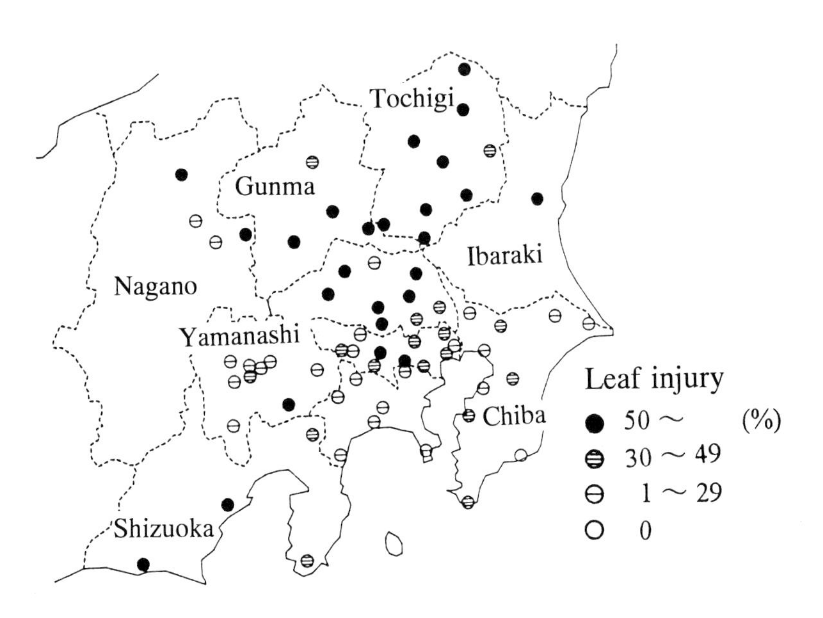

Fig. 4. Distribution of morning glory leaf injury in the Kanto district and its surrounding prefectures, Japan, in 1996 (from Department of Air Pollution, Environmental Protection Measures Promotion Headquarters, Kanto District Governors Association 1997)

injury occurrence sites to the total observation sites from 1974 to 1997 are shown in Fig. 5. During this period, the ratio remained approximately unchanged, suggesting that photochemical oxidant pollution showed no sign of improvement in the Kanto district and its surrounding prefectures.

5.3 Radish (*Raphanus sativus* L.)

Radish is also used as an indicator plant for O_3 because of its high sensitivity to this pollutant and its rapid growth and convenient compact size. Izuta et al. (1993) attemped to evaluate atmospheric environments polluted mainly by O_3 using the growth response of radish (cv. Comet) plants cultivated in small-sized open-top chambers (OTC). Using this method, the effect of ozone on radish plants can be

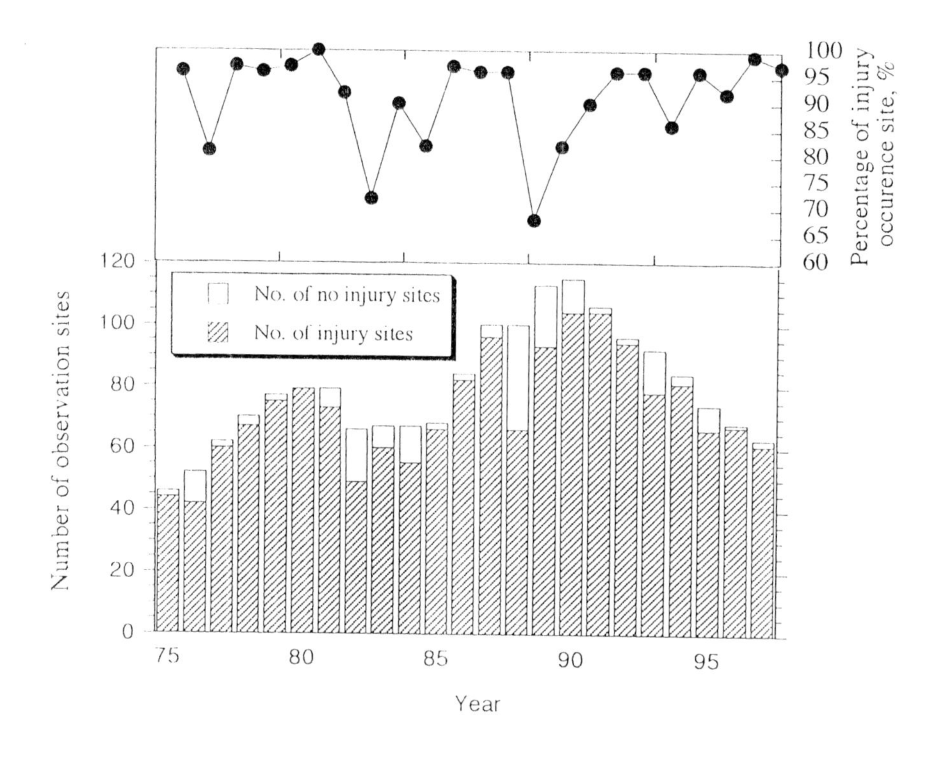

Fig. 5. Annual changes in injury occurrence of morning glory in the Kanto district and its surrounding prefectures from 1976 to 1997 (from Department of Air Pollution, Environmental Protection Measures Promotion Headquarters, Kanto District Governors Association 1998)

determined quantitatively by comparing the growth of plants cultivated in nonfiltered and in charcoal-filtered treatments. The OTC system consists of two acrylic chambers with 16 holes at the top, and a wooden box containing culture soil supplemented with fertilizer (Fig. 6). Each OTC, for the charcoal-filtered and nonfiltered treatments, is aerated by a small fan through a dust filter with and without activated charcoal, respectively, installed at the air inlet. The concentration of ozone in the OTC is reduced to less than 25% that of ambient air for the charcoal-filtered treatment, but is the same as that in ambient air for the nonfiltered treatment.

On the 3rd day after sowing the radish seeds, the chambers are placed on top of the culture soil in which the seeds had been sown. From the 3rd to the 7th day after sowing, all seedlings are grown in charcoal-filtered air for 5 days and, on the 8th day, the seedlings are selected for uniformity and thinned to 24 seedlings per treatment. After thinning, 8 seedlings are harvested from each treatment to obtain initial values of leaf area and total dry weight of a whole plant (initial sampling). From the 8th to the 14th day, the seedlings are grown in charcoal-filtered (charcoal-filtered treatment, CF) or nonfiltered (nonfiltered treatment, NF) air for 7 days. On the 15th day after sowing, 16 plants are harvested from each treatment for measurement of leaf area and dry weight of plant organs (final sampling).

A total of 17 such field experiments were conducted from 1987 to 1989 in Fuchu, Tokyo, Japan, which has a very high concentration of photochemical oxidants (Izuta et al. 1993). Although the total dry weight in the NF treatment was significantly reduced compared with that in CF in only 5 experiments, a linear

Fig. 6. Outline of small-sized open top chambers (OTC) for radish plants (courtesy Dr. T. Izuta)

relationship between the relative value of total dry weight per radish plant cultivated in NF [Rt = (average value of total dry weight in NF/average value of total dry weight in CF) × 100] and the average daily 8-h dose (8 A.M. to 4 P.M.) of ambient ozone during the 7-day evaluation period was obtained from the 17 field experiments, at an average daily air temperature above 20°C and an average daily cumulative solar radiation greater than 5 MJ m^{-2} day^{-1} (Fig. 7).

5.4 Clover

In northern Europe, tobacco has some disadvantages as a bioindicator of O_3 because it is not commercially grown, and it is sensitive to low temperatures and wind. Recently, studies using clover suggest that white clover (*Trifolium repens* L.) and red clover (*Trifolium pratense* L.) may, by showing typical visible symptoms, be suitable substitutes for tobacco as O_3 bioindicators. Becker et al. (1989) concluded that ambient concentrations of O_3 in Switzerland were high

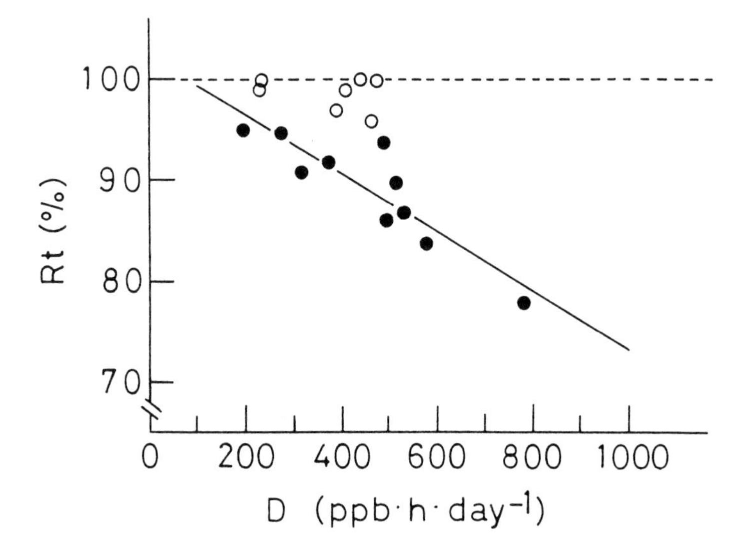

Fig. 7. Relationship between average daily 8-h dose (800-1600) of ozone (D, ppb-h day^{-1}) and relative value of total dry weight per plant (Rt, %) of radish plants cultivated in a nonfiltered OTC treatment (from Izuta et al. 1993, with permission). Open circles represent experiments with an average air temperature below 20°C at 900; black circles represent experiments with an average air temperature above 20°C at 900. The linear regression equation in experiments with an average air temperature above 20°C at 900 is represented by Rt = 0.029D + 102.1 (r = -0.894***)

enough to cause characteristic necrotic flecks on the leaves of sensitive cultivars of white clover. Luthy-Krause et al. (1989) found a relationship between the occurrence of necrotic spots on red clover and the O_3 concentration in a Swiss open-top chamber study. As the sensitivity to O_3 differs between clover species, Karlsson et al. (1995) investigated the O_3 sensitivity of three important clover species (*T. repens*, *T. pratense*, and *T. subterraneum*) cultivated in southwest Sweden. By evaluating visible foliar injury under Swedish conditions, *T. subterraneum* was found to be the most sensitive of the three species to O_3, whereas *T. pratense* was the most tolerant. The critical threshold level for ozone on *T. subterraneum* was found to be between 20 and 30 ppb.

6. Bioindicators for Peroxyacetyl Nitrate (PAN)

In the early 1940s, plants, such as romaine lettuce (*Lactuca sativa* L.), Swiss chard (*Beta chilensis* Hort.), and annual bluegrass (*Poa annua* L.) were identified as bioindicators for PAN pollution in the Los Angeles area of California even when PAN had not yet been chemically identified (Manning and Feder 1980). Characteristic PAN symptoms, such as glazing, bronzing, and silvering, occur on the lower surface of immature leaves. Petunia plants are also highly sensitive to PAN, are comparatively resistant to pests and diseases, and have a long growth period with a long duration of sensitivity to PAN and, as such, are suitable for detecting the existence of PAN. The sensitivity of petunia to PAN varies between cultivars and, in general, cultivars with white flowers are more sensitive to PAN than those with blue or red flowers.

6.1 Injury Symptoms on Petunia and the Relationship Between Foliar Injury and PAN Concentration

Typical visible symptoms on petunia induced by PAN are white or brown bronzing on the lower leaf surface extending to the upper leaf surfaces under severe injury (Fig. 8). The immature, younger leaves (2nd to 6th leaf positions from the top of the plant) are the most sensitive to PAN. According to the logarithmic equation between the extent of petunia foliar damaged and PAN concentration, obtained from PAN exposure experiments, the threshold PAN concentrations for foliar injury of white-flowered petunia (cv. white ensign) were 32, 14, and 7 ppb for exposure durations of 1, 3, and 8 h, respectively (Nouchi 1979).

Nouchi et al. (1984) conducted field surveys to clarify the relationship between PAN pollution and plant responses, based on visible foliar injury of field-grown petunia, *Petunia hybride* Vilm., at Yurakucho in central Tokyo. Visible injury on petunia cv. white ensign, a PAN-sensitive variety, and atmospheric PAN concentrations were recorded almost daily from April to November for 5 years (1976, 1977, 1978, 1982, and 1983). Foliar PAN injury occurred 15 times in 1976

Fig. 8. Petunia injury symptoms caused by ambient peroxyacetyl nitrate (PAN). More severe injury causes necrosis on upper surface of leaves

and 4-6 times in other years. It was found that the extent of PAN-induced injury increased when the daily PAN dose, calculated as a simple sum of the hourly mean PAN concentrations during the daytime hours (from 8 A.M. to 6 P.M.), exceeded 20 ppb-h (Table 1), or when the daily maximum PAN concentration was greater than 4 ppb (Table 2). These results indicated that a daily PAN dose of 20 ppb-h and a daily maximum PAN concentration of 4 ppb can be used as threshold PAN pollutant level for foliar damage of petunia plants under field conditions.

6.2 Long-Term Field Surveys of Petunia in the Kanto District and Its Surrounding Prefectures of Japan

Field surveys in the Kanto district and its surrounding prefectures, using three cultivars of petunia with white, blue, and red flowers (1983-1987) and two cultivars of petunia with white and blue flowers (1988-1997), were conducted annually from 1983 to 1997 in the last 10 days of July, as with the morning glory survey. Seeds of petunia plants were sown in pot soil at the beginning of May,

Table 1. Relationship between daily peroxyacetyl nitrate (PAN) dose and injury occurrence of "white ensign" petunia at Yurakucho, Tokyo (from Nouchi et al. 1984, with permission)

Daily PAN dose[a], ppb-h	Number of injuries	Number of polluted days	Injury occurrence, %
16-20	1	65	2
21-25	4	40	10
26-30	4	37	11
31-35	5	22	22
36-40	5	14	36
41-50	4	9	44
51-60	3	8	38
61-70	5	6	83
More than 71	3	3	100

[a])The sum of the hourly average concentration in daytime from 0800 to 1800.
Data obtained before April 30 and after November 1 were excluded from this table because PAN injury at these times did not occur every year.
Observation surveys were conducted from May 1 to October 30 for 5 years (1976, 1977, 1978, 1982, and 1983).

Table 2. Relationship between daily maximum PAN concentration and injury occurrence of "white ensign" petunia at Yurakucho, Tokyo (from Nouchi et al. 1984, with permission)

Daily maximum PAN concentration, ppb	Number of injuries	Number of polluted days	Injury occurrence, %
3	0	79	0
4	3	51	6
5	3	38	8
6	4	27	15
7	3	21	14
8	5	12	42
9	3	7	43
10-12	5	14	36
13-15	5	8	63
More than 16	2	2	100

Data obtained before April 30 and after November 1 were excluded from this table because PAN injury did not occur during these periods every year.
Observation surveys were conducted from May 1 to October 30 for 5 years (1976, 1977, 1978, 1982, and 1983).

transplanted, and then in the last 10 days of June three potted (surface area, 500 cm^2) plants per cultivar were placed at each observation site (11 in 1983 and 1984, 24 in 1985, 31-32 in 1986 and 1987, 44 in 1988, and 50-60 in 1989-1997).

The distribution of injury to white- and blue-flowered petunia in the Kanto district and its surrounding prefectures, at the end of July in 1996, is shown in Fig. 9. The distribution of petunia injury tended to expand with the increased number of observation sites from 1985 to 1997. The mean ratios of injury occurrence sites to the total number of observation sites for white (cv. titan white) and blue (cv. titan blue) flowers from 1985 to 1997 were 53% and 18%, respectively (Fig. 10). The sites with recognized injury occurrence for both white and blue titan cultivars were located in central Tokyo and Saitama Prefecture, sites with the highest photochemical oxidant concentrations.

7. Conclusion

If the concentration of an air pollutant is harmful to a plant, the plant will show an

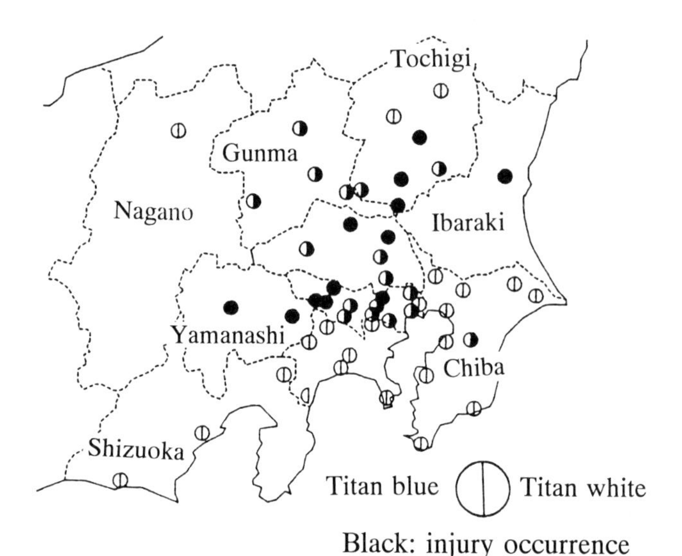

Fig. 9. Distribution of injury occurrence of petunia in the Kanto district and its surrounding prefectures, Japan, in 1996 (Department of Air Pollution, Environmental Protection Measures Promotion Headquarters, Kanto District Governors Association 1997)

appropriate response. Such plant responses, notably characteristic visible foliar injury, serve as an alert to the possible harmful effects of the pollutant to man and his environment. Hence, the bioindicator method reports directly whether the ambient concentration of an air pollutant is harmful to biological tissues. For this reason, ideal bioindicator plants should be sensitive to a specific air pollutant and respond proportionally to the pollutant concentration or dose, be native or adaptable to the region, and be resistant to pests and diseases. Although the bioindicator method is a cheap and easy means of environmental surveillance, compared with chemical and physical monitoring methods, it does not measure precise hourly pollutant fluctuations or produce easily comparable data as do the physical and chemical methods. Furthermore, although the bioindicator method does provide an integrated measure of the biological effects of air pollutants over a period of time, it cannot provide a direct quantitative measure of the amount of pollutant in the atmosphere, because the response of plants to a given pollutant dose will depend on other environmental factors such as temperature, soil water content, and light.

We should make greater efforts toward improving the bioindicator method (i.e., establishment of the dose-response relationship under various environmental conditions and quantification of injury degree) and into investigating the limits of

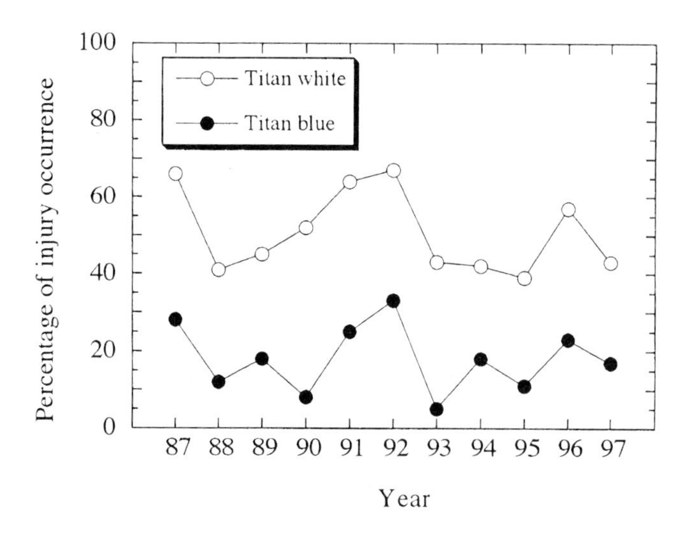

Fig. 10. Annual changes in injury occurrence of petunia in the Kanto district and its surrounding prefectures from 1987 to 1997 (from Department of Air Pollution, Environmental Protection Measures Promotion Headquarters, Kanto District Governors Association 1998)

this method. We can conclude that a more integrated approach, through a combination of physical and chemical methods together with indicator plants, is the most suitable means of monitoring ambient air protecting man and his environment. In Japan, the monitoring of ambient air using indicator plants, such as morning glory and petunia, has revealed that photochemical oxidant pollution has spread nationwide. The result further suggests that many other plant species must necessarily suffer not only visible foliar injury but also a reduction of growth or yield due to photochemical oxidants. Even after more than 30 years since the start of the study in Japan, using indicator plants, photochemical oxidant pollution has shown no sign of improvement, especially in the Kanto district and its surrounding prefectures. Since the global environmental crisis has now been recognized, it is important that we all understand the conditions of our environment. Bioindicators, which allow the dangers of air pollution to be clearly visualized, are important weapons in the fight against air pollution. It is necessary to continue monitoring environmental air pollution with bioindicators for a long time into the future in advanced countries, and equally so in large cities in developing countries where air pollution still increasing.

References

Abeles FB, Heggestad HE (1973) Ethylene: an urban air pollutant. J Air Pollut Control Assoc 23:517-521

Ashmore MR, Bell JNB, Reily CL (1978) A survey of ozone levels in the British isles using indicator plants. Nature 276:813-815

Ashmore MR, Bell JNB, Reily CL (1980) The distribution of phytotoxic ozone in the British isles. Environ Pollut Ser B 1:195-216

Becker K, Saurer M, Egger A, Fuhrer J (1989) Sensitivity of white clover to ambient ozone in Switzerland. New Phytol 112:235-243

Bytnerowicz A, Manning WJ, Grosjean D, Chmielewski W, Dmuchowski W, Grodzinska K, Godzik B (1993) Detecting ozone and demonstrating its phytotoxicity in forested areas of Poland: a pilot study. Environ Pollut 80: 301-305

Department of Air Pollution, Environmental Protection Measures Promotion Headquarters, Kanto District Governors Association (1997) Annual report in fiscal year 1996 of effects of photochemical smog on plants (in Japanese)

Department of Air Pollution, Environmental Protection Measures Promotion Headquarters, Kanto District Governors Association (1998) Annual report in fiscal year 1997 of effects of photochemical smog on plants (in Japanese)

DeSloover J, LeBlanc F (1968) Mapping of atmospheric pollution on the basis of lichen sensitivity. In: Misra R, Gopal B (eds) Proceedings of the symposium on recent advances in tropical ecology. Varanasi, India, pp 42-56

Hassan IA, Ashmore MR, Bell JNB (1995) Effect of ozone on radish and turnip under Egyptian field conditions. Environ Pollut 89:107-114

Heggestad HE (1991) Origin of Bel-W_3, Bel-C and Bel-B tobacco varieties and their use as indicators of ozone. Environ Pollut 74:264-291

Heggestad HE, Middeleton JT (1959) Ozone in high concentrations as cause of tobacco leaf injury. Science 129:208-210

Heggestad HE, Menser HA (1962) Leaf spot-sensitive tobacco strain Bel-W3, a biological indicator of the air pollutant ozone (abstract). Phytopathology 52:735

Izuta T, Funada S, Ohashi T, Miyake H, Totsuka T (1991) Effects of low concentrations of ozone on the growth of radish plants under different light intensities. Environ Sci 1:21-33

Izuta T, Miyake H, Totsuka T (1993) Evaluation of air-polluted environment based on the growth of radish plants cultivated in small-sized open-top chambers. Environ Sci 2:25-37

Karlsson GP, Sellden G, Skarby L, Pleijel H (1995) Clover as an indicator plant for phytotoxic ozone concentrations: visible injury in relation to species, leaf age and exposure dynamics. New Phytol 129:355-365

Larsen RI, Heck WW (1976) An air quality data analysis system interrelating effects, standards and needed source reductions. Part 3.Vegetation injury. J Air Pollut Control Assoc 26:325-333

LeBlanc F, De Sloover J (1970) Relation between industrialization and the distribution and growth of epiphytic lichens and mosses in Montreal. Can J Bot 48:1485-1496

LeBlanc F, Rao DN, Comeau G (1972) The epiphytic study of *Populus balsamifera* and its significance as air pollution indicator in Sudburry, Ontario. Can J Bot 50:519-528

LeBlanc F, Rao DN (1975) Effects of air pollutants on lichens and bryophytes. In: Mudd JB, Kozlowski TT (eds) Responses of plants to air pollution. Academic Press, New York, pp 237-272

Luthy-Krause B, Bleuler P, Landolt W (1989) Black poplar and red clover as bioindicators for ozone at a forest site. Angew Bot 63:111-118

Manning WJ, Feder WA (1980) Biomonitoring air pollutants with plants. Applied Science Publishers, London, p 142

Matsumaru T (1994) Monitoring of photochemical oxidants by plant indicators using morning glory and petunia plants. In: Proceedings of the international seminar for the simple measuring and evaluation method on air pollution, Chiba, March 2-3. Japan Society of Air Pollution, pp 11-26

Matsunaka S (1977) Utilization of morning glory as an indicator plant for photochemical oxidants in Japan. In: Kasuga S, Suzuki N, Yamada T, Kimura G, Inagaki K, Onoe K (eds) Proceedings of the Fourth International Clean Air Congress. Japan Union of Air Pollution Prevention Association, Tokyo, pp 91-94

Menser HA, Heggestad HE, Street OE (1963) Response of plants to air pollutants. II. Effects of ozone concentration and leaf maturity on injury to *Nicotiana tabacum*. Phytopathology 53:1304-1308

Nakagawa Y, Kobayashi T (1990) Estimation of air pollution based on the distribution and the component of epiphytic lichens by means of modified IAP method (in Japanese with English summary). Taiki Osen Gakkaishi (J Jpn Soc Air Pollut) 25:233-241

Nouchi I (1979) Effects of ozone and PAN concentrations and exposure duration on plant injury (in Japanese with English summary). Taiki Osen Gakkaishi (J Jpn Soc Air Pollut) 14:489-496

Nouchi I, Aoki K (1979) Morning glory as a photochemical oxidant indicator. Environ Pollut 18:289-303

Nouchi I, Ohashi T, Soufuku M (1984) Atmospheric PAN concentrations and foliar injury to petunia indicator plants in Tokyo (in Japanese with English summary). Taiki Osen Gakkaishi (J Jpn Soc Air Pollut) 19:392-402

Pleijel H, Ahlfors A, Skarby L, Pihl G, Sellden G, Sjodin A (1994) Effects of air pollutant emissions from a rural motorway on *Petunia* and *Trifolium*. Sci Total Environ 146/147:117-123

Posthumus AC (1983) Higher plants as indicators and accumulators of gaseous air pollution. Environ Monitoring Assessment 3:263-272

Posthumus AC (1984) Monitoring levels and effects of air pollutants. In: Treshow M (ed) Air pollution and plant life. Wiley, Chichester, pp 73-95

Taoda H (1972) Mapping of atmospheric pollution in Tokyo based upon epiphytic bryophytes. Jpn J Ecol 22:125-133

Tonneijck AEG, Posthumus AC (1987) Use of indicator plants for biological monitoring of effects of air pollution: the Dutch approach. VDI-Berichte 609:205-216

Tonneijck AEG, Bugter RJF (1991) Biological monitoring of ozone effects on indicator plants in the Netherlands: initial research on exposure-response functions. VDI-Berichte 901:613-624

Vasiloff GN, Smith ML (1974) A photocopy technique to evaluate fluoride injury on gladiolus in Ontario. Plant Dis Rep. 58:1091-1094

Yamamoto T, Suketa Y, Mikami E, Sato Y (1975) Environmental estimation of pollution by atmospheric fluoride using plant indicator (in Japanese with English summary). Nippon Nougeikagaku Kaishi (J Agric Chem Soc Jpn) 49:347-352

3 Phytomonitoring for Urban Environmental Management

Margaret Burchett, Rachid Mousine, and Jane Tarran

Centre for Ecotoxicology, Faculty of Science, University of Technology, Sydney (UTS), Westbourne St, Gore Hill, NSW 2065, Australia

1. Introduction

1.1 Why Use Plants as Bioindicators for Environmental Management?

It has been known for several decades that air pollution can adversely affect plant health, and the possible use of plants as passive monitors or indicators was early recognised (Bleasedale 1973; Harward and Treshow 1975; Roose et al. 1982). The quest is continuing for standardised indicator species that will show known, reliable dose - effect relationships with any gaseous pollutant or mixture; however, research has yet to yield the comprehensive data sets that would be ideal for the purpose. In the main we are still following the approach recommended by Posthumus (1985) as a fall-back option, namely, to study the effects themselves on naturally growing plants or cultivated crops. This is the approach that has been taken in the case study presented here, on the leaf responses to air pollution of three Australian native species commonly used in Sydney as street and park trees. The study, however, did have the benefit of exact parallel air quality data, which were supplied by the New South Wales Environment Protection Authority (NSW EPA).

The final aims of any pollution monitoring program are to establish the presence and levels of pollutants in an area and their effects on significant species

Air Pollution and Plant Biotechnology
–Prospects for Phytomonitoring and Phytoremediation–
Edited by K. Omasa, H. Saji, S. Youssefian, and N. Kondo
© Springer -Verlag Tokyo 2002

or the ecosystem (Manning and Feder 1980; Forbes and Forbes 1994; Krupa and Legge 1995). Ecotoxicological approaches to management and risk assessment now commonly use what is termed a triad approach (USEPA 1993; Pascoe and DalSoglio 1994; Wang and Freemark, 1994): testing for the presence and concentrations of chemicals; ecoepidemiological surveys to identify extent of harm; and toxicity testing to indicate cause - effect links and dose - response relationships. Criteria for suitable bioindicator species (Phillips 1980) include relative tolerance to pollution exposure; sedentary habit; abundant presence; ease of laboratory holding and testing; and the ability to accumulate some pollutants and hence show dose - response relationships. Many plant species fulfil these criteria (see the chapter by Nouchi, this volume). They are also useful ecosystem indicators, because harm to them has implications for the whole, and plant sampling has the added advantage that it is inexpensive and provides information on toxic effects that cannot be predicted from chemical testing. In addition, plants are more sensitive to air pollution than humans, and so can act as an early-warning indicator of deteriorating quality. Compare, for example, the standards cited in Table 1 with those for example of Pfleeger et al. (1993), and Bergmann et al. (1995), who report adverse symptoms in various plant species at ozone levels of 0.04 and 0.07 ppm, and a number of studies that have shown the adverse effects on plant metabolism of nitrogen oxide at 0.01 – 0.03 ppm and nitrogen dioxide at levels of 0.04 – 0.06 ppm (Bobbink 1988; Koziol et al. 1988; Dizengremel and Citerne 1988; Pell 1988).

Many studies have now been conducted on the responses of plants to air pollution (for example, see reviews by Hutchinson and Meema 1987; Bennett 1996). The results of chamber experiments are valuable in indicating causal links between pollutants and the onset of toxic symptoms. However, they often show different results from those obtained by field sampling (passive biomonitoring) because the latter reveals the integrated effects of pollution over lifespans under varying pollutant mixtures and results of interactions of pollution with other environmental variables. In general, two categories of plant are useful for monitoring air pollution; sensitive species in which visible symptoms indicate damage; and tolerant species that accumulate pollutants and demonstrate dose - response relationships (for example see Canas et al. 1997).

A further use for such plant field studies is in their application to bioremediation. This is a newly emerging technology, defined by Salt et al. in a recent review (1998) as "the use of green plants to remove pollutants from the environment or to render them harmless". Although considerably more attention has been paid to the phytoremediation of contaminated soils and waters (Watanabe 1997), the possibility of using plants to reduce air pollution has also been recognised (Raaschou-Nielson et al. 1995; Angold 1997; Beckett et al 1998; Roy and Sharma 1998; Salt et al. 1998). However, before plants can be used as a routine tool of urban environmental management and remediation, parallel epidemiological, experimental, and physicochemical investigations must be

Table 1. Air quality goals used by New South Wales Environment Protection Authority (NSW EPA 1996)

	Maximum concentration	Averaging time	Agency
Particles as PM_{10}	50 μg/m^3	1 year	USEPA
	150 μg/m^3	24 h	USEPA
	50 μg/m^3	*24 h*	*NEPC*
NO_2	0.16 ppm	1 h	NHMRC
	0.05 ppm	1 year	USEPA
	0.125 ppm	*1 h*	*NEPC*
Photochemical oxidant, as O_3	0.10 ppm	1 h	NHMRC
	0.10 ppm	*1 h*	*NEPC*
	0.08 ppm	*4 h*	*NEPC*
CO	87 ppm	15 min	WHO
	9 ppm	*8 h*	*NHMRC*
	9 ppm	*8 h*	*NEPC*
SO_2	0.25 ppm	1 h	NHMRC
	0.02 ppm	1 year	NHMRC
	0.12 ppm	1 h	WHO
	0.20 ppm	*1 h*	*NEPC*
	0.02 ppm	*1 year*	*NEPC*
Lead	1.4 μg/m^3	3 months	NHMRC
	0.5 μg/m^3	*1 year*	*NEPC*

Also shown (in italics) are the proposed Australian National Standards for air quality, to be achieved over next decade (National Environment Protection Council, NEPC, 1997). NHMRC, National Health and Medical Research Council (Australia); USEPA, United States Environmental Protection Agency; WHO, World Health Organization.

conducted. Few individual studies to date have encompassed this three-pronged approach.

1.2 Project Rationale

The project outlined here is presented as a case study of both the potential and the challenges of developing a phytoindicator system for management of urban air quality. The aim of the 3-year project (1995 - 1997) was to investigate the feasibility of developing the use of commonly planted street-tree species as bioindicators, and eventually as remediators, of air pollution in the Sydney region. The project included ecoepidemiological studies of three species for both local and regional responses, using leaf biochemical paramcters and heavy metal levels, and parallel field experiments with two of the species. The project was carried out with funding from the New South Wales Environmental Research Trust, and in co-operation with the New South Wales Environment Protection Authority (NSW EPA), which conducts a joint research centre with the University of Technology, Sydney (UTS), the Centre for Ecotoxicology.

2. Scope of Project

2.1 Context: Air Pollution and Human Health in Sydney Region

The largest Australian conurbation (~4 million) comprises Sydney (Fig. 1), Newcastle, 150 km north, and Wollongong, 100 km south, plus the coastal towns between. This strip is the site of heavy industry and coal mining. The region is susceptible to the development of poor air quality on some occasions (NSW EPA 1996), on some days approaching levels similar to those of Tokyo and New York (SEAC 1996). Parallel research surveys were undertaken, 1993 - 1995, by the NSW EPA (Metropolitan Air Quality Study, MAQS), which provided a description of regional air pollution dynamics, and the NSW Department of Health (Health and Air Research Program, HARP), which provided a description of regional air pollution dynamics and morbidity/mortality responses in the region (NSW Health Dept. 1996). Pollutants of greatest concern were found to be ozone (O_3); nitrogen oxides (NO_x), particularly nitric oxide (NO) and nitrogen dioxide (NO_2); fine particulates (PM_{10}, i.e., particulate matter of diameter $<10\ \mu m$); and reactive organic compounds (ROCs). Sulfur dioxide was not a major pollutant, and lead is of decreasing importance. Elevated carbon dioxide is considered to contribute more to national and global problems than regional ones. Monitoring is conducted by NSW EPA at 30 regional stations, of which 13 are in metropolitan Sydney (Fig. 1). All stations are situated away from heavy traffic routes and industrial installations.

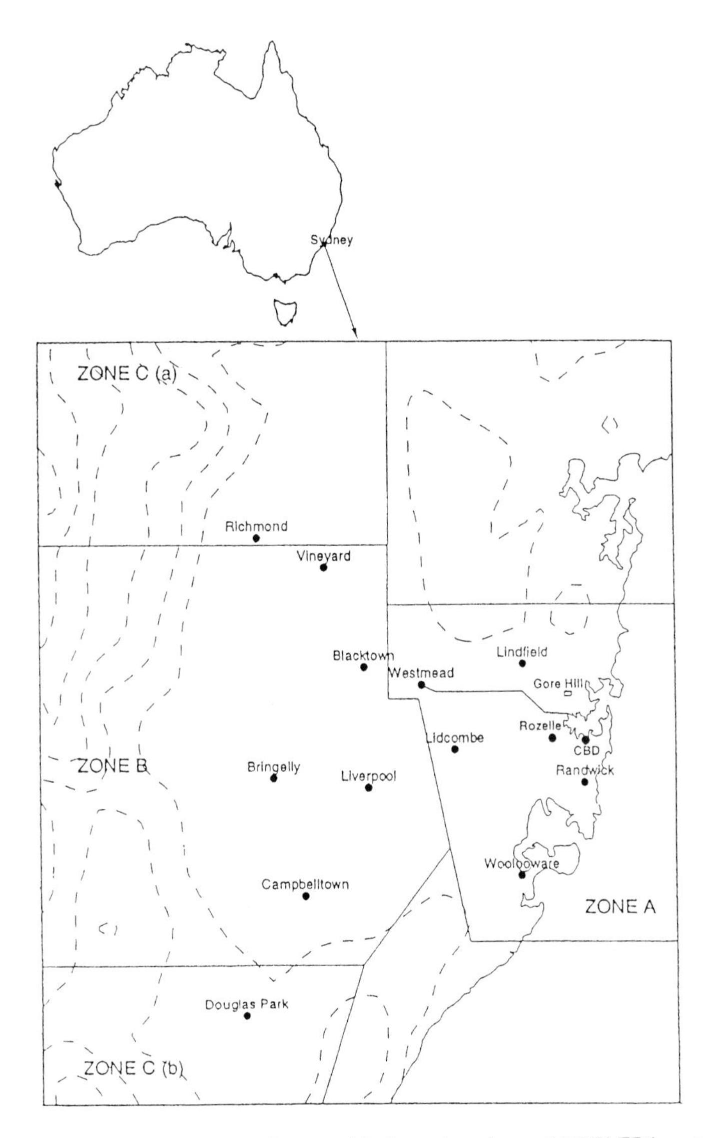

Fig. 1. Map showing Australia, Sydney, and in inset, locations of NSW EPA metropolitan air pollution monitoring stations and air pollution zones, identified on the basis of the Metropolitan Air Quality Study (MAQS) (1993-1995) (NSW EPA 1996). Under the influence mainly of easterly breezes to mountains west of Sydney, *Zone A* has relatively high NO_x, from fresh vehicle emissions and generally lower ozone levels; *Zone B* has mixed NO_x and ozone levels, with more fresh emissions, and ozone production from NO_x of both eastern and central areas; and *Zone C* is relatively low in NO_x, and higher in ozone, with photochemical reactions being completed and fewer fresh emissions. CBD, Central Business District

The health study showed that regional air pollution is associated annually with about 400 premature deaths; 250 hospital admissions with heart disease; 300 admissions from asthma; increases in respiratory illness and decreases in lung function; a 4% increase in winter deaths, when pollution is highest; a 1% increase in daily deaths for every 10 $\mu g/m^3$ increase in PM_{10} levels; and a similar relationship with higher NO_x levels (NSW Health Dept. 1996). The NSW EPA uses the air quality goals shown in Table 1 (which also includes the proposed National Standards for air to be achieved over the next decade; NEPC 1997).

It was concluded from the surveys that research was needed to assist in the development of new strategies for managing air quality in the Sydney region (NSW EPA 1996; NSW Health Dept. 1996). It was as part of the exploration for new strategies that this project was undertaken – for an effective and inexpensive means of air quality assessment which could augment physicochemical measurements, include an early-warning capacity, and lay the basis for the development of strategies for phytoremediation.

2.2 Project Design

Commonly used street-tree species were employed, because they could yield spatial and temporal regional comparisons and are presumably tolerant of air pollution and hence likely to show dose - response relationships. No previous coordinated studies of plants and regional air pollution have been carried out in Australia, although other studies have been conducted (for example, Greenhalgh and Brown 1982; Bridgman 1989; Murray and Wilson 1989; Monk and Murray 1995). The current project has included the following:

Epidemiological studies:
Preliminary identification of subregional zones of differing air quality; initial screening of species; and selection of the leaf parameters

Local traffic effects. Spatial and seasonal effects of traffic levels on leaf parameters

Regional effects. Long-term seasonal sampling at three primary sites, located in high-O_3, high-NO_x, and a mixed zone, respectively and shorter-term (1-2 year) sampling at subsidiary sites

Field experiments:
Potted seedlings of two species were placed at selected sites and analysed for the same parameters as used in the epidemiological study

3. Methodology: Ecoepidemiological

3.1 Site Selection Based on Air Pollution Data

Following the MAQS survey the NSW EPA divided the Sydney metropolitan area into three air pollution zones (Fig.1). Air in Zone A, which includes the Central Business District and main centres of population and industry, contained mainly fresh vehicle emissions, relatively high in NO_x and low in O_3. Zone B, the western suburbs, had mixed levels of NO_x, plus increased O_3 produced from photochemical reactions between NO_x and ROCs as air moved west under prevailing flows. New NO_x and ROCs are also produced in Zone B. Air in Zone C, to north and south of western Sydney, and backed by mountains, contained higher O_3 and lower NO_x, the photochemical reactions being largely completed, and with fewer new emissions (NSW EPA 1996). The three primary sites were selected around an EPA monitoring station in each zone. Subsidiary sites were selected later for comparative sampling:

Zone A, Gore Hill, about equidistant between Lindfield and Rozelle monitoring stations; three subsites: (a) disused cemetery, ie parkland, zero traffic, but adjacent to a highway (100 m removed); (b) highway, very heavy traffic road (50 - 60,000 vehicles per day); (c) side road, low traffic (500-2,000 vehicles per day); the three subsites were also used to study effects of local traffic loads

Subsidiary sites in Zone A: Lindfield, low traffic road; Rozelle, low traffic road; Lidcombe, parkland, zero to low traffic

Zone B, Bringelly, two subsites: (a) low traffic road next to monitoring station; (b) moderate to heavy traffic road within 2 km; both subsites also used for study of local traffic effects

Zone C, Richmond, experimental farm/parkland of University of Western Sydney, Hawkesbury, which houses the NSW EPA monitoring station; zero to low traffic

3.2 Screening of Species

Six species were used for initial screening, from which three were chosen for further study on the basis of their distribution and ease of biochemical analysis. The three are from the family Myrtaceae, the most dominant Australian woody family (evergreen broad-leafed):

Lophostemon confertus	Brush box (previously Tristania conferta)
Callistemon viminalis	Weeping red bottlebrush
Melaleuca quinquenervia	Paper bark, or broad-leafed tea-tree

3.3 Selection and Methods of Analysis of Leaf Parameters

The leaf biochemical parameters selected for seasonal sampling have all been shown to respond to air pollution stress and can be used as early warning biomarkers. A single survey was also made of the heavy metal content in soils and washed and unwashed leaves as a complementary estimate of exposure to air pollution.

3.3.1 Chlorophylls

All pollutants produce very reactive oxidative intermediates in plant tissues, and any increase in active oxygen destroys chlorophyll (Shimazaki et al. 1980; Rank 1997; see also chapters by K Omasa and A Polle, this volume). Reports of both decreased and increased chlorophylls in plants fumigated with NO_2, and decreases with raised O_3 and HF, were summarized by Koziol and Whatley as early as 1984. In field studies, acid gases and O_3 usually seem to cause decreases in chlorophylls (Singh et al. 1991; Ali 1993; Mousine and Aliev, 1994; Pandey and Agrawal, 1994). It was found in early chamber experiments, using SO_2 and O_3, that chlorophyll *a* and *b* levels could vary independently: hence, the chl*a* /chl/*b* ratios could sometimes be a useful bioindicator index (Ricks and Williams 1975; Knudson et al. 1977; Laurenroth and Dodd, 1981). Under field conditions, with prevailing air pollution, chlorophylls have often been found to vary together, i.e. without changing the chl*a*/chl*b* values (Rabe and Kreeb 1980; Aliev and Mousine 1992; Mousine and Aliev 1994). Ratios were computed in this study as a check on responses. Estimations were carried out using the method of Inskeep and Bloom (1985), following extraction over 48 hours in the dark in *N,N*-dimethylformamide (DMF).

3.3.2 Peroxidase Activity

An increase in peroxidase activity can be taken as reliable evidence of increased stress (Roy et al. 1992; Nast et al. 1993; Canas et al. (1997); Kangerjarvi et al. 1994; Puccinelli et al. 1998). Peroxidase activity was determined by means of a guaiacol assay.

3.3.3 Ascorbic Acid

Both increases and decreases in ascorbic acid levels have been found in plants under pollution stress. Field studies have generally shown a decrease in ascorbic acid with air pollution (for example, see Pandey and Agrawal 1994). A high-performance liquid chromatographic method (Finley and Duang 1981) was used to estimate ascorbic acid levels.

3.3.4 Heavy Metal Levels

Air-and soilborne heavy metals from vehicles may accumulate in plants, particularly along heavy traffic routes (for example, see Manfredi and Trenti 1994; Sanka et al. 1995; Garty et al. 1996; Bell and Ashenden 1997). Even where metal concentrations are not high enough to produce direct toxic effects, they may change biochemical parameters (Van Assche and Clijsters 1990; Ernst et al. 1992). Heavy metals were estimated by atomic absorption spectroscopy in soils and leaves of *L. confertus*, with different traffic loads, and from the three primary sites.

3.4 Data Analysis

From 6 to 15 trees per site or subsite were sampled every time. Leaves were sampled from two to three leafy branchlets taken from around the canopy at 1.5 - 2.5 m, from which two to three replicate batches were prepared. Data were analysed using the SYSTAT package for means, standard deviations, and standard errors. Results were subjected to a one-way ANOVA or a *t*-test for independent samples with unequal variance to test for significance of differences. From pilot trials it was found that if young mature leaves were used, individual variations in leaf parameters at the same site were only 5% -10%. Biological significance was not inferred, therefore, unless differences were greater than 15%, whether or not smaller differences were found to be statistically significant. Statistical significance is taken at $P < 0.05$ except in one or two cases, as indicated, where P values lie between 0.05 and 0.1.

4. Findings: Ecoepidemiological

4.1 Sydney Air Pollution During Study Period

NSW EPA data over the period (Figs. 2 - 4), showed that average monthly means for NO and NO_2 were, as predicted, markedly higher at stations in zone A than in zone B (Bringelly) or zone C (Richmond). There was an increasing westerly gradient of NO_x across zone A. However, levels of NO_x tended to be higher at Richmond than at Bringelly, which was not as predicted, although they were both much lower than in zone A. Pollution patterns in Sydney are variable, partly because of the unpredictable weather patterns (NSW EPA 1996). Levels of NO_x showed strong seasonal variation, higher in autumn/winter than in summer. Monthly means for O_3 (Fig. 4) were also not entirely as predicted; although they were significantly higher in zones B and C than in most of zone A, at Lindfield readings were always high, sometimes higher than in the two western zones. Some values for Rozelle are missing because of instrumentation problems; however,

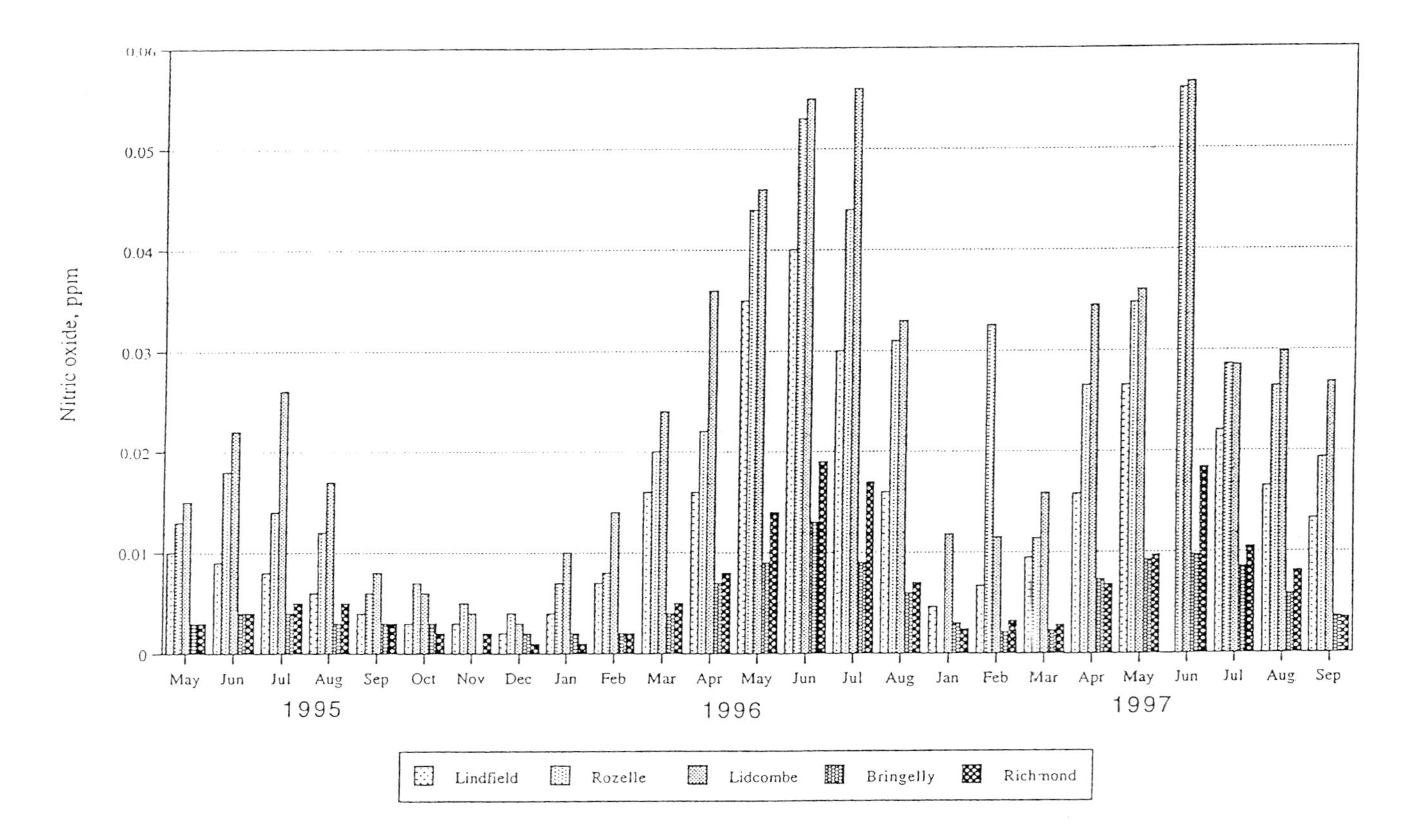

Fig. 2. Mean monthly levels of nitric oxide at NSW EPA air pollution monitoring stations relevant to ecoepidemiological sampling and experimentation

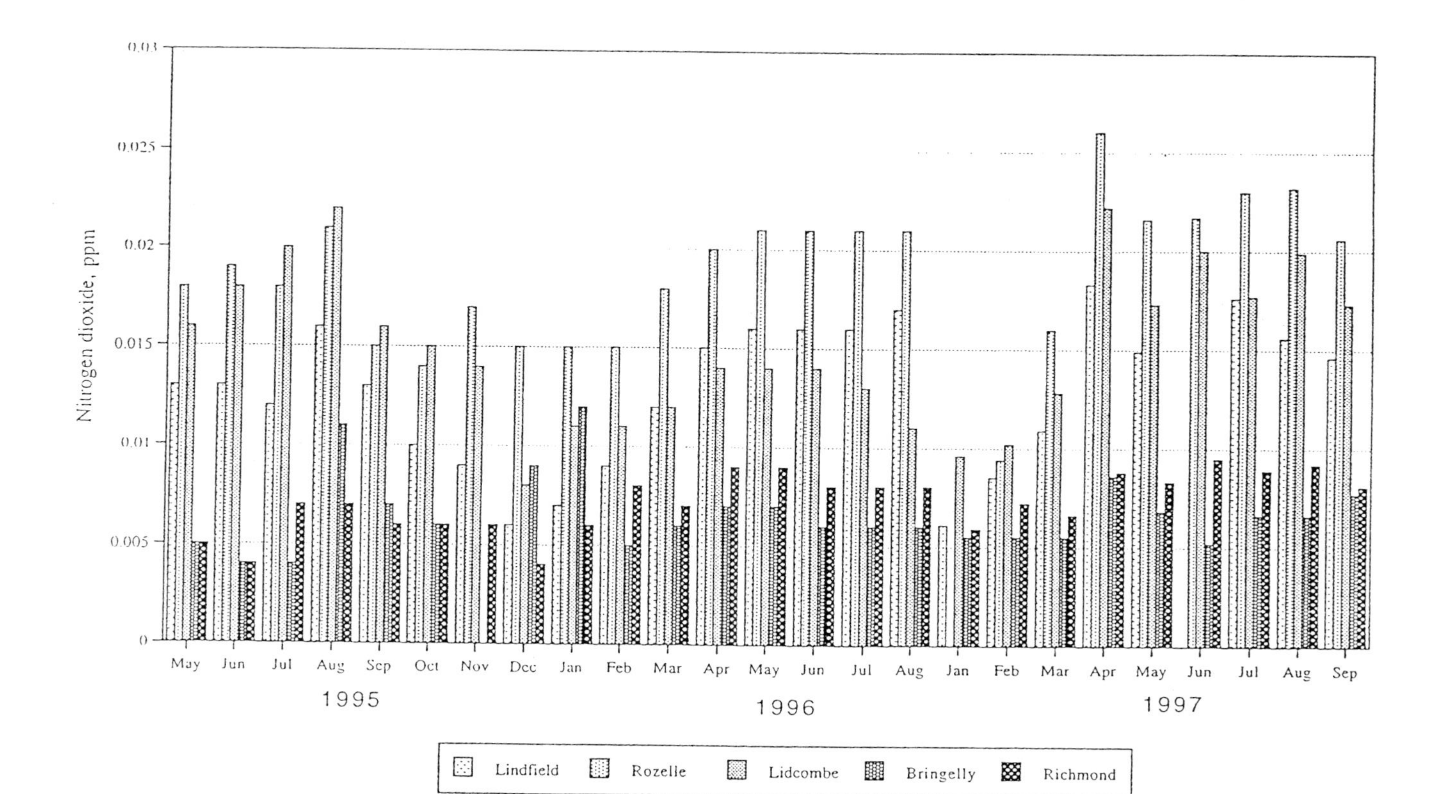

Fig. 3. Mean monthly levels of nitrogen dioxide at NSW EPA air pollution monitoring stations relevant to ecoepidemiological sampling and experimentation

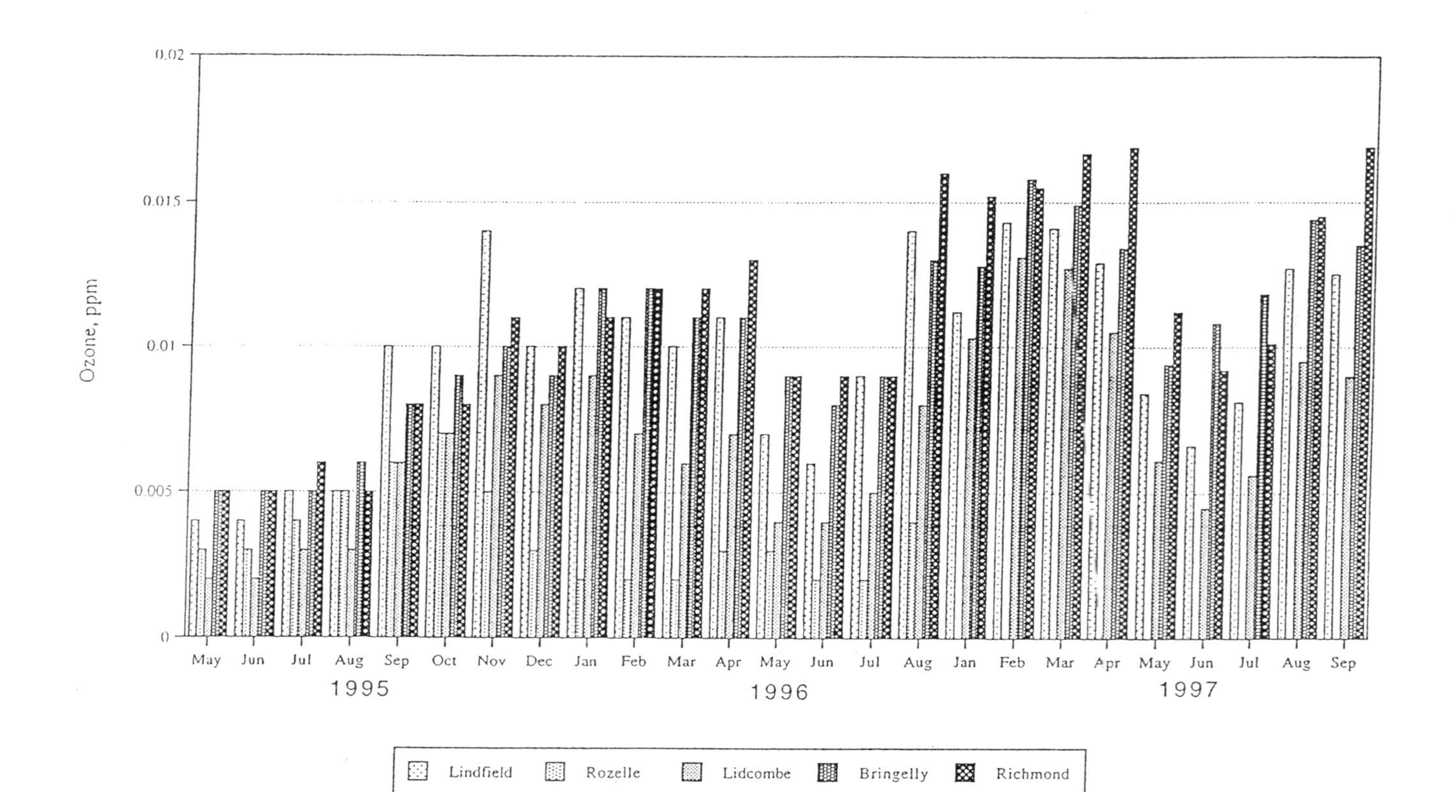

Fig. 4. Mean monthly levels of ozone at NSW EPA air pollution monitoring stations relevant to ecoepidemiological sampling and experimentation

from the data available it appears that levels of O_3 at Rozelle are generally similar to those at Lidcombe ($\pm$10%). Ozone levels were higher in summer than in winter, no doubt because of higher radiation intensities, which accelerate photochemical O_3 production.

4.2 Local Effects of Different Traffic Loads

Investigations for this component were carried out at Gore Hill with *L. confertus* and *M. quinquenervia*, and at Bringelly with *C. viminalis*.

4.2.1 Lophostemon confertus

At Gore Hill there were three stands of similar age (probably 30 - 50 years), all within 200 m of each other, on the highway; side road, and parkland. The plants were growing on the same soil and topography. Leaf parameters were sampled on a seasonal basis over 3 years on highway and parkland and over 18 months for the side road. Seasonal chlorophyll levels at the subsites are shown in Fig. 5 (A,B). Marked differences were found between chlorophyll *a* levels on the highway and those in either side road or parkland, and also between those of the side road and parkland. Differences in chlorophyll *b* between highway and other subsites were also significant, but not those between side road and parkland. There were resulting differences in total chlorophyll levels among subsites, but not in chl*a* /chl*b* ratios, although those on the highway were consistently lower than at the other subsites. In this species, increasing traffic levels produced increasing chlorophylls, indicating that this is an adaptive response to the air pollution levels, which were predominantly NO_x from fresh vehicle emissions.

Peroxidase activity levels are shown in Fig. 5C. Those levels from the highway were significantly higher than those of side road, which was higher than from parkland, again indicating higher stress with heavy traffic conditions. Less clear-cut results were obtained for ascorbic acid levels (Fig. 5.D), although those on the highway were consistently lower than those from the parkland, again indicating increased stress under heavy traffic conditions. Side road and highway levels were similar on the first two occasions, but higher on the side road on the third sampling.

Heavy metal levels in soil (top 10 cm) and leaves are shown in Table 2, along with results from the other primary sites, which are discussed later. Highway levels of all soil-borne metals were significantly higher than in side road or parkland; however, between the other subsites the zinc, lead, and iron were higher in parkland than the side road, and leaf levels, either unwashed and washed, were almost equal or higher in the parkland. The higher levels in the parkland are no doubt the result of its closer proximity to the highway. Differences between the highway and other subsites, for soil and leaf levels, indicate clearly that air-borne pollution is the major factor in determining metal contents of both.

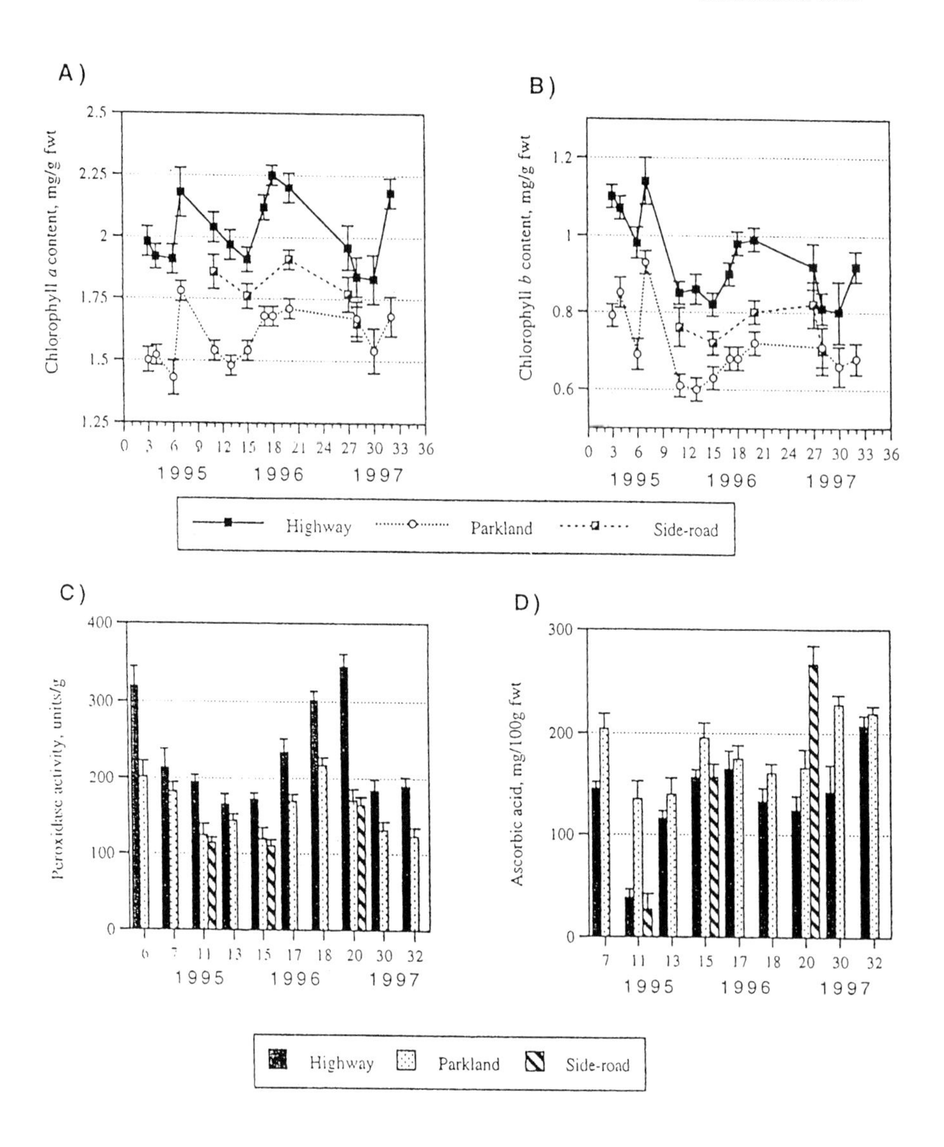

Fig. 5. A - D. Effects of local traffic levels, in highway, side road and parkland, Gore Hill, on leaf parameters of *Lophostemon confertus* A Chlorophyll *a* ; B Chlorophyll *b*; C Peroxidase activity; D Ascorbic acid

Table 2. Concentrations of heavy metals in soils and washed and unwashed leaves of *Lophostemon confertus* at subsites at Gore Hill with different traffic loads (μg/g dwt ± SE; ppt for soil iron)

Zone	Sub-site	Source	Copper	Zinc	Iron	Manganese
Zone A	**Gore Hill**					
	Highway - heavy traffic	Soil	46.2 ± 8.8	151.7 ± 28.4	36.1 ± 5.1 (ppt)	161.3 ±17.7
		Unwashed leaves	9.3 ± 0.7	56.2 ± 1.8	111.2 ± 3.4	75.3 ± 15.4
		Washed leaves	8.1 ± 0.8	48.5 ± 4.1	64.2 ± 6.8	65.3 ± 10.3
	Side road- low traffic	Soil	31.2 ± 7.3	84.1 ± 10.4	11.4 ± 1.1 (ppt)	97.4 ± 18.1
		Unwashed leaves	8.6 ± 0.5	31.1 ± 4.5	67.4 ± 10.6	61.1 ± 22.2
		Washed leaves	8.7 ± 0.4	32.8 ± 2.8	43.1 ± 1.4	70.9 ± 12.4
	Parkland - no traffic	Soil	27.2 ± 2.8	105.8 ±9.2	23.1 ± 2.1 (ppt)	80.0 ± 10.9
		Unwashed leaves	5.8 ± 0.6	35.1 ± 2.2	65.0 ± 4.9	71.6 ± 6.9
		Washed leaves	5.9 ± 0.6	33.8 ± 1.9	52.0 ± 2.1	58.2 ± 9.6
Zone B	**Bringelly**	Unwashed leaves	5.6 ± 0.3	21.0 ± 1.1	170.2 ± 1.8	118.1 ± 13.9
		Washed leaves	-	20.0 ± 1.9	73.0 ± 3.8	114.3 ± 12.3
Zone C	**Richmond**	Unwashed leaves	4.9 ± 0.2	42.9 ± 1.9	82.1 ± 1.6	419.2 ± 46.3
		Washed leaves	-	40.9 ± 1.6	57.1 ± 1.9	394.9 ± 42.9

4.2.2 *M. quinquenervia*

This species grew on highway and side road, but not in the parkland. In this species chlorophyll *a* and *b*, and therefore total chlorophyll, all significantly decreased on the highway compared with the side road (Fig. 6.A,B). There were no significant differences between chl*a*/ chl*b* ratios, although seasonal changes

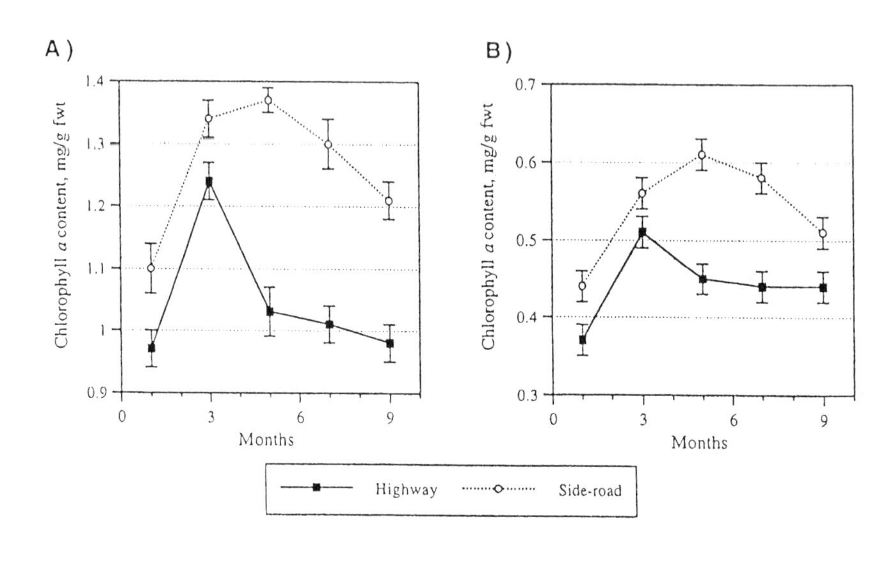

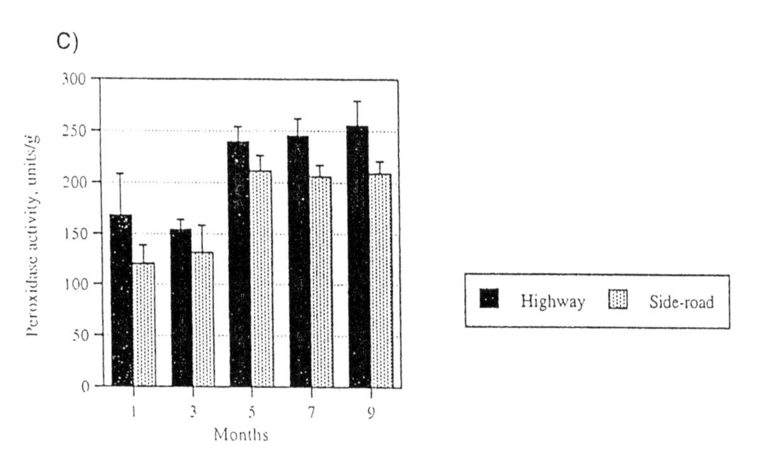

Fig. 6. A - C. Effects of local traffic levels, in highway and side road, Gore Hill, on leaf parameters of *Melaleuca quinquenervia* A Chlorophyll *a* ; B Chlorophyll *b*; C Peroxidase activity

occurred in both. Peroxidase levels were higher on the highway (Fig.6.C), indicating higher stress with heavier traffic. Ascorbic acid levels were too low (< 5 mg/100g) to be assessed for differences, although higher levels were observed at other sites (see next section).

4.2.3 *C. viminalis*

There were two subsites of this species: at Bringelly, on the side road with the monitoring station,with light traffic; and 2 km away, with moderate to heavy traffic. Significant differences in chlorophylls *a*, *b* , and total chlorophylls were found (Fig. 7.A,B), demonstrating that this species also shows increased chlorophyll content with increased pollution stress. Differences in chl*a*/chl*b* were not significant. An increase in peroxidase activity was also found with the heavier traffic (Fig. 7.C). Ascorbic acid results were again inconclusive (Fig. 7.D).

4.2.4 Summary of Local Traffic Effects

The results provide new information on the responses to local traffic effects of air pollution on these three Australian species. Both *L. confertus* and *C. viminalis* showed increased chlorophyll levels and increased peroxidase activity with heavier traffic conditions, whereas in *M. quinquenervia*, chlorophylls decreased as peroxidase activity increased. Ascorbic acid responses in all three species were less consistent. The results confirm the finding of other studies that local traffic loads are of major importance in the development of stress responses (Singh et al. 1991, 1995; Pfeffer 1994; Canas et al. 1997; Angold 1997; Bell and Ashenden 1997), and in the accumulation of airborne contaminants such as heavy metals (Munch 1993; Garcia and Millan 1994; Manfredi and Trenti 1994; Alfani et al. 1996; Garty et al. 1996). Hence, when the aim of a program is to develop a system of regional plant bioindicators of air pollution that can complement physicochemical monitoring to help ensure environmental protection, the presence of roads with different traffic loads must be taken into account at the design stage. The results also provide evidence of similarities and differences in response to the same stressors among different members from the same family (for differences among more closely related species, see Monk and Murray 1995; Carreras et al. 1996; Guidi et al. 1998). Such differences have the potential to help distinguish effects of individual pollutants across a region.

4.3 Regional Differences

Mean values for all samplings for all species and sites are presented, as a compressed summary of among-site differences. Results from both highway and parkland at Gore Hill are included because they represent local differences with respect to the heaviest traffic loads encountered in this study. It was again found that ascorbic acid levels were inconsistent with other results, so while the values are presented to provide comparisons, it must be concluded that for these three species the parameter is not a reliable indicator of air pollution effects.

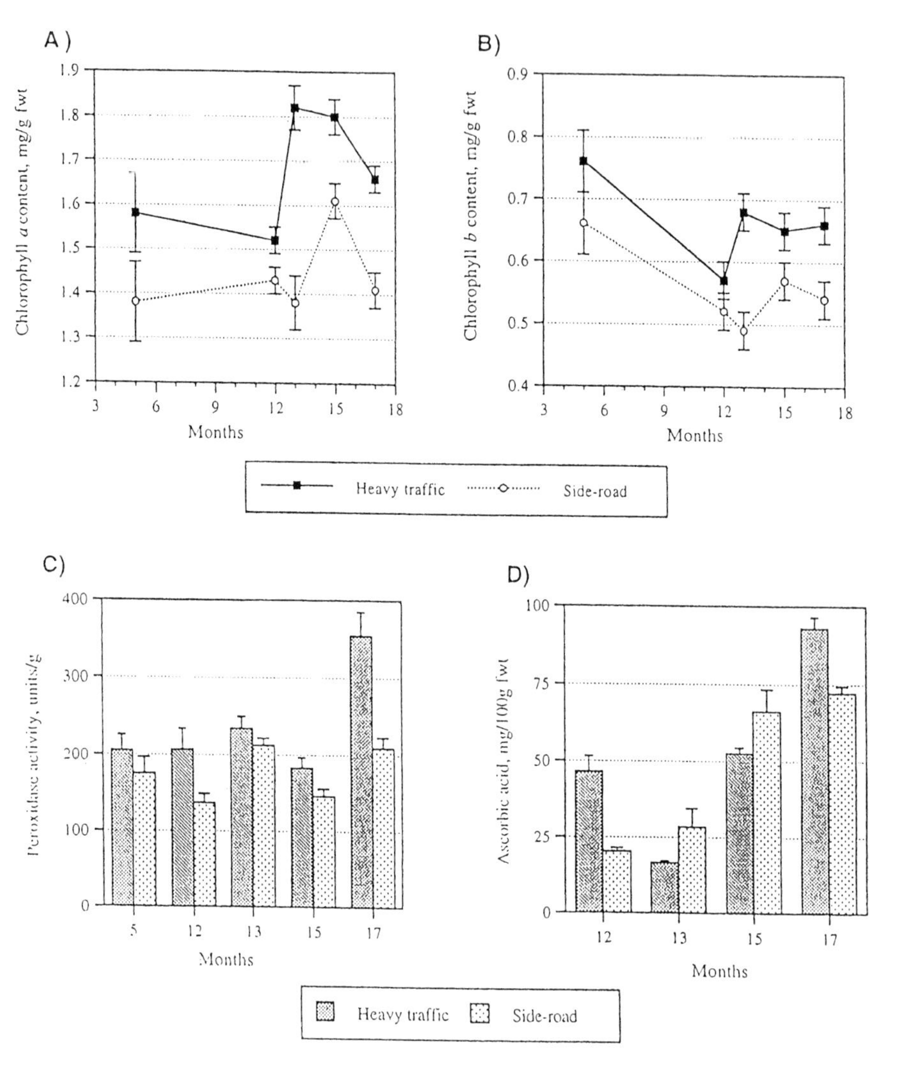

Fig.7. A - D. Effects of local traffic levels, side road and moderate to heavy traffic road, Bringelly, on leaf parameters of *Callistemon viminalis*. A Chlorophyll *a*; B Chlorophyll *b*; C Peroxidase activity; D Ascorbic acid

4.3.1 L. confertus

The sites fell into three groups with respect to chlorophyll *a* responses, (Fig. 8); levels at Gore Hill highway (zone A, highest NO_x) and Richmond (zone C, highest O_3) were significantly higher than at other sites; while those at Gore Hill parkland (zone A), Rozelle (zone A), and Bringelly (zone B) were higher than those at Lindfield and Lidcombe (zone A). Similar results were obtained for chlorophyll *b*. Peroxidase levels were significantly higher at Gore Hill highway than at all other sites, and the trends at Richmond also suggested higher levels (15% - 30%)

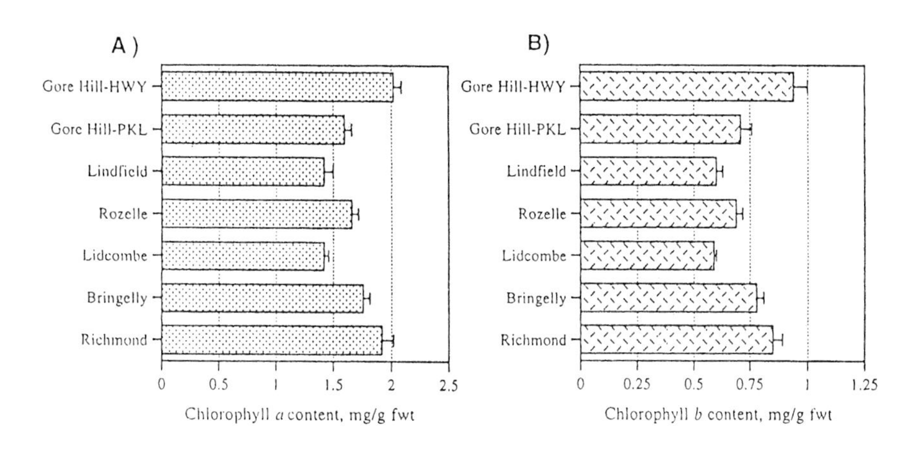

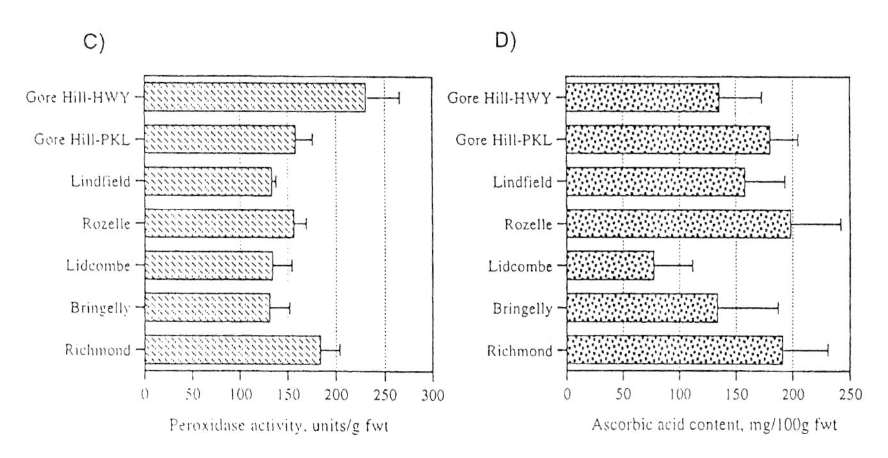

Fig. 8. A - D. Among-site differences across region, in leaf parameters of *L. confertus.*, A Chlorophyll *a* ; B Chlorophyll *b*; C Peroxidase activity; D Ascorbic acid

(P<0.07). There were no significant differences among other sites in peroxidase activity. A significant reduction in ascorbic acid was seen at Lidcombe compared with other sites, but trends were inconsistent with other parameters. Assuming that chlorophyll levels rose at all sites with increasing air pollution, these and the peroxidase levels together indicate that this species is susceptible to both high NO_x (highest at Gore Hill highway) and O_3 (highest at Richmond).

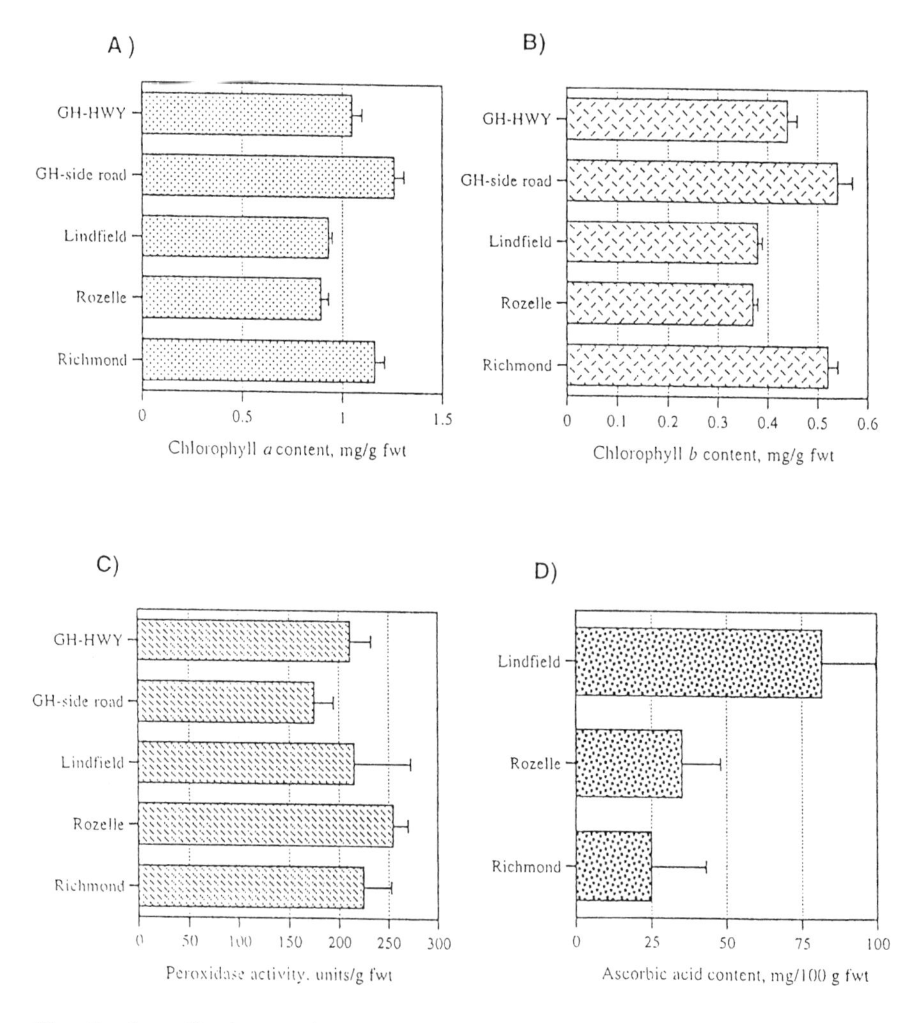

Fig. 9. A - C. Among-site differences across region, in leaf parameters of *M. quinquenervia*, A Chlorophyll *a* ; B Chlorophyll *b*; C Peroxidase activity

4.3.2 *M. quinquenervia*

In this species, chlorophylls decreased with increasing local air pollution. Across the region, chlorophylls *a* and *b* were lower at sites other than Gore Hill side road or Richmond, indicating greater stress at the other sites (Fig. 9. A,B). Peroxidase levels were also significantly elevated at all sites compared with Gore Hill side road or Richmond (Fig. 9.C), suggesting that in this species elevated NO_x concentrations were more important than O_3 levels in producing stress responses. Ascorbic acid levels (Fig. 9.D) at the three sites sampled showed significant differences but were inconsistent with other results.

4.3.3 *C. viminalis*

In this species, as in *L. confertus*, chlorophyll levels rose with air pollution at the local level, but again it is more problematic to interpret results at a regional level, because the limits of adaptive tolerance are not known. Chlorophyll *a* levels were lower at Lindfield and Richmond (both of which had higher O_3 levels) than at other sites (Fig. 10). Chlorophyll *b* was also lower at Lindfield than at other sites, but higher at Lidcombe and Rozelle (highest NO_x). Peroxidase activity was also significantly higher at Lidcombe than other sites and showed a tendency to be higher at Rozelle ($P<0.1$). The results point to this species also being more susceptible to high NO_x levels. Ascorbic acid levels were lower at Lidcombe and Richmond than at other sites, which is again inconsistent.

4.3.4 Heavy Metals

Heavy metal levels in unwashed and washed leaves from the three primary sites are shown in Table 2. Levels of copper, lead, and zinc were significantly lower at other sites compared with the Gore Hill highway, but otherwise there were few differences. The results indicate that, apart from local effects of traffic loads, there are roughly similar levels of air pollution across the region, albeit of varying composition. The levels are generally much lower than those reported for a number of overseas cities. The high levels of manganese at Richmond are thought to be the result of fertilizer use in adjacent experimental farm plots.

4.3.5 Summary of Regional Differences

Further seasonal sampling, at a greater number of sites, would be necessary to characterize precisely the responses of these species to different conditions of regional air quality. However, as a feasibility study the results are valuable in showing that significant and consistent differences in chlorophyll levels and ratios, and in peroxidase activity, occur in the three species at different sites across the region with known differences in air pollutant levels and mixtures. The results point to air pollution as having a significant effect on these species, even at the moderate levels normally experienced in the region. As the species show different patterns of response, more detailed fingerprinting could be useful for distinguishing the effects of individual pollutants more clearly.

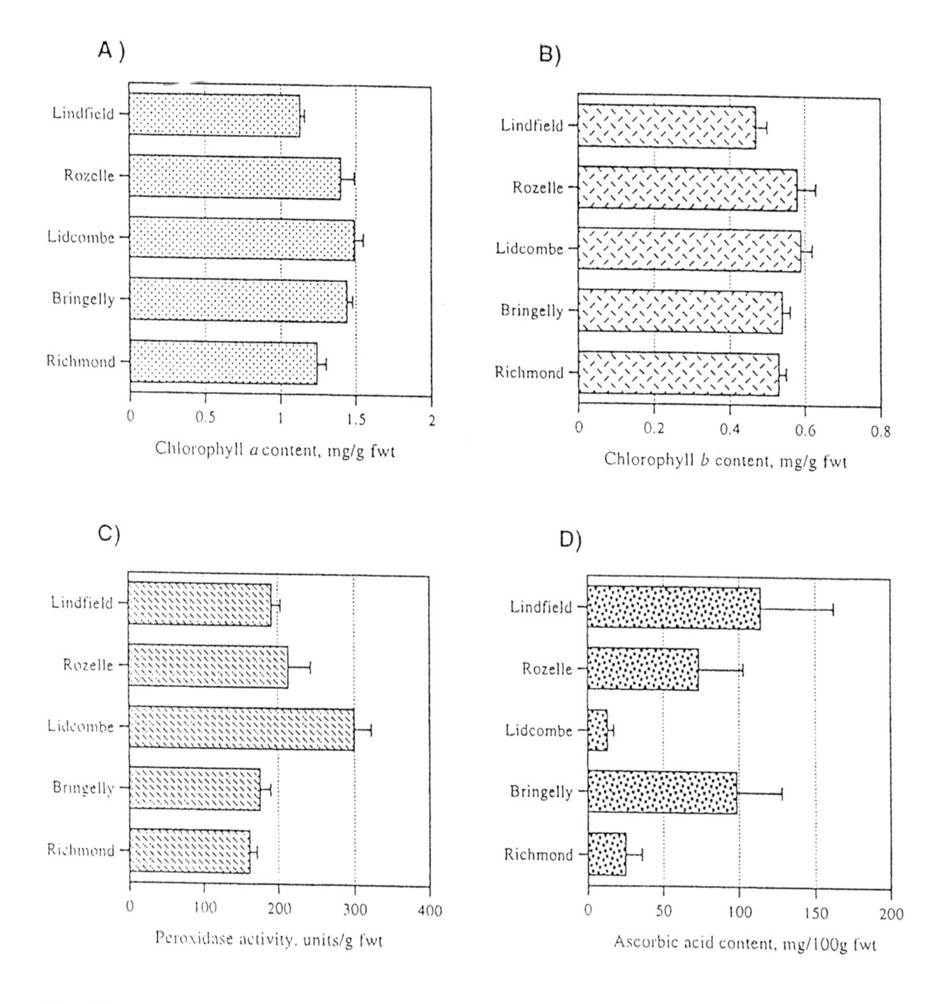

Fig. 10. A - D. Among-site differences across region, in leaf parameters of *C. viminalis* A Chlorophyll *a* ; B Chlorophyll *b*; C Peroxidase activity, D Ascorbic acid

5. Field Experiment

5.1 Methodology

Nine-month-old seedlings of *L. confertus* and *C. viminalis* (30 - 50 cm height) were obtained in May 1996 and replanted into 250-mm subirrigating pots with a standard native plant potting mixture. They were held in the open near Gore Hill, away from traffic, for 27 days to equilibrate to the new growth conditions, after which an initial harvest of one batch of each species was made (10/species) (results shown as 'Initial 1'). One batch was then retained under 'Initial Conditions' (they could not be called a control treatment because there was no control over air quality) until final harvest (results shown as 'Initial 2'). The remainder were placed (30/species/batch) within the EPA's monitoring stations at Lindfield (zone A north), Rozelle (zone A south), and at Bringelly (zone B), from which accurate air pollution data were obtained from the EPA. A first field harvest was made 105 - 109 days from field placement. At this time half the seedlings at Bringelly and Rozelle were exchanged, to test whether they would develop new responses in new conditions. A final harvest was made 156 - 162 days after field placement, including the batch that had remained under initial conditions (Initial 2). Leaves were analysed for the same parameters as for the epidemiological study. All pots were watered two to three times per week.

5.2 Experimental Findings

5.2.1 L. confertus

Chlorophyll *a* contents of seedlings at Bringelly (Fig. 11.A) rose significantly by the first field harvest (~20%) and were maintained to final harvest, but remained unchanged at Rozelle and Lindfield. Chlorophyll *b* remained unchanged at Bringelly but decreased at Rozelle and Lindfield (Fig. 11.B). As a result, chl*a* /chl*b* ratios increased at all sites. Plants moved from Bringelly to Rozelle showed a reduction in chlorophyll *a* of 18%, becoming equal with plants remaining at Rozelle. However, those transferred from Rozelle to Bringelly did not show a significant change in their chlorophyll *a* content. There were no significant changes in chlorophyll *b* in exchanged batches.

Peroxidase activity was higher at all sites at the first field harvest over initial values (Fig. 11.C), which may have been a response to a transfer to field conditions, made harder at Bringelly by some frost. By the final harvest, peroxidase activity at Bringelly and Lindfield was significantly reduced compared with the initial level before field placement (Initial 1 level) and in both cases were about half that at Rozelle. Seedlings transferred from Bringelly to Rozelle showed increased activity after transfer, to a level even higher than those remaining at Rozelle. Conversely, those transferred from Rozelle to Bringelly had decreased levels, equal to those remaining at Bringelly. Ascorbic acid levels rose at all sites above initial levels, at both harvests (Fig. 11.D). Plants moved from Rozelle to Bringelly showed a reduction in peroxidase activity compared with those

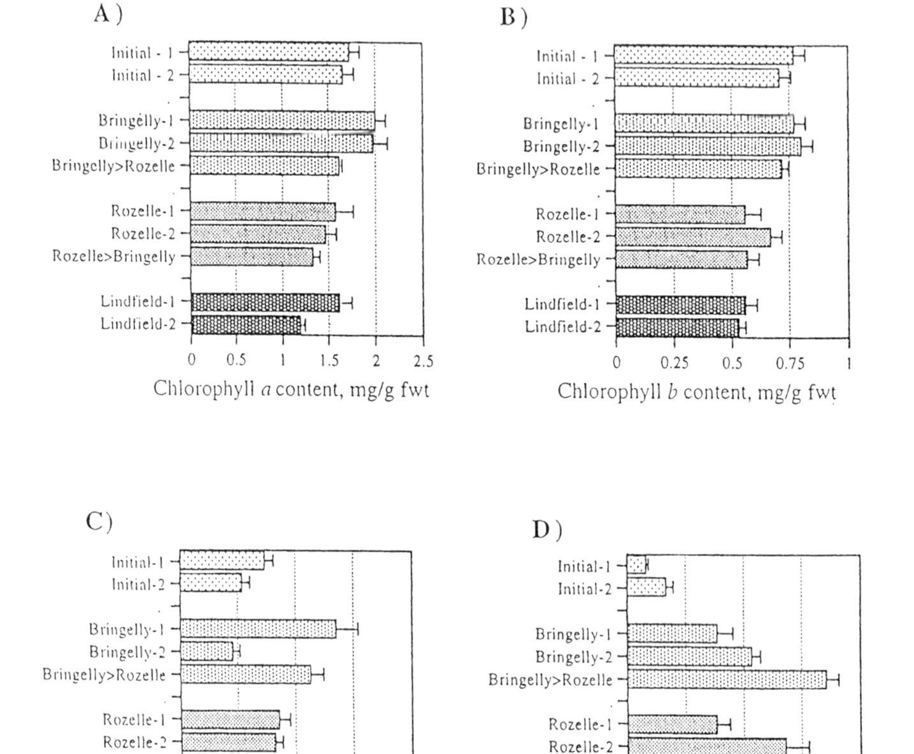

Fig. 11. A - D. Among-site differences in leaf parameters of potted seedlings of *L. confertus*. A Chlorophyll *a* ; B Chlorophyll *b*; C Peroxidase activity; D Ascorbic acid. Initial 1 refers to a harvest taken to establish initial levels of leaf parameters in the seedlings before the field placements. Initial 2 refers to the final harvest of the 'control' batch, i.e., those kept under initial conditions throughout, and harvested at the time of the final harvest of the field experiment as a whole

remaining at Rozelle whereas those transferred from Bringelly to Rozelle showed a rise of 40%.

The peroxidase results indicate that for these seedlings the most stressful site was Rozelle, where the O_3 was lower but NO_x levels higher than at the other two sites. With the mature trees, as discussed earlier, the results suggest that this

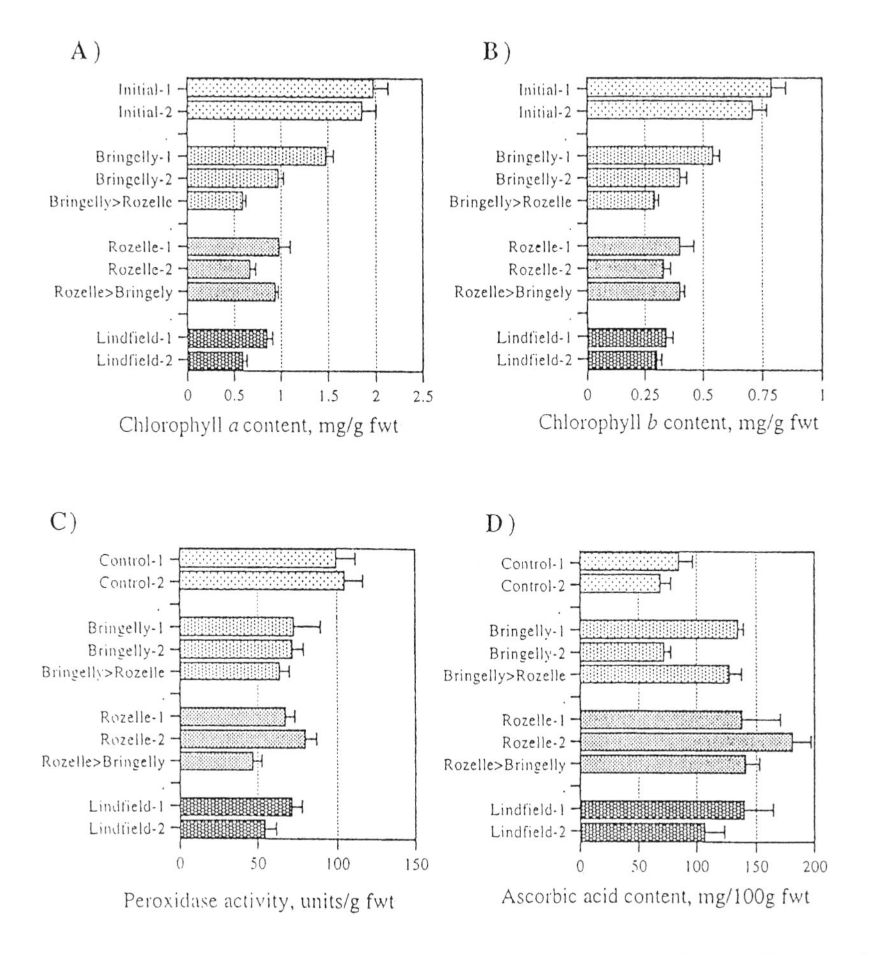

Fig. 12. A - D. Among-site differences in leaf parameters of potted seedlings of *C. viminalis.*, A Chlorophyll *a* ; B Chlorophyll *b*; C Peroxidase activity; D Ascorbic acid. Initial 1 refers to a harvest taken to establish initial levels of leaf parameters in the seedlings before the field placements. Initial 2 refers to the final harvest of the 'control' batch, i.e., those kept under initial conditions throughout, and harvested at the time of the final harvest of the field experiment as a whole

species is in fact susceptible to both high NO_x or high O_3. The results also suggest also that in seedlings, in contrast to mature plants, chlorophyll levels are reduced by increased stress.

5.2.2 *C. viminalis*

Chlorophyll *a* and *b* at all sites decreased from initial levels before field placement, and the decreases continued to final harvest (Fig. 12.A,B). Chlorophyll *a* at Bringelly was more than 30% higher than at the other two sites at first field harvest and still significantly higher at final harvest . Plants transferred from Bringelly to Rozelle showed a drop in chlorophyll *a* and *b* from the first field harvest levels (i.e., before transfer) to the final harvest, while those moved from Rozelle to Bringelly showed increases in both pigment levels, approaching those that had remained at Bringelly. Peroxidase levels decreased from initial, i.e. before-placement levels at all sites (Fig. 12.C), and there were no significant differences among sites. There was a significant difference, however, in seedlings moved from Rozelle to Bringelly, where there was a 40% decrease in peroxidase activity compared with those left at Rozelle. Thus, for seedlings of this species, as for the mature plants, it appears that chlorophyll levels may decrease with increased stress as indicated by rises in peroxidase activity. It also seems that Rozelle, with highest NO_x levels, was again the most stressful site. Ascorbic acid levels increased at all sites by 45% - 50 % in comparison with initial values, but were similar at all sites (Fig. 12.D), with differences emerging in exchanged plants.

5.2.3 *Summary of Findings of Field Experiment*

The results of the field experiment again show the development of consistent responses to air pollution in juveniles of the species tested, although with patterns somewhat different from those of mature plants. In both species the convergence in responses of plants exchanged between sites supports the view that air pollution is a major factor in bringing about the responses because plant material, growing medium, and watering regime, were all constants across sites.

6. Implications for Environmental Management

6.1 As Regional Bioindicators

Ecological/epidemiological sampling can reveal correlations between stressors and responses, but variations in soil conditions and microclimates among sites can confound interpretation of results. For example, average maximum temperatures in zones B and C are 2 – 3 degrees Celsius higher in summer (28 - 30°C), and winter minima 2 –3 degrees lower (2 - 4°C), than in zone A. However, the fact that the species are very commonly used in public plantings across the region, and exhibit the same structure, habit, and apparent health, indicate that edaphic and climatic variables are within the range of ecophysiological tolerance of the species concerned. Further, the results of the field experiment showed that seedlings faced with differences in air quality developed differences in leaf parameters that were consistent with, though different from, those in mature trees. The reciprocal changes in parameters brought about when seedlings were exchanged between sites indicate that air quality was a major factor in determining responses. Overall, the results point to the feasibility of using street-tree species for urban bioindication/biomonitoring of air quality. This approach is an alternative to that of Ali (1993), for example, or Klumpp et al. (1994), who used active biomonitoring, with seedlings of species with differing air pollution sensitivities, placed at various exposure sites across the urban areas of interest. As a tool of management, street-trees as passive phytomonitors have the advantage that they are continuously available and probably at no cost to the project. They can provide what is rarely contemplated in studies of plants and urban air pollution, that is, spatial and temporal comparisons over years, which would greatly increase our understanding of plant responses to air quality, and hence their value as biological indicators of environmental quality. These trees can also be used in a mixture of active and passive monitoring strategies (Tausz et al. 1996).

6.2 For Phytoremediation

The next phase of our project is to develop a foundation for the use of street/park trees for phytoremediation of local urban air pollution. If more air pollution can be absorbed by plants close to the source, an improvement in local air quality can be achieved and the number of exceedance days reduced. The aim appears feasible, since the following points are known: (a) air pollution is distributed unevenly at the local level, as evidenced by the foregoing results and other studies (Pfeffer, 1994; Raaschou-Nielsen et al. 1995); (b) plants absorb air pollutants, including heavy metals (see above), O_3 and NO_x etc (Singh et al. 1995; Bergmann et al. 1995; Alfani et al. 1996; Morikawa et al. 1998; Lea 1998; Wellburn 1998); (c) more than 50 potted species have been shown to be associated with reducing concentrations of volatile organics in test chambers (Wolverton, 1996). Studies in our laboratory have shown that several species used in interior plantscaping are instrumental in the removal of from two to five times the Worksafe Australia time-weighted allowable maximum concentrations of benzene and *n*-hexane within 24 h (Wood et al. 1997). It appears that rhizosphere microorganisms are involved in this removal, which also has implications for street-tree phytoremediation of air pollution.

We are currently measuring the deposition levels and characteristics of PM_{10} on leaves of a range of species using scanning electron microscopy, with heavy metal analysis as a further measure of levels of exposure to pollutants, in plants at different distances into parklands from heavy traffic routes (Nasrullah et al. 1994; Beckett et al. 1998; Roy and Sharma 1998). Chlorophyll fluorescence parameters are being used to assess stress responses along the gradient from the road. Another study is also being set up to examine the capacity of juveniles of the same species, in test chambers, to remove known doses of ROCs. Tolerant, absorptive species are required for buffer plantings along roads, around industrial sites and to protect schools, hospitals, and residential estates. However, mixed plantings should be designed, with the more sensitive species acting as early-warning bioindicators of deteriorating air quality.

References

Alfani A, Maisto G, Iovieno P, et al. (1996) Leaf contamination by atmospheric pollutants as assessed by elemental analysis of leaf tissue, leaf surface deposit and soil. J Plant Physiol 148:243-248

Ali EA (1993) Damage to plants due to industrial pollution and their use as bioindicators in Egypt. Environ Pollut 81:251-255

Aliev RR, Mousine RI (1992) Selection of wildlife plants for bioindication of air pollution. Uzb Biol Zh 5:25-29

Angold PG (1997) The impact of a road upon adjacent vegetation: effects on plant species composition. J Applied Ecol 34:409-417

Beckett KP, Freer-Smith PH, Taylor G (1998) Urban woodlands: their role in reducing effects of particulate pollution. Environ Pollut 99:347-360

Bell S, Ashenden TW (1997) Nitrogen dioxide pollution adjacent to rural roads. Water Air Soil Pollut 95:87-98

Bennett JP (1996) Floristic summary of plant species in the air pollution literature. Environ Pollut 92 (3):253-256

Bergmann E, Bender J, Weigel HJ (1995) Growth responses and foliar sensitivities of native herbaceous species to ozone exposures. Water Air Soil Pollut 85:1437-1442

Bleasedale JKA (1973) Effects of coal-smoke pollution gases on the growth of ryegrass (*Lolium perenne* L.) Environ Pollut 5:275-285

Bobbink R (1998) Impacts of tropospheric ozone and airborne nitrogenous pollutants on natural and semi-natural ecosystems: a commentary. New Phytol 139:161-168

Bridgman HA (1989) Acid rain studies in Australia and New Zealand. Arch Environ Contam Toxicol 18:137-146

Canas MS, Carreras HA, Orellana L, Pignata ML (1997) Correlation between environmental conditions and foliar chemical parameters in *Ligustrum lucidum* Ait. exposed to urban air pollutants. J Environ Manage 49: 167-181

Carreras HA, Canas MS, Pignata ML (1996) Differences in responses to urban air pollutants by *Ligustrum lucidum* Ait and *Ligustrum lucidum* Ait F tricolor (Rehd) Rehd. Environ Pollut 93 (2):211-218

Dizengrmel P, Citerne A (1988) Air pollution effect on mitochondria and respiration. In: Shulte-Hostede S, Darral NMD, Blank LW, Wellburn AR (eds), Air pollution and plant

metabolism. Elsevier, London, pp 169-188
Ernst WHO, Veklei J, Schat H (1992) Metal tolerance in plants. Acta Bot Neerl 41 (3):229-248
Finley JW, Duang EJ (1981) Resolution of ascorbic, dehyroascorbic and diketogulonic acids by paired-ion reverse-phase chromatography. Chromatography 207:449-453
Forbes VE, Forbes TL (1994) Ecotoxicology in theory and practice. Chapman and Hall, London
Garcia R, Millan E (1994) Heavy metal contents from road soils in Guipuzcoa (Spain). Sci Total Environ 146/147: 157-161
Garty J, Kauppi M, Kauppi A (1996) Accumulation of airborne elements from vehicles in transplanted lichens in urban sites. J Environ Qual 25 (2):265-272
Greenhalgh WJ, Brown GS (1982) In: Murray F (ed) Air monitoring and effects on vegetation and ecosystems. Academic Press Press, Sydney pp 125-138
Guidi LN, Nali C, Lorenzini G, Soldatini GF (1998) Photosynthetic response to ozone of two poplar clones showing different sensitivity. Chemosphere 36 (4-5):657-662
Harward M, Treshow M (1975) Impact of ozone on the growth and reproduction of understorey plants in the aspen zone of western USA. Environ Conserv 2:17-23
Hutchinson TC, Meema GHM (eds) (1987) Effects of atmospheric pollutants on forests, wetlands and agricultural systems. Springer-Verlag, Berlin
Inskeep WP, Bloom PR (1985) Extinction coefficients of chlorophyll *a* and *b* in *N,N*-dimethylformamide and 80% acetone. Plant Physiol 77:483-485
Kangerjarvi J, Talvinen J, Utriainen M, Karjalainen R (1994) Plant defence systems induced by ozone (review). Plant Cell Environ 17(7):783-794
Klumpp A, Klumpp G, Domingos M (1994) Plants as bioindicators of air pollution at the Serra do Mar near the industrial complex of Cubatao, Brazil. Environ Pollut 85:109 - 116
Knudson LL, Tibbits TW, Edwards GE (1977) Measurements of ozone injury by determination of leaf chlorophyll concentration. Plant Physiol 60:606-608
Koziol MJ, Whatley FR (eds) (1984) Gaseous pollutants and plant metabolism. Butterworth, London
Koziol MJ, Whatley FR, Shelvey JD (1988) An integrated view of the effects of gaseous air pollutants on plant carbohydrate metabolism. In: Shulte-Hostede S, Darral NMD, Blank LW, Wellburn AR (eds), Air pollution and plant metabolism. Elsevier, London, pp 148-168
Krupa SV, Legge AH (1995) Air quality and its possible impacts on the terrestrial ecosystems of the North American great plains: an overview. Environ Pollut 8:1213-1233
Laurenroth WK, Dodd JL (1981) Chlorophyll reduction in western wheat grass (*Agropyron smithii* Rydb) exposed to sulphur dioxide. Water Air Soil Pollut 15:309-315
Lea PJ (1998) Oxides of nitrogen and ozone: can our plants survive? New Phytol 139:25-26.
Manfredi VR, Trenti A (1994) Lead concentration in vegetable matrices from a Paduan urban area. Toxicol Environ Chem 41:169-174
Manning WJ, Feder WA (1980) Biomonitoring air pollutants with plants. Elsevier, London
Monk RJ, Murray F (1995) The relative tolerance of some eucalypt species to ozone exposure. Water Air Soil Pollut 85(3):1405-1411
Morikawa H, Higaki A, Nohno M, et al. (1998) More than a 600-fold variation in nitrogen dioxide assimilation among 217 plant taxa. Plant Cell Environ 21:180-190

Mousine RI, Aliev RR (1994) Standardisation of sulfur dioxide content in atmosphere of cotton plants. In, Assessment of industrial pollution effects on terrestrial plants. Hydrometeoizdat, Tashkent, Uzbekistan, pp 17-21

Munch D (1993) Concentration profiles of arsenic, cadmium, chromium, copper, lead, mercury, nickel, zinc, vanadium and polynuclear aromatic hydrocarbons (PAH) in forest soil beside an urban road. Sci Total Environ 138:47 - 55

Murray F (ed) (1982) Air monitoring and effects on vegetation and ecosystems. Academic Press, Sydney

Murray F, Wilson S (1989) The relationship between sulfur dioxide concentration and crop yield of five crops in Australia. Clean Air 23 (2):51-55

Nast W, Mortensen L, Fischer K, Fitting L (1993) Effect of air pollutants on the growth and antioxidative system of Norway spruce exposed in open-top chambers. Environ Pollut 80:85-90

Nasrullah N, Tatsumoto H, Misawa A (1994) Effect of roadside planting on suspended particulate matter concentration near road. Environ Technol 15:293-298

NEPC (National Environment Protection Council) (1997) Draft national protection measure and impact statement for ambient air quality. NEPC Service Corp. Adelaide, p 101

NSW EPA (1993 - 1995) Quarterly air quality monitoring reports, part A: EPA Data. EPA, Chatswood, NSW

NSW EPA (1996) Metropolitan air quality study - outcomes and implications for managing air quality. EPA 96/20. Chatswood, NSW

NSW Health Dept (1996) Proceedings of health and urban air quality in NSW conference, Sydney, June 3-4, vol I, II. NSW Health Dept, Gladesville, NSW

Pandey J, Agrawal M (1994) Evaluation of air pollution phytotoxicity in a seasonally dry tropical urban environment using three woody perennials. New Phytol 126: 53-61

Pascoe GA, DalSoglio JA (1994) Planning and implementation of a comprehensive ecological risk assessment at the Milltown Reservoir - Clark Fork River Superfund site, Montana, Environ Toxicol Chem 13 (12):1943-1956

Pell EJ (1988) Secondary metabolism and air pollutants. In: Shulte-Hostede S, Darral NMD, Blank LW, Wellburn AR (eds), Air pollution and plant metabolism. Elsevier, London, pp 22-237

Pfeffer HU (1994) Ambient air concentrations of pollutants at traffic-related sites in urban areas of North Rhine - Westphalia, Germany. Sci Total Environ 146/147:263-273

Pfleeger TG, Ratsch HC, Shimabuku RA (1993) A review of terrestrial plants as biomonitors. In: Gorsuch JW, Dwyer FJ, Ingersoll CG, La Point TW (eds) Environmental toxicology and risk assessment, Vol 2. ASTM STP, Philadelphia, pp 317-330

Phillips DJH (1980) Quantitative aquatic biological indicators. Applied Science, London

Posthumus AC (1985) Plants as indicators for atmospheric pollution. In: Nurnberg HW (ed) Pollutants and their ecotoxicological significance. Wiley, London, pp 55-65

Puccinelli P, Anselmi N, Bragaloni M (1998) Peroxidases: suitable markers of air pollution in trees from urban environments. Chemospher 36 (4-5):889 –894

Raaschou-Nielsen O, Nielsen ML, Gehl J (1995) Traffic-related air pollution: exposure and health effects in Copenhagen street cleaners and cemetery workers. Arch Environ Health 50 (3):207-213

Rabe R, Kreeb KH (1980) Bioindication of air pollution by chlorophyll destruction in plant leaves. Oikos 34:163-167

Rank B (1997) Oxidative stress responses and photosystem 2 efficiency in trees in urban areas. Photosynthetica 33 (3-4):467-481

Ricks GR, Williams RJ (1975) Effects of atmospheric pollution on deciduous woodland, part 3. Effects on photosynthetic pigments of leaves of *Quercus petraea* Mattushka Leibl. Environ Pollut 8:97-106

Roose ML, Bradshaw AD, Roberts TM (1982) Evolution of resistance to gaseous air pollutants. In: Unsworth MH, Ormond, DP (eds), Effects of gaseous pollutants in agriculture and horticulture. Butterworth, London, pp 379-409

Roy S, Inantola R, Hanninen O (1992) Peroxidase activity in lake macrophytes and its relation to pollution tolerance. Environ Exp Bot 32:457-464

Roy RK, Sharma SC (1998) Bioremediation of urban pollution by orientation of landscaping. Ind J Environ health 40 (2):203-208

Salt DE, Smith RD, Raskin I (1998) Phytoremediation. Annu Rev Plant Physiol Plant Mol Biol 49:643-668

Sanka M, Strnad M, Vondra J, Paterson E (1995) Sources of soil and plant contamination in an urban environment and possible assessment methods. Int J Environ Anal Chem 59:327-343

SEAC (State of Environment Advisory Committee) (1996) State of environment, Australia 1996. CSIRO, Collingwood, Australia

Shimazaki KL, Sakaki T, Kondo N, Sugahara K (1980) Active oxygen participation in chlorophyll destruction and lipid peroxidation in SO_2 exposed leaves of spinach. Plant Cell Physiol 21:1193-1204

Singh SK, Rao ND, Agrawal M, Pandey J, Narayan D (1991) Air pollution tolerance index of plants. J Environ Manage 32:45-55

Singh N, Yunus M, Srivastava K, et al (1995) Monitoring of auto exhaust pollution by roadside plants. Environ Monit Assess 34:13-25

Tausz M, Batic F, Grill D (1996) Bioindication of forest sites - concepts, practice and outlook. Phyton 36 (3):7-14

USEPA (1993) A review of ecological risk assessment: case studies from a risk assessment perspective. Risk assessment forum. EPA/ 630/ R-92/005, May 1993, pp 1-2

Van Assche F, Clijster J (1990) Effects of metals on enzyme activity in plants. Plant Cell Environ13:195-206

Wang W, Freemark K (1994) The use of plants for environmental monitoring and assessment. Ecotoxicol Environ Saf 30:289-301

Watanabe ME (1997) Phytoremediation on the brink of commercialization. Environ Sci Technol News 31 (4):182-186

Wellburn AR (1998) Atmospheric nitrogenous compounds and ozone-is NO_x fixation by plants a possible solution? New Phytol 139:5-9

Wolverton BC (1996) Eco-friendly house plants. Weidenfeld and Nicolson, London

Wood RA, Orwell RL, Burchett MD (1997) Rates of absorption of VOCs by commonly used indoor plants. Proceedings healthy buildings/IAQ '97, Washington, DC, 27 Sept - 2 Oct, vol 1, pp 59-64

4
Effects of Air Pollutants on Lipid Metabolism in Plants

Takeshi Sakaki

Department of Bioscience and Technology, School of Engineering, Hokkaido Tokai University, Minami-sawa 5-1-1-1, Minami-ku, Sapporo 005-8601, Japan

1. Introduction

Ozone (O_3), nitrogen oxides (NO_x), peroxyacetyl nitrate (PAN), and sulfur dioxide (SO_2) are major global air pollutants, causing serious vegetative damage and forest decline. To clarify both the acute and chronic phytotoxic mechanisms of these pollutants, their physiological and biochemical effects on plants have been extensively studied during the past several decades. According to ultrastructural observations of plant cells injured by these pollutants, cellular membrane systems are affected by the pollutants (Thomson 1975; Huttunen and Soikkeli 1984), and membrane permeability is also seen to change after treatment with SO_2 (Malhotra and Hocking 1976) and O_3 (Heath and Castillo 1988). It is now generally accepted that cellular membranes are among the primary sites of pollutant attack and, since lipids are important membrane components and play essential roles in maintaining membrane structure and function, many workers have examined the effects of pollutants on lipids to clarify the mechanisms of their phytotoxicity (Mudd et al. 1984; Heath 1984; Sakaki 1998).

Previous reports concerning lipids and air pollution may be divided into two groups. The first group deals with the oxidation of unsaturated fatty acids in membrane lipids, due primarily to the potential reactivity of O_3, NO_2, and PAN with these fatty acids (Hippeli and Elstner 1996). More recent studies rather address the oxidation of these lipids by active oxygen species (AOS), which have

Air Pollution and Plant Biotechnology
–Prospects for Phytomonitoring and Phytoremediation–
Edited by K. Omasa, H. Saji, S. Youssefian, and N. Kondo
© Springer -Verlag Tokyo 2002

been demonstrated to be generated by O_3, SO_2, and NO_2 treatment. The second group of studies concerns the metabolic alteration of membrane lipids in plant cells, which can be classified into three major groups: glycerolipids, sphingolipids, and sterols. To date, studies have been limited to those on glycerolipids and sterols because of the ease of analysis and the considerable amount of information available on the metabolic pathways, especially for glycerolipids. Since some of the glycerolipid classes are distributed in specific intracellular organelles, their analysis provides additional information on the intracellular sites affected by the pollutants.

In this chapter, I review the oxidative and metabolic alterations in leaf lipids induced by air pollutants with respect to the biochemical mechanisms of phytotoxicity. The results from molecular biology studies of lipids reported so far are also discussed.

2. Leaf Glycerolipids and Their Metabolism

Glycerolipids in green leaf tissues are classified into glyco-, phospho-, and neutral lipids (Harwood 1980). In healthy leaves, glycolipids consist of galactolipids (monogalactosyldiacylglycerol, MGDG; digalactosyldiacylglycerol, DGDG) and sulfolipid (sulfoquinovosyldiacylglycerol, SQDG). The glycolipids are exclusively located in the envelope and thylakoid membranes of chloroplasts (Joyard et al. 1991). Phospholipids in leaf cells consist of phosphatidylcholine (PC), phosphatidylethanolamine (PE), phosphatidylglycerol (PG), phosphatidylinositol (PI), phosphatidylserine (PS), cardiolipin, and phosphatidic acid (PA). Except for the majority of PG located in chloroplasts and a minor amount of PC in chloroplast envelopes, these phospholipids are principally extrachloroplast membrane lipids. Cardiolipin is restricted to inner mitochondrial membranes, but other phospholipids are distributed in extrachloroplast membranes. Green leaves usually contain trace amounts of free fatty acid (FFA) and neutral glycerolipids, such as triacylglycerol (TG) and diacylglycerol (DG). The glycerolipids in leaves, especially MGDG and DGDG, are characterized by their acylation with a high proportion of polyunsaturated fatty acids (Table 1) and their susceptibility to oxidation. Mature glycerolipids are synthesized in three major steps: synthesis of palmitate (16:0) and oleate (18:1) in chloroplasts, acylation of the fatty acids to glycerol 3-phosphate to form PA, and metabolism of the head groups to final glycerolipid classes with lipid-linked desaturation of the fatty acid residues (Browse and Somerville 1991). Compared to these biosynthetic pathways, the catabolic pathways for glycerolipids have not been studied in detail.

The content of glycerolipids and the proportion of fatty acid species acylated to each lipid class are usually constant in leaves of the same plant species. However, they vary considerably under stress conditions, such as low temperature and salinity stress (Kuiper 1985), and air pollution as described below. Leaf age is also an important factor determining lipid and fatty acid composition (Harwood

Table 1. Content and fatty acid composition of major glycerolipids in mature green leaves of spinach[a]

Lipids[b]	Content	Fatty acid composition (mol%)[c]						
	(nmol/cm^2 leaf area)	16:0	t16:1	16:3	18:0	18:1	18:2	18:3
MGDG	93	2	0	25	tr	1	2	70
DGDG	50	13	0	4	1	3	3	75
SQDG	13	47	0	0	3	5	9	35
PG	20	18	36	0	2	3	6	34
PC	26	20	0	0	1	25	23	31
PE	11	34	0	0	3	12	31	19
PI	6	40	0	0	6	10	21	22

[a]Spinach plants (*Spinacia oleracea* L. cv. New Asia) were cultivated in a glasshouse for 6 weeks. Glycerolipid content and their fatty acid composition in the spinach leaves were determined as described (Sakaki et al. 1985).
[b]MGDG, monogalactosyldiacylglycerol; DGDG, digalactosyldiacylglycerol; SQDG, sulfoquinovosyldiacylglycerol; PG, phosphatidylglycerol; PC, phosphatidylcholine; PE, phosphatidylethanolamine; PI, phosphatidylinositol. More than 90% of fatty acids ocurring in the leaves are acylated to these seven glycerolipids.
[c]16:0, palmitate; t16:1, 3-*trans*-hexadecenoate; 16:3, hexadecatrienoate; 18:0, stearate; 18:1, oleate; 18:2, linoleate; 18:3, α-linolenate; tr, trace.

1980). In addition to glycerolipids, green leaves contain free sterol and three types of sterol derivatives: steryl glycoside, acylated steryl glycoside, and steryl ester. The content of these lipids is also affected by various stresses (Kuiper 1985), including air pollution.

3. Lipid Oxidation by Air Pollutants

3.1 Participation of Active Oxygen Species in Lipid Oxidation

The pollutants O_3, PAN, NO_2, and SO_2 can all chemically react with the double bonds of unsaturated fatty acids (Hippeli and Elstner 1996), and therefore have the potential to oxidize the unsaturated lipids. In fact, when O_3 was bubbled into suspensions of unicellular algae (Frederick and Heath 1975), isolated chloroplasts (Mudd et al. 1971b), and microsomes (Pauls and Thompson 1981), malondialdehyde, one of the oxidation products of polyunsaturated fatty acids, accumulated while levels of unsaturated fatty acids decreased. The characteristics of the reaction of O_3 with membrane lipids demonstrated by these bubbling experiments resemble those with pure fatty acid and lipid molecules.

When intact plant leaves were fumigated with O_3 (Tomlinson and Rich 1970; Sakaki et al. 1983), PAN (Nouchi and Toyama 1988) or SO_2 (Peiser and Yang 1979; Shimazaki et al. 1980; Tanaka and Sugahara 1980), malondialdehyde also accumulated in the leaves. The observations that the levels of unsaturated fatty acids were decreased by fumigation with O_3 (Sakaki et al. 1985) and PAN (Nouchi and Toyama 1988), and that ethane, a decomposition product of fatty acid hydroperoxides, was emitted by SO_2 (Peiser and Yang 1979), throw little doubt about the oxidative damage of leaf lipids by these pollutants. However, in contrast to the bubbling experiments with O_3, lipid oxidation in the fumigated leaves is attributable to the reaction with AOS, leading to lipid peroxidation, rather than to a direct reaction with O_3. Sakaki et al. (1983, 1985) have shown that an accumulation of malondialdehyde in leaves after O_3 fumigation is light- and O_2 dependent but O_3 independent, and that superoxide anions (O_2^-) generated in leaf cells are responsible for the accumulation. This is also the case with other pollutants. Shimazaki et al. (1980) have demonstrated the involvement of O_2^- in the formation of malondialdehyde in SO_2-fumigated leaves, and the toxicity of NO_2 has been ascribed, at least in part, to that of O_2^- (Shimazaki et al. 1992). Thus, these pollutants induce oxidative stress, mediated by AOS, on leaf cells like other environmental stresses (Bowler et al. 1992). Further studies on the detection and distribution of hydroperoxides in lipid classes may well identify intracellular sites affected by AOS.

Air pollutants function by enhancing the production of AOS and/or by damaging the antioxidant systems of leaf cells. SO_2 entering leaf cells moves into chloroplasts, where it is aerobically oxidized in the light to sulfate with the enhanced production of O_2^- and hydroxyl radicals ($OH^\cdot$) (Asada and Kiso 1973). The SO_2 may inactivate superoxide dismutase, further contributing to the increased O_2^- production (Shimazaki et al. 1980). On the other hand, O_3 appears to enhance O_2^- production mainly by inactivating the antioxidant systems, such as by oxidation of ascorbates and inactivation of superoxide dismutases, ascorbate peroxidases, and catalases (Sakaki et al. 1983; Tanaka et al. 1985).

Although O_2^- does not react with polyunsaturated fatty acids, the protonated form (HO_2) does. More importantly, O_2^- produces $OH^\cdot$ and singlet molecular oxygen ($^1O_2$), which are highly reactive to acids, via the Haber-Weiss reaction and interaction with cellular components, respectively (Elstner 1982). It is probable that these AOS are the actual direct inducers of lipid peroxidation in leaves treated with pollutants. Not only unsaturated lipids, but also sterols, are oxidized by these AOS (Smith 1987).

3.2 Mechanisms of Cellular Protection Against Lipid Oxidation

To cope with the threat of AOS damage, leaves exposed to air pollutants are known to enhance their antioxidant systems, which include superoxide dismutase, ascorbate peroxidase, and glutathione reductase (see Aono, Polle, in this volume). Recently, transgenic plants with enhanced levels of antioxidant enzymes have been generated, and shown to be resistant to air pollutants (Aono et al. 1991, 1993; Van Camp et al. 1994). Although the levels of lipid peroxidation have not been measured in these transgenic plants, it is reasonable to assume that they may be diminished to nontoxic levels as a result of AOS removal. Genetic manipulation of the levels of low molecular weight antioxidants, such as ascorbate and tocopherol, will also be useful for improving the resistance of membrane lipids to oxidative stress.

4. Metabolic Alteration of Lipids by Air Pollutants

4.1 Lipid Changes by Air Pollutants

4.1.1 Ozone

Upon fumigation of plants with O_3 at concentrations sufficient to induce acute injuries (0.1-0.5 ppm) within a day, various leaf lipids, including membrane-constituting ones, change drastically (Fig. 1, Table 2) before the onset of fatty acid oxidation (Sakaki et al. 1985, 1994). Thus lipid molecules embedded in membranes are clear targets of O_3. The lipids most greatly reduced in amount include galactolipids (Sakaki et al. 1985, 1994; Nouchi and Toyama 1988) and free sterol (Tomlinson and Rich 1971; Trevathan et al. 1979; Whitaker et al. 1990). Phospholipids and sulfolipids are more stable to O_3 than galactolipids, and occasionally the amounts of some phospholipid classes, such as PC, are increased by O_3 (Fong and Heath 1981; Sakaki et al. 1985, 1994; Nouchi and Toyama1988). PA generally accumulates in O_3-treated leaves (Sakaki et al. 1985; Nouchi and Toyama 1988). Besides these lipids, TG (Sakaki et al. 1985, 1994), FFA (Mackay et al. 1987), sterol derivatives (Tomlinson and Rich 1971; Whitaker et al. 1990), and oligogalactolipids, such as tri- and tetragalactosyldiacylglycerol (Sakaki et al. 1990b), accumulate in O_3-treated leaves. Similar patterns of lipid change have been observed in leaves treated with lower concentrations of O_3 (0.05-0.1 ppm) for several days (Carlsson et al. 1994, 1996).

Such lipid changes are undoubtedly toxic to plant cells. Because the contents of membrane lipids reflect the developments of intracellular membrane systems where the lipids are located, galactolipid decreases suggest that chloroplast membranes disintegrate and disappear in response to O_3. This finding is consistent with morphological observations (Thomson 1975; Huttunen and Soikkeli 1984) and early damage to the photosynthetic apparatus (Heath 1980). The decreased amount of galactolipids closely correlates with chlorophyll loss in several plant species exposed to O_3 (Table 2, Fig. 2). Loss of chloroplast lipids and the resultant membrane disintegration would induce instability of membrane components such as chlorophyll. In addition, newly accumulated lipids may affect membrane structure and function. For example, FFA induces the swelling of thylakoid membranes (Okamoto et al. 1977) and inhibits photosynthesis (McCarty and Jagendorf 1965).

Another set of reports describes lipid changes in conifer needles fumigated in

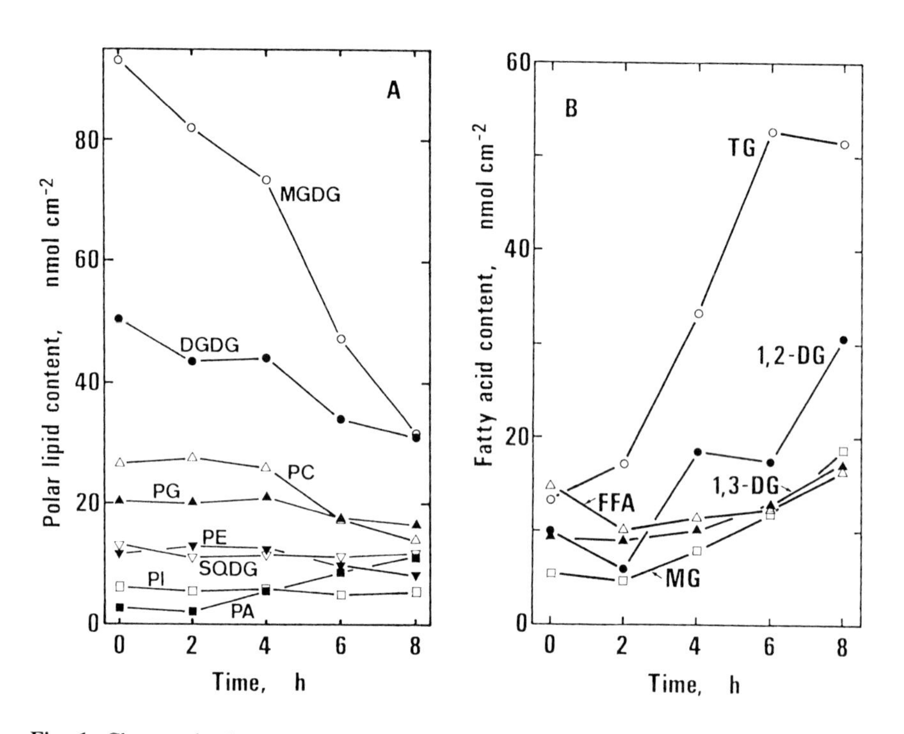

Fig. 1. Changes in the content of glyco- and phospholipid classes (A) and neutral lipid classes (B) in spinach leaves fumigated with O_3 (0.5 ppm, v/v). MGDG, monogalactosyldiacylglycerol; DGDG, digalactosyldiacylglycerol; PC, phosphatidylcholine; PG, phosphatidylglycerol; PE, phosphatidylethanolamine; SQDG, sulfoquinovosyldiacylglycerol; PI, phosphatidylinositol; PA, phosphatidic acid; TG, triacylglycerol; DG, diacylglycerol; FFA, free fatty acid; MG, monoacylglycerol. (Data from Sakaki et al. 1985)

Table 2. Loss of galactolipids and chlorophyll *a* in leaves of various plant species fumigated with O_3[a]

Plant species[b]	O_3[b]	Galactolipids[c] (nmol/cm^2 leaf area)		Chlorophyll *a*[c] (μg/cm^2 leaf area)
		MGDG	DGDG	
Lettuce	-	60.3	44.9	29.2
	+	53.7	38.2	27.8 (25.7)
Tobacco	-	79.5	52.5	40.7
	+	60.3	31.8	38.9 (36.8)
Broad bean	-	93.9	78.8	50.2
	+	67.8	59.2	48.7 (46.5)
Maize	-	63.4	41.5	-
	+	43.4	37.5	-
Radish	-	76.0	40.1	41.7
	+	47.1	28.4	41.7 (28.3)
Kidney bean	-	62.6	33.8	30.4
	+	25.5	21.2	31.4 (16.3)
Spinach	-	81.0	51.3	44.2
	+	31.6	30.2	46.3 (17.7)

[a]Each value is the average of two samples (T. Sakaki, unpublished results).
[b]Plants of lettuce (*Lactuca sativa* L.), tobacco (*Nicotiana tabacum* L.), broad bean (*Vicia faba* L.), maize (*Zea mays* L.), radish (*Raphanus sativus* L.), kidney bean (*Phaseolus vulgaris* L.), and spinach (*Spinacia oleracea* L.) were grown in glasshouses and fumigated with O_3 (0.5 ppm, v/v) for 6 h (Sakaki et al. 1994).
[c]Contents of galactolipids and chlorophyll *a* in leaves were determined as described (Sakaki et al. 1983). Values in parentheses are the chlorophyll *a* contents after O_3-treated plants were kept in the light for a further 20 h (Sakaki et al. 1983).

open-top chambers with very low concentrations of O_3 (<0.05 ppm) for months or years (Fangmeier et al. 1990; Wolfenden and Wellburn 1991; Wellburn et al. 1994). These experiments, designed to estimate the impact of ambient O_3 levels on field-grown plants, demonstrate decreased unsaturation levels of total lipids and MGDG.

4.1.2 Sulfur Dioxide and Peroxyacetyl Nitrate

Khan and Malhotra (1977) showed decreases in the content and unsaturation levels of galactolipids and sulfolipid in pine needles incubated in an aqueous solution of SO_2. The authors attributed these lipid changes to the inhibition of their

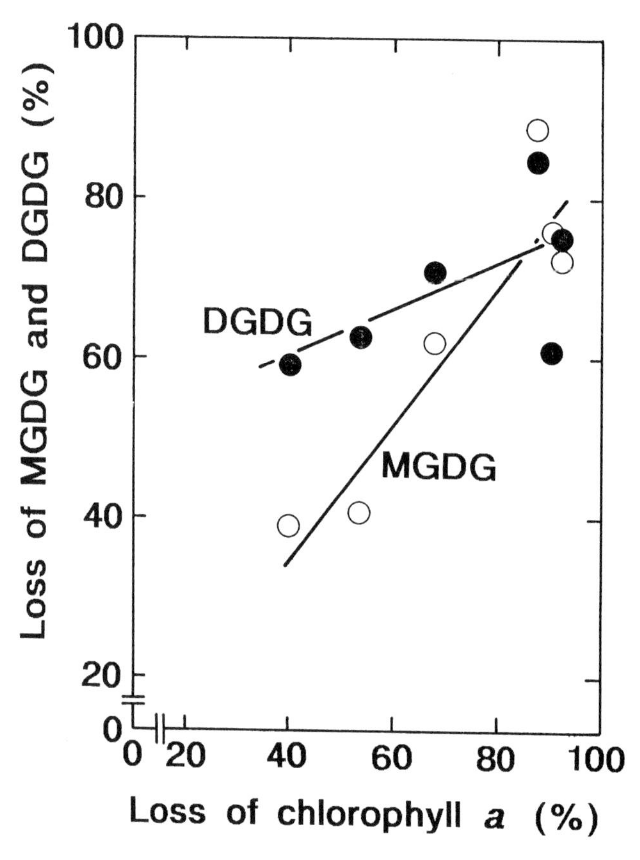

Fig. 2. Relationship between loss of MGDG and DGDG during 6 h of O_3 fumigation and subsequent loss of chlorophyll *a* after 20 h of illumination. The values represent each of the plant species shown in Table 2

biosynthesis (Malhotra and Khan 1978). However, as they treated the needles with aqueous SO_2 under high light intensities for 22 h, oxidative destruction of unsaturated lipids by AOS may well have contributed to the lipid decreases. In contrast, when barley plants were fumigated with low concentrations of SO_2 (0.04 and 0.117 ppm) for 48 days, the degree of unsaturation in the galactolipids remained constant despite a sharp loss of the lipids (Navari-Izzo et al. 1992). These authors also found increases in TG, FFA, and PA, and a decrease in free sterol, following SO_2 fumigation (Navari-Izzo et al. 1991, 1992). The characteristics of the lipid responses to gaseous SO_2 resemble those to O_3.

As with O_3 and SO_2, PAN was found to cause greater decreases in the amounts of galactolipids than phospholipids in kidney bean leaves fumigated with 0.1 ppm for 1 h (Nouchi and Toyama 1988). PA again accumulated in the treated leaves. Treatment with PAN for longer periods decreased the phospholipid contents as well as galactolipids and unsaturated fatty acids.

4.2 Metabolic Changes in Lipids by Air Pollutants

4.2.1 Effects on Lipid Biosynthesis

Several steps of lipid biosynthesis in plants are reported to be affected by O_3. Mudd et al. (1971a,b) showed inhibition of lipid synthesis from acetyl-CoA and of oligogalactolipid synthesis from UDP-galactose in isolated chloroplasts bubbled with O_3. Peters and Mudd (1982) measured the activities of several enzymes of lipid synthesis in rat lung microsomes bubbled with O_3, and demonstrated that glycerol 3-phosphate acyltransferase was the most susceptible to O_3. This group emphasized that oxidation of sulfhydryl groups by O_3 was responsible for the inactivation of the enzymes. However, these results are rather inconsistent with those obtained from intact plants fumigated with O_3. Sakaki et al. (1990c) have shown that galactolipid: galactolipid galactosyltransferase, the enzyme responsible for oligogalactolipid synthesis, is fully active in O_3-fumigated spinach leaves. Hellgren et al. (1995) designed experimental systems to estimate separately the synthesis and degradation activities of lipids in pea leaves fumigated with O_3, and demonstrated enhancement of galactolipid degradation with little inhibition of the synthesis by O_3. These conflicting results may be attributable to the experimental conditions; bubbling of O_3 to membrane suspensions is close to the conditions of chemical reactions and does not always reflect fumigation conditions of intact plants (see Section 3.1).

Desaturation of glycerolipid-linked fatty acids is another step of lipid biosynthesis. Although the fatty acid composition of most glycerolipid classes changes little upon fumigation with O_3, at least for short periods of time, one of the exceptions is PC, which becomes more unsaturated by accumulating α-linolenate (18:3) (Whitaker et al. 1990; Sakaki et al. 1994). This is partly due to reacylation to PC of 18:3, which is hydrolyzed from galactolipids (see next section). In addition, chloroplast ω-3 fatty acid desaturase is reportedly induced after wounding of tobacco leaves, resulting in an increase in the levels of 18:3-PC (Hamada et al. 1996). During development of O_3-induced injuries in leaf cells, the desaturase might be induced and desaturate fatty acids in PC.

Malhotra and Khan (1978) examined the effects of aqueous SO_2 on lipid synthesis from acetate in pine needles, and demonstrated decreased synthesis of all the glyco-, phospho-, and neutral lipid classes measured. They speculated that the early steps of lipid synthesis, such as of fatty acids, would be affected by SO_2. The decreased amounts of various glycerolipids observed in barley seedlings fumigated with SO_2 (Navari-Izzo et al. 1992) might be caused by the inhibition of these early steps.

4.2.2 Metabolic Conversion of Lipids

The majority of lipid changes induced by O_3 can be now explained by the induction of new metabolic steps as depicted in Fig. 3 (Sakaki et al. 1985, 1990a,b). The first step of this metabolic pathway is the stimulation of galactolipase activity by O_3, resulting in the hydrolysis of galactolipids to produce FFA, mainly 18:3. FFA formed in the chloroplasts reaches envelope membranes, where FFA is on one hand converted to acyl-CoA outside the chloroplasts and on the other affects enzyme activities for galactolipid metabolism (Sakaki et al. 1990c). FFA in the envelopes decreases MGDG biosynthesis by inhibiting UDP-galactose:diacylglycerol galactosyltransferase, whereas it increases the production of DG and oligogalactolipids from MGDG by enhancing the activity of galactolipid:galactolipid galactosyltransferase. Finally, TG is synthesized from acyl-CoA and DG derived from MGDG. Part of acyl-CoA is also incorporated into PC, thus increasing 18:3-PC (Sakaki et al. 1994). Collectively, operation of the pathway results in a large loss of galactolipids with an accumulation of TG, oligogalactolipids, FFA, and 18:3-PC.

Although it is unknown how O_3 enhances galactolipase activity, the production of FFA in leaves is not specific to O_3 but rather a general response of plants to various environmental stresses (see Sakaki et al. 1994). Under chilling of leaves, FFA also accumulates in chloroplasts and has deleterious effects on photosynthetic functions (Garstka and Kaniuga 1991). This effect is similar to the decreases in MGDG and accumulation of FFA and TG in response to SO_2 (Navari-Izzo et al. 1991), suggesting operation of the same pathway with SO_2. Although FFA is a substrate for vital molecules in plants (see next section), FFA itself is toxic to membrane structure and functions, and plant cells must metabolize it to nontoxic levels. Exogenously applied FFA has been shown to be rapidly incorporated into various lipids, including TG and PC, in intact leaf tissues (Sakaki et al. 1990b). Thus, the pathway from FFA to TG and PC is ready to operate in case FFA is generated spontaneously in leaf cells.

Besides the increase in 18:3-PC, another characteristic response of phospholipids is an accumulation of PA following O_3 treatment. PA also accumulates in leaves treated with SO_2 or PAN. Since the PA increased by O_3 is rich in polysunsaturated fatty acids such as 18:3 (Sakaki et al. 1985), it cannot be the accumulated intermediate of lipid biosynthesis, but is rather produced from mature phospholipids through the action of phospholipase D. This enzyme is widely distributed in plant species and its activity is elevated in leaves subjected to various stresses (Wang 1993). Thus, the production of PA also belongs to the general stress-response changes in lipid metabolism. No study has examined the response of phospholipid hydrolysis, through the action of phospholipases A_1, A_2, or B, to air pollutants.

Free sterol is drastically decreased by O_3 and SO_2. Because steryl glycoside and acylated steryl glycoside accumulate concomitantly with a decrease in free sterol following O_3 treatment (Tomlinson and Rich 1971; Whitaker et al. 1990), part of the free sterol would be metabolized to these derivatives. The physiological relevance of such sterol metabolism is unknown.

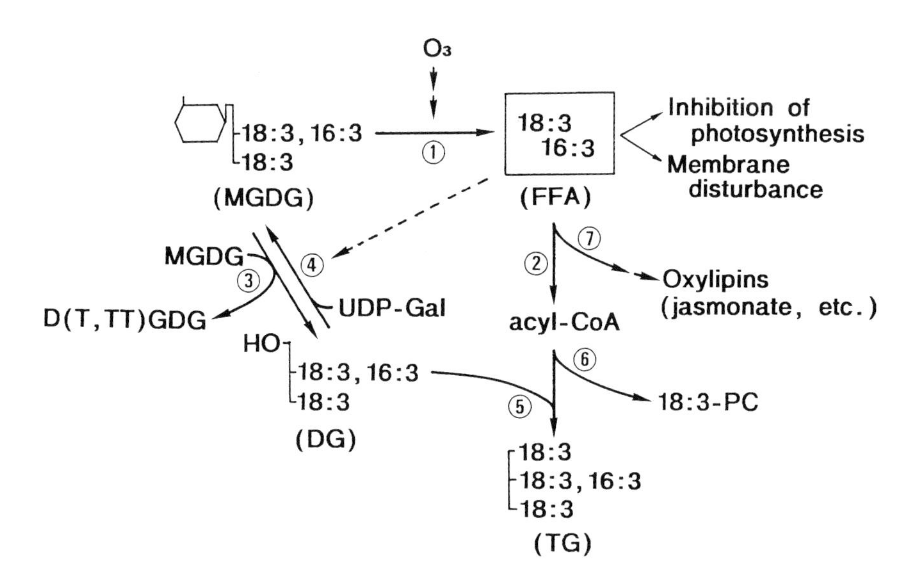

Fig. 3. Scheme showing metabolic changes in leaf lipids induced by O_3 fumigation. *1*, galactolipase; *2*, acyl-CoA synthetase; *3*, galactolipid:galactolipid galactosyltransferase; *4*, UDP-galactose:diacylglycerol galactosyltransferase; *5*, diacylglycerol acyltransferase; *6*, acyl-CoA:lyso-PC acyltransferase; *7*, lipoxygenase, D(T,TT)GDG, di-, tri-, and tetragalactosyldiacylglycerols. The *broken line* indicates the galactolipid-metabolizing enzymes affected by FFA

4.2.3 Effects on Lipid Degradation

Air pollutants, especially O_3, are often considered to induce premature senescence in leaves. During normal senescence processes, chloroplast thylakoid membranes disappear with concomitant loss of galactolipids and chlorophylls (Thomas 1986). To some extent, these structural and biochemical changes resemble those in leaves treated with O_3. Because galactolipids are metabolized to soluble sugars as well as to CO_2 during senescence (Wanner et al. 1991), both β-oxidation and the glyoxylate cycle appear to be involved in the metabolism. If this is also the case with O_3, then part of the FFA liberated from galactolipids by galactolipase might be metabolized through the pathway.

Lipoxygenase is a key enzyme for another step of FFA metabolism. While this enzyme produces toxic fatty acid hydroperoxides from FFA, the hydroperoxides formed would be rapidly metabolized (Blee and Joyard 1996) via the so-called lipoxygenase pathway to various substances, including volatiles and jasmonate (Vick 1993). Vick (1993) further suggested important roles of this enzyme in the chemical signaling of stresses through the production of jasmonate. Maccarrone et al. (1992) showed induction of lipoxygenase in O_3-treated soybean seedlings, but Sharma and Davis (1994) could not detect its induction in O_3-fumigated *Arabidopsis* leaves. Operation of the lipoxygenase pathway in pollutant-stressed plants will be a necessary subject of future study, with respect to the metabolism of FFA and the signal transduction of air pollutants.

The induction of glutathione *S*-transferase by fumigation with O_3 has also been reported (Price et al. 1990; Sharma and Davis 1994; Conklin and Last 1995). This enzyme has been proposed to function in the detoxification of various xenobiotic substances, including lipid hydroperoxides (Marrs 1996). Future work is necessary to establish the exact role of this enzyme in the metabolism of lipid hydroperoxides and its protective functions against pollutants.

5. Conclusions and Prospects for Biotechnology

As discussed here, lipid constituents in membranes, especially those in chloroplasts, are drastically affected by air pollutants through their oxidation by AOS and changes in metabolism. Thus, lipids and their metabolic processes are candidates for both the monitoring of air pollution and improving plant resistance against them with biotechnology. Although much progress on the biochemical mechanisms of lipid changes has been made in the past decade, I still cannot help but emphasize only similarities in lipid changes among the pollutants, such as decreases in galactolipids and increases in TG and PA. In the future, more detailed experiments on lipids, including missing components like sphingolipids, will identify the specific responses to each pollutant, and will allow their use for pollutant monitoring. In this respect, there is an increasing number of reports on epicuticular wax as affected by air pollution (for example, see Percy et al. 1992).

Because the wax on the leaf surface is the first component to be exposed to air pollutants, a promising field is to find specific wax responses to each pollutant.

New approaches to the study of the metabolism of AOS and lipids using molecular biology techniques are now in progress (see Aono, Polle, in this volume; Budziszewski et al. 1996). Many genes encoding antioxidative and lipid-metabolizing enzymes have been cloned and characterized, and some of them have been introduced into foreign plants to produce transgenic organisms. Unlike the case of transgenic plants expressing antioxidant enzymes, those with altered lipid-metabolizing enzymes have been mainly studied from agricultural aspects and have yet to be applied to investigations of responses to air pollutants. Since the pollutants destroy unsaturated fatty acids, transgenic plants with increased levels of fatty acid synthesis and desaturation may be expected to show enhanced levels of resistance. A notable example is the enhanced chilling tolerance shown by a cyanobacterium transformed with a desaturase gene (Wada et al. 1990). In addition, because lipoxygenase and glutathione *S*-transferase might function to remove FFA and their hydroperoxides, respectively, correlations between the expression of these enzymes and subsequent resistance against air pollution is a worthy area of future study. Conversely, if enzymes toxic to membrane lipids, such as galactolipase and phospholipase D, could be genetically suppressed, such transgenic plants could be expected to be resistant to O_3 and other air pollutants.

References

Aono M, Kubo A, Saji H, Natori T, Tanaka K, Kondo N (1991) Resistance to active oxygen toxicity of transgenic *Nicotiana tabacum* that expresses the gene for glutathione reductase from *Escherichia coli*. Plant Cell Physiol 32:691-697

Aono M, Kubo A, Saji H, Tanaka K, Kondo N (1993) Enhanced tolerance to photooxidative stress of transgenic *Nicotiana tabacum* with high chloroplastic glutathione reductase activity. Plant Cell Physiol 34:129-135

Asada K, Kiso K (1973) Initiation of aerobic oxidation of sulfite by illuminated spinach chloroplasts. Eur J Biochem 33:253-257

Blee E, Joyard J (1996) Envelope membranes from spinach chloroplasts are a site of metabolism of fatty acid hydroperoxides. Plant Physiol 110:445-454

Bowler C, Van Montagu M, Inze D (1992) Superoxide dismutase and stress tolerance. Annu Rev Plant Physiol Plant Mol Biol 43:83-116

Browse J, Somerville C (1991) Glycerolipid synthesis: biochemistry and regulation. Annu Rev Plant Physiol Plant Mol Biol 42:467-506

Budziszewski GJ, Croft KPC, Hildebrand DF (1996) Uses of biotechnology in modifying plant lipids. Lipids 31:557-569

Carlsson AS, Hellgren LI, Sellden G, Sandelius AS (1994) Effects of moderately enhanced levels of ozone on the acyl lipid composition of leaves of garden pea (*Pisum sativum*). Physiol Plant 91:754-762

Carlsson AS, Wallin G, Sandelius AS (1996) Species- and age-dependent sensitivity to ozone in young plants of pea, wheat and spinach: effects on acyl lipid and pigment content and metabolism. Physiol Plant 98:271-280

Conklin PL, Last RL (1995) Differential accumulation of antioxidant mRNAs in *Arabidopsis thaliana* exposed to ozone. Plant Physiol 109:203-212

Elstner EF (1982) Oxygen activation and oxygen toxicity. Annu Rev Plant Physiol 33:73-96

Fangmeier A, Kress LW, Lepper P, Heck WW (1990) Ozone effects on the fatty acid composition of loblolly pine needles (*Pinus taeda* L.). New Phytol 115:639-647

Fong F, Heath RL (1981) Lipid content in the primary leaf of bean (*Phaseolus vulgaris*) after ozone fumigation. Z Pflanzenphysiol 104:109-115

Frederick PE, Heath RL (1975) Ozone-induced fatty acid and viability changes in *Chlorella*. Plant Physiol 55:15-19

Garstka M, Kaniuga Z (1991) Reversal by light of deleterious effects of chilling on oxygen evolution, manganese and free fatty acid content in tomato thylakoids is not accompanied by restoration of the original membrane conformation. Physiol Plant 82:292-298

Hamada T, Nishiuchi T, Kodama H, Nishimura M, Iba K (1996) cDNA cloning of a wounding-inducible gene encoding a plastid ω-3 fatty acid desaturase from tobacco. Plant Cell Physiol 37:606-611

Harwood JL (1980) Plant acyl lipids: structure, distribution, and analysis. In: Stumpf PK (ed) The biochemistry of plants, vol 4. Lipids: structure and function. Academic Press, New York, pp 1-55

Heath RL (1980) Initial events in injury to plants by air pollutants. Annu Rev Plant Physiol 31:395-431

Heath RL (1984) Air pollutant effects on biochemicals derived from metabolism: organic, fatty and amino acids. In: Koziol MJ, Whatley FR (eds) Gaseous air pollutants and plant metabolism. Butterworths, London, pp 275-290

Heath RL, Castillo FJ (1988) Membrane disturbances in response to air pollutants. In: Schulte-Hostede S, Darrall NM, Blank LW, Wellburn AR (eds) Air pollution and plant metabolism. Elsevier, London, pp 55-75

Hellgren LI, Carlsson AS, Sellden G, Sandelius AS (1995) *In situ* leaf lipid metabolism in garden pea (*Pisum sativum* L.) exposed to moderately enhanced levels of ozone. J Exp Bot 46:221-230

Hippeli S, Elstner EF (1996) Mechanisms of oxygen activation during plant stress: biochemical effects of air pollutants. J Plant Physiol 148:249-257

Huttunen S, Soikkeli S (1984) Effects of various gaseous pollutants on plant cell ultrastructure. In: Koziol MJ, Whatley FR (eds) Gaseous air pollutants and plant metabolism. Butterworths, London, pp 117-127

Joyard J, Block MA, Douce R (1991) Molecular aspects of plastid envelope biochemistry. Eur J Biochem 199:489-509

Khan AA, Malhotra SS (1977) Effects of aqueous sulphur dioxide on pine needle glycolipids. Phytochemistry 16:539-543

Kuiper PJC (1985) Environmental changes and lipid metabolism of higher plants. Physiol Plant 64:118-122

Maccarrone M, Veldink GA, Vliegenthart JFG (1992) Thermal injury and ozone stress affect soybean lipoxygenases expression. FEBS Lett 309:225-230

Mackay CE, Senaratna T, McKersie BD, Fletcher RA (1987) Ozone induced injury to cellular membranes in *Triticum aestivum* L. and protection by the triazole S-3307. Plant Cell Physiol 28:1271-1278

Malhotra SS, Hocking D (1976) Biochemical and cytological effects of sulphur dioxide on plant metabolism. New Phytol 76:227-237

Malhotra SS, Khan AA (1978) Effects of sulphur dioxide fumigation on lipid biosynthesis in pine needles. Phytochemistry 17:241-244

Marrs KA (1996) The functions and regulation of glutathione *S*-transferases in plants. Annu Rev Plant Physiol Plant Mol Biol 47:127-158

McCarty RE, Jagendorf AT (1965) Chloroplast damage due to enzymatic hydrolysis of endogenous lipids. Plant Physiol 40:725-735

Mudd JB, McManus TT, Ongun A, McCullogh TE (1971a) Inhibition of glycolipid biosynthesis in chloroplasts by ozone and sulfhydryl reagents. Plant Physiol 48:335-339

Mudd JB, McManus TT, Ongun A (1971b) Inhibition of lipid metabolism in chloroplasts by ozone. In: Englund HM, Beery WT (eds) Proceedings of the 2nd international clean air congress. Academic Press, New York, pp 256-260

Mudd JB, Banerjee SK, Dooley MM, Knight KL (1984) Pollutants and plant cells: effects on membranes. In: Koziol MJ, Whatley FR (eds) Gaseous air pollutants and plant metabolism. Butterworths, London, pp 105-116

Navari-Izzo F, Quartacci MF, Izzo R (1991) Free fatty acids, neutral and polar lipids in *Hordeum vulgare* exposed to long-term fumigation with SO_2. Physiol Plant 81:467-472

Navari-Izzo F, Quartacci MF, Izzo R, Pinzino C (1992) Degradation of membrane lipid components and antioxidant levels in *Hordeum vulgare* exposed to long-term fumigation with SO_2. Physiol Plant 84:73-79

Nouchi I, Toyama S (1988) Effects of ozone and peroxyacetyl nitrate on polar lipids and fatty acids in leaves of morning glory and kidney bean. Plant Physiol 87:638-646

Okamoto T, Katoh S, Murakami S (1977) Effects of linolenic acid on spinach chloroplast structure. Plant Cell Physiol 18:551-560

Pauls KP, Thompson JE (1981) Effects of in vitro treatment with ozone on the physical and chemical properties of membranes. Physiol Plant 53:255-262

Peiser GD, Yang SF (1979) Ethylene and ethane production from sulfur dioxide-injured plants. Plant Physiol 63:142-145

Percy KE, Jensen KF, McQuattie CJ (1992) Effects of ozone and acidic fog on red spruce needle epicuticular wax production, chemical composition, cuticular membrane ultrastructure and needle wettability. New Phytol 122:71-80

Peters RE, Mudd JB (1982) Inhibition by ozone of the acylation of glycerol 3-phosphate in mitochondria and microsomes from rat lung. Arch Biochem Biophys 216:34-41

Price A, Lucas PW, Lea PJ (1990) Age-dependent damage and glutathione metabolism in ozone fumigated barley: a leaf section approach. J Exp Bot 41:1309-1317

Sakaki T (1998) Photochemical oxidants: toxicity. In: De Kok LJ, Stulen I (eds) Responses of plant metabolism to air pollution and global change. Backhuys, Leiden, pp 117-129

Sakaki T, Kondo N, Sugahara K (1983) Breakdown of photosynthetic pigments and lipids in spinach leaves with ozone fumigation: role of active oxygens. Physiol Plant 59:28-34

Sakaki T, Ohnishi J, Kondo N, Yamada M (1985) Polar and neutral lipid changes in spinach leaves with ozone fumigation: triacylglycerol synthesis from polar lipids. Plant Cell Physiol 26:253-262

Sakaki T, Saito K, Kawaguchi A, Kondo N, Yamada M (1990a) Conversion of monogalactosyldiacylglycerols to triacylglycerols in ozone-fumigated spinach leaves. Plant Physiol 94:766-772

Sakaki T, Kondo N, Yamada M (1990b) Pathway for the synthesis of triacylglycerols from monogalactosyldiacylglycerols in ozone-fumigated spinach leaves. Plant Physiol 94:773-780

Sakaki T, Kondo N, Yamada M (1990c) Free fatty acids regulate two galactosyltransferases in chloroplast envelope membranes isolated from spinach leaves. Plant Physiol 94:781-787

Sakaki T, Tanaka K, Yamada M (1994) General metabolic changes in leaf lipids in response to ozone. Plant Cell Physiol 35:53-62

Sharma YK, Davis KR (1994) Ozone-induced expression of stress-related genes in *Arabidopsis thaliana*. Plant Physiol 105:1089-1096

Shimazaki K, Sakaki T, Kondo N, Sugahara K (1980) Active oxygen participation in chlorophyll destruction and lipid peroxidation in SO_2-fumigated leaves of spinach. Plant Cell Physiol 21:1193-1204

Shimazaki K, Yu SW, Sakaki T, Tanaka K (1992) Differences between spinach and kidney bean plants in terms of sensitivity to fumigation with NO_2. Plant Cell Physiol 33:267-273

Smith LL (1987) Cholesterol autooxidation 1981-1986. Chem Phys Lipids 44:87-125

Tanaka K, Sugahara K (1980) Role of superoxide dismutase in defense against SO_2 toxicity and an increase in superoxide dismutase activity with SO_2 fumigation. Plant Cell Physiol 21:601-611

Tanaka K, Suda Y, Kondo N, Sugahara K (1985) O_3 tolerance and the ascorbate-dependent H_2O_2 decomposing system in chloroplasts. Plant Cell Physiol 26:1425-1431

Thomas H (1986) The role of polyunsaturated fatty acids in senescence. J Plant Physiol 123:97-105

Thomson WW (1975) Effects of air pollutants on plant ultrastructure. In: Mudd JB, Kozlowski TT (eds) Responses of plants to air pollution. Academic Press, New York, pp 179-194

Tomlinson H, Rich S (1970) Lipid peroxidation, a result of injury in bean leaves exposed to ozone. Phytopathology 60:1531-1532

Tomlinson H, Rich S (1971) Effect of ozone on sterols and sterol derivatives in bean leaves. Phytopathology 61:1404-1405

Trevathan LE, Moore LD, Orcutt DM (1979) Symptom expression and free sterol and fatty acid composition of flue-cured tobacco plants exposed to ozone. Phytopathology 69:582-585

Van Camp W, Willekens H, Bowler C, Van Montagu M, Inze D, Reupold-Popp P, Sandermann H Jr, Langebartels C (1994) Elevated levels of superoxide dismutase protect transgenic plants against ozone damage. Bio/Technology 12:165-168

Vick BA (1993) Oxygenated fatty acids of the lipoxygenase pathway. In: Moore TS Jr (ed) Lipid metabolism in plants. CRC Press, Boca Raton, pp 167-191

Wada H, Gombos Z, Murata N (1990) Enhancement of chilling tolerance of a cyanobacterium by genetic manipulation of fatty acid desaturation. Nature 347:200-203

Wang X (1993) Phospholipases. In: Moore TS Jr (ed) Lipid metabolism in plants. CRC Press, Boca Raton, pp 505-525

Wanner L, Keller F, Matile P (1991) Metabolism of radiolabelled galactolipids in senescent barley leaves. Plant Sci 78:199-206

Wellburn AR, Robinson DC, Thomson A, Leith ID (1994) Influence of episodes of summer O_3 on delta-5 and delta-9 fatty acids in autumnal lipids of Norway spruce [*Picea abies* (L.) Karst]. New Phytol 127:355-361

Whitaker BD, Lee EH, Rowland RA (1990) EDU and ozone protection: foliar glycerolipids and steryl lipids in snapbean exposed to O_3. Physiol Plant 80:286-293

Wolfenden J, Wellburn AR (1991) Effects of summer ozone on membrane lipid composition during subsequent frost hardening in Norway spruce [*Picea abies* (L.) Karst]. New Phytol 118:323-329

5
Effects of Ethylene on Plant Responses to Air Pollutants

Nobuyoshi Nakajima

Biodiversity Conservation Research Project, National Institute for Environmental Studies, Onogawa 16-2, Tsukuba, Ibaraki 305-8506, Japan

1. Introduction

Ozone (O_3) and sulfur dioxide (SO_2) induce various forms of damage to plants. In the case of O_3, its effects differ depending on whether exposure is acute (high concentration) or chronic (repetitive low concentration; see the chapter by I. Nouchi, this volume). In many plant species, ethylene production is one of the earliest plant responses to these pollutants, under both exposure conditions, and the extent of pollutant -induced leaf injury has been shown to be associated with the rate of ethylene production (Kangasjärvi et al. 1994). Furthermore, recent physiological studies has shown that hormonal action of ethylene promotes leaf damage and senescence under ozone exposure (Bae et al. 1996; Miller et al. 1999). Ethylene is a gaseous plant hormone that induces a variety of physiological phenomena, such as leaf epinasty, abscission, fruit ripening, hook opening, lateral expansion of cells, and (leaf) senescence. The rate of ethylene production increases in response to various environmental stimuli, such as wounding, fungal infection, irradiation, water logging, and air pollution (Abeles et al.1992). In some cases, stress-induced ethylene induces physiological or morphological changes that make the plants more stress tolerant; however, its role in these processes is only partially understood. Extensive research into the ethylene biosynthetic pathway and of the genes encoding enzymes of this pathway has been carried out since 1979. Ethylene is synthesized from *S*-adenosyl-L-methionine (SAM) via 1-

Air Pollution and Plant Biotechnology
–Prospects for Phytomonitoring and Phytoremediation–
Edited by K. Omasa, H. Saji, S. Youssefian, and N. Kondo
© Springer -Verlag Tokyo 2002

aminocyclopropane-1-carboxylic acid (ACC). The formation of ACC from SAM, which is catalyzed by ACC synthase (ACS; EC 4.4.1.14), is often a rate-limiting step, and an increased activity of ACS is seen to precede the rise in ethylene production (Yang and Hoffmann 1984). Several genes encoding ACSs are known to exist in a plant genome. For example, eight cDNAs encoding ACS have been isolated from tomato, and their transcripts of these genes are found to accumulate in response to various stimuli (Kende 1993). When the concentration of ACC in the tissue becomes high, the formation of ethylene from ACC, which is catalyzed by ACC oxidase (ACO), limits the rate of ethylene production. This effect is often observed when ACS has already been induced to a high level by stimuli such as wounding. Genes encoding ACO have also been shown to constitute a multigene family; however, the physiological significance of these different isozymes is not yet fully understood (Kende 1993).

In addition to studies on ethylene biosynthesis, research on ethylene signaling has made extensive progress over the last decade. Components of the ethylene-responsive pathway have been identified by molecular genetic studies of *Arabidopsis thaliana*. By 1997, more than 20 mutants displaying abnormal responses to ethylene had been isolated. Several complementary DNAs (cDNAs) that can compensate for some of these mutant phenotypes have been isolated and characterized. The deduced amino acid sequences encoded by these cDNAs have specific protein kinase motifs, suggesting that ethylene sensing involves a protein kinase cascade (Kieber 1997). Progress in our understanding of the basic concepts of ethylene biosynthesis and signaling, as just described, are expected to significantly enhance researches into the molecular mechanisms by which air pollutants, especially O_3, induce ethylene production and subsequent foliar damage and senescence.

This review summarizes recent progress in the molecular mechanisms regulating air pollutant-induced ethylene synthesis and considers the function of ethylene in the appearance of leaf damage.

2. Ozone-Induced Ethylene Synthesis

It has been demonstrated that, under acute exposure condition, O_3-induced ethylene synthesis is regulated by the level of ACS. The activity of ACS and the level of ACC are found to increase just before ethylene production during O_3 exposure, and an inhibitor of ACS completely prevents ethylene production (Bae et al. 1996; Schlagnhaufer et al. 1997).

Three cDNAs (*LE-ACS1A, LE-ACS2, LE-ACS6*), encoding O_3-induced ACSs, have been isolated from acute O_3-exposed tomato. Transcripts for *LE-ACS1A* and *LE-ACS6* accumulate within 1 h after the start of $0.2 \cdot l^{-1}$ of O_3 exposure, reach a maximum level at 2 h and then rapidly decline. Transcripts for *LE-ACS2* start to accumulate 2 h after the start of O_3 exposure and then gradually decline (Tuomainen et al. 1997; Nakajima et al., unpublished data). These isozymes were also determined to be elicitor-inducible in suspension cultured tomato cells (Oetiker et al. 1997), and wound-inducible in tomato fruits (Tatsuki and Mori 1999; Lincoln et al. 1993). The expression patterns of these isozymes during acute O_3 exposure were similar to those induced by elicitor treatment and wounding (Oetiker et al. 1997). However, levels of transcripts for PinII, a wound-inducible proteinase inhibitor, did not increase by acute O_3 exposure, suggesting that the mechanism of ACS induction differs from that of wound signaling (Nakajima et al., unpublished data).

Other O_3-inducible ACS cDNAs have also been identified from potato (*ST-ACS4, ST-ACS5*) and *Arabidopsis* (*A+ACS6*) (Schlagnhaufer et al. 1995, 1997; Vahala et al. 1998). The extent of nucleotide sequence homology and the patterns of *A+ACS6, ST-ACS4*, and *ST-ACS5* expression suggest that they may have identical functions to those of *LE-ACS1A, LE-ACS2*, and *LE-ACS6*, respectively.

It has also been demonstrated that the activity of O_3-inducible ACSs may be regulated by protein phosphorylation/dephosphorylation under acute O_3 condition. Treatment with K-252a, a protein kinase inhibitor, was found to reduce the level of ACS activity after acute O_3 exposure, while calyculin A, an inhibitor of protein phosphatase, increased the level of enzyme activity (Tuomainen et al. 1997). However, it is not clear whether phosphorylation/dephosphorylation affects only ACS enzyme activities or whether gene expression of O_3-inducible ACS is also activated.

Acute O_3 exposure was found to result in three fold higher levels of ethylene production in tomato plants applied with excessive amounts of ACC than in plants without ACC application (Bae et al. 1996), suggesting that O_3-induced ethylene production is also partially regulated at the step of ACC conversion into ethylene.

The question arises as to whether chronic O_3 exposure also activates the same pattern of isozyme expression induced by acute O_3 conditions. However, as only one case, the induction of *A+ACS6* by both acute and chronic O_3 exposures, has been reported (Vahala et al. 1998, Miller et al. 1999), further investigations with other species are required. The rate of O_3-induced ethylene production may also be affected by polyamines, which can act as radical scavengers in plant cells. It is known, for example, that the levels of polyamines increased on O_3 exposure in some O_3-tolerant plants without a concomitant induction of ethylene production (Langebartels et al. 1991; Nagireddy et al. 1993), suggesting that polyamines could inhibit O_3-induced ethylene synthesis. In higher plants, polyamines are synthesized by decarboxylation of arginine to form agmatin, a rate-limiting step catalyzed by arginine decarboxylase (ADC). Agmatin is metabolized to putrescin, spermidine, and spermine using SAM, a precursor common to both ethylene and polyamine synthesis. It is conceivable, therefore, that a high level of polyamine synthesis could limit the supply of SAM required for ethylene production. In fact, putrescin, spermidine, and spermine have been shown to act as noncompetitive inhibitors of ACS in vitro (Hyodo and Tanaka 1986). A strong correlation between the level of ADC activity under O_3 exposure and the polyamine content has been observed in an O_3 -tolerant tobacco cultivar (Langebartels et al. 1991),

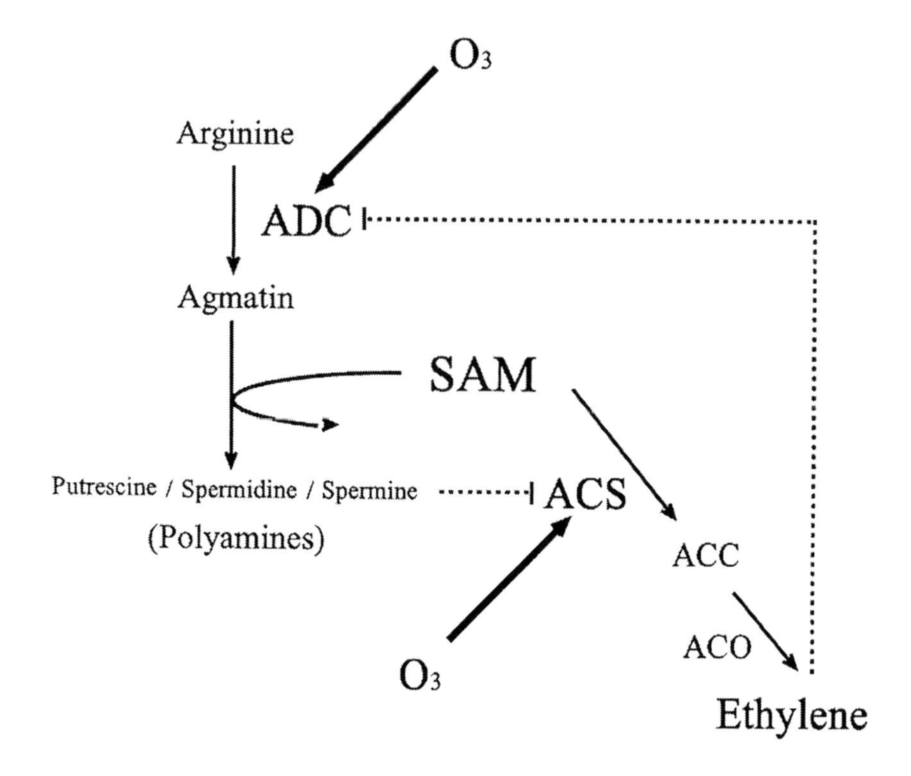

Fig. 1. Biosynthetic pathway of ethylene and polyamines in higher plants. *SAM*, S-adenosyl-L-methionine; ACC, 1-aminocyclopropane-1-carboxylic acid; ACD, Arginine decarboxylase; ACS, ACC synthase; ACO, ACC oxidase. Ozone may induce ADC and ACS (*large arrows*). Polyamines and ethylene may inhibit each other's synthesis by suppression of ADC induction and inhibition of ACS activity

and application of an ADC inhibitor was found to enhance O_3-induced visible damage (Bamford et al. 1989). Furthermore, ethylene is reported to be a negative regulator of ADC induction in pea seedlings (Apelbaum et al. 1985). These findings together suggest that ethylene and polyamines inhibit each other's synthesis, so that the rate of polyamine synthesis could affect plant susceptibility to O_3.

3. Effects of Ethylene Under Acute Ozone Exposure

When plants are exposed to acute O_3 conditions, ethylene production is promoted within 30 min, and leaf epinasty is observed 1 h after the start of O_3 exposure (Craker 1971); this is one of the earliest observed symptoms and is probably induced by ethylene. Within 6 h of exposure, expression of many of the genes related to phenylpropanoid synthesis and carose metabolism is induced (Rosemann et al. 1991; Schraudner et al. 1992; Ernst et al. 1992; Galiano et al. 1993), followed by appearance of necrotic lesions on the surface of leaves. Accumulating evidence suggests that the extent of ethylene production determines plant sensitivity to O_3. Exposure of a range of plant species and cultivars to varying O_3 concentrations demonstrated that the rate of ethylene production, as a function of the O_3 concentration, correlated with the extent of visible injury (Tingey et al. 1976). Similarly, pretreatment of plants with an inhibitor of ethylene biosynthesis reduced the extent of O_3-induced foliar damage (Mehlhorn and Wellburn 1987; Mehlhorn et al. 1991; Wenzel et al. 1995; Bae et al. 1996). Furthermore, an O_3-sensitive cultivar of tobacco (Bel W3) was found to rapidly produce ethylene in response to O_3, while the rate of ethylene production in an O_3-insensitive cultivar (Bel B) was considerably lower (Langebartels et al. 1991).

One hypothesis for the effects of ethylene under acute O_3 exposure is that O_3 produces oxygen free radicals that react with volatile compounds, such as ethylene and terpenes, and so create reactive aldehydes that lead to leaf injury (Mehlhorn and Wellburn 1987; Elstner 1987). Observations using electron spin resonance spectroscopy have demonstrated the formation of free radicals in O_3-exposed plants (Mehlhorn et al. 1990), and the formation of reactive aldehydes is supported by the identification of an organic hydroperoxide, 1-hydroxymethyl hydroperoxide (HEHP), in O_3-exposed California poppy leaves (Hewitt et al. 1990). However, it is not clear if a correlation exists between the amount of HEHP and the extent of leaf injury, or whether HEHP is really produced by the interaction between O_3 and ethylene in plants.

Another possible explanation is that, under O_3 exposure, ethylene acts as a plant hormone. This idea is supported by the observation that treatment with 2,5-norbornadiene, an inhibitor of the hormonal action of ethylene, enhances O_3-induced ethylene production but reduces the extent of O_3-induced leaf injury (Bae et al. 1996). Tyron, a free radical scavenger, strongly reduces O_3-induced leaf damage but has no inhibitory effects on the evolution of ethylene from tomato leaves, suggesting that both ethylene and free radicals may be necessary to cause O_3-induced leaf damage (Bae et al. 1996).

Although ethylene production during O_3 exposure triggers foliar damage, pretreatment with ethylene is found to reduce the extent of O_3-induced damage. Mehlhorn (1990) reported that pretreatment of mung bean and pea seedlings with ethylene increased both their tolerance to O_3 and their activities of ascorbate peroxidase, an enzyme involved in the scavenging of active oxygen species in plant cells. The authors postulated that plants normally respond to O_3 with an increased production of ethylene and initiate the peroxidative reaction before ascorbate peroxidase is induced in damaged cells. However, pretreatment with ethylene increased their ascorbate peroxidase activities to levels that were sufficient to protect the plants from subsequent O_3-induced damage.

4. Effects of Ethylene Under Chronic Ozone Exposure

When plants are subjected to chronic O_3 exposure, necrosis of the leaves is rarely observed. Most plant species respond to this type of exposure with accelerated senescence of the leaves and chlorosis (Pell et al. 1997). In normal senescing leaves, degradation of ribulose bisphosphate carboxylase/oxygenase (RuBisCo) protein begins to export nitrogen and carbon to growing region of the plant. Under chronic O_3 exposure, a decline in the net photosynthetic rate was observed in potato plants with a correlated loss of RuBisCo activity. However, neither the photosynthetic electron transport system nor the composition of thylakoid proteins was affected (Dann and Pell 1989). Glick et al. (1995) reported that transcript levels of rbcS, which encodes the small subunit of RuBisCo, decreased five fold with the increased ethylene production in potato leaves under chronic O_3 exposure. This response paralleled the decrease in transcript levels for chloroplastic glyceraldehyde-3-phosphate dehydrogenase and chlorophyll *a*/*b*-binding protein in photosystem II, and the increase in transcripts for cytosolic glyceraldehyde-3-phosphate dehydrogenase, suggesting that chronic O_3 exposure affects transcription of some genes and decreases transcripts for rbcS to reduce RuBisCo activity.

The reduction in rbcS transcripts by O_3 treatment was not completely suppressed by aminooxyacetic acid, an inhibitor of ACS. This finding suggests that ethylene may act not as a trigger but as an accelerator of O_3-induced leaf senescence (Schlagnhaufer et al. 1995). A recent study has reported that transcripts for some, but not all, senescence-associated genes accumulated with chronic O_3 exposure, suggesting that chronic O_3 induces some part of the senescence process (Miller et al. 1999).

5. Sulfur Dioxide-Induced Ethylene Production

Although it is a generally accepted phenomenon, there are only a limited number of reports on sulfur dioxide (SO_2)-induced ethylene production (Peiser and Yang 1979). The amount of ethylene produced depends on the SO_2 concentration and duration of the exposure (Bressan et al. 1979). Bae et al. (1995) showed that pretreatment with aminoethoxyvinylglycine, a competitive inhibitor of ACS, reduced both SO_2-induced ethylene production and the extent of leaf injury in tomato plants. These results suggest that ethylene might also trigger the appearance of leaf injury in SO_2-exposed plants, and that the role of ethylene is similar to that in O_3 exposed plants. Furthermore, SO_2 induces expression of the same ACSs (*LE-ACS1A, LE-ACS2*, and *LE-ACS6*) as are induced by acute O_3-exposure in tomato leaves, although their expression patterns differ. These findings suggest that different mechanisms of ACS induction by SO_2 and O_3 exist (Nakajima et al. unpublished data).

6. Conclusion

Recent progress in molecular biology and advances in biochemical research have partly revealed the mechanisms by which O_3 and SO_2 enhance ethylene production. These pollutants induce the transcription of genes encoding specific isozymes of ethylene biosynthesis, but they appear to induce ethylene production through different signaling pathways. Future studies should focus on the mechanisms by which these pollutants control transcription of genes encoding enzymes of ethylene biosynthesis, especially DNA sequences and proteins regulating ACS and ACO transcription. Understanding these mechanisms will open up the possibility that we can control the rate of pollutant-induced ethylene synthesis and so regulate the O_3 susceptibility of plants.

There is also evidence supporting the hypothesis that the hormonal action of ethylene may be involved in the induction of O_3-induced leaf damage. Under acute conditions, O_3 triggers leaf necrosis, but it induces senescence of leaves under chronic conditions. Although physiological differences between acute and chronic conditions appear to be highly complex, studies with the mutants that have lost their capacity for ethylene signaling may allow us to identify the genes responsible for this difference.

References

Abeles FB, Morgan PW, Saltveit ME Jr. (1992) Ethylene in plant biology, 2nd edn. Academic Press, New York

Apelbaum A, Goldlust A, Icekson I (1985) Control by ethylene of arginine decarboxylase activity in pea seedlings and its implication for hormonal regulation of plant growth. Plant Physiol 79:635-640

Bae GY, Kondo N, Nakajima N, Ishizuka K (1995) Ethylene production in tomato plants by SO_2 in relation to leaf injury. J Jpn Soc Atmos Environ 30:367-373

Bae GY, Nakajima N, Ishizuka K, Kondo N (1996) The role in ozone phytotoxicity of the evolution of ethylene upon induction of 1-aminocyclopropane-1-carboxylic acid synthase by ozone fumigation in tomato plants. Plant Cell Physiol 37:129-134

Bamford AJR, Borland AM, Lea PL, Mansfield TA (1989) The role of arginine decarboxylase in modulating the sensivity of barley to ozone. Environ Pollut 61:95-106

Bressan RA, LeCureux L, Wilson LG, Filner P (1979) Emission of ethylene and ethane by leaf tissue exposed to injurious concentration of sulfur dioxide or bisulfite ion. Plant Physiol 63:924-930

Craker L (1971) Ethylene production from ozone injured plants. Environ Pollut 1:299-304

Dann MS, Pell EJ (1989) Decline of activity and quantity of ribulose bisphosphate carboxylase/oxygenase and net photosynthesis in ozone-treated potato foliage. Plant Physiol 91: 427-432

Elstner EF (1987) Ozone and ethylene stress. Nature 328: 482

Ernst D, Schraudner M, Langebartels C, Sandermann H Jr. (1992) Ozone-induced changes of mRNA levels of β-1,3-glucanase, chitinase and 'pathogenesis-related' protein 1b in tobacco plants. Plant Mol Biol 20:673-682

Galiano H, Cabane M, Eckerskorn C, Lottspeich F, Sandermann H Jr., Ernst D (1993) Molecular cloning, sequence analysis and elicitor-/ozone-induced accumulation of cinnamyl alcohol dehydrogenase from Norway spruce (*Picea abies L.*) Plant Mol Biol 23:145-156

Glick RE, Schlagnhaufer CD, Arteca RN, Pell E (1995) Ozone-induced ethylene emission accelerates the loss of ribulose-1,5-bisphosphate carboxylase/oxygenase and nuclear-encoded mRNAs in senescing potato leaves. Plant Physiol 109:891-898

Hewitt CN, Kok GL, Fall R (1990) Hydroperoxides in plants exposed to ozone mediate air pollution damage to alkene emitters. Nature 344:56-58

Hyodo H, Tanaka K (1986) Inhibition of 1-aminocyclopropane-1-carboxylic acid synthase activity by polyamines their related compounds and metabolites of *S*-adenosylmethionine. Plant Cell Physiol 3:391-398

Kangasjärvi J, Talvinen J, Utriainen M, Karjalainen R (1994) Plant defense systems induced by ozone. Plant Cell Environ 17:783-794

Kende H (1993) Ethylene biosynthesis. Annu Rev Plant Physiol Plant Mol Biol 44:283-307

Kieber JJ (1997) The ethylene response pathway in *Arabidopsis*. Annu Rev Plant Physiol Plant Mol Biol 48:277-296

Langebartels C, Kerner K, Leonard S, Schraudner M, Trost M, Heller W, Sandermann H Jr. (1991) Biochemical plant responses to ozone. Plant Physiol 95:882-889

Lincoln JE, Campbell AD, Oetiker J, Rottmann WH, Oeller PW, Shen NF, Theologis A (1993) LE-ACS4, a fruit ripening and wound-induced 1-aminocyclopropane-1-carboxylate synthase gene of tomato (*Lycopersicon esculentum*). J Biol Chem 268:19422-19430

Mehlhorn H (1990) Ethylene-promoted ascorbate peroxidase activity protects plants against hydrogen peroxide, ozone and paraquat. Plant Cell Environ 13:971-976

Mehlhorn H, Wellburn AR (1987) Stress ethylene formation determines plant sensitivity to ozone. Nature 327:417-418

Mehlhorn H, Tabner BJ, Wellburn AR (1990) Electron spin resonance evidence for the formation of free radicals in plants exposed to ozone. Physiol Plant 79:377-383

Mehlhorn H, O'Shea JM, Wellburn AR (1991) Atmospheric ozone interacts with stress ethylene formation by plant to cause visible plant injury. J Exp Bot 42:17-24

Miller JD, Arteca RN, Pell EJ (1999) Senescence-associated gene expression during ozone-induced leaf senescence in *Arabidopsis*. Plant Physiol 120:1015-1023

Nagireddy G, Arteca RN, Dai YR, Flores HE, Negm FB, Pell EJ (1993) Changes in ethylene and polyamines in relation to mRNA leaves of the large and small subunits of ribulose bisphosphate carboxylase/oxygenase in ozone-stressed potato foliage. Plant Cell Environ 16:819-826

Oetiker JH, Olson DC, Shiu OY, Yang SF (1997) Differential induction of seven 1-aminocyclopropane-1-carboxylate synthase genes by elicitor in suspension culture of tomato (*Lycopersicon esculentum*). Plant Mol Biol 34:275-286

Peiser GD, Yang SF (1979) Ethylene and ethane production from sulfur dioxide-injured plants. Plant Physiol 63:142-145

Pell EJ, Schlagnhaufer CD, Arteca RN (1997) Ozone-induced oxidative stress: mechanisms of action and reaction. Physiol. Plant 100:264-273

Rosemann S, Heller W, Sandermann H Jr. (1991) Biochemical plant responses to ozone. Plant Physiol 97:1280-1286

Schlagnhaufer CD, Glick RE, Arteca RN, Pell EJ (1995) Molecular cloning of an ozone-induced 1-aminocyclopropane-1-carboxylate synthase cDNA and its relationship with a loss of rbcS in potato (*Solanum tuberosum L.*) plants. Plant Mol Biol 28: 93-103

Schlagnhaufer CD, Glick RE, Arteca RN, Pell EJ (1997) Sequential expression of two 1-aminocyclopropane-1-carboxylate synthase genes in response to biotic and abiotic stress in potato (*Solanum tuberosum L.*) leaves. Plant Mol Biol 35:683-688

Schraudner M, Ernst D, Langebartels C, Sandermann H Jr. (1992) Biochemical plant responses to ozone. Plant Physiol 99:1321-1328

Tatsuki M, Mori H (1999) Rapid and transient expression of 1-aminocyclopropane-1-carboxylate synthase isogenes by touch and wound stimuli in tomato. Plant Cell Physiol 40: 709-715

Tingey DT, Standley C, Field RW (1976) Stress ethylene evolution: a measure of ozone effects on plants. Atmos Environ 10:969-974

Tuomainen J, Betz C, Kangasjärvi J, Ernst D, Yin ZH, Langebartels C, Sandermann H (1997) Ozone induction of ethylene emission in tomato plants: regulation by differential accumulation of transcripts for the biosynthetic enzymes. Plant J 12:1151-1162

Vahala J, Schlagnhaufer CD, Pell EJ (1998) Induction of an ACC synthase cDNA by ozone in light-grown *Arabidopsis thaliana* leaves. Physiol Plant 103: 45-50

Wenzel AA, Schlautmann H, Jones CA, Kuppers K, Mehlhorn (1995) Aminoethoxyvinylglycine, cobalt and ascorbic acid all reduce ozone toxicity in mung bean by inhibition of ethylene biosynthesis. Physiol Plant 93:286-290

Yang SF, Hoffman NE (1984) Ethylene biosynthesis and its regulation in higher plants. Annu Rev Plant Physiol 35:155-189

6
Effects of Air Pollutants on Gene Expression in Plants

Akihiro Kubo

Environmental Biology Division, National Institute for Environmental Studies, Onogawa 16-2, Tsukuba, Ibaraki 305-8506, Japan

1. Introduction

Air pollutants exert various effects on plants, which then show various responses to the pollutants (Alscher and Wellburn 1994; Yunus and Iqbal 1996). Among the effects and responses, those at the level of gene expression are rapidly becoming one of the best understood. For instance, several reviews on the effects of ozone (O_3) on gene expression in plants (Kangasjärvi et al. 1994; Pell et al. 1994, 1997; Schraudner et al. 1996, 1997; Sandermann 1996; Sharma and Davis 1997) and one on the molecular effects of sulfur dioxide (SO_2) in plants (Okpodu et al. 1996) have appeared. Overall, reports on this subject area relating to O_3 outnumber those concerning other air pollutants. This chapter summarizes the effects of phytotoxic air pollutants on gene expression in higher plants, and discusses the application of this research field to environmental biotechnology.

Gene expression proceeds as illustrated in Fig. 1, and each step of the process can be detected by the currently established methods shown. The first step of gene expression (transcription) results in the synthesis of transcripts, such as mRNA. Transcription rate of specific genes affects mRNA levels of the genes. The transcription rate can be determined by nuclear run-on transcription assay. The level of mRNA of specific genes can be determined directly by northern blotting, etc. Translation is the second step, resulting in the synthesis of proteins, such as enzymes. Translation rate of specific proteins reflects mRNA levels of the

---

Air Pollution and Plant Biotechnology
–Prospects for Phytomonitoring and Phytoremediation–
Edited by K. Omasa, H. Saji, S. Youssefian, and N. Kondo
© Springer -Verlag Tokyo 2002

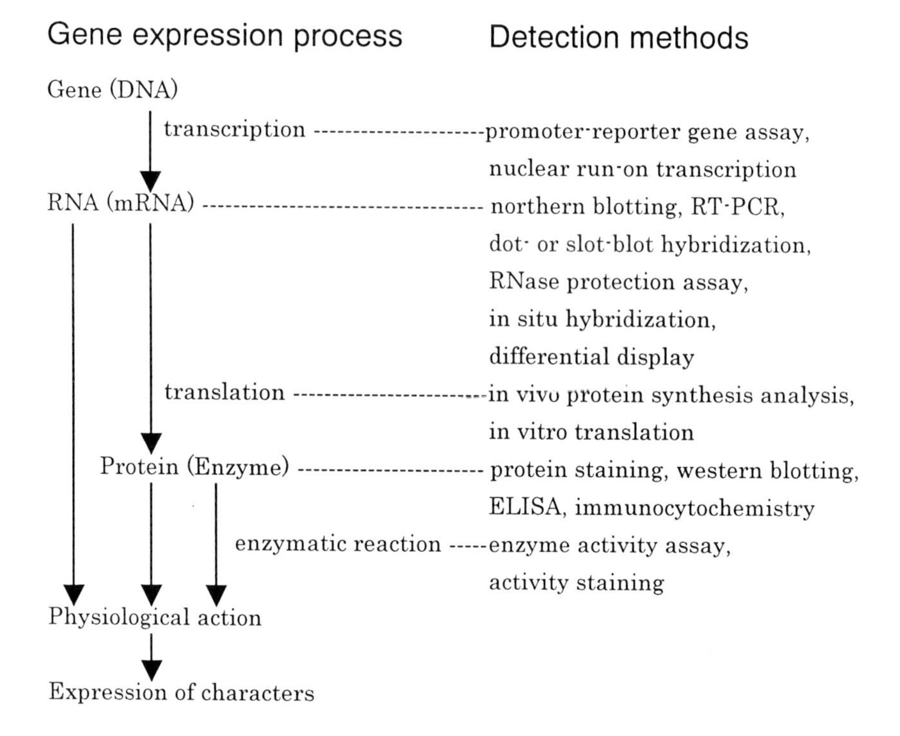

Fig. 1. The process of gene expression and current methods of detecting expression at each step of the process. This chapter covers papers using the detection methods that are written with bold letters in this figure

corresponding genes. The translation rate can be determined both in vivo and in vitro. The amount of specific proteins can be determined by protein staining, western blotting, and so on.

This review is a compilation of reports concerning changes in the relative levels of specific transcripts, using either direct or indirect detection methods. Changes in protein levels and enzymatic activities have been reported in some detail elsewhere, but are not included here since they reflect several mechanisms other than just transcriptional regulation.

2. Effects of Air Pollutants on Gene Expression

2.1 Detection at the Translation Step

We can estimate the levels of cellular mRNA that can be translated into proteins by in vivo protein synthesis analysis and in vitro translation. Neither method requires cloned DNA, but both do provide information on many and unspecified genes by electrophoretic analysis of labeled proteins. However, it is difficult to subsequently identify or clone the genes that have been found to show differential expression by these methods.

2.1.1 In Vivo Protein Synthesis Analysis

In vivo protein synthesis analysis is the method of detecting newly synthesized proteins in vivo by radiolabeling. Schmitt and Sandermann (1990) found two unidentified proteins synthesized in *Picea abies* needles exposed to O_3 and acid mist. Pino et al. (1995) reported both O_3-induced increases and decreases in the accumulation of some newly synthesized proteins in leaves of *Zea mays*. Such changes were observed in the soluble, membrane, and thylakoid membrane fractions of leaves in the absence of visible leaf injury. One of these thylakoid membrane-associated proteins, which showed reduced synthesis in response to O_3, was identified by immunoprecipitation as the D1 protein of photosystem II. Ranieri et al. (1997) also found reduced synthesis of the D1 protein in visibly uninjured leaves of *Helianthus annuus* fumigated with O_3. They also observed O_3-enhanced synthesis of two unidentified thylakoid proteins using two-dimensional fluorography. These changes under no-visible injury indicate that phytomonitoring of O_3 by detection of such changes could be more sensitive than that by detection of visible injuries.

2.1.2 In Vitro Translation

In vitro translation is the method of detecting proteins synthesized in vitro using extracted mRNA as the template and radiolabeled amino acids. The amount of specific polypeptides synthesized reflects the level of specific transcripts translatable. Eckey-Kaltenbach et al. (1994b), using two-dimensional fluorography, showed both increases and decreases in the amount of several in vitro translation products using poly $(A)^+$ RNA extracted from leaves of *Petroselium crispum* fumigated with 200 ppb O_3. They further demonstrated the accumulation of translatable pathogenesis-related PR1-1 mRNA in O_3-exposed plants by hybrid-selected in vitro translation. In addition, O_3-induced increase and decrease in the amounts of some in vitro translation products were observed in visibly uninjured *Pinus sylvestris* needles (Großkopf et al. 1994) and in *Lycopersicon esculentum* leaves (Kirtikara and Talbot 1996).

2.2 Detection at the Transcription Step

Measurements of the transcription rate of specific genes, despite being more laborious than measurements of the mRNA levels, reveal the direct effects of air pollutants on gene transcription. The transcription rate can be measured by nuclear run-on transcription, which is the method of detecting specific transcripts elongating in the isolated nuclei. Using the nuclear run-on transcription assay, Schlagnhaufer et al. (1995) demonstrated O_3-induced transcriptional activation of the gene encoding a 1-aminocyclopropane-1-carboxylate (ACC) synthase in *Solanum tuberosum* leaves. Glick et al. (1995), by the same method, showed reduced transcription rate of the genes encoding the small subunit of ribulose-1,5-bisphosphate carboxylase/oxygenase (Rubisco) and 28S rRNA in O_3-exposed leaves of *Solanum tuberosum*. They further suggested that the Rubisco small subunit gene expression is regulated by both transcription and mRNA degradation rates, since the relatively small decrease in transcription of this gene did not account for the large drop in the mRNA levels.

The promoter of genes regulates transcription. The activity of the promoter can be monitored in the transgenic plants, in which a reporter gene connected to the promoter had been introduced. The promoter activity indirectly reflects the transcription rate. Using the promoter-reporter gene assay, Schubert et al. (1997) and Grimmig et al. (1997) studied the stilbene synthase gene from *Vitis vinifera* in transgenic tobacco plants, and showed O_3-induced activation of this promoter and histochemical localization of the promoter activity. Their work also revealed an O_3-responsive region of this promoter.

2.3 Detection of Transcripts

The level of mRNA of specific genes is important in the process of gene expression because it affects translation rates to synthesize specific proteins. It is determined by both transcription and mRNA degradation rates. Measurements of mRNA levels of specific genes are achieved mainly by hybridization of a labeled probe (DNA or RNA complementary to the mRNA) to the mRNA or by reverse transcription-polymerase chain reaction (RT-PCR). The increasing availability of such suitable probes and PCR primers is accelerating studies of transcript accumulation, and numerous papers have appeared since 1992 concerning changes in mRNA levels of specific genes in response to air pollutants. In many cases, such changes in mRNA levels could be detected in plants fumigated with air pollutants at concentrations not causing visible injury, or before visible injury was apparent. These reports cover herbaceous and woody plants, including dicots, monocots, and gymnosperms. Recently developed DNA chip or microarray technology (Ramsay 1998) will enable us to carry out large-scale analysis of mRNA levels in plants exposed to air pollutants.

2.3.1 Effects of Ozone

Tables 1–4 show the reports on increases and Tables 5 and 6 the reports on decreases of mRNA levels by O_3 exposure. The proteins (enzymes) in Tables 1–6

Table 1. Increases in higher plant mRNA levels in response to ozone: pathogenesis-related enzymes (proteins)

Enzyme (protein)	Plant species	Reference
Lipoxygenase	*Glycine max*	Maccarrone et al. (1992)
	Lens culinaris	Maccarrone et al. (1997)
β-1,3-Glucanase	*Nicotiana tabacum*	Schraudner et al. (1992)
		Ernst et al. (1992, 1996)
		Bahl et al. (1995)
Chitinase	*Nicotiana tabacum*	Ernst et al. (1992)
		Bahl et al. (1995)
PR-1b	*Nicotiana tabacum*	Ernst et al. (1992)
PR1-1 and PR2	*Petroselinum crispum*	Eckey-Kaltenbach et al. (1994a,b)
Hydroxyproline-rich glycoprotein (Extensin)	*Petroselinum crispum*	Eckey-Kaltenbach et al. (1994a,b)
	Fagus sylvatica	Schneiderbauer et al. (1995)
	Pinus sylvestris	Schneiderbauer et al. (1995)
	Picea abies	Schneiderbauer et al. (1995)
	Pisum sativum	Sävenstrand et al. (2000)
Eli 16	*Petroselinum crispum*	Eckey-Kaltenbach et al. (1994a,b)
AtOZI1	*Arabidopsis thaliana*	Sharma and Davis (1995)
PR1	*Arabidopsis thaliana*	Sharma et al. (1996)
		Rao et al. (2000)
		Overmyer et al. (2000)
PR1-3 and PR1-4	*Petroselinum crispum*	Eckey-Kaltenbach et al. (1997)
PR-10	*Betula pendula*	Pääkkönen et al. (1998)
PR-4A (1.4 kb mRNA),	*Pisum sativum*	Sävenstrand et al. (2000)
Disease-resistance response protein	*Pisum sativum*	Sävenstrand et al. (2000)
Pre-hevein-like protein	*Pisum sativum*	Sävenstrand et al. (2000)
Allen oxide synthase	*Arabidopsis thaliana*	Rao et al. (2000)

have been tentatively classified into groups according to their known characters. Ozone enhances the mRNA levels for, among others, pathogenesis-, antioxidant- and protein metabolism-related enzymes (proteins) and enzymes for secondary metabolism and ethylene biosynthesis. In situ hybridization, which is the method of detecting mRNA histochemically, revealed that O_3-induced β-1,3-glucanase mRNA accumulation was distributed uniformly throughout cross sections of *Nicotiana tabacum* leaves except for necrotic areas (Ernst et al. 1996). In contrast, by the same method, ozone-induced stilbene synthase and cinnamyl alcohol dehydrogenase mRNAs were revealed to accumulate preferentially in mesophyll cells of *Pinus sylvestris* needles (Zinser et al. 1998).

On the other hand, O_3 reduces mRNA levels for photosynthesis enzymes (proteins) (Table 5) and some antioxidant enzymes (Table 6). In most of the reports describing reduced mRNA levels of a particular gene, an increase in the mRNA level of another gene was demonstrated in the same experiment, indicative

Table 2. Increases in higher plant mRNA levels in response to ozone: antioxidant and related enzymes (proteins)

Enzyme (protein)	Plant species	Reference
Mn superoxide dismutase	*Nicotiana tabacum*	Hérouart et al. (1993)
Cu/Zn superoxide dismutase (cytosolic)	*Nicotiana tabacum*	Hérouart et al. (1993)
		Willekens et al. (1994)
	Arabidopsis thaliana	Sharma and Davis (1994)
		Conklin and Last (1995)
		Kliebenstein et al. (1998)
	Populus tremuloides	Akkapeddi et al. (1999)
Cu/Zn superoxide dismutase (plastidic)	*Arabidopsis thaliana*	Rao and Davis (1999)
Guaiacol peroxidase	*Petroselinum crispum*	Eckey-Kaltenbach et al. (1994a,b)
	Arabidopsis thaliana	Sharma and Davis (1994)
	Ipomoea batatas	Kim et al. (1999)
Glutathione S-transferase	*Arabidopsis thaliana*	Sharma and Davis (1994, 1995)
		Conklin and Last (1995)
Glutathione S-transferase 1 (cytosolic)	*Arabidopsis thaliana*	Sharma et al. (1996),
		Rao and Davis (1999)
		Clayton et al. (1999)
		Rao et al. (2000)
		Overmyer et al. (2000)
Glutathione S-transferase 2	*Arabidopsis thaliana*	Overmyer et al. (2000)
Glutathione peroxidase	*Nicotiana tabacum*	Willekens et al. (1994)
		Schraudner et al. (1998)
	Nicotiana plumbaginifolia	Willekens et al. (1994)
Glutathione peroxidase (plastidic)	*Arabidopsis thaliana*	Rao and Davis (1999)
Catalase 2 and 3	*Nicotiana tabacum* and *N. plumbaginifolia*	Willekens et al. (1994)
Catalase 1 and 3	*Arabidopsis thaliana*	Overmyer et al. (2000)
Salicylic acid-binding catalase	*Nicotiana tabacum*	Örvar et al. (1997)
Ascorbate peroxidase (cytosolic)	*Nicotiana tabacum*	Willekens et al. (1994)
		Örvar et al. (1997)
	Arabidopsis thaliana	Kubo et al. (1995)
		Conklin and Last (1995)
		Rao and Davis (1999)
		Rao et al. (2000)
Glutathione reductase (cytosolic)	*Brassica campestris*	Lee et al. (1998)
Metallothionein-like protein	*Picea abies*	Buschmann et al. (1998)
Copper chaperon	*Arabidopsis thaliana*	Himelblau et al. (1998)
		Miller et al. (1999)

Table 3. Increases in higher plant mRNA levels in response to ozone: enzymes for secondary metabolism and proteins related to protein metabolism

Enzyme (protein)	Plant species	Reference
Enzymes for secondary metabolism		
Cinnamyl alcohol dehydrogenase	*Picea abies*	Galliano et al. (1993)
	Pinus sylvestris	Zinser et al. (1998)
Phenylalanine ammonia-lyase	*Petroselinum crispum*	Eckey-Kaltenbach et al. (1994a,b)
	Arabidopsis thaliana	Sharma and Davis (1994)
		Sharma et al. (1996)
	Nicotiana tabacum	Bahl et al. (1995)
	Betula pendula	Tuomainen et al. (1996)
		Pääkkönen et al. (1998)
	Populus maximowizii × *P. trichocarpa*	Koch et al. (1998)
	Pisum sativum	Brosché and Strid (1999b)
4-Coumaroyl-CoA ligase	*Petroselinum crispum*	Eckey-Kaltenbach et al. (1994a,b)
Chalcone synthase	*Petroselinum crispum*	Eckey-Kaltenbach et al. (1994a,b)
	Nicotiana tabacum	Bahl et al. (1995)
	Pisum sativum	Brosché and Strid (1999b)
3-Hydroxy-3-methylglutaryl-CoA-synthase (1.95-kb mRNA)	*Pinus sylvestris*	Wegener et al. (1997a)
Stilbene synthase	*Vitis vinifera*	Schubert et al. (1997)
	Pinus sylvestris	Zinser et al. (1998)
		Chiron et al. (2000)
O-Methyltransferase	*Populus maximowizii* × *P. trichocarpa*	Koch et al. (1998)
Pinosylvin methyltransferase	*Pinus sylvestris*	Chiron et al. (2000)
Proteins related to protein metabolism		
Proteinase inhibitor	*Atriplex canescens*	No et al. (1997)
Thiol protease,	*Atriplex canescens*	No et al. (1997)
Polyubiquitin	*Pinus sylvestris*	Wegener et al. (1997b)
	Pisum sativum	Brosché and Strid (1999b)
Trypsin inhibitor (wound-inducible)	*Populus maximowizii* × *P. trichocarpa*	Koch et al. (1998)
Ribosomal protein S26	*Pisum sativum*	Brosché and Strid (1999a)
Protease regulator ERD1 (early response to dehydration)	*Arabidopsis thaliana*	Miller et al. (1999)

Table 4. Increases in higher plant mRNA levels in response to ozone: enzymes for ethylene biosynthesis and other proteins

Enzyme (protein)	Plant species	Reference
Enzymes for ethylene biosynthesis		
1-Aminocyclopropane-1-carboxylate synthase	*Solanum tuberosum*	Schlagnhaufer et al. (1995, 1997)
		Glick et al. (1995)
1-Aminocyclopropane-1-carboxylate synthase (LE-ACS2)	*Lycopersicon esculentum*	Tuomainen et al. (1997)
1-Aminocyclopropane-1-carboxylate synthase (ACS6)	*Arabidopsis thaliana*	Vahala et al. (1998)
		Miller et al. (1999)
		Overmyer et al. (2000)
1-Aminocyclopropane-1-carboxylate oxidase	*Lycopersicon esculentum*	Tuomainen et al. (1997)
	Oryza sativa	Ohki et al. (1999)
S-Adenosylmethionine synthetase (SAM3)	*Lycopersicon esculentum*	Tuomainen et al. (1997)
Others		
Short-chain alcohol dehydrogenase	*Picea abies*	Bauer et al. (1993)
	Arabidopsis thaliana	Miller et al. (1999)
	Pisum sativum	Brosché and Strid (1999b)
Glyceraldehyde-3-phosphate dehydrogenase C (cytosolic)	*Solanum tuberosum*	Glick et al. (1995)
Phosphoribosylanthranilate transferase (plastidic)	*Arabidopsis thaliana*	Conklin and Last (1995)
Glycine-rich protein	*Atriplex canescens*	No et al. (1997)
Small heat shock protein	*Petroselinum crispum*	Eckey-Kaltenbach et al. (1997)
Phosphate translocator (mitochondrial)	*Betula pendula*	Kiiskinen et al. (1997)
Porin-like protein	*Picea abies*	Buschmann et al. (1998)
Blue copper-binding protein	*Arabidopsis thaliana*	Richards et al. (1998)
		Miller et al. (1999)
pEARLI2, 5 (aluminium-induced)	*Arabidopsis thaliana*	Richards et al. (1998)
SAG18, 20, 21 (senescence-associated)	*Arabidopsis thaliana*	Miller et al. (1999)
UOD1 (1.0 kb mRNA)	*Pisum sativum*	Sävenstrand et al. (2000)
Lipid transfer protein	*Pisum sativum*	Sävenstrand et al. (2000)
Leucine-rich repeat protein	*Pisum sativum*	Sävenstrand et al. (2000)
Vegetative storage protein	*Arabidopsis thaliana*	Rao et al. (2000)

Table 5. Decreases in higher plant mRNA levels in response to ozone: photosynthesis enzymes (proteins)

Enzyme (protein)	Plant species	Reference
Ribulose-1,5-bisphosphate carboxylase/oxygenase large subunit	*Solanum tuberosum*	Reddy et al. (1993)
		Eckardt and Pell (1994)
		Glick et al. (1995)
	Populus maximowizii × *trichocarpa*	Brendley and Pell (1998)
Ribulose-1,5-bisphosphate carboxylase/oxygenase small subunit	*Solanum tuberosum*	Reddy et al. (1993)
		Eckardt and Pell (1994)
		Schlagnhaufer et al. (1995)
		Glick et al. (1995)
	Nicotiana tabacum	Bahl and Kahl (1995)
		Torsethaugen et al. (1997)
	Arabidopsis thaliana	Conklin and Last (1995)
		Miller et al. (1999)
	Betula pendula	Tuomainen et al. (1996)
10-kDa protein of the water-evolving complex of photosystem II	*Nicotiana tabacum*	Bahl and Kahl (1995)
Chlorophyll a/b-binding protein	*Nicotiana tabacum*	Bahl and Kahl (1995)
	Arabidopsis thaliana	Conklin and Last (1995)
		Rao and Davis (1999)
		Miller et al. (1999)
	Solanum tuberosum	Glick et al. (1995)
	Pinus sylvestris	Wegener et al. (1997a)
		Zinser et al. (1998)
	Pisum sativum	Brosché and Strid (1999b)
Glyceraldehyde-3-phosphate dehydrogenase A and B (plastidic)	*Solanum tuberosum*	Glick et al. (1995)
Carbonic anhydrase	*Nicotiana tabacum*	Örvar et al. (1997)

of decreases in specific mRNA levels rather than a general reduction in total mRNA contents. Consequently, the reduced mRNA levels in response to O_3 appear to reflect the regulated expression of specific genes by plants to cope with stress conditions imposed by O_3.

2.3.2 Effects of Air Pollutants Other than Ozone

Some reports have shown that the mRNA levels of specific enzymes (proteins) either increase (Table 7) or decrease (Table 8) in response to air pollutants other than O_3. The number of reports describing the effects of SO_2 on gene expression is far less than for O_3. All changes in transcript levels in response to SO_2 seem to be similar to those with O_3, although there are a few reports concerning other than antioxidant enzymes to date. As with O_3, some of these reports show SO_2-

Table 6. Decreases in higher plant mRNA levels in response to ozone: antioxidant enzymes and others

Enzyme (protein)	Plant species	Reference
Antioxidant enzymes		
Fe superoxide dismutase	*Nicotiana tabacum*	Hérouart et al. (1993)
		Willekens et al. (1994)
	Arabidopsis thaliana	Conklin and Last (1995)
		Kliebenstein et al. (1998)
	Nicotiana plumbaginifolia	Willekens et al. (1994)
Catalase 1	*Nicotiana plumbaginifolia*	Willekens et al. (1994)
Glutathione reductase (plastidic)	*Arabidopsis thaliana*	Conklin and Last (1995)
		Rao and Davis (1999)
Cu/Zn superoxide dismutase (plastidic)	*Arabidopsis* thaliana	Kliebenstein et al. (1998)
		Overmyer et al. (2000)
Glutathione peroxidase (cytosolic)	*Arabidopsis thaliana*	Rao and Davis (1999)
Others		
3-Hydroxy-3-methylglutaryl-CoA-synthase (1.2 kb mRNA)	*Pinus sylvestris*	Wegener et al. (1997a)
Amine oxidase	*Lens culinaris*	Maccarrone et al. (1997)
Actin	*Pinus sylvestris*	Zinser et al. (1998)
UOD1 (3.7 kb mRNA)	*Pisum sativum*	Sävenstrand et al. (2000)

induced or reduced accumulation of different mRNAs in the same experiment, indicative of regulated gene expression. In contrast, all the reports concerning the effects of automobile exhaust on transcript accumulation showed only a decrease in mRNA levels. Therefore, the overall mRNA content may be reduced by exhaust gas. However, a combination of O_3 and exhaust gas could elevate some mRNA levels more than O_3 alone (Bahl et al. 1995).

3. Biological Significance of Gene Expression in Response to Air Pollutants

As mentioned above, expression of specific genes in response to air pollutants seems to be an active response by the plants. Therefore, it may be considered as a defense response to these pollutants. Tolerance to pollutants would be improved by induction of the genes that are the limiting factors in the tolerance mechanisms. Some of the antioxidant enzymes have been shown to be involved in tolerance to

Table 7. Increases in higher plant mRNA levels in response to air pollutants other than ozone alone

Gas	Enzyme (protein)	Plant species	Reference
SO_2			
	Glutathione peroxidase,	*Nicotiana plumbaginifolia*	Willekens et al. (1994)
	Catalase 2	*Nicotiana plumbaginifolia*	Willekens et al. (1994)
	Ascorbate peroxidase (cytosolic)	*Arabidopsis thaliana*	Kubo et al. (1995)
	Thiol protease	*Atriplex canescens*	No et al. (1997)
	Glycine-rich protein	*Atriplex canescens*	No et al. (1997)
SO_2 and NO_2			
	Cu/Zn superoxide dismutase (plastidic and cytosolic)	*Pinus sylvestris*	Karpinski et al. (1992)
O_3 and exhaust gas			
	Phenylalanine ammonia-lyase	*Nicotiana tabacum*	Bahl et al. (1995)
	Chalcone synthase	*Nicotiana tabacum*	Bahl et al. (1995)
	β-1,3-Glucanase	*Nicotiana tabacum*	Bahl et al. (1995)
	Chitinase	*Nicotiana tabacum*	Bahl et al. (1995)
	Mn superoxide dismutase	*Nicotiana tabacum*	Bahl et al. (1995)
	Cu/Zn superoxide dismutase (cytosolic)	*Nicotiana tabacum*	Bahl et al. (1995)

air pollutants by the analysis of transgenic plants with altered levels of such enzymes (see the chapter by M. Aono, this volume). Accumulation of transcripts for antioxidant enzymes in plants exposed to air pollutants (Tables 2 and 7) is assumed to be performed to scavenge active oxygen species (AOS) generated in the plants by air pollutants. It is unknown why mRNA levels of some antioxidant genes, such as the ironsuperoxide dismutase (FeSOD) gene, are reduced in their expression in response to O_3 or SO_2 (Tables 6 and 8). Overproduction of *Arabidopsis* FeSOD conferred paraquat tolerance on transgenic maize (Van Breusegem et al. 1999).

Plant responses to O_3 resemble those against pathogen attack (Kangasjärvi et al. 1994; Sharma and Davis 1997; Sandermann et al. 1998). The accumulation of transcripts encoding pathogenesis-related (PR) enzymes (proteins) in response to O_3 (see Table 1) raises some important questions about their regulation and functions. Are these proteins effective for protecting the plants against O_3? Are they induced by mistake because of similarities in action between the pathogen and O_3, or do various stresses induce a common set of defense genes without regard to the effect of the individual member of the gene group? The biological significance of the induction of these pathogenesis-related genes is poorly understood. Sharma

Table 8. Decreases in higher plant mRNA levels in response to air pollutants other than ozone

Gas	Enzyme (protein)	Plant species	Reference
SO_2			
	Cu/Zn superoxide dismutase (plastidic)	*Pisum sativum*	Madamanchi et al. (1994)
	Catalase 1	*Nicotiana plumbaginifolia*	Willekens et al. (1994)
	Fe superoxide dismutase	*Nicotiana plumbaginifolia*	Willekens et al. (1994)
Exhaust gas			
	Ribulose-1,5-bisphosphate carboxylase/oxygenase small subunit	*Nicotiana tabacum*	Bahl and Kahl (1995)
	10-kDa protein of the water-evolving complex of photosystem II	*Nicotiana tabacum*	Bahl and Kahl (1995)
	Chlorophyll a/b-binding protein	*Nicotiana tabacum*	Bahl and Kahl (1995)
	Chalcone synthase	*Nicotiana tabacum*	Bahl et al. (1995)
	Mn superoxide dismutase	*Nicotiana tabacum*	Bahl et al. (1995)
	Cu/Zn superoxide dismutase (cytosolic)	*Nicotiana tabacum*	Bahl et al. (1995)

et al. (1996) reported that *Arabidopsis npr1* mutant plants, which are defective in the expression of systemic acquired resistance, were not obviously more susceptible to O_3 treatment, although they were unable to accumulate PR1 mRNA. Instead, high accumulation of PR1 mRNA may correlate with O_3-induced hypersensitive response-like cell death (Rao et al. 2000). Schubert et al. (1997), using transgenic tobacco plants, demonstrated that the O_3-responsive promoter region of the stilbene synthase gene (*Vst*1) from *Vitis vinifera* differed from its pathogen-responsive sequence. This result indicates that induction of the *Vst*1 gene in O_3-fumigated plants was not the result of simple activation of the signaling pathway for defense against pathogens.

Most of the identified O_3-induced genes for secondary metabolism are involved in the phenylpropanoid pathway (see Table 3), which is induced by various biotic and abiotic stresses and which is thought to have various roles (Dixon and Paiva 1995; Solecka 1997). However, a flavonoid-deficient *Arabidopsis* mutant, *tt5*, did not differ significantly from the wild type in growth retardation under O_3-exposed conditions (Rao et al. 1995).

Ozone induces various other genes, including those inducible by stresses, such as heat shock, wounding, or dehydration, and those encoding ethylene biosynthesis enzymes (see Table 3 and 4). Ethylene is a plant hormone that is generated under various stress conditions and which affects the sensitivity of plants to O_3 (see the chapter by N. Nakajima, this volume). Analysis of air pollutant-induced changes in gene expression has indicated close relationships between plant responses to air pollutant stress and those to some other stresses.

Ozone-fumigated plants downregulate photosynthesis by suppressing expression of photosynthetic genes in both the electron transport and carbon dioxide fixation systems (see Table 5). This response is thought to be part of the process for accelerated foliar senescence or programmed cell death (Pell et al. 1997). Miller et al. (1999) reported that transcript levels for eight of 12 senescence-associated genes characterized showed induction by O_3 in *A. thaliana*. The physiological significance of downregulated photosynthesis as a defense response against O_3 is unknown. *Rbc*S-antisense transformed tobacco plants, expressing reduced quantities of ribulose-1,5-bisphosphate carboxylase/oxygenase (Rubisco), exhibited increased O_3 sensitivity by their morphological changes in the leaf tissue (Wiese and Pell 1997).

4. Application of the Detection of Gene Expression to Environmental Biotechnology

Generally speaking, changes in mRNA levels are faster and larger than those in the corresponding protein levels and enzymatic activities. Besides, changes in mRNA levels are often transient, and so may be suitable for monitoring the relatively short-term responses of plants to pollutants. However, attention should be paid to the daily periodicity of the mRNA levels of some genes, such as the *Arabidopsis APX1* gene (Kubo et al. 1995). If it is necessary to determine relatively long-term or accumulated effects of pollutants on plants, it may be more suitable to determine specific protein levels or enzymatic activities. Sandermann (2000) proposed the use of molecular biomarkers, comprising certain stress transcripts, proteins and metabolites, to investigate ozone/biotic disease interactions because biochemical changes induced in a high ozone episode may persist until a subsequent biotic attack period.

Many genes are affected in their expression by more than one stress, and it is unclear whether genes exist whose expression is affected solely by one air pollutant but by no other stresses. Willekens et al. (1994) reported that O_3, SO_2, and ultraviolet-B (UV-B) had similar effects on accumulation of mRNA for antioxidant genes in *Nicotiana plumbaginifolia*. To apply the analysis of gene expression to the diagnosis of plants and phytomonitoring, it is essential to clarify the stress specificity of the pollutant-induced or repressed genes. Brosché and Strid (1999a) reported that expression of the pea ribosomal protein S26 gene can be used as a molecular marker to differentiate between O_3 and UV-B radiation, since O_3 increased the level of mRNA for this gene, whereas UV-B reduced the level. The level of mRNA for disease-resistance response protein 230, extensin, and UOD1 (3.7 kb mRNA) were also shown to be differentially regulated by O_3

and UV-B radiation in opposite ways in pea (Sävenstrand et al. 2000). To construct reporter genes specific for O_3, the O_3-responsive promoter region (Schubert et al. 1997; Ernst et al. 1999) could be used.

A small damage-associated RNA of about 450 nucleotides was found in spruce needles in the forest (Beuther et al. 1988; Köster et al. 1988). The levels of mRNA for the small subunit of Rubisco, subunit II of photosystem I, and metallothionein-like protein in spruce needles were shown to be higher in damaged trees in a forest with a high air pollution input than in symptomless trees in a forest with low input (Etscheid et al. 1993, 1999). For analysis of field-grown plants, the potentially high background of genetic variation must be considered. One further prerequisite is, of course, to simplify the biochemical analysis of plants in the field.

A possible role of air pollutant-induced genes is to provide plants with some level of tolerance to the pollutant. Of the numerous genes that may be involved in providing tolerance to air pollutants, those that show induced expression may be the most effective when expression is induced. Therefore, identification of the air pollutant-induced genes will give us useful information for the generation of air pollutant-tolerant transgenic plants or transgenic plants that have a high ability to detoxify air pollutants for phytoremediation. On the other hand, transgenic plants with reduced expression of specific genes for defense against air pollutants, by such as antisense strategy, may be useful as indicator plants for the air pollutant because of their high sensitivity. Some transgenic plants, transformed with sense or antisense DNA of antioxidant genes, have been shown to have altered sensitivity to air pollutants (see the chapter by M. Aono, this volume).

5. Conclusions

I have reviewed the effects of air pollutants on gene expression in higher plants, especially at the mRNA level. Transcripts of some genes increase, whereas others decrease, in response to air pollutants. Such changes appear to be the result of regulated gene expression by plants to cope with the stress imposed by the pollutants. Many genes that are affected in their expression by air pollutants are also affected by other environmental or biotic stresses. It is unclear whether genes exist in which expression is influenced specifically by one particular air pollutant. Studies on the regulation of gene expression in response to air pollutants constitute an important basis of environmental biotechnology related to air pollution.

Field-grown plants live under stress conditions caused by various factors, including air pollutants. Such stresses are not independent, and their interrelationships are becoming clearer through investigations into the stress mechanisms. Indeed without such investigations, the close relationship between O_3 and pathogen infection would not have been identified. Moreover, the involvement of AOS in almost all stresses is becoming increasingly obvious as the result of studies on the stress mechanisms. A more detailed understanding of the relationships among various stresses will be obtained through elucidation of the signal transduction pathways from perception of the stress to gene expression. Such studies will provide important information for the understanding of field-grown plants living under stress conditions imposed by various stressors, including air pollutants.

Acknowledgment. This work was supported, in part, by Special Coordination Funds for Promoting Science and Technology from the Science and Technology Agency of Japan.

References

Alscher RG, Wellburn AR (eds) (1994) Plant responses to the gaseous environment. Chapman & Hall, London

Akkapeddi AS, Noormets A, Deo BK, et al (1999) Gene structure and expression of the aspen cytosolic copper/zinc-superoxide dismutase (PtSodCcl). Plant Sci 143:151-162

Bahl A, Kahl G (1995) Air pollutant stress changes the steady-state transcript levels of three photosynthesis genes. Environ Pollut 88:57-65

Bahl A, Loitsch SM, Kahl G (1995) Transcriptional activation of plant defence genes by short-term air pollutant stress. Environ Pollut 89:221-227

Bauer S, Galliano H, Pfeiffer F, et al (1993) Isolation and characterization of a cDNA clone encoding a novel short-chain alcohol dehydrogenase from Norway spruce (*Picea abies* L. Karst). Plant Physiol 103:1479-1480

Beuther E, Köster S, Loss P, et al (1988) Small RNAs originating from symptomless and damaged spruces (*Picea* spp.) I. Continuous observation of individual trees at three different locations in NRW. J Phytopathol 121:289-302

Brendley BW, Pell EJ (1998) Ozone-induced changes in biosynthesis of Rubisco and associated compensation to stress in foliage of hybrid poplar. Tree Physiol 18:81-90

Brosché M, Strid Å (1999a) The mRNA-binding ribosomal protein S26 as a molecular marker in plants: molecular cloning, sequencing and differential gene expression during environmental stress. Biochim Biophys Acta 1445:342-344

Brosché M, Strid Å (1999b) Cloning, expression, and molecular characterization of a small pea gene family regulated by low levels of ultraviolet B radiation and other stresses. Plant Physiol 121:479-487

Buschmann K, Etscheid M, Riesner D, et al (1998) Accumulation of a porin-like mRNA and a metallothionein-like mRNA in various clones of Norway spruce upon long-term treatment with ozone. Eur J For Pathol 28:307-322

Chiron H, Drouet A, Lieutier F, et al (2000) Gene induction of stilbene biosynthesis in Scots pine in response to ozone treatment, wounding, and fungal infection. Plant Physiol 124:865-872

Clayton H, Knight MR, Knight H, et al (1999) Dissection of the ozone-induced calcium signature. Plant J 17:575-579

Conklin PL, Last RL (1995) Differential accumulation of antioxidant mRNAs in *Arabidopsis thaliana* exposed to ozone. Plant Physiol 109:203-212

Dixon RA, Paiva NL (1995) Stress-induced phenylpropanoid metabolism. Plant Cell 7:1085-1097

Eckardt NA, Pell EJ (1994) O_3-induced degradation of Rubisco protein and loss of Rubisco mRNA in relation to leaf age in *Solanum tuberosum* L. New Phytol 127:741-748

Eckey-Kaltenbach H, Ernst D, Heller W, et al (1994a) Biochemical plant responses to ozone IV. Cross-induction of defensive pathways in parsley (*Petroselinum crispum* L.) plants. Plant Physiol 104:67-74

Eckey-Kaltenbach H, Großkopf E, Sandermann H Jr, et al (1994b) Induction of pathogen defence genes in parsley (*Petroselinum crispum* L.) plants by ozone. Proc R Soc Edinburgh 102B:63-74

Eckey-Kaltenbach H, Kiefer E, Grosskopf E, et al (1997) Differential transcript induction of parsley pathogenesis-related proteins and of a small heat shock protein by ozone and heat shock. Plant Mol Biol 33:343-350

Ernst D, Schraudner M, Langebartels C, et al (1992) Ozone-induced changes of mRNA levels of β-1,3-glucanase, chitinase and 'pathogenesis-related' protein 1b in tobacco plants. Plant Mol Biol 20:673-682

Ernst D, Bodemann A, Schmelzer E, et al (1996) β-1,3-Glucanase mRNA is locally, but not systemically induced in *Nicotiana tabacum* L. cv. Bel W3 after ozone fumigation. J Plant Physiol 148:215-221

Ernst D, Grimmig B, Heidenreich B, et al (1999) Ozone-induced genes: mechanisms and biotechnological applications. In: Smallwood MF, Calvert CM, Bowles DJ (eds) Plant responses to environmental stress. BIOS Scientific Publishers, Oxford, pp 33-41

Etscheid M, Buschmann K, Köhler R, et al (1993) Differential screening in a cDNA-library from spruce for clones associated with forest decline reveals accumulation of ribulose-1,5-bisphosphate carboxylase small subunit mRNA. J Phytopathol 137:317-343

Etscheid M, Klümper S, Riesner D (1999) Accumulation of a metallothionein-like mRNA in Norway spruce under environmental stress. J Phytopathol 147:207-213

Galliano H, Cabané M, Eckerskorn C, et al (1993) Molecular cloning, sequence analysis and elicitor-/ozone-induced accumulation of cinnamyl alcohol dehydrogenase from Norway spruce (*Picea abies* L.). Plant Mol Biol 23:145-156

Glick RE, Schlagnhaufer CD, Arteca RN, et al (1995) Ozone-induced ethylene emission accelerates the loss of ribulose-1,5-bisphophate carboxylase/oxygenase and nuclear-encoded mRNAs in senescing potato leaves. Plant Physiol 109:891-898

Grimmig B, Schubert R, Fischer R, et al (1997) Ozone- and ethylene-induced regulation of a grapevine resveratrol synthase promoter in transgenic tobacco. Acta Physiol Plant 19:467-474

Großkopf E, Wegener-Strake A, Sandermann H Jr, et al (1994) Ozone-induced metabolic changes in Scots pine: mRNA isolation and analysis of in vitro translated proteins. Can J For Res 24:2030-2033

Hérouart D, Bowler C, Willekens H, et al (1993) Genetic engineering of oxidative stress resistance in higher plants. Philos Trans R Soc Lond B 342:235-240

Himelblau E, Mira H, Lin S-J, et al (1998) Identification of a functional homolog of the yeast copper homeostasis gene *ATX1* from Arabidopsis. Plant Physiol 117:1227-1234

Kangasjärvi J, Talvinen J, Utriainen M, et al (1994) Plant defence systems induced by ozone. Plant Cell Environ 17:783-794

Karpinski S, Wingsle G, Karpinska B, et al (1992) Differential expression of CuZn-superoxide dismutases in *Pinus sylvestris* needles exposed to SO_2 and NO_2. Physiol Plant 85:689-696

Kiiskinen M, Korhonen M, Kangasjärvi J (1997) Isolation and characterization of cDNA for a plant mitochondrial phosphate translocator (*Mpt1*): ozone stress induces *Mpt1* mRNA accumulation in birch (*Betula pendula* Roth). Plant Mol Biol 35:271-279

Kim K-Y, Huh G-H, Lee H-S, et al (1999) Molecular characterization of cDNAs for two anionic peroxidases from suspension cultures of sweet potato. Mol Gen Genet 261:941-947

Kirtikara K, Talbot D (1996) Alteration in protein accumulation, gene expression and ascorbate-glutathione pathway in tomato (*Lycopersicon esculentum*) under paraquat and ozone stress. J Plant Physiol 148:752-760

Kliebenstein DJ, Monde R-A, Last RL (1998) Superoxide dismutase in Arabidopsis: an eclectic enzyme family with disparate regulation and protein localization. Plant Physiol 118:637-650

Koch JR, Scherzer AJ, Eshita SM, et al (1998) Ozone sensitivity in hybrid poplar is correlated with a lack of defense-gene activation. Plant Physiol 118:1243-1252

Köster S, Beuther E, Riesner D (1988) Small RNAs originating from symptomless and damaged spruces (*Picea abies* L., Karst.) II. Investigation of different trees from two differently exposed forest sections in the Hils area. J Phytopathol 121:303-312

Kubo A, Saji H, Tanaka K, et al (1995) Expression of *Arabidopsis* cytosolic ascorbate peroxidase gene in response to ozone or sulfur dioxide. Plant Mol Biol 29:479-489

Lee H, Jo J, Son D (1998) Molecular cloning and characterization of the gene encoding glutathione reductase in *Brassica campestris*. Biochim Biophys Acta 1395:309-314

Maccarrone M, Veldink GA, Vliegenthart JFG (1992) Thermal injury and ozone stress affect soybean lipoxygenases expression. FEBS Lett 309:225-230

Maccarrone M, Veldink GA, Vliegenthart JFG, et al (1997) Ozone stress modulates amine oxidase and lipoxygenase expression in lentil (*Lens culinaris*) seedlings. FEBS Lett 408:241-244

Madamanchi NR, Donahue JL, Cramer CL, et al (1994) Differential response of Cu,Zu superoxide dismutases in two pea cultivars during a short-term exposure to sulfur dioxide. Plant Mol Biol 26:95-103

Miller JD, Arteca RN, Pell EJ (1999) Senescence-associated gene expression during ozone-induced leaf senescence in Arabidopsis. Plant Physiol 120:1015-1023

No E-G, Flagler RB, Swize MA, et al (1997) cDNAs induced by ozone from *Atriplex canescens* (saltbush) and their response to sulfur dioxide and water-deficit. Physiol Plant 100:137-146

Ohki T, Matsui H, Nagasaka A, et al (1999) Induction by ozone of ethylene production and an ACC oxidase cDNA in rice (*Oryza sativa* L.) leaves. Plant Growth Regul 28:123-127

Okpodu CM, Alscher RG, Grabau EA, et al (1996) Physiological, biochemical and molecular effects of sulfur dioxide. J Plant Physiol 148:309-316

Örvar BL, McPherson J, Ellis BE (1997) Pre-activating wounding response in tobacco prior to high-level ozone exposure prevents necrotic injury. Plant J 11:203-212

Overmyer K, Tuominen H, Kettunen R, et al (2000) Ozone-sensitive Arabidopsis *rcd1* mutant reveals opposite roles for ethylene and jasmonate signaling pathways in regulating superoxide-dependent cell death. Plant Cell 12:1849-1862

Pääkkönen E, Seppänen S, Holopainen T, et al (1998) Induction of genes for the stress proteins PR-10 and PAL in relation to growth, visible injuries and stomatal conductance in birch (*Betula pendula*) clones exposed to ozone and/or drought. New Phytol 138:295-305

Pell EJ, Eckardt NA, Glick RE (1994) Biochemical and molecular basis for impairment of photosynthetic potential. Photosynth Res 39:453-462

Pell EJ, Schlagnhaufer CD, Arteca RN (1997) Ozone-induced oxidative stress: mechanisms of action and reaction. Physiol Plant 100:264-273

Pino ME, Mudd JB, Bailey-Serres J (1995) Ozone-induced alterations in the accumulation of newly synthesized proteins in leaves of maize. Plant Physiol 108:777-785

Ramsay G (1998) DNA chips: state-of-the art. Nature Biotechnology 16:40-44

Ranieri A, Tognini M, Tozzi C, et al (1997) Changes in the thylakoid protein pattern in sunflower plants as a result of ozone fumigation. J Plant Physiol 151:227-234

Rao MV, Davis KR (1999) Ozone-induced cell death occurs via two distinct mechanisms in *Arabidopsis*: the role of salicylic acid. Plant J 17:603-614

Rao MV, Paliyath G, Ormrod DP (1995) Differential response of photosynthetic pigments, rubisco activity and rubisco protein of *Arabidopsis thaliana* exposed to UVB and ozone. Photochem Photobiol 62:727-735

Rao MV, Lee H, Creelman RA, et al (2000) Jasmonic acid signaling modulates ozone-induced hypersensitive cell death. Plant Cell 12:1633-1646

Reddy GN, Arteca RN, Dai Y-R, et al (1993) Changes in ethylene and polyamines in relation to mRNA levels of the large and small subunits of ribulose bisphosphate carboxylase/oxygenase in ozone-stressed potato foliage. Plant Cell Environ 16:819-826

Richards KD, Schott EJ, Sharma YK, et al (1998) Aluminum induces oxidative stress genes in *Arabidopsis thaliana*. Plant Physiol 116:409-418

Sandermann H Jr (1996) Ozone and plant health. Annu Rev Phytopathol 34:347-366

Sandermann H Jr (2000) Ozone/biotic disease interactions: molecular biomarkers as a new experimental tool. Environ Pollut 108:327-332

Sandermann H Jr, Ernst D, Heller W, et al (1998) Ozone: an abiotic elicitor of plant defence reactions. Trends Plant Sci 3:47-50

Sävenstrand H, Brosché M, Ängehagen M, et al (2000) Molecular markers for ozone stress isolated by suppression subtractive hybridization: specificity of gene expression and identification of a novel stress-regulated gene. Plant Cell Environ 23:689-700

Schlagnhaufer CD, Glick RE, Arteca RN, et al (1995) Molecular cloning of an ozone-induced 1-aminocyclopropane-1-carboxylate synthase cDNA and its relationship with a loss of *rbcS* in potato (*Solanum tuberosum* L.) plants. Plant Mol Biol 28:93-103

Schlagnhaufer CD, Arteca RN, Pell EJ (1997) Sequential expression of two 1-aminocyclopropane-1-carboxylate synthase genes in response to biotic and abiotic stresses in potato (*Solanum tuberosum* L.) leaves. Plant Mol Biol 35:683-688

Schmitt R, Sandermann H Jr (1990) Biochemical response of Norway spruce (*Picea abies* (L.) Karst.) towards 14-month exposure to ozone and acid mist: Part II—effects on protein biosynthesis. Environ Pollut 64:367-373

Schneiderbauer A, Back E, Sandermann H Jr, et al (1995) Ozone induction of extensin mRNA in Scots pine, Norway spruce and European beech. New Phytol 130:225-230

Schraudner M, Ernst D, Langebartels C, et al (1992) Biochemical plant responses to ozone III. Activation of the defense-related proteins β-1,3-glucanase and chitinase in tobacco leaves. Plant Physiol 99:1321-1328

Schraudner M, Langebartels C, Sandermann H Jr (1996) Plant defence systems and ozone. Biochem Soc Trans 24:456-461

Schraudner M, Langebartels C, Sandermann H (1997) Changes in the biochemical status of plant cells induced by the environmental pollutant ozone. Physiol Plant 100:274-280

Schraudner M, Moeder W, Wiese C, et al (1998) Ozone-induced oxidative burst in the ozone biomonitor plant, tobacco Bel W3. Plant J 16:235-245

Schubert R, Fischer R, Hain R, et al (1997) An ozone-responsive region of the grapevine resveratrol synthase promoter differs from the basal pathogen-responsive sequence. Plant Mol Biol 34:417-426

Sharma YK, Davis KR (1994) Ozone-induced expression of stress-related genes in *Arabidopsis thaliana*. Plant Physiol 105:1089-1096

Sharma YK, Davis KR (1995) Isolation of a novel *Arabidopsis* ozone-induced cDNA by differential display. Plant Mol Biol 29:91-98

Sharma YK, Davis KR (1997) The effects of ozone on antioxidant responses in plants. Free Radical Biol Med 23:480-488

Sharma YK, León J, Raskin I, et al (1996) Ozone-induced responses in *Arabidopsis thaliana*: the role of salicylic acid in the accumulation of defense-related transcripts and induced resistance. Proc Natl Acad Sci USA 93:5099-5104

Solecka D (1997) Role of phenylpropanoid compounds in plant responses to defferent stress factors. Acta Physiol Plant 19:257-268

Torsethaugen G, Pitcher LH, Zilinskas BA, et al (1997) Overproduction of ascorbate peroxidase in the tobacco chloroplast does not provide protection against ozone. Plant Physiol 114:529-537

Tuomainen J, Pellinen R, Roy S, et al (1996) Ozone affects birch (*Betula pendula* Roth) phenylpropanoid, polyamine and active oxygen detoxifying pathways at biochemical and gene expression level. J Plant Physiol 148:179-188

Tuomainen J, Betz C, Kangasjärvi J, et al (1997) Ozone induction of ethylene emission in tomato plants: regulation by differential accumulation of transcripts for the biosynthetic enzymes. Plant J 12:1151-1162

Vahala J, Schlagnhaufer CD, Pell EJ (1998) Induction of an ACC synthase cDNA by ozone in light-grown *Arabidopsis thaliana* leaves. Physiol Plant 103:45-50

Van Breusegem F, Slooten L, Stassart J-M, et al (1999) Overproduction of *Arabidopsis thaliana* FeSOD confers oxidative stress tolerance to transgenic maize. Plant Cell Physiol 40:515-523

Wegener A, Gimbel W, Werner T, et al (1997a) Molecular cloning of ozone-inducible protein from *Pinus sylvestris* L. with high sequence similarity to vertebrate 3-hydroxy-3-methylglutaryl-CoA-synthase. Biochim Biophys Acta 1350:247-252

Wegener A, Gimbel W, Werner T, et al (1997b) Sequence analysis and ozone-induced accumulation of polyubiquitin mRNA in *Pinus sylvestris*. Can J For Res 27:945-948

Wiese CB, Pell EJ (1997) Influence of ozone on transgenic tobacco plants expressing reduced quantities of Rubisco. Plant Cell Environ 20:1283-1291

Willekens H, Van Camp W, Van Montagu M, et al (1994) Ozone, sulfur dioxide, and ultraviolet B have similar effects on mRNA accumulation of antioxidant genes in *Nicotiana plumbaginifolia* L. Plant Physiol 106:1007-1014

Yunus M, Iqbal M (eds) (1996) Plant response to air pollution. Wiley, Chichester

Zinser C, Ernst D, Sandermann H Jr (1998) Induction of stilbene synthase and cinnamyl alcohol dehydrogenase mRNAs in Scots pine (*Pinus sylvestris* L.) seedlings. Planta 204:169-176

7
Biotechnology for Phytomonitoring

Hikaru Saji

Environmental Biology Division, National Institute for Environmental Studies, Onogawa 16-2, Tsukuba, Ibaraki 305-8506, Japan

1. Introduction

The various forms of foliar damage caused by air pollutants to plants, including easily visible chlorotic and necrotic symptoms, have been successfully used to detect and monitor these pollutants in the natural environment (see the chapters by I. Nouchi, this volume). The use of plants for such monitoring (phytomonitoring) has both advantages and disadvantages over methods based on physicochemical analyses (see the chapters by I. Nouchi and by M. Burchett et al.). Plant materials suitable for such a purpose must demonstrate high sensitivity, specificity, and reproducibility of the symptoms in addition to allowing easy detection and measurement of the degree of damage. For example, sensitive strains of plants such as tobacco, morning glory, and clover have been used for phytomonitoring (see the chapter by I. Nouchi). Excellent strains, such as ozone-sensitive Bel-W3 tobacco, have also been developed by traditional breeding and selection (Heggestad 1991).

However, such conventional breeding methods are time consuming and also limited by the availability of the genetic variation that can be transferred from closely related cross-hybridizable species. In contrast, the development of modern biotechnology has made it possible to introduce genes from any organism, as well as artificially synthesized or processed genes, into plant genomes in a relatively short period. The potential to genetically modify plants has been rapidly increasing because of the advances in manipulation techniques and with the

Air Pollution and Plant Biotechnology
–Prospects for Phytomonitoring and Phytoremediation–
Edited by K. Omasa, H. Saji, S. Youssefian, and N. Kondo
© Springer -Verlag Tokyo 2002

increasing number of genes being isolated and characterized from different organisms. Although the number of identified genes involved in the sensitivity or specificity of plant responses to air pollutants is still limited, there will undoubtedly be a steady increase in these numbers as studies in plant molecular biology are actively pursued in numerous laboratories around the world.

Various biochemical responses to air pollutants can be detected before any visible symptoms appear on plants, as described in several other chapters in this volume. For example, the expression of many genes in plants is altered by exposure to air pollutants (see the chapter by A. Kubo). As some of these changes in gene expression are detected at pollutant concentrations that are lower than those required to induce visual symptoms, these changes may also be used as sensors of air pollutants. In this chapter, I discuss the current status and prospects of biotechnology and the problems to be addressed for its application to the phytomonitoring of air pollutants.

2. Use of Biotechnology to Generate Plants with Altered Sensitivity to Air Pollutants

2.1 Selection of Artificially Mutated Plants

The large variations between plants in physiological traits, such as their sensitivity to air pollutants, are the result of the differential accumulation of naturally occurring mutations in the genomes over long evolutionary periods. Such mutations, which have been effective in changing various plant traits, are being used in combination with conventional breeding methods to create new strains or cultivars of crops with the desired characteristics. Chemical mutagens, such as ethylmethane sulfonate, or physical mutagens, such as ionizing radiation, are also commonly used to artificially induce mutations in plants (Lightner and Caspar 1998). With the development of molecular genetics, which combines classical genetics and modern molecular biology, numerous mutants have been isolated from the model plant species, *Arabidopsis thaliana* (ABRC 1995), and used for basic physiological, biochemical, and molecular biological studies.

Several reports of mutant plant lines with different sensitivities to air pollutants have been made. The cultivar Kentucky bluegrass (*Poa pratensis* L.), which was originally selected from mutated plants for its resistance to rust (*Puccinia* spp.), was also found to be tolerant to ozone (Wilton et al. 1972). Barley mutants resistant to nitrogen dioxide (Lea et al. 1994) and those with increased resistance or sensitivity to a mixture of sulfur dioxide and nitrogen dioxide (Kasana and Lea 1994) have also been isolated. However, the most extensively studied plant mutants with altered sensitivities to air pollutants appear to be a set of *A. thaliana* mutants obtained by Conklin et al. (1996, 2000). By screening mutated plants on the basis of their sensitivity to ozone, several mutant lines were obtained with higher sensitivities to ozone than wild-type plants. One of these mutants, *vtc1* (originally named *soz1*), in addition to being highly ozone sensitive, was found to possess an ascorbate content that was only 30% that of wild-type plants. Because ascorbate is a major redox substance in plant cells, its deficiency, as in the *vtc1* mutant, may result in various pleiotropic effects, especially sensitivity to oxidative stress conditions. As *vtc1* is also found to be sensitive to sulfur dioxide and ultraviolet B irradiation, its use for phytomonitoring may be limited to the detection of a broad range of oxidative stress conditions rather than the specific detection of air pollutants.

Such results are indicative of the potential use of artificially induced mutagenesis in the generation of new plant lines with altered sensitivity to air pollutants. The main advantage of this approach is that plants with such altered sensitivity may be obtained without any understanding of the gene(s) involved. The disadvantage, however, is that only alterations of genes already existing in the genome of a plant are possible; the genetic resources existing in other plants or organisms cannot be utilized by this method.

With the recent advances in genome analyses, especially in international collaborative projects on plants such as *A. thaliana* and rice, it is becoming increasingly feasible to identify and isolate mutated genes from mutants of these plants (Martínez-Zapater and Salinas 1998). Such mutants thus serve as ideal materials for the isolation of genes that affect specific plant characteristics and which can be used for genetic modification of various plant species as well as for fundamental studies into plant physiology and molecular biology (Fig. 1). Several laboratories around the world are actively engaged in the isolation of *Arabidopsis* mutants with altered sensitivities to air pollutants (Overmyer et al. 1998).

2.2 Genetic Manipulation

2.2.1 Background

As described recent developments in biotechnology have made it possible to introduce genes from any organism, or those artificially synthesized or processed, into various plant genomes. The resulting plants, often referred to as transgenic plants, have so far been mainly used to clarify the function of plant genes or gene products in basic studies of plant physiology and biochemistry (Allen 1995). Steadily increasing numbers of researchers, however, have begun to apply these molecular techniques to agriculture and other bioindustries (Lindsey 1998). Recent successes in conferring agronomically important traits, such as herbicide and pathogen resistance, on crops with these techniques suggest that genetically modified plants with traits ideally suited for detecting air pollutants are within reach. To fully appreciate the potential application of these techniques to

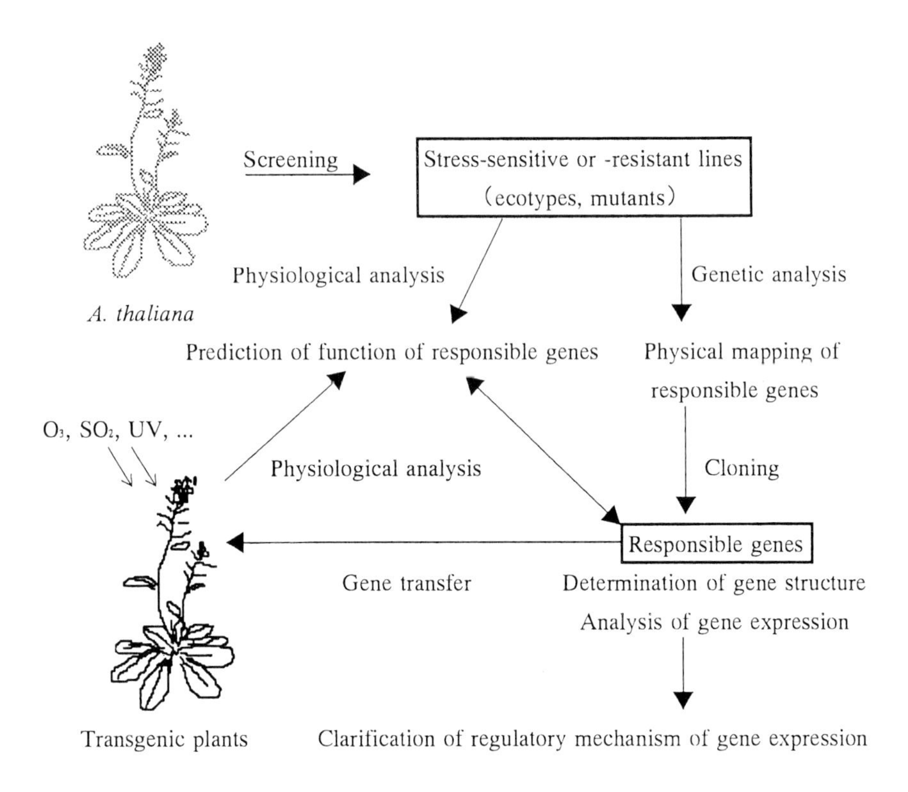

Fig. 1. Molecular genetic studies on plant stress resistance using *Arabidopsis thaliana*

phytomonitoring, however, it is important to briefly review (1) the currently available methods and strategies for the genetic modification of plants and (2) the information regarding, and accessibility to, the genes to be manipulated.

2.2.2 Methods for Manipulating Genes in Plants

Transfer of Foreign Genes into Plants. Based on the discovery of the tumor-inducing (Ti) plasmid in the soil bacterium, *Agrobacterium tumefaciens*, and its role in transferring into plant genomes a DNA fragment (T-DNA) containing genes involved in the tumorigenesis of plant cells, a method to artificially introduce genes into the plant genome was established. This method, which has been routinely used to introduce genes into a wide variety of plant species, especially dicotyledonous plants (Hooykaas 1989), consists of several steps (Horsch et al. 1985). The gene to be introduced (transgene) is first appropriately processed and inserted into the T-DNA region of a Ti plasmid vector. The resulting construct is then introduced into an *A. tumefaciens* strain, which is then used to infect culture tissues of the host plant. Finally, the infected tissues are cultured under appropriate conditions to regenerate transgenic plants. Numerous variations of this method have now been successfully employed, for example, germinating seeds (Feldmann and Marks 1987) or cut shoot ends (Chang et al. 1994) of *Arabidopsis thaliana* have been infected with *Agrobacterium tumefaciens* harboring the desired gene construct, and transgenic seeds were obtained by further growth of these plants without going through the tissue culture step.

Another widely used method to introduce genes into plants is by the "particle bombardment" or "microprojectile-mediated transformation" method. In this method, DNA to be introduced is adsorbed to small gold or tungsten particles (microprojectiles), which are then bombarded into plant cells with an instrument called a particle gun (Klein et al. 1987). In principle, this method is applicable to any plant species but is of special value for introducing genes into plants, such as cereals, in which *Agrobacterium*-mediated transformation is more challenging (Christou 1997).

Expression of Foreign Genes in Plants. When genes such as those encoding enzymes or regulatory factors are introduced into plant cells, it is essential that their spatial and temporal expression and the localization of their products are appropriately regulated. By placing regulatory DNA fragments, such as gene "promoters," next to the introduced gene before introduction into plant cells, the genes can be expressed either constitutively, or in particular tissues (Ficker et al. 1997), or in response to specific developmental cues (Pickardt et al. 1998) and/or environmental signals (Blume and Grierson 1997; Kurata et al. 1998). Moreover, by modifying the gene itself, for example, by introducing short DNA sequences that encode transit or signal peptides, the gene products can be targeted to specific subcellular organelles such as chloroplasts and mitochondria (Philip et al. 1998; Shen et al. 1997).

In most research, genes have been connected to regulatory fragments, such as cauliflower mosaic virus 35S promoter, and transferred into plant cells to be expressed constitutively (Benfey and Chua 1990). However, it seems better to appropriately confine, both timely and spatially, the expression of the transferred gene, especially when the gene expression has a side effect on plants. Although overexpression of the cDNA encoding a stress-responsive transcription factor, DREB1A, in transgenic *Arabidopsis* plants resulted in improved tolerance to drought, salt loading, and freezing, use of the 35S promoter to drive expression of *DREB1A* also resulted in severe growth retardation under normal growing conditions. In contrast, expression of *DREB1A* from the stress-inducible *rd29A* promoter gave rise to minimal effects on plant growth while providing an even greater tolerance to stress conditions than did expression of the gene from the 35S promoter (Kasuga et al. 1999). It is also essential that the product of the transferred gene is targeted to proper subcellular loci: the difference in effect of transgene expression between different subcellular targeting of the gene products has sometimes been reported (see the chapter by M. Aono, this volume).

Inhibition of Expression of Endogenous Genes in Plants. Although overexpression of endogenous genes or introduced foreign genes in plants is often an effective means of altering specific plant traits, inhibition of the endogenous genes is an alternative means of modifying plant characteristics. If the gene to be inhibited has already been isolated and is available, the most commonly used method has been to introduce an antisense-orientated form of the DNA into the plant (Bourque 1995). The antisense DNA, which may represent the whole or a fragment of transgene of interest, is connected in a reverse orientation to the regulatory DNA regions, such as a promoter, as described here. Although the gene used to produce this antisense effect does not have to originate from the same host plant species, a high level of DNA sequence homology with the corresponding endogenous host gene may be essential to inhibit expression of the endogenous gene.

Another approach to obtain plants in which the expression of certain genes is inhibited is to generate and select plants in which the endogenous genes have been inactivated by insertional mutagenesis. Many *Arabidopsis* lines that have some DNA fragment (tag) such as T-DNA or a transposon, inserted at various sites of the genome, have been generated (Feldmann 1991). If the insertion site in one of these tagged lines happens, by chance, to correspond to the site where an endogenous gene exists, the function of the gene is expected to be lost or modified.

2.2.3 Genes to Be Manipulated

Once a gene involved in plant responses to one or more air pollutants is identified and isolated, the gene can be transferred to plants and either (over)expressed or used to inhibit the corresponding gene(s), as described. However, the biggest obstacle to this approach at present, the lack of suitable genes for manipulation, reflects the extent of progress in basic physiological analysis of plant responses to air pollutants. There are numerous black boxes in our understanding of the mechanisms by which air pollutants induce plant damage and those by which plants protect themselves from pollutants. Therefore, at present, methods of genetic manipulation and molecular genetic studies with mutants are used primarily to analyze, rather than to utilize, these responses.

It has been reported that several low molecular weight bioactive substances are produced in response to air pollutants and that these are involved in damage to, or protection of, plants (Langebartels et al. 1991; Sharma et al. 1996; see also the chapters by N. Nakajima and by M. Aono, this volume)(Fig. 2). Therefore, there are many genes worth considering as potential targets for genetic manipulation to alter the plant's sensitivity to air pollutants. Several researchers have already reported some success. Overproduction of enzymes involved in the scavenging of active oxygen species have been shown to enhance plant resistance to ozone or sulfur dioxide (see the chapter by M. Aono, this volume), and overexpression of genes for sulfur metabolism enhanced plant resistance to sulfur dioxide (see the chapter by S. Youssefian). Expression of a bacterial gene encoding salicylate

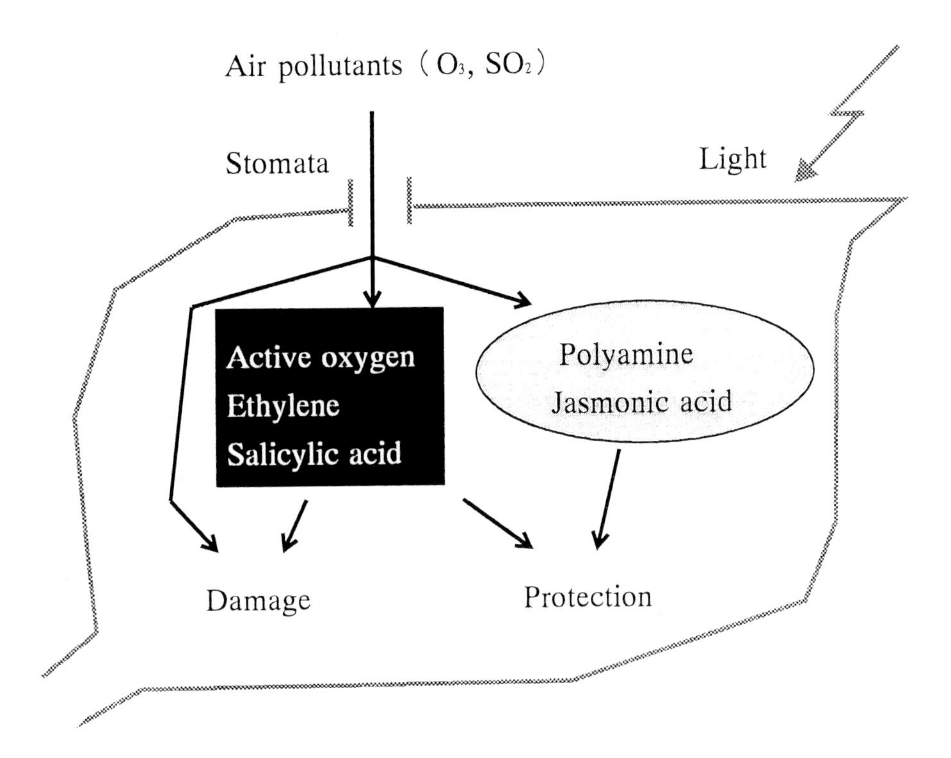

Fig. 2. Production and hypothetical effects of low molecular weight bioactive substances in plants exposed to air pollutants

hydroxylase and the consequent inhibition of salicylic acid accumulation in plant cells was shown to enhance ozone sensitivity in *A. thaliana* (Sharma et al. 1996), although it reduced sensitivity in tobacco (Örvar et al. 1997).

Such studies suggest that it may not be too long before we can efficiently control the sensitivity of plants to air pollutants by genetic manipulation. Where the difficulties will lie, however, will be in how to alter the plant's sensitivity only to air pollutants. As all the substances shown in Fig. 2 are involved in plant responses not only to air pollutants but also to other stress factors, such as ultraviolet light, pathogens, and wounding (Klessig and Malamy 1994; Yalpani et al. 1994; Conconi et al. 1996), the manipulation of genes involved in these pathways would also be expected to affect plant responses to these stress factors. However, the effects of these substances may vary with the particular stress factor. For example, transgenic tobacco plants generated in our laboratory to overproduce glutathione reductase exhibit enhanced tolerance to sulfur dioxide but not to ozone, even though active oxygen species appear to be involved in the damage induced by both these gases (see the chapter by M. Aono, this volume). Therefore, many studies with transgenic plants are necessary to clarify the roles of various genes in the resistance of plants to each of many stress factors.

3. Molecular Sensors of Air Pollutants

As described, many biochemical responses to air pollutants can be detected before the appearance of visual symptoms in plants. Such responses include changes in lipid metabolism, increases in the concentrations of various bioactive substances, such as ethylene (see Fig. 2), and changes in the expression of numerous genes (see chapters by T. Sakaki, by N. Nakajima, and by A. Kubo, this volume). These biochemical responses may well be useful as molecular sensors of air pollutant levels. However, the greatest obstacle to this approach appears again to be the lack of specificity in these responses to air pollutants because almost all the previously reported biochemical changes are also observed in response to other stress factors, such as ultraviolet light and pathogens (Kangasjärvi et al. 1994).

Despite such limitations, there does appear to exist some difference in response to some of these stress factors. For example, the patterns of induction in expression of genes encoding ethylene-synthesizing enzymes are much different between ozone and sulfur dioxide treatments (see the chapter by N. Nakajima, this

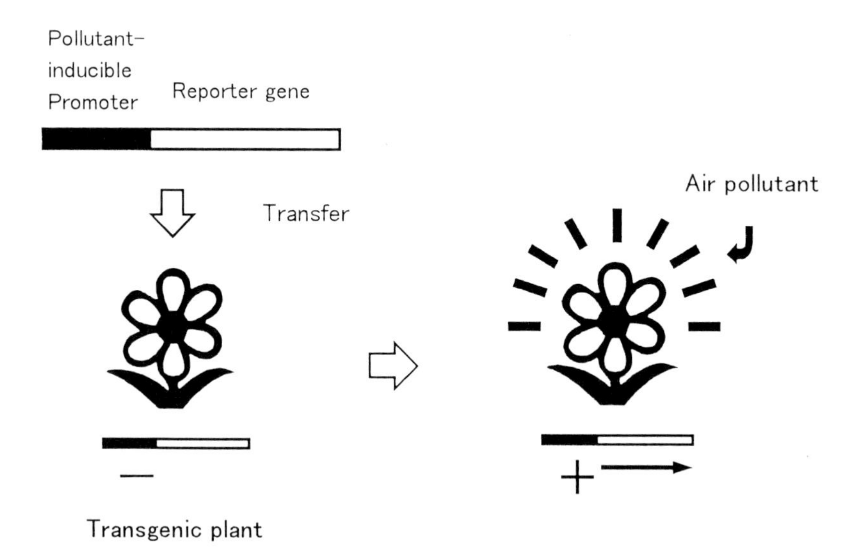

Fig. 3. A detector plant, transgenic for a chimeric gene consisting of an air pollutant-responsive promoter and a fluorescence-emitting reporter gene

volume). Clearly, further studies are necessary to clarify the specificity of this and other biochemical responses to air pollutants.

When such specific-response genes have been identified, their regulatory DNA regions, including promoter regions, which generally control the specificity of gene induction, can be isolated and used as a sensor of air pollutants in the following way: the pollutant-inducible promoter is attached to reporter genes, such as those encoding pigment-synthesizing enzymes like β-glucuronidase (Jefferson et al. 1987) and fluorescence-emitting protein as green fluorescent protein (Sheen et al. 1995), and then introduced into the sensor plant. The transgenic plant thus obtained can be used to detect air pollutants by specific induction of the reporter gene (Fig. 3).

Even if we cannot find a gene whose expression is induced specifically by air pollutants, specific detection of air pollutants may be possible by use of the combination of several stress-responsive genes. Furthermore, there is a rapid advancement in methods to measure expressions of many genes at once: the transcript imaging by hybridization of total RNA to high-density oligonucleotide arrays (Lockhart et al. 1996; Wodicka et al. 1997). Therefore, it may not be long before a novel method to detect various stress factors, including air pollutants, will be developed based on the profile of expression of many genes in a model plant species.

4. Conclusions

The responses of plants to air pollutants, such as the appearance of easily visible chlorotic foliar symptoms and changes in expression of certain genes, can be used to detect and monitor air pollutants in the natural environment. The potential use of plants for such monitoring (phytomonitoring) is expected to be largely enhanced by modern biotechnology. Various methods and strategies have been rapidly developed to analyze and manipulate genes in plants. Although only a limited number of genes have been isolated and characterized that are involved in the sensitivity or specificity of plant responses to air pollutants and which are potential targets for genetic manipulation, the number will undoubtedly increase with the advancement of plant physiological and molecular biological studies.

References

ABRC (1995) Seed and DNA stock list. Arabidopsis Biological Resource Center, The Ohio State University, Columbus, Ohio, USA

Allen RD (1995) Dissection of oxidative stress tolerance using transgenic plants. Plant Physiol 107:1049-1054

Benfey PN, Chua NH (1990) The cauliflower mosaic virus 35S promoter: combinatorial regulation of transcription in plants. Science 250:959-966

Blume B, Grierson D (1997) Expression of ACC oxidase promoter-GUS fusions in tomato and *Nicotiana plumbaginifolia* regulated by developmental and environmental stimuli. Plant J 12:731-746

Bourque JE (1995) Antisense strategies for genetic manipulations in plants. Plant Sci 105:125-149

Chang SS, Park SK, Kim BC, Kang BJ, Kim DU, Nam HG (1994) Stable genetic transformation of *Arabidopsis thaliana* by *Agrobacterium* inoculation *in planta*. Plant J 5:551-558

Christou P (1997) Rice transformation: bombardment. Plant Mol Biol 35:197-203

Conconi A, Smerdon MJ, Howe GA, Ryan, CA (1996) The octadecanoid signalling pathway in plants mediates a response to ultraviolet radiation. Nature 383:826-829

Conklin PL, Saracco SA, Norris SR, Last RL (2000) Identification of ascorbic acid-deficient *Arabidopsis thaliana* mutants. Genetics 154:847-856

Conklin PL, Williams EH, Last RL (1996) Environmental stress sensitivity of an ascorbic acid-deficient Arabidopsis mutant. Proc Natl Acad Sci USA 93:9970-9974

Feldmann KA (1991) T-DNA insertion mutagenesis in Arabidopsis: mutational spectrum. Plant J 1:71-82

Feldmann KA, Marks MD (1987) *Agrobacterium*-mediated transformation of germinating seeds of *Arabidopsis thaliana*: a non-tissue culture approach. Mol Gen Genet 208:1-9

Ficker M, Wemmer T, Thompson RD (1997) A promoter directing high level expression in pistils of transgenic plants. Plant Mol Biol 35:425-431

Heggestad HE (1991) Origin of Bel-W3, Bel-C and Bel-B tobacco varieties and their use as indicators of ozone. Environ Pollut 74:264-291

Hooykaas PJJ (1989) Transformation of plant cells via Agrobacterium. Plant Mol Biol 13:327-336

Horsch RB, Fry JE, Hoffmann NL, Eichholtz D, Rogers SG, Fraley RT (1985) A simple and general method for transferring genes into plants. Science 227:1229-1231

Jefferson RA, Kavanagh TA, Bevan MW (1987) GUS fusions: β-glucuronidase as a sensitive and versatile gene fusion marker in higher plants. EMBO J 6:3901-3907

Kangasjärvi J, Talvinen J, Utriainen M, Karjalainen R (1994) Plant defence systems induced by ozone. Plant Cell Environ 17:783-794

Kasana MS, Lea PJ (1994) Growth responses of mutants of spring barley to fumigation with SO_2 and NO_2 in combination. New Phytol 126:629-636

Kasuga M, Liu Q, Miura S, Yamaguchi-Shinozaki K, Shinozaki K (1999) Improving plant drought, salt, and freezing tolerance by gene transfer of a single stress-inducible transcription factor. Nature Biotechnol 17:287-291

Klein TM, Wolf ED, Wu R, Sanford JC (1987) High-velocity microprojectiles for delivering nucleic acids into living cells. Nature 327:70-73

Klessig DF, Malamy J (1994) The salicylic acid signal in plants. Plant Mol Biol 26:1439-1458

Kurata HI, Takemura T, Furusaki S, Kado CI (1998) Light-controlled expression of a foreign gene using the chalcone synthase promoter in tobacco BY-2 cells. J Ferment Bioeng 86:317-323

Langebartels C, Kerner K, Leonardi S, Schraudner M, Trost M, Heller W (1991) Biochemical plant responses to ozone. I. Differential induction of polyamine and ethylene biosynthesis in tobacco. Plant Physiol 95:882-889

Lea PJ, Wolfenden J, Wellburn AR (1994) Influences of air pollutants upon nitrogen metabolism. In: Alscher RG, Wellburn, AR (eds) Plant responses to the gaseous environment: molecular metabolic and physiological aspects. Elsevier, Barking, Essex, pp 279-300

Lightner J, Caspar T (1998) Seed mutagenesis of Arabidopsis. In: Martínez-Zapater JM, Salinas J (eds) *Arabidopsis* protocols. Humana Press, Totowa, NJ, pp 91-104

Lindsey K (1998) Transgenic plant research. Harwood, Amsterdam

Lockhart DJ, Dong H, Byrne MC, Follettie MT, Gallo MV, Chee MS, Mittmann M, Wang C, Kobayashi M, Horton H, Brown EL (1996) Expression monitoring by hibridization to high-density oligonucleotide arrays. Nat Biotechnol 14:1675-1680

Martínez-Zapater JM, Salinas J (1998) *Arabidopsis* protocols. Methods in molecular biology, vol 82, Humana Press, Totowa, NJ, pp 277-351

Örvar B, McPherson J, Ellis BE (1997) Pre-activating wounding response in tobacco prior to high-level ozone exposure prevents necrotic injury. Plant J 11:203-212

Overmyer K, Tuominen H, Kettunen R et al (2000) Ozone-sensitive Arabidopsis *rcd1* mutant reveals opposite roles for ethylene and jasmonate signaling pathways in regulating superoxide-dependent cell death. Plant Cell 12: 1849-1862

Philip R, Darnowski DW, Sundararaman V, Cho MJ, Vodkin LO (1998) Localization of β-glucuronidase in protein bodies of transgenic tobacco seed by fusion to an amino terminal sequence of the soybean lectin gene. Plant Sci 137:191-204

Pickardt T, Ziervogel B, Schade V, Ohl L, Bäumlein H, Meixner M (1998) Developmental-regulation and tissue-specific expression of two different seed promoter GUS-fusions in transgenic lines of *Vicia narbonensis*. J Plant Physiol 152:621-629

Sharma YK, León J, Raskin I, Davis KR (1996) Ozone-induced responses in *Arabidopsis thaliana*: the role of salicylic acid in the accumulation of defense-related transcripts and induced resistance. Proc Natl Acad Sci USA 93:5099-5104

Sheen J, Hwang S, Niwa Y, Kobayashi H, Galbraith DW (1995) Green-fluorescent protein as a new vital marker in plant cells. Plant J 8:777-784

Shen B, Jensen RG, Bohnert HJ (1997) Increased resistance to oxidative stress in transgenic plants by targeting mannitol biosynthesis to chloroplasts. Plant Physiol 113:1177-1183

Wilton AC, Murray JJ, Heggestad HE, Juska FV (1972) Tolerance and susceptibility of Kentucky bluegrass (*Poa pratensis* L.) cultivars to air pollution: in the field and in an ozone chamber. J Environ Qual 1:112-114

Wodicka L, Dong H, Mittmann M, Ho MH, Lockhart DJ (1997) Genome-wide expression monitoring in *Saccharomyces cerevisiae*. Nat Biotechnol 15: 1359-1367

Yalpani N, Enyedi AJ, León J, Raskin I (1994) Ultraviolet light and ozone stimulate accumulation of salicylic acid, pathogenesis-related proteins and virus resistance in tobacco. Planta 193:372-376

II. Resistant Plants and Phytoremediation

8
Absorption of Organic and Inorganic Air Pollutants by Plants

Kenji Omasa[1], Kazuo Tobe[2], and Takayuki Kondo[3]

[1] Department of Biological and Environmental Engineering, Graduate School of Agricultural and Life Sciences, The University of Tokyo, Yayoi 1-1-1, Bunkyo-ku, Tokyo 113-8657, Japan
[2] Laboratory of Intellectual Fundamentals for Environmental Studies, National Institute for Environmental Studies, Onogawa 16-2, Tsukuba, Ibaraki 305-8506, Japan
[3] Air Quality Section, Toyama Prefectural Environmental Science Research Center, Nakataikouyama 17-1, Kosugi, Toyama 939-0363, Japan

1. Introduction

Plant leaves possess metabolic processes that can biochemically transform many of the absorbed air pollutants (Koziol and Whatley 1984; Schulte-Hostede et al. 1987; Yunus and Iqbal 1996; Sandermann et al. 1997; De Kok and Stulen 1998), thereby allowing the leaves to continuously absorb the gases without being saturated with them. Therefore, plants are thought to play key roles in determining the fate of atmospheric pollutants of both man-made and natural origins.

Many air pollutants have deleterious effects on leaf tissues and can cause changes in stomatal opening (Heath 1980; Omasa et al. 1985, 1990; Winner et al. 1988; Darrall 1989; Mansfield and Pearson 1996), thus reducing the capacity of plants to act as effective sinks for the pollutants. Both the capacity for foliar absorption and the susceptibility to the pollutants vary widely between species and also between particular growth environments (NIES 1980, 1984, 1992; Yunus and Iqbal 1996; Sandermann et al. 1997; Matyssek et al. 1997; De Kok and Stulen 1998). Therefore, to assess the role of vegetation as a sink for air pollutants, it is

---

Air Pollution and Plant Biotechnology
–Prospects for Phytomonitoring and Phytoremediation–
Edited by K. Omasa, H. Saji, S. Youssefian, and N. Kondo
© Springer -Verlag Tokyo 2002

important for us to evaluate, over a wide range of species, the efficacy by which the leaves absorb these pollutants and the extent to which the leaves are adversely affected by the exposure.

Gas diffusion models have been widely used to analyze the exchange of water vapor and CO_2 between the atmosphere and plant leaves (Monteith 1973; Jones 1992). Hill, Bennett and colleagues (Hill 1971; Bennett and Hill 1973, 1975; Bennett et al. 1973) modified these models so as to describe the foliar absorption of atmospheric pollutant gases. Since then, considerable effort has focused on the quantitative analyses of foliar absorption of pollutant gases (for example, O'Dell et al. 1977; Omasa 1979; Hosker and Lindberg 1982; Parkhurst 1994; Kondo et al. 1996a; Omasa et al. 2000a,b). Based on the gas diffusion model, it is possible to separately determine stomatal and nonstomatal components of gas absorption rates and to infer the foliar capacity for metabolism of the absorbed pollutant gases, whether inorganic or organic.

In this chapter, the characteristics of foliar absorption of more than 20 organic and inorganic air pollutants are discussed based on a simple gas diffusion model. In addition, the difference between species for stomatal control of gas absorption and susceptibility of plants to air pollutants are briefly described to select species suitable for phytoremediation.

2. A Simple Gas Diffusion Model for Analyzing Gas Absorption by Plant Leaves

Carbon dioxide (CO_2) in the atmosphere is absorbed by plant leaves through the stomata and used for photosynthesis in leaf cells. In a similar manner, atmospheric pollutants that can be metabolized inside the leaf tissue are absorbed through the stomata. Consequently, a concentration gradient of these gases is generated from the ambient air to the interior of the leaf (Fig. 1) and acts as a factor that regulates foliar gas absorption. Also, the evaporation of water through the stomata to the atmosphere (transpiration) is driven by the water vapor density gradient between the inside and outside of the leaf, although the direction of water vapor flux is opposite to that of CO_2 and pollutant gases. Therefore, the analysis of transpiration and gas absorption using a gas diffusion model aids us to understand the gas absorption phenomena taking place in the leaves.

The gas and water vapor fluxes between ambient air and air at the gas-liquid interface in the stomatal cavity are regulated by two limiting sites, the leaf boundary layer and the stomata (Fig. 1). Resistance to the flow of gas or water vapor at the leaf boundary layer (r_b; leaf boundary layer resistance) depends on the thickness of the leaf boundary layer; the resistance and the thickness decrease with increasing wind velocity at the leaf surface, and are also dependent on leaf size and shape. Stomatal resistance to gas flow (r_s) depends on the density and size of the stomata and the degree of stomatal opening. Consequently, the total resistance (r) can be described as the sum of the two resistance components:

$$r = r_b + r_s \tag{1}$$

The reciprocals of these resistances (r, r_b, and r_s) are referred to as the conductance. Thus, Eq. 1 can be rewritten

$$1/g = 1/g_b + 1/g_s \tag{2}$$

where g, g_b, and g_s are the total leaf conductance, the leaf boundary layer conductance, and the stomatal conductance, respectively.

The rate of water evaporation through the cuticle is usually very small compared with that through the stomata (Kramer and Boyer 1995). However, the cuticular deposition (adsorption and decomposition) of some pollutant gases may not be negligible and needs to be considered when modeling foliar gas absorption. Therefore, on a unit leaf area basis, the rates of cuticular gas deposition, total foliar gas absorption (to be exact, the total amount of absorption, adsorption, and decomposition), and transpiration can be represented by α, Q, and E, respectively, with the transpiration rate (E) and the stomatal gas absorption rate (Q - α) being approximated by Fick's law of diffusion:

$$E = g^{W}\,(w_i - w_o) \tag{3}$$

and

$$Q - \alpha = g^{G}(c_o - c_i) \tag{4}$$

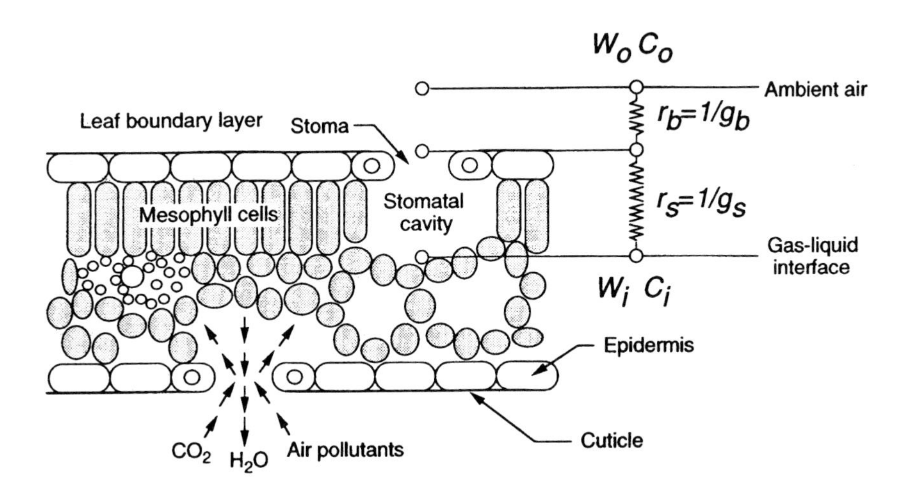

Fig. 1. Pathways for diffusion of gases and water vapor between ambient air and the interior of a leaf

where w_i and c_i, respectively, are the saturated water vapor density and the gas concentration in the air at the gas-liquid interface of the stomatal cavity, indicating the capacity to scavenge the gas in the leaves, and w_o and c_o, respectively, are the water vapor density and the gas concentration of the ambient air surrounding the leaf (Fig. 1). The respective W and G superscripts on g represent conductance for water vapor and the gas.

The total foliar gas absorption per unit gas concentration (q) is obtained from Eq. 4 as:

$$q = Q/c_o$$
$$= (1 - c_i/c_o)\, g^G + \alpha/c_o \qquad (5)$$

The ratio of g^G to g^W can be regarded as a constant, k. That is:

$$k = g^G / g^W \qquad (6)$$

Thus, Eq. 5 can be rearranged as follows:

$$q = k\,(1 - c_i/c_o)\, g^W + \alpha/c_o \qquad (7)$$

The leaf boundary layer conductance (g_b) and the stomatal conductance (g_s) can be regarded to be proportional to $D^{2/3}$ and D, respectively, where D denotes the diffusion coefficient (Monteith 1973). Therefore, the quotients g_b^G / g_b^W and g_s^G / g_s^W (k_b and k_s, respectively) are given by

$$k_b = g_b^G / g_b^W$$
$$= (D^G/D^W)^{2/3} \qquad (8)$$

and

$$k_s = g_s^G / g_s^W$$
$$= D^G/D^W \qquad (9)$$

where the W and G superscripts on D indicate the diffusion coefficients for water vapor and pollutant gas, respectively.

The ratio of the diffusion coefficients can, in turn, be approximated by

$$D^G / D^W = (M^G / M^W)^{-1/2} \qquad (10)$$

where M^W and M^G are the molecular weights of water vapor and pollutant gas, respectively. This relationship indicates that the lower the molecular weight of a pollutant, the more diffusive the pollutant is through both the leaf boundary layer and the stomata.

From the relationships in Eqs. 2, 6, 8, and 9, it is deduced that the value k is a constant between k_s and k_b. Therefore, if $M^W < M^G$, as holds for many pollutant gases, then

$$k_s < k < k_b \tag{11}$$

or, from Eqs. 8-10

$$(M^G / M^W)^{-1/2} < k < (M^G / M^W)^{-1/3} \tag{12}$$

Equation 7 indicates that by plotting q versus g^W and calculating the range of k from Eq. 12, c_i and α can be estimated from the gradient and the y-intercept of this plot, respectively. For this estimation, it is necessary to measure E, Q, w_o, and c_o, and to obtain w_i, which can be equated with saturated water vapor density at the leaf temperature (Farquhar and Raschke 1978).

It is clear from Eq. 7 that, at a given c_o, the stomatal absorption of a gas is governed by g^G (= $1/g_b^G + 1/g_s^G$) and c_i. Because g_b is usually large and does not depend on physiological factors, the major changes in g^G can be attributed to changes in g_s^G, which in turn is determined by the density and size of the stomata on the leaves, as well as the degree of stomatal opening. Usually, to evaluate the conductance or resistance of gases through the stomata, g_s^W (instead of g_s^G) is utilized, and g_s^W is merely referred to as 'stomatal conductance.' Consequently, the regulation of stomatal gas absorption can be mainly attributed to two factors, g_s^W and c_i. Therefore, the importance of each of these two parameters, stomatal conductance for water vapor and gas concentration in the air at the gas-liquid interface of the stomatal cavity on foliar gas absorption are discussed in the following sections.

3. Analysis of Foliar Absorption of Pollutant Gases by the Gas Diffusion Model

Hill (1971) reported that O_3, NO_2, SO_2, Cl_2, HF, and peroxyacetyl nitrate (PAN) were absorbed continuously and at considerably high rates by an artificial alfalfa canopy, whereas absorption of NO and CO by the canopy was considerably smaller. He also reported that the absorption rates of all these gases by the canopy were proportional to the ambient gas concentration and were governed by the degree of stomatal opening. Subsequently, many researchers have focused on the foliar absorption of inorganic air pollutants by plants, and the results of these studies have been occasionally reviewed (for example, Omasa 1979; Garsed 1984; Smith 1984; Lange et al. 1989; De Kok and Stulen 1998). Recently, studies of foliar gas absorption analysis have been extended to organic air pollutants.

Consequently, it was found that four kinds of C_1-C_4 aldehydes (Mutters et al. 1993; Kondo et al. 1995, 1996a 1998) as well as phenol (Kondo et al. 1996b, 1999) could be continuously absorbed by leaves through the stomata. More recently, Omasa et al. (2000b) measured the foliar absorption of seven organic pollutants (acetone, acetonitrile, acrolein, methyl ethyl ketone [MEK], isobutyl methyl ketone [IBMK], chloroform, and benzene) by two woody species (*Populus nigra* and *Camellia Sasanqua*). Of these seven pollutants, acrolein was absorbed by both the tested species through the stomata, and MEK absorption was detected only in *C. Sasanqua*, whereas absorption of the other five pollutants could not be detected in either of the two species. Such finding suggests differences between species in absorption of different organic pollutants.

To investigate the mechanism of gas absorption by leaves, the gas diffusion model described in the previous section was applied to the results of foliar gas absorption measurements. Figure 2 shows the relationship between the

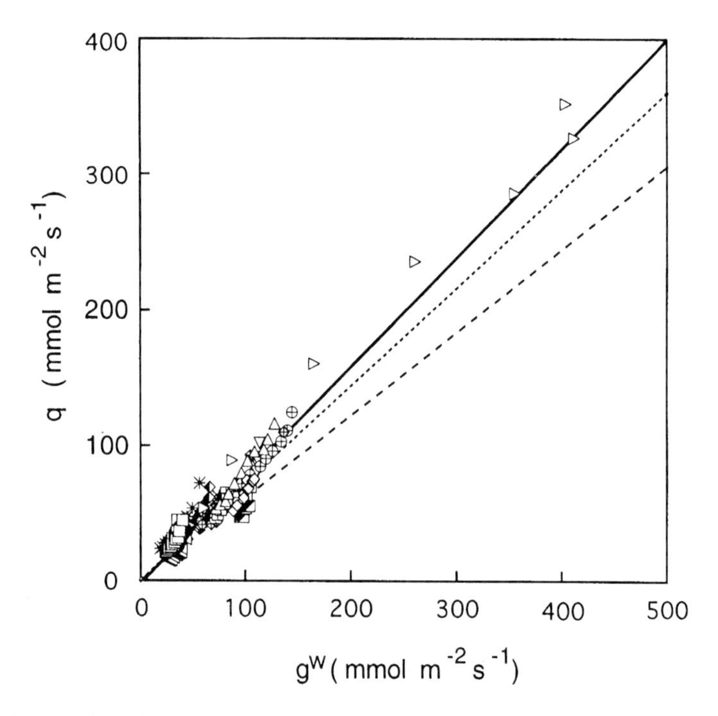

Fig. 2. Relationship between the normalized O_3 absorption rate (q) and total leaf conductance of water vapor diffusion (g^W) for 15 woody species which were exposed to 0.5-ppmv O_3. Different symbols present measurements of different species. The regression line is shown by a solid line, and lines $q = k_b\, g^W$ and $q = k_s\, g^W$ are also indicated by dotted and broken lines, respectively. See Fig. 8 for the reference of the species names. The equation of the regression line is $y = 0.804x - 0.002$ ($r = 0.972$). (from Omasa et al. 2000a)

normalized O_3 absorption rate (q; see Eq. 7) and total leaf conductance of water vapor diffusion (g^W) for 15 different woody species exposed to 0.5 ppmv O_3. The y-intercept of this regression line is close to zero ($\alpha \fallingdotseq 0$), indicating that the cuticular deposition (adsorption and decomposition) of O_3 was negligibly small. Based on Eq. 7, the slope of the regression line should have a value between k_s and k_b (Eq. 11). Although the slope of the regression line is slightly larger than k_b, it would be reasonable to expect that c_i is close to zero, indicating all absorbed O_3 is very rapidly scavenged in the leaf tissues (Polle 1998). The error may be due to either the incompleteness of the model applied or experimental errors. Because most of the changes in g^W can be regarded to result from changes in stomatal opening, this result would indicate that, for a wide range of species, foliar O_3 absorption is predominantly through the stomata and is strongly governed by stomatal conductance.

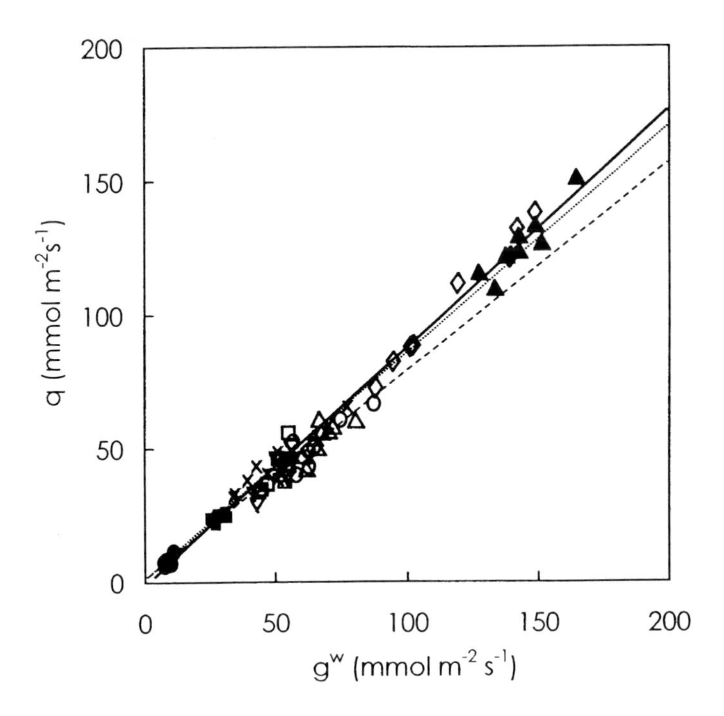

Fig. 3. Relationship between q and g^W in 9 woody species exposed to 0.07 ppmv formaldehyde. Different symbols represent measurements of different species (▲, Lombardy popular; ◇, Locust; △, Japanese maple; ○, Japanese elm; □, Maidenhair tree; X, Camellia; ▽, Bamboo-leafed oak; ■, Himalayan cedar; ●, Japanese black pine). The regression line is shown by a solid line, and lines $q = k_b g^W$ and $q = k_s g_w$ are also indicated by dotted and broken lines, respectively. The equation of the regression line is $y = 0.894x - 3.1$ ($r = 0.983$). (from Kondo et al. 1996a)

From gas absorption measurements of coniferous forests, it has been suggested that cuticular deposition also plays an important role in removing O_3 from the atmosphere (Rondon et al. 1993; Coe et al. 1995; Grant and Richter 1995). These reports indicated that the foliar gas absorption rate was approximately threefold that absorbed only through the stomata, and that the rate of cuticular gas deposition showed diurnal changes, being maximal at midday. However, other O_3 absorption studies (Omasa et al. 1979, 2000a; Laisk et al. 1989; Matyssek et al.

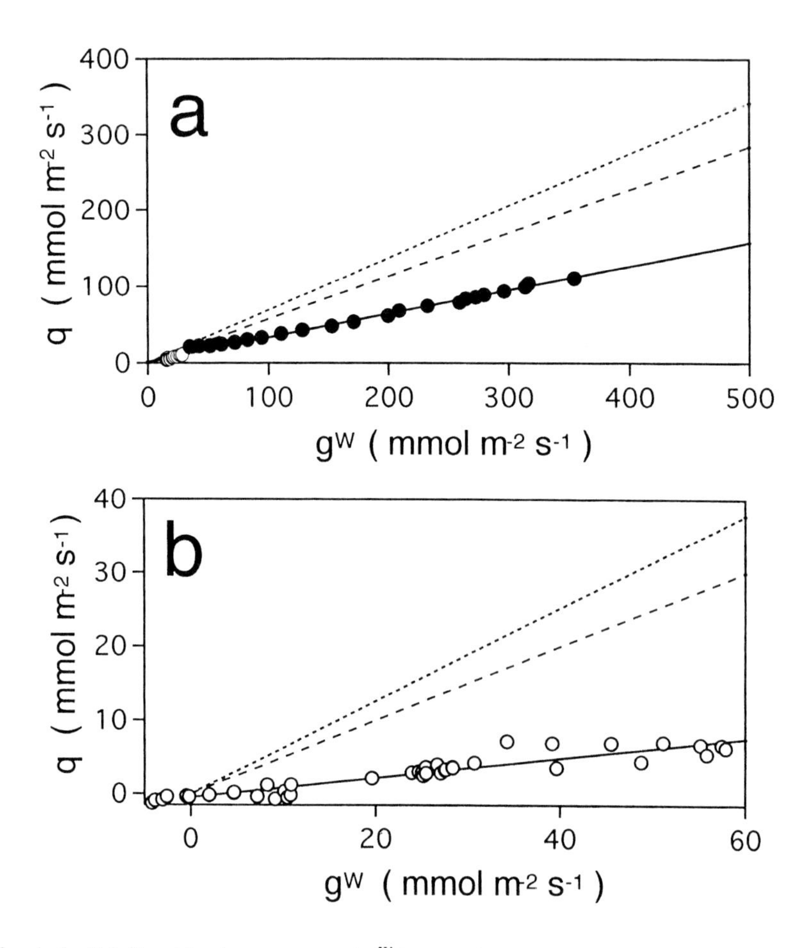

Fig. 4a,b. Relationships between q and g^W in *Populus nigra* (closed circles) and *Camellia Sasanqua* (open circles) exposed to 1 ppmv acrolein (**a**) or MEK (**b**). The regression lines are shown by solid lines, and lines $q = k_b g^W$ and $q = k_s g^W$ are also indicated by dotted and broken lines, respectively. Equations of the regression lines are **a** (acrolein): $y = 0.312x + 1.479$ ($r = 0.997$); **b** (MEK): $y = 0.134x - 0.478$ ($r = 0.938$). Note difference in axes scales in **a** and **b**. (from Omasa et al. 2000b)

1997; Polle 1998) do not support these findings, which may have been affected by inconstancies in environmental parameters that are inevitable in field experiments.

For organic pollutants, the absorption rate varied with species of plants and gases. The relationship between the normalized formaldehyde absorption rate (q) and total leaf conductance of water vapor diffusion (g^W) for nine different woody species is shown in Fig. 3. The results indicate that both c_i and α were close to zero, as for O_3, suggesting that this pollutant is absorbed through the stomata and rapidly scavenged in the leaf tissues. This result is supported by reports demonstrating that formaldehyde is rapidly metabolized to sugars, amino acids, and other metabolites in leaf tissues and used for plant growth (Krall and Tolbert 1957; Girard et al. 1989; Giese et al. 1994).

The relationships of the absorption rates (q) of acrolein and MEK with total leaf conductance (g^W) suggest positive linear correlations, with y-intercepts of almost zero, for both pollutants (Fig. 4). These findings indicate that foliar absorption of these gases is also regulated by stomatal conductance. However, the slopes of the regression lines of both acrolein and MEK are considerably smaller than k_s which, in consideration of Eqs. 7 and 11, suggests that c_i/c_o is not close to zero, indicating limitations of metabolic capacity of these gases in the leaf tissues.

Table 1 shows the results of model analysis of several organic and inorganic pollutant gases. For SO_2 (Omasa and Abo 1978; Black and Unsworth 1979), NO_2 (Omasa et al. 1979; Natori et al. 1981; Neubert et al. 1993), O_3 (Omasa et al. 1979, 2000a; Laisk et al. 1989), PAN (Nouchi 1980), and formaldehyde (Kondo et al. 1996a), c_i/c_o s are almost zero, which indicates that these absorbed gases are very quickly scavenged from the air at the surface of the mesophyll cells, under favorable growing conditions. Hence, the foliar absorption rates of these gases are governed solely by their gas-phase diffusion (i.e., mostly by stomatal conductance). However, a part of the sulfur derived from SO_2 is emitted as H_2S (see Kondo; De Kok et al., this volume), and also NO_2 absorption depends on ascorbate concentrations in leaf apoplast (see Yoneyama et al., this volume). For NH_3, c_i/c_o is between 0.05 and 0.1 at concentrations above the compensation point (0.4 - 15 ppbv), but $c_i > c_o$ at concentrations below the compensation point, which indicates that NH_3 is emitted from the leaf into the atmosphere through the stomata (Farquhar et al. 1980; Geßler and Rennenberg 1998). For NO and CO, c_i/c_o is above 0.9, suggesting that stomatal absorption is very small, thus indicating that metabolic capacity of these gases is very small.

For some organic gases such as C_2-C_4 aldehydes and phenol, c_i/c_o is found to range between 0 and 1. For these gases, therefore, stomatal absorption appears to play an important role in foliar gas absorption, but the relating slow rates of metabolism of the gas in leaf tissues may act as a rate-limiting step. Consequently, for these pollutants, both stomatal conductance and metabolic rate determine the foliar gas absorption rate. For other organic pollutant gases (for example, acetone, chloroform, benzene, and trichloroethylene), foliar gas absorption could not be detected and so c_i/c_o is assumed to be unity (Table 1).

It has been reported that some organic pollutants absorbed by leaves exist in the

Table 1. Solubility in water of gases and c_i/c_o calculated from foliar gas absorption measurements under favorable growing conditions

Gas	solubility in water (mol kg^{-1})*)	calculated c_i/c_o	c_o and literature
CO_2	0.037	0.5-0.9 (in light condition)	350 ppmv; Zeiger *et al.* (1987)
CO	0.00083	≅ 1	0.5 ppmv; estimated from Hill (1971)
NO_2	decomposes	≥ 0	0.1-6 ppmv; Omasa *et al.* (1979) and others
NO	0.0021	0.9-0.95	0.02-0.1 ppmv; calculated from Neubert *et al.* (1993)
NH_3	33	0.05-0.1	0.05 ppmv; Farquhar *et al.* (1980)
O_3	0.011	≥ 0	0.2-0.9 ppmv; Omasa *et al.* (1979,2000a) and others
SO_2	1.6	≥ 0	0.25-1.5 ppmv; Omasa and Abo (1978) and others
Peroxyacetyl nitrate (PAN)	low	≥ 0	0.01-0.08 ppmv; Nouchi (1980)
Formaldehyde	18	≥ 0	0.05-1 ppmv; Kondo *et al.* (1996a)
Acetaldehyde	∞	0.4-0.5	0.1 ppmv; calculated from Kondo *et al.* (1998)
Propanal	3.8	0.35-0.45	0.1 ppmv; calculated from Kondo *et al.* (1998)
Butanal	0.56	0.3-0.5	0.1 ppmv; calculated from Kondo *et al.* (1998)
Pentanal	low	0.3-0.5	0.1 ppmv; calculated from Kondo *et al.* (1998)
Acrolein	4.6	0.2-0.6	1 ppmv; Omasa *et al.* (2000b)
Acetone	∞	≅ 1	1 ppmv; Omasa *et al.* (2000b)
Methyl ethyl ketone (MEK)	5.0	0.7-0.8	1 ppmv; Omasa *et al.* (2000b)
Methyl isobutyl ketone (MIBK)	0.017	≅ 1	1 ppmv; Omasa *et al.* (2000b)
Phenol	0.71	0.45-0.6	0.05 ppmv; calculated from Kondo *et al.* (1999)
Acetonitrile	∞	≅ 1	1 ppmv; Omasa *et al.*(2000b)
Benzene	0.0023	≅ 1	1 ppmv; Omasa *et al.* (2000b)
Chloroform	0.0069	≅ 1	1 ppmv; Omasa *et al.*(2000b)
Trichloroethylene	0.0076	≅ 1	0.05 ppmv; Kondo *et al.*, unpublished data

*): Data from Richardson (1992-1994), Hill (1971) and other literatures.

leaves without being decomposed (Bacci et al. 1990a; Keymeulen et al. 1993; Simonich and Hites 1994a,b; Ockenden et al. 1998). In such cases, the foliar and ambient air pollutant levels will eventually become equilibrated, drastically limiting the capacity of the leaves to remove these pollutants from the atmosphere. Models describing the foliar absorption of such organic pollutants have been proposed (Bacci et al. 1990b; Riederer 1990; Trapp et al. 1990; Peterson et al. 1991; Deinum et al. 1995), and are used to estimate the partitioning and transport of organic pollutants between the atmosphere and different leaf parts.

Hill (1971) and Bennett et al. (1973) reported that stomatal absorbency of a gas is determined by its solubility in water and the effectiveness of its decomposition or metabolization in the leaf tissues. Among the pollutants listed in Table 1, no conspicuous relationship could be found between the solubility of a pollutant in water and its foliar absorbency. For example, although the solubility of O_3 in water is relatively low, the c_i/c_o for O_3 was close to zero. On the other hand, the foliar absorption of acetone or acetonitrile could not be detected, even though they are both highly soluble in water. These results imply that solubility in water is not a critical factor determining the foliar absorbency of a pollutant. However, there is a possibility that the foliar absorption of CO, chloroform, benzene, and trichloroethylene (Table 1) is limited by their solubility because their gas-liquid-phase transfer into the leaves is obstructed.

The fact that all the aldehydes (including acrolein) noted in Table 1 are absorbed by leaves indicates that leaves have a high capacity to metabolize aldehydes, although this capacity varies between the different kinds of aldehydes. However, different ketones show different levels of foliar absorbency (Table 1). For example, MEK absorbance is found in *C. Sasanqua* but is not detectable in *P. nigra* (Omasa et al. 2000b). Furthermore, although MEK and acetone have similar chemical structures and are both highly soluble in water, the leaves show no absorbance of acetone. Such variation may result from interspecific differences in the activity of metabolic enzymes or in the characteristics of the metabolic processes.

4. Stomatal Control of Gas Absorption and Susceptibility of Plants to Air Pollutants

As described, stomatal conductance is a critical factor in determining the foliar absorbency of pollutant gases that are decomposed or metabolized inside the leaves. Figure 5a-c shows the mean stomatal conductance of deciduous and evergreen species, under sunlight and adequate water conditions of midday in summer in the field. The large variation in stomatal conductance between species is indicative of the variation in their gas absorption capacities. In general, stomatal conductance is found to be greater in deciduous than in evergreen species. Because the absorption of a gas by trees in the field depends on their leaf area index, trees with a large number of leaves, as well as higher stomatal conductance,

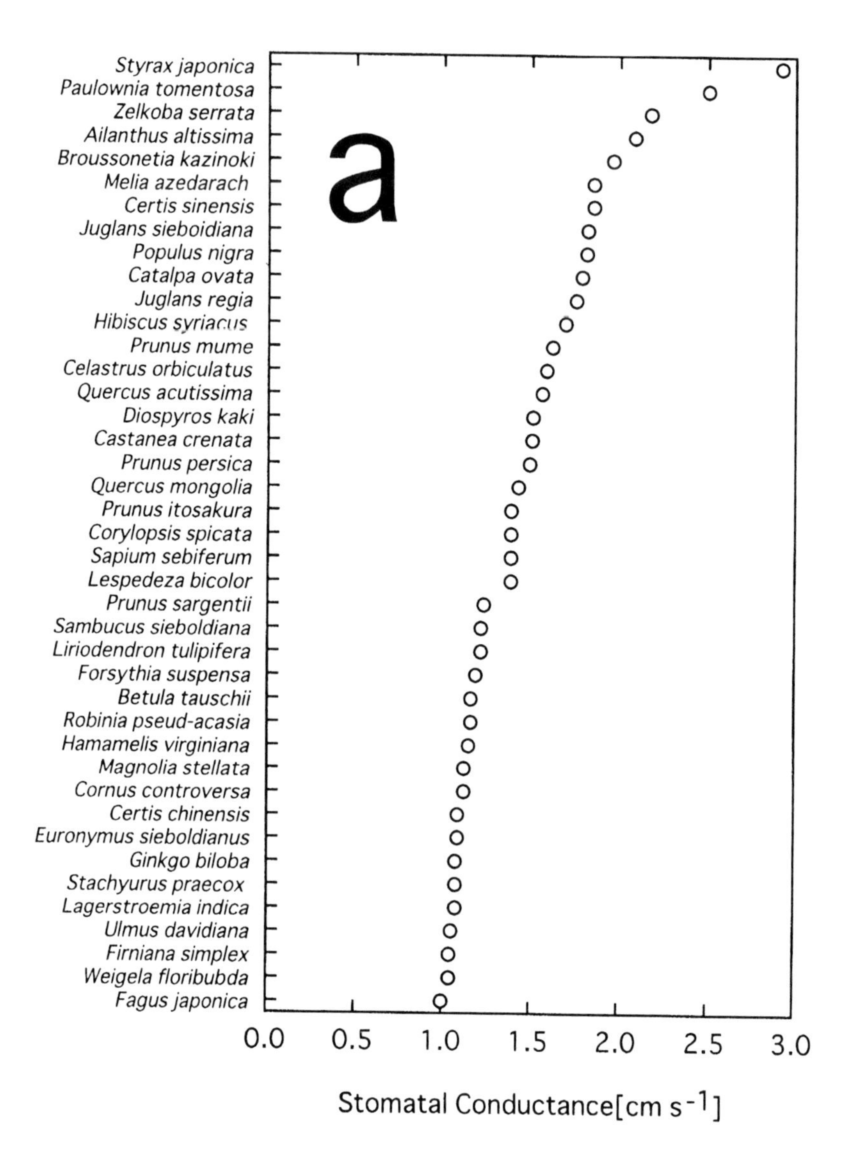

Fig. 5a,b,c. Stomatal conductance of different tree species. **a,** deciduous tree species with larger stomatal conductance; **b**, deciduous tree species with smaller stomatal conductance; **c**, evergreen tree species. Note differences in horizontal axis scales in **a**, **b** and **c**. Measurements were made in parks and experimental farm in Tsukuba, Japan during the daytime in summer. Each point represents the mean of five replications. (Modified from unpublished data and Fujinuma et al. 1985)

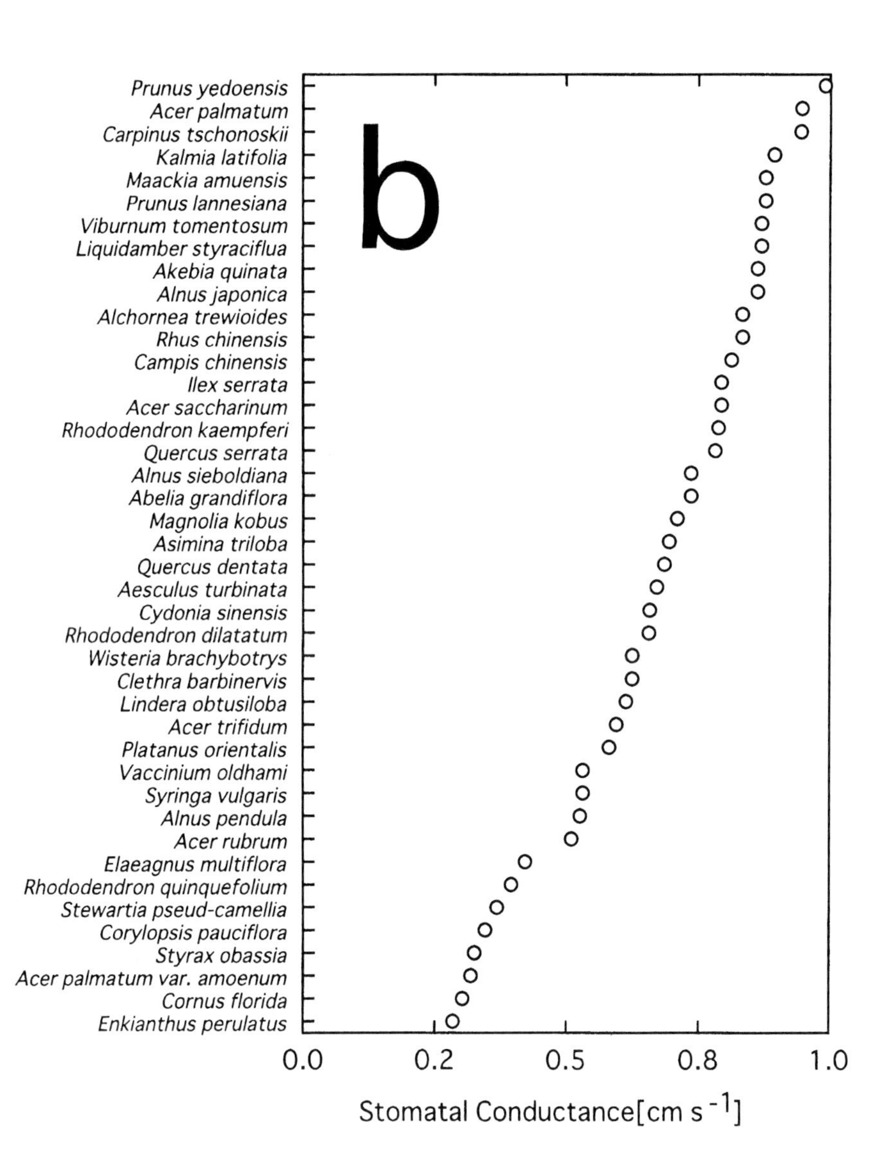

Fig. 5a,b,c. Continued

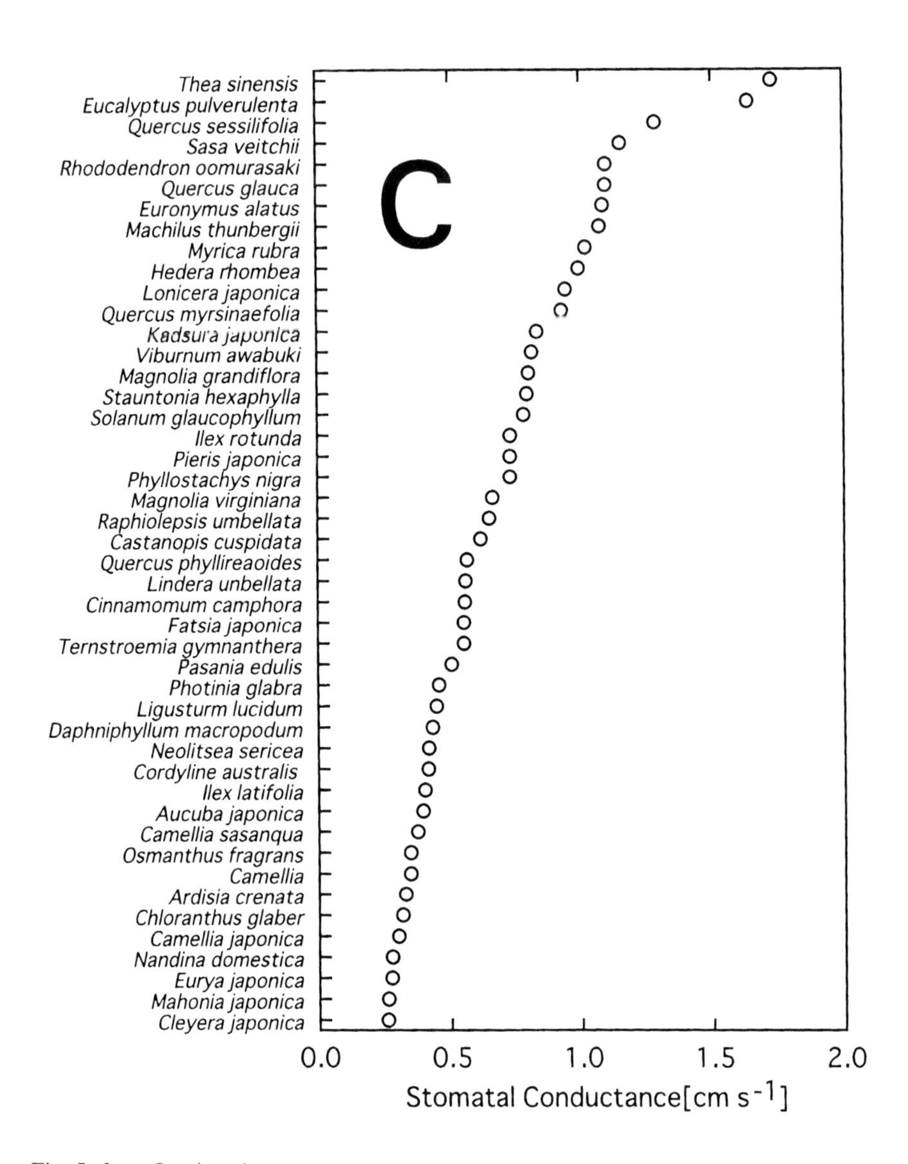

Fig. 5a,b,c. Continued

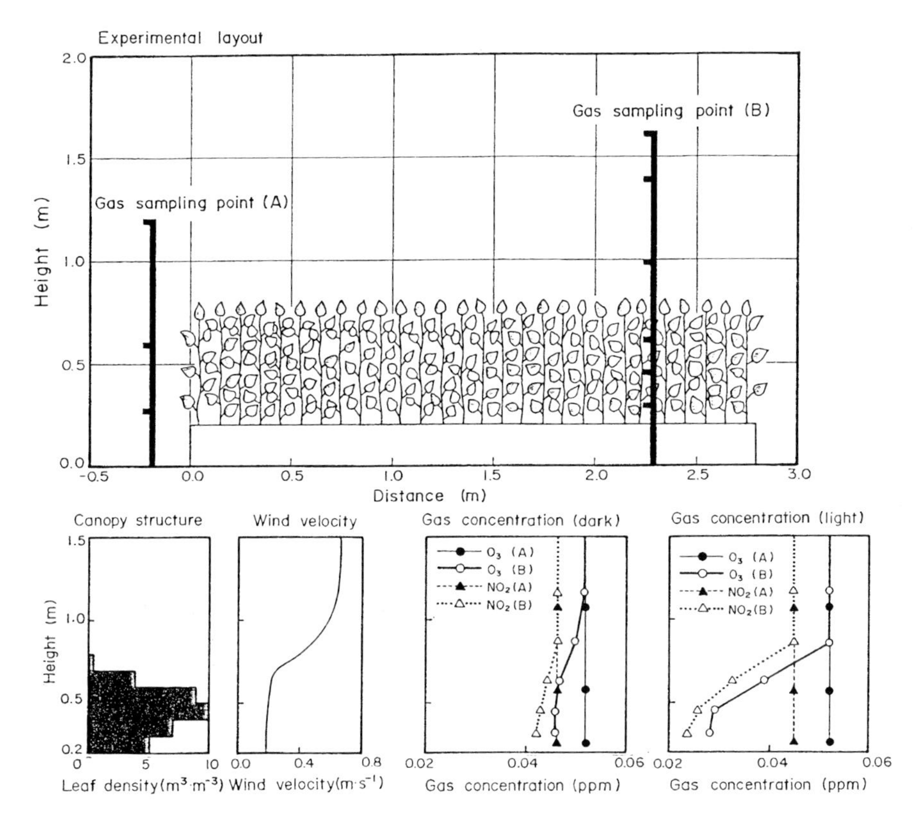

Fig. 6. Vertical profiles of gas concentrations in a wind tunnel at upstream and downstream sides of the air flux. Approximately 200 poplar saplings were exposed to a mixture of NO_2 and O_3 in either light (40 klx) or dark conditions. Air temperature and humidity were maintained at 25°C and 60% RH, respectively. (from Aiga et al. 1984)

are perhaps the most effective for purification of pollutant gases.

To qualitatively examine the capacity of a plant community to absorb pollutant gases, about 200 poplar (*Populus nigra*) seedlings were exposed to a mixture of O_3 and NO_2 in an environment-controlled wind tunnel (Aiga et al. 1984), in which air flow proceeded from sampling point A to point B (Fig. 6). *Populus nigra* has great potential for trees valid for the purification of pollutant gases because it has a large number of leaves with a high level of stomatal conductance, as well as a rapid growth rate. The results from this study demonstrated that, under light conditions (light intensity, 40 klx), concentrations of both gases at sampling point B were considerably lower than those at point A at heights where leaf density was high. However, in the dark, differences in the gas concentrations between points A and B were considerably smaller than in the light, reflecting the fact that the stomata of most plant species open in response to light and close in the dark. Indeed, environmental conditions are known to effect changes in the stomatal aperture of leaves (Zeiger et al. 1987; Jones 1992), and decrease the capacity of the leaves to absorb gases. Light is the most important factor affecting the diurnal changes in stomatal opening, although soil water deficit and reduced air humidity also result in decreased stomatal apertures (Matyssek et al. 1997). Therefore, to be effective scavengers of atmospheric pollutants, plants should not suffer from exposure to other environmental stresses.

Exposure to SO_2 or O_3 has been reported to cause changes in stomatal opening of plant leaves (for example, Winner et al. 1988; Michael et al. 1998; Omasa et al. 2000a). However, there is no general agreement on the effect of these gases on stomatal responses: some researchers observed stomatal opening was brought about by exposure to SO_2 (Mansfield and Majernik 1970; Black and

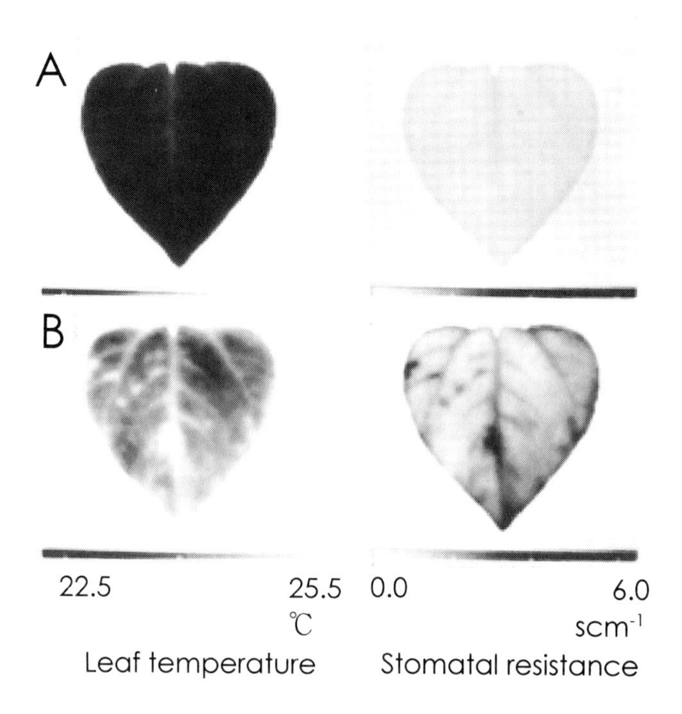

Fig. 7. Changes in spatial distributions of stomatal resistance evaluated from leaf temperature images of a sunflower plant during exposure to 1.2 ppmv O_3. A, just prior exposure; B, 30 min after exposure. Air temperature and humidity were maintained at 25.0ºC and 62%RH, respectively. The distribution of shortwave and longwave radiation and boundary layer resistance on the leaf surface were maintained constant. (from Omasa et al. 1981b)

Unsworth 1980) or O_3 (Evans and Ting 1974), while others reported that stomata were closed by exposure to SO_2 (Omasa and Abo 1978) or O_3 (Omasa et al. 1979, 2000a; Temple 1986). Meanwhile, Omasa et al. (1981a,b, 1992) have found spatial differences in stomatal responses to these pollutants at sites all over the leaf, based on the stomatal resistance images evaluated from leaf temperature images (see Fig. 7). Furthermore, by directly observing the responses of stomata

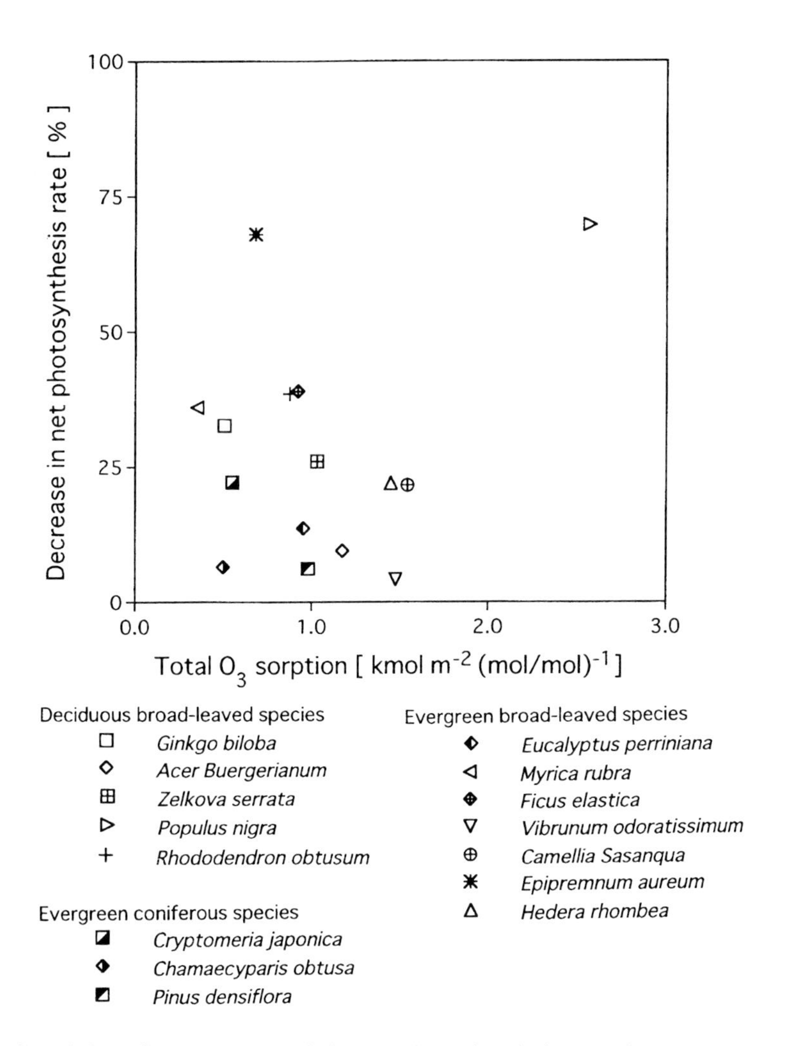

Fig. 8. Relationship between cumulative gas absorption during 4 h (from 1 h to 5 h after gas exposure) (horizontal axis) and the decrease in CO_2 absorption rate (net photosynthesis rate) during the same in 15 different woody species exposed to 0.5-ppmv O_3. Data are indicated on a unit leaf area basis. (from Omasa et al. 2000a)

on an attached leaf using a remote-control light microscope, Omasa et al. (1985, 1990) showed that stomata closed when the leaf was exposed to concentrations of 0.1 ppmv SO_2, NO_2, or O_3, but that stomatal opening varied with changes in the pressure balance between guard cells and epidermal cells caused by the water-soaking of epidermal cells. Organic pollutants are also found to change stomatal apertures when present at high concentrations (Omasa et al. 2000b).

Many pollutant gases entering plant leaves have deleterious effects on cell organelles and metabolic processes when their atmospheric concentrations are higher than threshold levels (see other chapters in this volume). Consequently, they can cause lesions on the leaf surface and a decrease in plant productivity, thus decreasing the capacity of the plants to act as sinks for these pollutants. The absorption rates of O_3 (Turner et al. 1972; Thorne and Hanson 1976), SO_2 (Jensen and Kozlowski 1975; Bressan et al. 1978; Caput et al. 1978; Furukawa et al. 1980), and NO_2 (Okano et al. 1988, 1989) have, in many species and cultivars, been reported to correlate with their susceptibility to the pollutant.

In addition, there are reports of differences in pollutant sensitivities between various species (Natori and Totsuka 1984; Okano et al. 1990; NIES 1992; Omasa et al. 2000a) and varieties (Furukawa et al. 1984; NIES 1992). For example, when 15 woody species were exposed to O_3, both the total O_3 absorbed and the decrease in CO_2 absorption rate (net photosynthetic rate) varied widely between species, but there was no correlation between total O_3 absorption and the decrease in

Fig. 9. Relationship between total absorption of O_3 during a 60-min exposure period (A) and visible injury one day later (B). Visible injury did not appear during the exposure. Environmental conditions were as described in Fig.7. (from Omasa et al. 1981b)

photosynthesis (Fig. 8) (Omasa et al. 2000a). By analyzing the leaf temperature images (see Fig. 9), Omasa et al. (1981b) have also shown that differences in the degree of visible injury and stomatal responses to O_3 are not dependent on the total O_3 absorption at a specific site. Furthermore, Okano et al. (1990) suggested that interspecific differences in the competence of metabolic processes to decompose PAN determine their resistance to this pollutant. Therefore, whether plants exposed to pollutants can maintain their absorption rate for a longer period depends on their resistance to the air pollutants. Species that show both higher stomatal conductance and lower sensitivities to atmospheric pollutants and other environmental stresses will be the most competent to ameliorate the atmospheric pollutants.

5. Conclusion

The foliar absorption of pollutant gases that are effectively metabolized inside the leaves (for example, SO_2, NO_2, O_3, and formaldehyde) is dependent on their rate of metabolism, which continually maintains a concentration gradient between the atmosphere and the interior of the leaves. Foliar absorption of this type of pollutant is governed mainly by stomatal conductance. On the other hand, many pollutants not only cause changes in stomatal opening but also have deleterious effects on the physiological functions of plants, thus reducing their capacity to effectively scavenge the pollutants. Therefore, to utilize plants for the remediation of atmospheric pollutants, the most suitable plant species or cultivars will be those with high stomatal conductance and lower sensitivities to the pollutants. In addition, the susceptibility of the plant species to other environmental stresses (especially water stress), and their ease of management, as well as the species type (for example, deciduous versus evergreen, with variation in leaf size, leaf density, and plant height), should be important parts of consideration depending on the environment in which they are to be introduced.

References

Aiga I, Omasa K, Matsumoto S (1984) Phytotrons in the National Institute for Environmental Studies. Res Rep Natl Inst Environ Stud Jpn 66: 133-155

Bacci E, Calamarl D, Gaggi C, et al (1990a) Bioconcentration of organic chemical vapors in plant leaves: Experimental measurements and correlation. Environ Sci Technol 24: 885-889

Bacci E, Cerejeira MJ, Gaggi C, et al (1990b) Bioconcentration of organic chemical vapors in plant leaves: The azalea model. Chemosphere 21: 525-535

Bennett JH, Hill AC (1973) Absorption of gaseous air pollutants by a standardized canopy. J Air Pollut Control Assoc 23: 203-206

Bennett JH, Hill AC (1975) Interactions of air pollutants with canopies of vegetation. In:

Mudd JB, Kozlowski TT (eds) Responses of plants to air pollution. Academic Press, New York, pp 273-306
Bennett JH, Hill AC, Gates DM (1973) A model for gaseous pollutant sorption by leaves. J Air Pollut Control Assoc 23: 927-962
Black VJ, Unsworth MH (1979) Resistance analysis of sulfur dioxide fluxes to *Vacia faba*. Nature 282:68-69
Black VJ, Unsworth MH (1980) Stomatal responses to sulfur dioxide and vapour pressure deficit. J Exp Bot 31:667-677
Bressan RA, Wilson LG, Filner P (1978) Mechanisms of resistance to sulfur dioxide in the *Cucurbitaceae*. Plant Physiol 61: 761-767
Caput C, Belot Y, Auclair D, Decourt N (1978) Absorption of sulfur dioxide by pine needles leading to acute injury. Environ Pollut 16: 3-15
Coe H, Gallagher MW, Choularton TW, et al (1995) Canopy scale measurements of stomatal and cuticular O_3 uptake by sitka spruce. Atmos Environ 29: 1413-1423
Darrall NM (1989) The effects of air pollutants on physiological processes in plants. Plant Cell Environ 12:1-30
Deinum G, Baart AC, Bakker DJ, et al (1995) The influence of uptake by leaves on atmospheric deposition of vapor-phase organics. Atmos Environ 29: 997-1005
De Kok LJ, Stulen I (eds) (1998) Responses of plant metabolism to air pollution and global change. Backhuys, Leiden
Evans LS, Ting IP (1974) Ozone sensitivity of leaves: relationship to leaf water potential, gas transfer resistance, and anatomical characteristics. Am J Bot 61: 592-597
Farquhar GD, Raschke K (1978) On the resistance to transpiration of the sites of evaporation within the leaf. Plant Physiol 61: 1000-1005
Farquhar GD, Firth PM, Wetselaar R, Weir B (1980) On the gaseous exchange of ammonia between leaves and the environment: determination of the ammonia compensation point. Plant Physiol 66: 710-714
Fujinuma Y, Machida T, Okano K, Natori T, Totsuka T (1985) Screening of air-filtering plants - Interspecific difference in characteristics of leaf diffusive resistance among broad-leaved tree species-. Res Rep Natl Inst Environ Stud Jpn 82: 13-28 (in Japanese)
Furukawa A, Isoda O, Iwaki H, et al (1980) Interspecific difference in resistance to sulfur dioxide. Res Rep Natl Inst Environ Stud Jpn 11: 113-126
Furukawa A, Katase M, Ushijima T, et al (1984) Inhibition of photosynthesis of poplar species and sunflower by O_3. Res Rep Natl Inst Environ Stud Jpn 65: 77-87
Garsed SG (1984) Uptake and distribution of pollutants in the plant and residence time of active species. In: Koziol MJ, Whatley FR (eds) Gaseous air pollutants and plant metabolism. Butterworths, London, pp 83-103
Geßler A, Rennenberg H (1998) Atmospheric ammonia: mechanisms of uptake and impacts on N metabolism of plants. In: De Kok LJ, Stulen I (eds) Responses of plant metabolism to air pollution and global change. Backhuys, Leiden, pp 81-94
Giese M, Bauerdoranth U, Langebartels C, et al (1994) Detoxification of formaldehyde by the spider plant (*Chlorophytum comosum* L.) and by soybean (*Glycine max* L.) cell-suspension cultures. Plant Physiol 104: 1301-1309
Girard F, Jolivet P, Belot Y (1989) Fixation and incorporation of atmospheric C-14-formaldehyde by sunflower leaves. C R Acad Sci 309: 447-452
Grant L, Richter A (1995) Dry deposition to pine of sulphur dioxide and ozone at low concentration. Atmos Environ 29: 1677-1683
Heath RL (1980) Initial events in injury to plants by air pollutants. Ann Rev Plant Physiol

31: 395-431
Hill AC (1971) Vegetation: A sink for atmospheric pollutants. J Air Pollut Control Assoc 21: 341-346
Hosker RP Jr, Lindberg SE (1982) Review: Atmospheric deposition and plant assimilation of gases and particles. Atmos Environ 16: 889-910
Jensen KF, Kozlowski TT (1975) Absorption and translocation of sulfur dioxide by seedlings of four forest tree species. J Environ Qual 4: 379-382
Jones HG (1992) Plants and microclimate, 2nd edn. Cambridge University Press, Cambridge
Keymeulen R, Schamp N, Van Langenhove H (1993) Factors affecting airborne monocyclic aromatic hydrocarbon uptake by plants. Atmos Environ 27A:175-180
Kondo T, Hasegawa K, Uchida R, et al (1995) Absorption of formaldehyde by oleander (*Nerium indicum*). Environ Sci Technol 29: 2901-2903
Kondo T, Hasegawa K, Uchida R, et al (1996a) Absorption of atmospheric formaldehyde by deciduous broad-leaved, evergreen broad-leaved, and coniferous tree species. Bull Chem Soc Jpn 69: 3673-3679
Kondo T, Hasegawa K, Kitagawa C, et al (1996b) Absorption of atmospheric phenol by evergreen broad-leaved tree species. Chem Lett 11: 997-998
Kondo T, Hasegawa K, Uchida R, et al (1998) Absorption of atmospheric C_2-C_5 aldehyde by various tree species and their tolerance to C_2-C_5 aldehyde. Sci Total Environ 224: 121-132
Kondo T, Hasegawa K, Kurokawa H, et al (1999) Absorption of atmospheric phenol by various tree species and their tolerance to phenol. Toxicol Environ Chem 69: 183-200
Koziol MJ, Whatley FR (eds) (1984) Gaseous air pollutants and plant metabolism. Butterworths, London
Krall AR, Tolbert NE (1957) A comparison of the light dependent metabolism of carbon monoxide by barley leaves with that of formaldehyde, formate and carbon dioxide. Plant Physiol 32: 321-326
Kramer PJ, Boyer JS (1995) Water relations of plants and soils. Academic Press, San Diego, pp 201-256
Laisk A, Kull O, Moldau H (1989) Ozone concentration in leaf intercellular air space is close to zero. Plant Physiol 90: 1163-1167
Lange OL, Heber ED, Schulze ED, et al (1989) Atmospheric pollutants and plant metabolism. In: Schulze ED, Lange OL, Oren R (eds) Forest decline and air pollution. Springer, Berlin, pp 238-273
Mansfield TA, Majernik O (1970) Can stomata play a role in protecting plants against air pollutants? Environ Pollut 1: 149-154
Mansfield TA, Pearson M (1996) Disturbances in stomatal behavior in plants exposed to air pollution. In: Yunus M, Iqubal M (eds) Plant response to air pollution. Wiley, Chichester, pp 179-193
Matyssek R, Havranek WM, Wieser G, Innes JL (1997) Ozone and the forests in Austria and Switzerland. In: Sandermann H, Wellburn AR, Heath RL (eds) Forest decline and ozone. Springer, Berlin, pp 95-134
Michael F, Heath J, Mansfield TA (1998) Disturbances in stomatal behaviour caused by air pollutants. J Exp Bot 49: 461-469
Monteith JL (1973) Principles of environmental physics. Arnold, London
Mutters RG, Mandre M, Bytnerowicz A (1993) Formaldehyde exposure affects growth and metabolism of common bean. Air Waste 43: 113-116

Natori T, Totsuka T (1984) Effects of mixed gas on transpiration rate of several woody plants. 1. Interspecific difference in the effects of mixed gas on transpiration rate. Res Rep Natl Inst Environ Stud Jpn 65: 45-53

Natori T, Omasa K, Abo F, et al (1981) Effects of fumigation periods and light condition during NO_2 fumigation on plant's factors controlling NO_2 sorption rate. Res Rep Natl Inst Environ Stud Jpn 28: 123-132 (in Japanese with English summary)

Neubert A, Kley D, Wildt J, et al (1993) Uptake of NO, NO_2 and O_3 by sunflower (*Helianthus annuus* L.) and tobacco plants (*Nicotina tabacum* L.): dependence on stomatal conductivity. Atmos Environ 27A: 2137-2145

NIES (National Institute for Environmental Studies) (1980) Studies on the effects of air pollutants on plants and mechanisms of phytotoxicity. Res Rep Natl Inst Environ Stud Jpn 265p

NIES (1984) Studies on effects of air pollutant mixtures on plants. Part 1 & 2. Res Rep Natl Inst Environ Stud Jpn 163p & 155p

NIES (1992) Studies on development of indicator plants for evaluation of atmospheric environment by biotechnology. Special Res Rep Natl Inst Environ Stud Jpn pp 13-18 (in Japanese)

Nouchi I (1980) A study on the foliar absorption rate of ozone and PAN. Ann Rep Tokyo Metro Res Inst Environ Prot pp 77-83 (in Japanese)

Ockenden WA, Steinnes E, Parker C, et al (1998) Observations on persistent organic pollutants in plants: Implications for their use as passive air samplers and for POP cycling. Environ Sci Technol 32: 2721-2726

O'Dell RA, Taheri M, Kabel RL (1977) A model for uptake of pollutants by vegetation. J Air Pollut Control Assoc 27: 1104-1109

Okano K, Machida T, Totsuka T (1988) Absorption of atmospheric NO_2 by several herbaceous species: estimation by the ^{15}N dilution method. New Phytol 109: 203-210

Okano K, Machida T, Totsuka T (1989) Differences in ability of NO_2 absorption in various broad-leaved tree species. Environ Pollut 58: 1-17

Okano K, Tobe K, Furukawa A (1990) Foliar uptake of peroxyacetyl nitrate (PAN) by herbaceous species varying in susceptibility to this pollutant. New Phytol 114:139-145

Omasa K (1979) Sorption of air pollutants by plant communities – Analysis and modelling of phenomena –. Res Rep Natl Inst Environ Stud. 10: 367-385 (in Japanese)

Omasa K (1990) Image instrumentation methods of plant analysis. In: HF Linskens and JF Jackson (eds) Modern methods of plant analysis. New Series, vol. 11. Springer, pp 203-243

Omasa K, Abo F (1978) Studies of air pollutant sorption by plants. (I) Relation between local SO_2 sorption and acute visible leaf injury. J Agr Met 34: 51-58 (in Japanese with English summary) [also Res Rep Natl Inst Environ Stud 11: 181-193 (1980) (English translation)]

Omasa K, Croxdale JG (1992) Image analysis of stomatal movements and gas exchange. In: Häder D-P (ed) Image analysis in biology. CRC Press, Boca Raton, pp 171-193

Omasa K, Abo F, Natori T, et al (1979) Studies of air pollutant sorption by plants. (II) Sorption under fumigation with NO_2, O_3 or $NO_2 + O_3$. J Agr Met 35: 77-83 (in Japanese with English summary) [also Res Rep Natl Inst Environ Stud 11: 213-224 (1980) (English translation)]

Omasa K, Hashimoto Y, Aiga I (1981a) A quantitative analysis of the relationships between SO_2 or NO_2 sorption and its acute effects on plant leaves using image instrumentation. Environ Control Biol 19: 59-67

Omasa K, Hashimoto Y, Aiga I (1981b) A quantitative analysis of the relationships between O_3 sorption and its acute effects on plant leaves using image instrumentation. Environ Control Biol 19: 85-92

Omasa K, Hashimoto Y, Kramer PJ, et al (1985) Direct observation of reversible and irreversible stomatal responses of attached sunfluxer leaves to SO_2. Plant Physiol 79: 153-158

Omasa K, Aiga I, Kondo J (1990) Remote-control light microscope system. In: Hashimoto Y, Kramer PJ, Nonami H, et al (eds) Measurement techniques in plant science. Academic Press, San Diego, pp 387-401

Omasa K, Tobe K, Hosomi M, et al (2000a) Experimental studies on O_3 sorption mechanism of green area – Analysis of O_3 sorption rates of plants and soils. Environ Sci 13: 33-42 (in Japanese with English summary)

Omasa K, Tobe K, Hosomi M, et al (2000b) Absorption of ozone and seven organic pollutants by *Populus nigra* and *Camellia Sasanqua*. Environ Sci Technol 34:2498-2500

Parkhurst DF (1994) Diffusion of CO_2 and other gases inside leaves. New Phytol 126: 449-479

Peterson S, Mackay D, Bacci E, et al (1991) Correlation of the equilibrium and kinetics of leaf-air exchange of hydrophobic organic chemicals. Environ Sci Technol 25: 866-871

Polle A (1998) Photochemical oxidants: uptake and detoxification mechanisms. In: De Kok LJ, Stulen I (eds) Responses of plant metabolism to air pollution and global change. Backhuys, Leiden, pp 93-116

Richardson ML (ed) (1992-1994) The dictionary of substances and their effects, 7 vols. Royal Society of Chemistry, London

Riederer M (1990) Estimating partitioning and transport of organic chemicals in the foliage/atmosphere system: discussion of a fugacity-based model. Environ Sci Technol 24: 829-837

Rondon A, Johansson C, Grant L (1993) Dry deposition of nitrogen dioxide and ozone to coniferous forests. J Geophys Res 98: 5159-5172

Sandermann H, Wellburn AR, Heath RL (eds) (1997) Forest decline and ozone. Springer, Berlin

Schulte-Hostede S, Darrall NM, Blank LW, et al (eds) (1987) Air pollution and plant metabolism. Elsevier, London

Simonich SL, Hites RA (1994a) Importance of vegetation in removing polycyclic aromatic hydrocarbons from the atmosphere. Nature 370: 49-51

Simonich SL, Hites RA (1994b) Vegetation –atmospheric partitioning of polycyclic aromatic hydrocarbons. Environ Sci Technol 28: 939-943

Smith WH (1984) Pollutant uptake by plants. In: Treshow M (ed) Air pollution and plant life. Wiley, Chichester, pp 417-450

Temple PJ (1986) Stomatal conductance and transpirational responses of field-grown cotton to ozone. Plant Cell Environ 9: 315-321

Thorne L, Hanson GP (1976) Relationship between genetically controlled ozone sensitivity and gas exchange rate in *Petunia hybrida* Vilm. J Am Soc Hortic Sci 101: 60-63

Trapp S, Matthies M, Scheunert I, et al (1990) Modeling the bioconcentration of organic chemicals in plants. Environ Sci Technol 24: 1246-1252

Turner NC, Rich S, Tomlinson H (1972) Stomatal conductance, fleck injury, and growth of tobacco cultivars varying in ozone tolerance. Phytopathology 62: 63-67

Winner WE, Gillespie C, Shen W-S, et al (1988) Stomatal responses to SO_2 and O_3. In:

Schulte-Hostede S, Darrall NM, Blank LW, et al (eds) Air pollution and plant metabolism. Elsevier, London, pp 255-271

Yunus M, Iqbal M (eds) (1996) Plant response to air pollution. Wiley, Chichester

Zeiger E, Farquhar GD, Cowan IR (eds) (1987) Stomatal function. Stanford University Press, Stanford

9
Uptake, Metabolism, and Detoxification of Sulfur Dioxide

Noriaki Kondo

Department of Biological Sciences, Graduate School of Science, The University of Tokyo, Hongo 7-3-1, Bunkyo-ku, Tokyo 113-0033, Japan

1. Introduction

Sulfur dioxide (SO_2) is a major air pollutant that is artificially produced by fossil fuel combustion, mainly in the industrialized areas of both developed and developing countries, and also results from volcanic emission, biogenic emissions, etc. SO_2 can also form aerosols, sulfate particles, in the atmosphere by photochemical reactions. Aerosols can then be incorporated into clouds and/or transported over long distances, thus causing severe acid precipitation as wet and dry depositions in surrounding countries.

Both acidic dry depositions such as SO_2 and sulfate particles and acidic wet depositions cause severe damage to vegetation, aquatic ecosystems, and concrete structures as well as human health. High concentrations of deposited SO_2 not only can result in visible foliar damage (Shimazaki et al. 1980) but can also decrease the plant photosynthetic rate (Kropff 1987; Hogetsu and Shishikura 1994). In some plant species, SO_2 causes rapid stomatal closure, thereby decreasing the photosynthetic rate, but it may also rapidly limit photosynthetic activity even without stomatal closure (Furukawa et al. 1980b; Price and Long 1989; Barton et al. 1980). In contrast, plant responses to ozone (O_3), the major photochemical oxidant, are relatively slow and gradual, tending to reduce photosynthetic rates, following stomatal closure, after 30 min to an hour after the start of O_3 fumigation (Furukawa et al. 1984). Chronic doses of SO_2 that cause no visible damage can

Air Pollution and Plant Biotechnology
–Prospects for Phytomonitoring and Phytoremediation–
Edited by K. Omasa, H. Saji, S. Youssefian, and N. Kondo
© Springer -Verlag Tokyo 2002

also inhibit the growth of plants (Ashenden 1978; Mejstřík 1980) and accelerate senescence of leaves (Martin and Thimann 1972a,b), although those sometimes promote plant growth under low sulfur fertility (Cowling and Lockyer 1978; Clarke and Murray 1990). SO_2 also promotes the leakage of solutes from leaves (Navari-Izzo et al. 1989), and the dehydration of leaves (Neighbour et al. 1988; Lucas 1990). Therefore, the effects of drought stress, which decrease protein content and photosynthetic rate, are amplified by SO_2 (Pierre and Queiroz 1988; Cornic 1987).

The acute injury, the visible foliar damage caused by high concentrations of SO_2, depends on the absorption rate through the stomata (Thomas and Hill 1935; Furukawa et al. 1980a). Therefore, stomatal density and aperture size are important factors determining the extent of SO_2-induced injury. In addition, stomata also respond to environmental conditions, such as light intensity, carbon dioxide (CO_2) concentration, and exposure to SO_2. Thus, when fumigated with SO_2, the response of some plant species is to close their stomata (Robinson et al. 1998) and thereby suppress the entrance of the pollutant into the leaves, a mechanism by which the level of foliar damage is reduced (Kondo and Sugahara 1978; Furukawa et al. 1980a).

SO_2, absorbed by leaves predominantly through stoma, dissolves in the aqueous phase of the cell surface (apoplast) or in the cytoplasm to produce bisulfite and sulfite ions. These sulfites are highly phytotoxic (Thomas et al. 1943), and can also give rise to secondary toxicants, such as bisulfite addition product, α-hydroxysulfonate (Zelitch 1957), which inhibits photosynthesis. In addition, sulfite enhances the generation of reactive oxygen species (ROS) (Asada and Kiso 1973) and ethylene (Bae et al. 1995), which also act as secondary phytotoxicants, inducing injuries that include chlorophyll bleaching and destruction of cell membranes (Shimazaki et al. 1980; Aono et al. 1993; Bae et al. 1995).

More than 20 years ago, it was pointed out that the resistance of plants to SO_2 might be controlled primarily by biochemical factors (Bressan et al. 1978). Indeed, it is now clear that plants possess various metabolic systems that limit damage by bisulfite and sulfite, the most important being their oxidation or reduction to less damaging forms. In addition, plants possess an intricate system of antioxidants and enzymes that scavenge and detoxify the highly reactive ROS. The phytotoxicity of SO_2 and sulfite, and the mechanisms employed by plants to detoxify them are outlined next.

2. Absorption of Sulfur Dioxide

2.1 Stomatal Responses to SO_2

It is assumed that plants absorb SO_2 predominantly by the stomata (Laisk et al.

1988), whereas the absorption through the cuticle is distinctively small, because it has been reported that gas exchange of water vapor between mature leaves and ambient air may occur solely through stomata (Omasa et al. 1983). SO_2 absorption by plants from ambient air follows the physical diffusion law. Hence, the absorption rate is determined by stomatal conductance (resistance), dependent stomatal density and aperture size, and the difference of SO_2 concentrations between leaf surface and the surface of cells surrounding stomatal cavity. As SO_2 absorbed through stomata is dissolved in aqueous layer of cell surface, to instantly form sulfite and bisulfite under normal conditions as mentioned later, SO_2 concentration in the aqueous layer is always actually near zero (Omasa et al. 1980). Therefore, SO_2 absorption rate is predominantly determined by stomatal conductance and SO_2 concentration in the surrounding air. In general, plant species tolerant to acute SO_2 injury show higher stomatal resistance to SO_2 than susceptible species and thereby possess a lower rate of SO_2 absorption (Kimmerer and Kozlowski 1981; Ayazloo et al. 1982). It has been reported that stomata of some plant species respond rapidly to SO_2 fumigation (Furukawa et al. 1979; Omasa et al. 1985). However, stomatal responses actually depend on environmental conditions (Jensen and Roberts 1986) and the plant species. For example, in moist air, the stomata of *Vicia faba* normally open wider in the presence of SO_2 at concentrations of 0.25 ppm and above (Majernik and Mansfield 1970), whereas in relatively low humidity, the stomata close in response to the same SO_2 range (Mansfield and Majernik 1970). Furthermore, Menser and Heggestad (1966) have reported that *Vicia faba* stomata open when fumigated with SO_2 while tobacco stomata close under the same conditions.

Kondo and Sugahara (1978) suggested that such variations in stomatal responses to SO_2 were dependent on the abscisic acid (ABA) content of leaves. Hence, the stomata of plants with high ABA concentrations closed immediately after the start of SO_2 fumigation, with the extent of stomatal closure being dependent on the ABA content. This type of stomatal response was reversible, so that stomatal aperture was slowly restored after terminating SO_2 fumigation (Kondo and Sugahara 1984). The ABA content in tomato, peanut, tobacco (*Nicotiana tabacum*), rice, and adlay were found to be high, while contents in perilla, spinach, radish, tobacco (*N. glutinosa*), *Vicia faba*, maize, and sorghum were low (Kondo and Sugahara 1984; Kondo et al. 1980b). SO_2 fumigation had little effect on ABA contents of leaves (Kondo and Sugahara 1984). ABA is known to induce stomatal closure and stored mainly in plastids of leaves and roots. When leaves are fumigated with SO_2, ABA is assumed to be transported to guard cells by the acidic effect of SO_2 (Kondo et al. 1980b) or by pH lowering of stroma (Cowan et al. 1982), resulting from inhibited photosynthetic electron flow (Shimazaki and Sugahara 1979a,b).

However, SO_2-induced stomatal opening cannot be explained by ABA content. Black and Black (1979) found that low concentrations of SO_2 cause severe damage to *Vicia faba* epidermal cells adjacent to guard cells (a pair of guard cells constitutes a stoma) and induce the stomatal opening in *Vicia*. It is assumed that the stomata opened as a result of losing turgor pressure of the adjacent epidermal cells.

The ABA-dependent, SO_2-induced stomatal closure is independent of stomatal closure in response to high concentrations of CO_2. Hence, these effects on stomata are additive (Kondo 1987). In the case of maize, the stomata close immediately after the start of SO_2 fumigation, irrespective of its low ABA content (Kondo et al. 1980b), with stomatal aperture size being rapidly restored after termination of fumigation. Stomata of maize are especially sensitive to CO_2 concentrations, closing immediately after transfer to high CO_2 concentrations. However, this stomatal closure is not enhanced by additional SO_2 fumigation (Kondo 1987). SO_2 is known to rapidly inhibit photosynthetic CO_2 absorption (Furukawa et al. 1980b), which raises CO_2 concentrations in the leaves. Thus, the SO_2-induced closure of maize stomata is assumed to result from the rise in CO_2 concentrations in the leaves.

Generally, plants that have stomata which rapidly and largely close in response to SO_2 are tolerant to acute injury (Kondo and Sugahara 1978; Furukawa et al. 1980a; Ayazloo et al. 1982). In many cases, when SO_2 and nitrogen dioxide (NO_2) are given in combination, plants are severely injured (Ashenden and Mansfield 1978), although NO_2 alone has little effects on plants. Amundson and Weinstein (1981), however, reported that exposure to a combination of SO_2 and NO_2 produces less injury to soybean leaves than the same single dose of SO_2. In this case, the decrease in stomatal aperture size was dramatic in plants exposed to the SO_2 and NO_2 combination, but only marginal in plants exposed to SO_2 and NO_2 alone. However, the mechanism of the synergistic effects of SO_2 and NO_2 on stomatal closure has yet to be clarified.

2.2 Sulfite Incorporation into Cells

SO_2 absorbed by plant leaves dissolves in the aqueous phase of the cell surface (apoplast) to form an aqueous SO_2 solution, from which it then yield bisulfite (HSO_3^-) and sulfite (SO_3^{2-}) ions, and protons (H^+). Generation of H^+ lowers the pH of the solution. In the aqueous solution, SO_2 establishes the following equilibria at 24°C (Malhotra and Hocking 1976):

$$SO_2 + H_2O \rightleftarrows [SO_2 \cdot H_2O] \quad (1)$$

$$[SO_2 \cdot H_2O] \rightleftarrows HSO_3^- + H^+ , \ pK_1 = 1.76 \quad (2)$$

$$HSO_3^- \rightleftarrows SO_3^{2-} + H^+ , \ pK_2 = 7.20 \quad (3)$$

Near neutral pH, HSO_3^- and SO_3^{2-} exist at approximately equal amounts according to Eq. 3, whereas aqueous SO_2 ($SO_2 \cdot H_2O$) increases at lower pH.

Mesophyll cells take up more sulfite when the surrounding pH is lower. According to formulas 1, 2, and 3, sulfite is thought to pass across cell membranes mainly in the form of SO_2 ($SO_2 \cdot H_2O$) (Sakaki and Kondo 1984). However, the incorporation of sulfite into mesophyll cells was suggested to depend on carrier-mediated anion transport at physiological pH (Pfanz et al. 1987a). At least a part of sulfite is oxidized to sulfate in the apoplast (Takahama et al. 1992) and then incorporated into mesophyll cells. Sulfite absorbed by mesophyll cells is taken up by chloroplasts and this, in part, depends on a phosphate translocator of the chloroplast inner envelope membranes (Hampp and Ziegler 1977).

3. Phytotoxicity of Sulfur Dioxide

3.1 Acidic Effects

The ABA-dependent, SO_2-induced stomatal closure has been suggested to be caused by the lowering of apoplastic pH as a result of SO_2 absorption (Kondo et al. 1980b). Indeed, the pH of the cytoplasm is known to decline following SO_2 exposure (Pfanz et al. 1987b). Since acidic substances tend to accumulate in the chloroplastic stroma, which has a high pH, SO_2 and bisulfite easily build up in chloroplasts. According to Eqs. 2 and 3, when SO_2 and/or bisulfite are absorbed by chloroplasts, H^+ is generated and the pH of the stroma is lowered. This pH lowering results in the inhibition of photosynthetic carbon fixation because some enzymes of the Calvin cycle are active only under alkaline conditions. Photosynthesis inhibition by low SO_2 concentrations is thought to be particularly dependent on this pH decline (Pfanz et al. 1987b).

3.2 Sulfite Effects

Photosynthesis of intact chloroplasts was found to be inhibited by in vitro sulfite treatment of the chloroplasts (Silvius et al. 1975). Libera et al. (1975) reported that CO_2 fixation in isolated spinach chloroplasts is inhibited by sulfite in a fully competitive manner with respect to inorganic carbon added to the medium. Furthermore, from in vitro experiments, it was reported that sulfite inhibits the activities of some enzymes involved in carbon fixation, by competiting for the carbon dioxide or bicarbonate binding sites. Such enzymes include ribulose-1,5-bisphosphate carboxylase (RuBPC) of spinach chloroplasts (Ziegler 1972), malate dehydrogenase in extracts of *Zea mays* (Ziegler 1974), and phosphoenolpyruvate carboxylase (PEPC) in extracts of *Zea mays* and spinach (Ziegler 1973; Mukerji and Yang 1974). Furthermore, Gezeilus and Hällgren (1980) reported that sulfite inhibited RuBPC in extracts of pine leaves by binding carbonate in a noncompetitive fashion. This result suggests that these enzymes possess some binding sites to which sulfite binds irreversibly, as well as a carbon dioxide or bicarbonate binding site.

Sakaki and Kondo (1985), using *Vicia* mesophyll cell protoplasts, subsequently demonstrated a parallel relationship between the rate of sulfite absorption and the rate of inhibition of photosynthesis. In this case, carbon fixation was inhibited in a noncompetitive manner by exogenously added 1 mM sulfite and in a mixed manner by 10 mM sulfite (Sakaki and Kondo 1984). Sulfite could suppress the absorption of carbonate by such protoplasts, which was at least partly attributed to the inhibition of carbonic anhydrase activity. In addition, at the high sulfite concentrations, hydrogen peroxide, secondarily produced, might participate in the sulfite-induced inhibition of carbon fixation (Tanaka et al. 1982b).

Another serious mode of attack of sulfite on plants is through its ability to cleave disulfide linkages of enzymes and structural proteins (Bailey and Cole 1959).

$$RSSR + SO_3^{2-} \rightleftarrows RSSO_3^- + RS^- \quad (4)$$

Because the tertiary structure of many enzymes is dependent on the integrity of their disulfide bonds, disruption of this unit can deactivate the enzymes. Sugahara et al. (1980) found that the structure of water-soluble chlorophyll proteins from *Chenopodium album* was altered by in vitro treatment with sulfite, possibly by disruption of the disulfide bonds. For the structural changes of this protein, preillumination of the samples was required, so that precedent conformational changes of the protein, which is driven by light energy absorbed by chlorophylls covalently linked to the protein, were an apparent prerequisite to sulfite attack. Shimazaki et al. (1984a) found that sulfite could inhibit the photochemical system (PS) II reaction center in spinach leaves. The sulfite-induced inhibition of PS II occurs under light illumination, even in dim light, but not under total darkness. However, it is not yet clear whether the cleavage of disulfide linkages by sulfite is involved in the inhibition of photosynthesis inhibition mentioned earlier.

3.3 Production of Secondary Toxicants

Another reaction of aqueous SO_2 with biological substances is the formation of bisulfite addition products. The bisulfite ion reacts reversibly with most aldehydes, methylketones, and unhindered cyclic ketones:

$$R_1(R_2)CO + HSO_3^- \rightleftarrows R_1(R_2)C(OH)SO_3^- \quad (5)$$

For example, the formation of glycolate bisulfite in plant leaves exposed to SO_2 has been confirmed (Tanaka et al. 1972). Bisulfite compounds inhibit enzymes, such as PEPC (Osmond and Avadhani 1970) and glycolate oxidase (Zelitch 1957), so that photosynthesis is inhibited (Lüttge et al. 1972).

A part of SO_2-derived sulfite in plant leaves is reduced to hydrogen sulfide (H_2S) and released into the atmosphere (Wilson et al. 1978; Hällgren and Frederiksson 1982). Although H_2S emission is generally regarded as a process for detoxification of SO_2, H_2S is, in itself, toxic (see the chapter by De Kok et al., this volume), inhibiting the NADH-oxidizing enzymes (Maas and De Kok 1988) and the activity of the photosynthetic electron transport system (Maas et al. 1988). Short-term exposure of plants to SO_2 is found to reversibly inhibit the photosynthetic water- splitting enzyme system (Shimazaki et al. 1984b), which is also inhibited by H_2S (Oren et al. 1979).

In addition, plants generate small amounts of ROS and ethylene under normal conditions; their production is enhanced by sulfite, a feature that is most critical for the phytotoxic effects of SO_2 (Okpodu et al. 1996). Sulfite is oxidized to sulfate by the chain reaction with superoxide radicals (O_2^-), which are formed in illuminated chloroplasts by the reducing power of PS I. During this reaction process, a large amount of O_2^- and sulfite radicals is generated (Asada and Kiso 1973), which give rise to the production of various other highly reactive species, including hydrogen peroxide (H_2O_2), hydroxyl radicals ($OH\cdot$), and singlet oxygen ($^1O_2$) (see the chapter by Morita and Tanaka, this volume). H_2O_2 inhibits SH enzymes of the Calvin cycle (Kaiser 1979; Tanaka et al. 1982b). Shimazaki et al. (1980) showed that the SO_2-induced breakdown of chlorophyll and membrane lipids could be dramatically suppressed by treatment with antioxidants, such as 1,2-dihydroxybenzene-3,5-disulfonate (tiron) or hydroquinone, suggesting that O_2^- and $^1O_2$ were responsible for the peroxidative destruction of cell components such as chlorophyll and membrane lipids. Peiser et al. (1982) also indicated that free radicals produced during sulfite oxidation were responsible for peroxidation of membrane lipids. ROS are also known to induce the damage to DNA by DNA cleavage or by the formation of 8-hydroxy-deoxyguanosine, an oxidized form of deoxyguanosine, although SO_2 has not yet been shown to directly induce these effects in plants. However, the current general consensus is that acute injury caused by SO_2 is mediated mainly through the SO_2-induced production and action of ROS (see the chapter by Morita and Tanaka, this volume).

The phytohormone ethylene is known to be transiently emitted from plants exposed to SO_2, and while this formation of ethylene is intensely related to SO_2-induced plant injury (Peiser and Yang 1979; Bressan et al. 1979), ethylene itself has no detrimental effects on plants. Ethylene is similarly emitted transiently following O_3 fumigation, and the suppression of ethylene formation by pretreatment with an ethylene biosynthesis inhibitor could depress O_3-induced visible injury (Mehlhorn and Wellburn 1987; Bae et al. 1996; see the chapter by Nakajima, this volume). The SO_2-induced injury could also be noticeably alleviated by inhibiting ethylene formation (Bae et al. 1995). In contrast, pretreatment with ethylene has been shown to make plants more tolerant to O_3 (Mehlhorn 1990). Therefore, ethylene possibly plays a dual role; as a promoter of damage when supplied together with SO_2 or O_3, but as a suppressor of damage when applied alone. Ethylene emission in response to these air pollutants may also mediate signals for the induction of plant defense responses. Indeed, plants previously exposed to SO_2 become more tolerant to subsequent SO_2 exposure (Tanaka and Sugahara 1980), although it is not clear whether ethylene is involved in acquisition of the tolerance. However, this acquired tolerance does not seem to be the case for O_3 as plants maintain their O_3 sensitivity even after O_3 fumigation. At present, there is no convincing hypothesis that can explain the mechanisms that are involved in ethylene-induced injury and tolerance in plants.

3.4 Effects on Photosynthate Partitioning

Exposure of plants to SO_2 often results in an increase in the shoot:root dry weight ratio (Murray 1985), probably because of the changes in photosynthate partitioning between organs. Such partitioning is determined by phloem loading, transport, and unloading. Indeed, both phloem loading and transport are distinctly inhibited by SO_2 in wheat seedlings, but not so clearly in maize seedlings (Minchin and Gould 1986; Gould et al. 1988). The incorporation of externally supplied sucrose into castor bean cotyledons was found to be promoted by low concentrations of sulfite (<1.0 mM), although it was inhibited by higher sulfite concentrations (Lorenc-Plucińska and Ziegler 1987). As respiration promoted by SO_2 exposure (Koziol and Jordan 1978; Black and Unsworth 1979), the promotion of sucrose uptake by sulfite may have been dependent on the promotion of respiration by sulfite. At least a part of the inhibition of phloem loading by SO_2 could conceivably have therefore resulted from the promotion of sucrose incorporation into the cotyledons.

On the other hand, Maurousset and Bonnenmain (1990) showed that sulfite treatment decreased sucrose uptake by parenchyma and veins of broadbean leaves, which was suggested to at least partly result from depolarization of the transmembrane potential difference (PD). Maurousset et al. (1992) also suggested that the sulfite-induced depolarization of transmembrane PD was due to an indirect inhibition of the plasma membrane H^+-ATPase activity following the decrease of the available level of ATP. Thus, SO_2-induced inhibition of phloem loading seems to be caused by complex mechanisms.

4. Metabolism of Sulfur Dioxide

4.1 Acid Neutralization

As presented in formulas 1, 2, and 3, the dissolution of SO_2 in water generates H^+, which thus lowers the solution pH (Pfanz et al. 1987b). Plants possess specific mechanisms to neutralize such pH changes (Thomas and Runge 1992). Using a buffering function known as pH stat, plant cells can modulate their pH through the synthesis of malic acid, catalyzed by the activity of PEPC, which is reversibly controlled by the pH of the cytoplasm. Furthermore, SO_2 is known to enhance the production of the polyamine putrescine in pea plants supplied with ammonium as a nitrogen source (Priebe et al. 1978). Polyamines are considered to play a buffering role to neutralize SO_2-derived H^+. Indeed, chronic fumigation with SO_2 increases the content of malate and basic amino acids, such as arginine, lysine, and hisitidine, as well as levels of the polyamines putrescine and spermine, in kidney bean leaves (Pierre and Queiroz 1981), presumably to reinforce the buffering capacity of the cells. It has also been reported that SO_2 fumigation of pea plants enhances the release of H^+ from the roots, with the subsequent absorption of cations, such as K^+, from the soil (Kaiser et al. 1993). Although H^+ release from roots apparently neutralizes cellular pH in plants exposed to SO_2, these plants were unable to release H^+ into K^+-deficient soil.

4.2 Sulfite Metabolism

4.2.1 Sulfite Decrease

The inhibition of photosynthesis by sulfite can be restored by light illumination but not in the dark, probably because of the decrease in cellular sulfite content as a result of light-driven sulfur metabolism (Veljovic-Jovanovic et al. 1993). In SO_2-fumigated plants, the residence time of sulfite in leaves is reported to differ between SO_2-tolerant and -sensitive cultivars of soybean (Miller and Xerikos 1979) and pea (Alscher et al. 1987). Thus, rapid metabolism of sulfite is considered to constitute the main form of resistance to SO_2 toxicity. This resistance is achieved either through the oxidation of sulfite to sulfate or through its reduction to sulfide. Most of generated sulfate is transported into vacuoles and stored there (Cram 1983; Kaiser et al. 1989), while the sulfide can be used to synthesize cysteine (see Fig. 1) or be partly emitted into the atmosphere as H_2S (Hällgren and Frederiksson 1982).

Plants also possess another metabolic pathway where sulfite can be emitted as SO_2 (Veeranjaneyulu et al. 1994). Sulfite, at acidic pH, equilibrates with bisulfite and SO_2, as shown by Eqs. 1, 2, and 3. Such reactions can proceed in favor of SO_2 generation in acidic medium if the accumulated sulfite is neither oxidized nor reduced. However, the mechanism by which SO_2 is produced and emitted has yet to be explained.

4.2.2 Reduction of Sulfite

Exposure of plants to SO_2 increases their thiol content (Miszalski and Ziegler 1979; Maas et al. 1987), especially the water-soluble thiols glutathione and cysteine (Grill et al. 1979), and also the sulfhydryls of proteins (Grill et al. 1980). Similarly, in wheat leaves exposed to chronic SO_2 doses, the accumulation of glutathione and cysteine was observed and accompanied by an increase in the reduced/oxidized glutathione (GSH/GSSG) ratio (Soldatini et al. 1992), which is probably involved in acquisition of tolerance to SO_2-induced oxidative stress. It has been suggested that glutathione synthesis is a plant response to environmental stresses, and that glutathione plays important roles in plant resistance to oxidative stress conditions (Tanaka et al. 1985; see the chapter by Youssefian, this volume).

A part of the sulfite formed in plant leaves exposed to SO_2 is reduced to sulfide

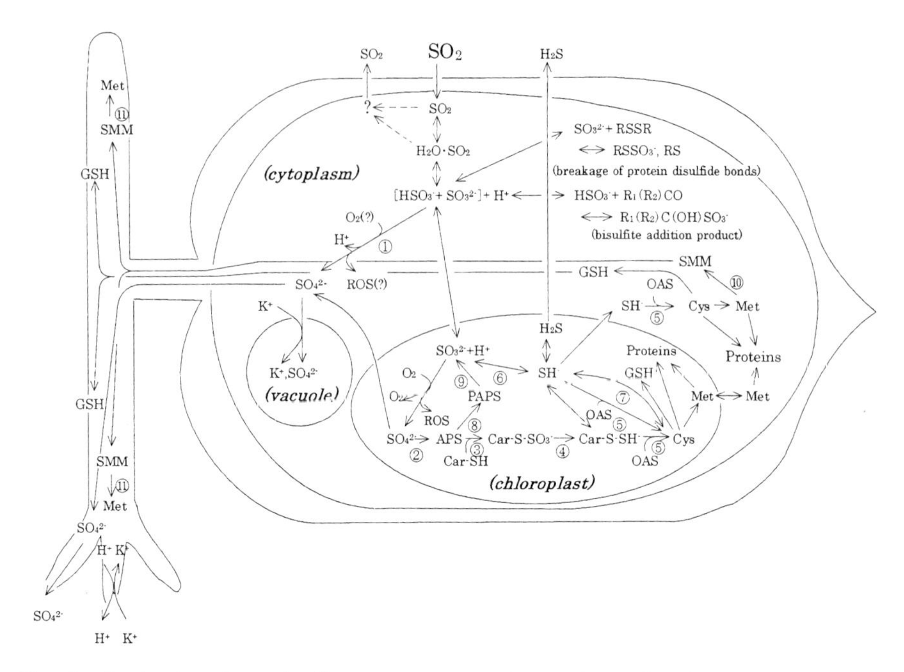

Fig. 1. A simple schematic diagram of a tentative pathway for the metabolism and translocation of sulfur dioxide absorbed by plant leaves. *APS*, adenosine 5'-phosphosulfate; *GSH*, reduced glutathione; *OAS*, *O*-acetylserine; *PAPS*, 3'-phosphoadenosine 5'-phosphosulfate; *ROS*, reactive oxygen species; *SMM*, *S*-methylmethionine; ①, sulfite oxidase (?); ②, ATP sulfurylase; ③, APS sulfotransferase; ④, thiosulfonate reductase; ⑤, *O*-acetylserine (thiol) lyase; ⑥, sulfite reductase; ⑦, L-cysteine desulfhydrase; ⑧, APS kinase; ⑨, PAPS reductase; ⑩, *S*-adenosylMet:Met S-methyltransferase; ⑪, homocysteine *S*-methyltransferase

by a ferredoxin-dependent sulfite reductase (Takahashi et al. 1996) in illuminated chloroplasts (Schiff and Hodson 1973), and is either used for cysteine biosynthesis or is released into the atmosphere as H_2S (Hällgren and Frederiksson 1982). Although this emission of H_2S may play some role in the response of plants to SO_2, the quantity of H_2S emitted corresponds to, at most, only 10% of the total sulfur derived from SO_2 (Sekiya et al. 1982b), and so it is unclear if this represents a true detoxification pathway for SO_2. Rather, H_2S emission is a possible indicator of sulfur reduction to generate cysteine from sulfite. The biosynthesis of cysteine from *O*-acetylserine and sulfide or carrier-bound sulfide is catalyzed by *O*-acetylserine (thiol) lyase. The detailed metabolism of H_2S in plants is described by De Kok et al. (this volume). However, when cysteine accumulates excessively in the leaves, sulfate and sulfite as well as glutathione are generated and a part of the excess sulfur is released into the atmosphere as H_2S

(Sekiya et al. 1982a), suggesting that H_2S emission is also a means of removing excessive sulfur from plants.

When lichens were exposed to SO_2, they also emitted H_2S, but photosynthetic activity did not participate in this emission of H_2S (Gries et al. 1997), suggesting that in the cytosol something resembling a sulfite reductase, which uses some reducing power other than ferredoxin, is operating in lichens.

The produced cysteine is then utilized to synthesize glutathione and proteins. On the contrary, cysteine synthesis seems to actively proceed only when synthesis of glutathione and proteins is also active (Dittrich et al. 1992). It has been reported that SO_2 promoted wheat growth when it was given simultaneously with high concentrations of CO_2, although it inhibited growth even at low concentrations (Deepak and Agrawal 1999). It seems possible that protein synthesis activated by high CO_2 promotes SO_2-driven cysteine synthesis, resulting in growth promotion induced by SO_2 as a fertilizer.

4.2.3 Oxidation of Sulfite

Most sulfite derived from SO_2 is oxidized to sulfate in plant cells (Weigl and Ziegler 1962). In some cases, 70% to 90% of absorbed ambient SO_2 was oxidized to sulfate (Slovik et al. 1995). Sulfate is estimated to be approximately 30 fold less toxic than sulfite (Thomas et al. 1943). Thus, the oxidation of sulfite to sulfate is assumed to be a form of detoxification reaction. Indeed, plants with the highest rates of sulfite oxidation showed the highest tolerance to SO_2, while plants with relatively low sulfite oxidation rates showed lower levels of SO_2 tolerance (Miller and Xerikos 1979; Ayazloo et al. 1982). Photosynthesis in some bryophytes is found to be suppressed by sulfite treatment. However, in the case of species of bryophytes that contain transfer metals in their cell walls, exogenous sulfite in the bathing solution decreased promptly by the oxidation to sulfate. In addition, such bryophytes showed tolerance to external sulfite (Baxter et al. 1989, 1991). These results are consistent with the model that oxidation of sulfite is a detoxification mechanism.

Peroxidase (Fridovich and Handler 1961; Klebanoff 1961; Yang 1967), cytochrome oxidase (Fridovich and Handler 1961), and ferredoxin NADP reductase (Nakamura 1970) are reportedly able to oxidize sulfite together with the generation of oxygen radicals. As oxygen radicals are toxic, sulfite oxidation catalyzed by these enzymes seems unlikely to be detoxification reaction. In animals, the activity of hepatic sulfite oxidase, which can oxidize sulfite using molecular oxygen without generation of ROS, is closely related to their SO_2 resistance (Cohen et al. 1973), although such enzymes have not been identified in plants. However, plants, especially castor bean and kidney bean, have been shown to possess heat-labile sulfite-oxidizing activities (Kondo et al. 1980a). These high molecular weight active substances, unlike hepatic sulfite oxidase, did not reduce cytochrome *c* during sulfite oxidation and, unlike peroxidase, cytochrome oxidase, and ferredoxin-NADP reductase, which require Mn^{2+}, ferrocytochrome, and

NADP, respectively, did not require the presence of any cofactor during sulfite oxidation. Takahama et al. (1992) reported that an apoplastic peroxidase can oxidize sulfite, but the oxidation reaction is slow and depends on the rate of apoplastic H_2O_2 generation. The active substances found by us remain to be precisely characterized.

Most of the sulfate formed by oxidation of sulfite is accumulated and stored in the vacuoles (Cram 1983; Kaiser et al. 1989), which may play a role in removing excess sulfur from the cytoplast. However, transportation to the vacuoles is an energy-consuming process using ATP. In needles of Norway spruce, sulfuric acid (SO_4^{2-}) derived from SO_2 is neutralized by K^+, which accumulates in vacuoles together with SO_4^{2-}. As a result, however, needles appear to become deficient in K^+ (Slovik et al. 1996).

In leaves of soybean plants fumigated with SO_2, the derived SO_4^{2-} was found to be excreted into the nutrient solution from the roots (Garsed and Read 1977). However, Kaiser et al. (1993) could not observe such excretion from either pea or barley roots, although a large part of the SO_4^{2-}, derived from SO_2, is transported to the roots in pea (Kaiser et al. 1993). Thus, it is questionable whether such excretion is a common feature among various plant species.

4.3 Sulfate Metabolism

The metabolism of sulfate to other substances is a light-promoted reaction (Rothermel and Alscher 1985). Light energy may be necessary for ATP formation and the production of reducing substances, such as NADPH and reduced ferredoxin, which are required for sulfate reduction. Sulfate is used in the sulfur assimilatory reductive pathway for the synthesis of the amino acid cysteine, from which methionine, glutathione, and proteins are subsequently produced. *S*-Methylmethionine (SMM) is produced from methionine. SMM and glutathione play an important role in sulfur transport in plants (Bourgis et al. 1999). On the other hand, SO_2 inhibits the transport of sulfate absorbed by roots to shoots. The sulfate transport seems to be regulated by SO_2-derived glutathione (Hershbach et al. 1995). An outline of the sulfate reduction pathway, which has been previously proposed, is as follows (Schiff and Hodson 1973). Adenosine 5'-phosphosulfate (APS) is generated from sulfate and ATP using ATP sulfurylase. The SO_3^- of APS is then transferred to a carrier (car) by adenosine 5'-phosphosulfate sulfotransferase (APSSTase), to generate car-S-SO_3^-, which is then further reduced to car-S-SH^- by sulfite reductase. The thiol of car-S-SH^- is then incorporated into *O*-acetylserine to generate cysteine. At present, two possible pathways, an 'APS-bound' pathway, as mentioned previously, and a 'PAPS-free' pathway have been proposed (Hell 1997). In the latter pathway, APS is converted to 3'-phosphoadenosine 5'-phosphosulfate (PAPS) by an APS kinase without combining with any carrier. PAPS is then reduced to sulfite by PAPS reductase. The sulfur assimilatory pathway, which has been recently reviewed by Leustek and Saito

(1999), is not completely resolved but is known to be well regulated by the levels of transcripts and activities of several of the biosynthetic enzymes, as well as by the substrates and products of the pathway.

It was reported that SO_2 suppresses APSSTase activity in the green leaves of beans (Wyss and Brunold 1980) and spruce trees (Tschanz et al. 1986). Low concentrations of SO_2, which do not cause any visible foliar damage, do change the activities of various foliar enzymes, such as *iso*-citrate dehydrogenase, malic enzyme, aspartate aminotransferase, glutamate dehydrogenase, and peroxidase (Pierre and Queiroz 1981, 1982). It is probable that these metabolic changes reflect an adaptive response to SO_2, maintaining a constant sulfur balance in the plant cells (Pierre and Queiroz 1982). The decline of APSSTase activity caused by SO_2 may also be a similar form of adaptive response, suppressing the overproduction of molecules such as cysteine and glutathione.

5. Conclusion

SO_2 absorbed through open stomata dissolves in the aqueous media of the apoplastic cell surface and produces bisulfite and/or sulfite. Not only are these substances phytotoxic, but they also reversibly generate more toxic substances, such as bisulfite addition products, in the leaves. A part of sulfite is rapidly reduced to sulfide, which either is used to produce cysteine or is released into the atmosphere as H_2S. Moreover, most sulfite is oxidized to sulfate, which actively accumulates in the vacuoles. If sulfite persists for an appreciable length of time in plant cells, the plant will be subjected to severe stress and possible damage. Therefore, the reduction or oxidation of sulfite is assumed to represent the detoxification of bisulfite and sulfite.

During the process of oxidation of sulfite to sulfate in illuminated chloroplasts, a substantial amount of O_2^- is generated by chain reactions, which, in turn, gives rise to other highly reactive species, especially H_2O_2, $OH\cdot$, and $^1O_2$. The H_2O_2 can inhibit SH enzymes of the Calvin cycle, whereas if not kept in check, O_2^- and $^1O_2$ destroy chlorophyll and cell membranes. Therefore, this oxidation reaction may not represent a detoxification mechanism for sulfite derived from SO_2 but rather a toxicant-generating process. However, because plants are equipped with antioxidants and other scavenging systems that detoxify these ROS, they are able to avoid severe damage under normal environmental conditions. Indeed, it has been shown that SO_2-induced ROS levels rapidly decrease and return to the original levels after termination of fumigation (Tanaka et al. 1982a).

In contrast to its toxicity, sulfite may, under sulfur-deficient conditions, serve as a sulfur nutrient that promotes growth. Thus, SO_2 in the atmosphere has both positive and negative effects on plants (Murray and Wilson 1990). Global environmental conditions recently have been changing rapidly, as demonstrated by global warming and the increase in ultraviolet radiation. When atmospheric CO_2 levels dramatically increase in the future, as has been predicted, the positive effects of SO_2 may become more apparent. However, high temperatures, drought, and enhanced ultraviolet radiation, which will result from these global changes, may enhance the formation of ROS. Thus, even at the present atmospheric levels of SO_2, enhanced formation of ROS may result and thereby promote further plant damage.

At present, the development of plants tolerant to SO_2 is essential to maintain or enhance food production and to conserve global ecosystems. Such SO_2-tolerant plants have been successfully generated by genetic manipulations (see the chapters by Youssefian, by Aono, and by Endo and Ebinuma, this volume).

SO_2 is known to causes a decline in photosynthetic activity, to affect or promote stomatal closure, and to enhance respiration and, as a result, to retard plant growth. At the cellular level, SO_2 disturbs metabolic events through processes including acidification of the cytoplasm and chloroplasts, the consumption of energy for metabolism of surplus sulfite, and the induced deficiency of cations, probably resulting in senescence acceleration. We are still uncertain of the exact factors that are crucial for the chronic effects of SO_2. We are in need of greater understanding of the chronic damage caused by subnecrotic fumigation with SO_2 to take necessary measures against chronic damage by SO_2. In addition, it is essential to examine whether transgenic plants that are tolerant to acute SO_2 injury are also tolerant to chronic effects of SO_2.

References

Alscher R, Bower JL, Zipfel W (1987) The basis for different sensitivities to SO_2 in two cultivars of pea. J Exp Bot. 38:99-108

Amundson RG, Weinstein LH (1981) Joint action of sulfur dioxide and nitrogen dioxide on foliar injury and stomatal behavior in soybean. J Environ Qual 10:204-206

Aono M, Kubo A, Saji H, Tanaka K, Kondo N (1993) Enhanced tolerance to photooxidative stress of transgenic *Nicotiana tabacum* with high chloroplastic glutathione reductase activity. Plant Cell Physiol 34:129-135

Asada K, Kiso K (1973) Initiation of aerobic oxidation of sulfite by illuminated spinach chloroplasts. Eur J Biochem 33:253-257

Ashenden TW (1978) Growth reductions in cocksfoot (*Dactylis glomerata* L.) as a result of SO_2 pollution. Environ Pollut 15:161-166

Ashenden TW, Mansfield TA (1978) Extreme pollution sensitivity of grasses when SO_2 and NO_2 are present in the atmosphere together. Nature 273:142-143

Ayazloo M, Garsed SG, Bell JNB (1982) Studies on the tolerance to sulphur dioxide of grass populations in polluted areas. II. Morphological and physiological investigations. New Phytol 90:109-126

Bae GY, Kondo N, Nakajima N, Ishizuka K (1995) Ethylene production in tomato plants by SO_2 in relation to leaf injury (in Japanese with English abstract). J Jpn Soc Atmos Environ 30:367-373

Bae GY, Nakajima N, Ishizuka K, Kondo N (1996) The role in ozone phytotoxicity of the evolution of ethylene upon induction of 1-aminocyclopropane-1-carboxylic acid synthase by ozone fumigation in tomato plants. Plant Cell Physiol 37:129-134

Bailey JL, Cole RD (1959) Studies on the reaction of sulfite with proteins. J Biol Chem 234:1733-1739

Barton JR, McLaughlin SB, McConathy RK (1980) The effects of SO_2 on components of leaf resistance to gas exchange. Environ Pollut Ser A 21:255-265

Baxter R, Emes MJ, Lee JA (1989) The relationship between extracellular metal accumulation and bisulphate tolerance in *Sphangnum cuspidatum* Hoffm. New Phytol 111:463-472

Baxter R, Emes MJ, Lee JA (1991) Short term effects of bisulphite on pollution-tolerant and pollution sensitive populations of *Sphangnum cuspidatum* Ehrh. (ex. Hoffm.). New Phytol 118:425-431

Black CR, Black VJ (1979) The effects of low concentrations of sulphur dioxide on stomatal conductance and epidermal cell survival in field bean (*Vicia faba* L.). J Exp Bot 30:291-298

Black VJ, Unsworth MN (1979) Effects of low concentrations of sulphur dioxide on net photosynthesis and dark respiration. J Exp Bot 30:473-483

Bourgis F, Roje S, Nuccio ML, Fisher DB, Tarczynski MC, Li C, Herschbach C, Rennenberg H, Pimenta MJ, Shen T-L, Gage DA, Hanson AD (1999) *S*-Methylmethionine plays a major role in phloem sulfur transport and is synthesized by a novel type of methyltransferase. Plant Cell 11:1485-1497

Bressan RA, Wilson LG, L, Filner P (1978) Mechanisms of resistance to sulfur dioxide in the Cucurbitaceae. Plant Physiol 61:761-767

Bressan RA, LeCureux L, Wilson LG, L, Filner P (1979) Emission of ethylene and ethane by leaf tissue exposed to injurious concentrations of sulfur dioxide or bisulfite ion. Plant Physiol 63:924-930

Clarke K, Murray F (1990) Stimulatory effects of SO_2 on growth of *Eucalyptus rudis* Endl. New Phytol 115:633-637

Cohen HJ, Drew RT, Johnson JL, Rajagopalan KV (1973) Molecular basis of the biological function of molybdenum. The relationship between sulfite oxidase and the acute toxicity of bisulfite and SO_2. Proc Natl Acad Sci USA 70:3655-3659

Cornic G (1987) Interaction between sublethal pollution by sulphur dioxide and drought stress. The effect on photosynthetic capacity. Physiol Plant 71:115-119

Cowan IR, Raven JA, Hartung W, Farquhar GD (1982) A possible role for abscisic acid in coupling stomatal conductance and photosynthetic carbon metabolism in leaves. Aust J Plant Physiol 9:489-498

Cowling DW, Lockyer DR (1978) The effect of SO_2 on *Lolium perenne* L. grown at different levels of sulphur and nitrogen nutrition. J Exp Bot 29: 257-265

Cram WJ (1983) Sulphate accumulation is regulated at the tonoplast. Plant Sci Lett 31:329-338

Deepak SS, Agrawal M (1999) Growth and yield responses of wheat plants to elevated levels of CO_2 and SO_2, singly and in combination. Environ Pollut 104:411-419

Dittrich APM, Pfanz H, Heber U (1992) Oxidation and reduction of sulfite by chloroplasts and formation of sulfite addition compounds. Plant Physiol 98:738-744

Fridovich I, Handler P (1961) Detection of free radicals generated during enzymic oxidations by the initiation of sulfite oxidation. J Biol Chem 236: 1836-1840

Furukawa A, Isoda O, Iwaki H, Totsuka T (1979) Interspecific difference in responses of transpiration to SO_2. Environ Control Biol 17:153-159

Furukawa A, Isoda O, Iwaki H, Totsuka T (1980a) Interspecific difference in resistance to sulfur dioxide. In: Studies on the effects of air pollutants on plants and mechanisms of

phytotoxicity. Res. Rep. Natl. Inst. Environ. Stud. Jpn No.11, pp 113-126
Furukawa A, Natori T, Totsuka T (1980b) The effect of SO_2 on net photosynthesis in sunflower leaf. In: Studies on the effects of air pollutants on plants and mechanisms of phytotoxicity. Res. Rep. Natl. Inst. Environ. Stud. Jpn No.11, pp 113-126
Furukawa A, Katase M, Ushijima T, Totsuka T (1984) Inhibition of photosynthesis of poplar species and sunflower by O_3. In: Studies on effects of air pollutant mixtures on plants, Part 1. Res. Rep. Natl. Inst. Environ. Stud. Jpn No.65, pp 77-87
Garsed SG, Read DJ (1977) Sulphur dioxide metabolism in soybean, *Glycine max* var. biloxi. I. The effects of light and dark on the uptake and translocation of $^{35}SO_2$. New Phytol 78:111-119
Gezeilus K, Hällgren J-E (1980) Effect of SO_3^{2-} on the activity of ribulose bisphosphate carboxylase from seedlings of *Pinus sivestris*. Physiol Plant 49:354-358
Gould RP, Minchin PEH, Young PC (1988) The effects of sulphur dioxide on phloem transport in two cereals. J Exp Bot 39:997-1007
Gries C, Romagni JG, Nash TH III, Kuhn U, Kesselmeier J (1997) The relation of H_2S release to SO_2 fumigation of lichens. New Phytol 136:703-711
Grill D, Esterbauer H, Klösch U (1979) Effect of sulphur dioxide on glutathione in leaves of plants. Environ Pollut Ser A 19:187-194
Grill D, Esterbauer H, Scharner M, Felgitsh C (1980) Effect of sulphur dioxide on protein-SH in needles of *Picea abies*. Eur J For Pathol 10:263-267
Hällgren J-E, Frederiksson S-A (1982) Emission of hydrogen sulfide from sulfur dioxide-fumigated pine trees. Plant Physiol 70:456-459
Hampp R, Ziegler I (1977) Sulfate and sulfite translocation via the phosphate translocator of the inner envelope membrane of chloroplasts. Planta 137:309-312
Hell R (1997) Molecular physiology of plant sulfur metabolism. Planta 202:138-148
Herschbach C, De Kok LJ, Rennenberg H (1995) Net uptake of sulphate and its transport to the shoot in tobacco plants fumigated with H_2S or SO_2. Plant Soil 175:75-84
Hogetsu T, Shishikura M (1994) Effects of sulfur dioxide and ozone on intact leaves and isolated mesophyll cells of groundnut plants (*Arachis hypogaea* L.). J Plant Res 107:229-235
Jensen KF, Roberts BR (1986) Changes in yellow poplar (*Liriodendron tulipifera*) stomatal resistance with sulfur dioxide and ozone fumigation. Environ Pollut Ser A 41:235-246
Kaiser G, Martinoia E, Schröppel-Meier G, Heber U (1989) Active transport of sulfite into the vacuole of plant cells provides halotorelance and can detoxify SO_2. J Plant Physiol 133:756-763
Kaiser WM (1979) Reversible inhibition of the Calvin cycle and activation of oxidative pentose phosphate cycle in isolated intact chloroplasts by hydrogen peroxide. Planta 145:377-382
Kaiser WM, Höfler M, Heber U (1993) Can plants exposed to SO_2 excrete sulfuric acid through the roots? Physiol Plant 87:61-67
Kimmerer TW, Kozlowski TT (1981) Stomatal conductance and sulfur uptake of five clones of *Populus tremuloides* exposed to sulfur dioxide. Plant Physiol 67:990-995
Klebanoff SJ (1961) The sulfite-activated oxidation of reduced pyrimidine nucleotides by peroxidase. Biochim Biophys Acta 48:93-103
Kondo N (1987) Changes in transpiration rate caused by air pollutants and contents of phytohormones. In: Studies on the role of vegetation as a sink of air pollutants. Res. Rep. Natl. Inst. Environ. Stud. No.108, pp 187-197 (in Japanese)
Kondo N, Sugahara K (1978) Changes in transpiration rate of SO_2-resistant and -sensitive

plants with SO_2 fumigation and the participation of abscisic acid. Plant Cell Physiol 19:365-373

Kondo N, Sugahara K (1984) Effects of air pollutants on transpiration rate in relation to abscisic acid content. In: Studies on effects of air pollutant mixtures on plants, Part 1. Res. Rep. Natl. Inst. Environ. Stud. Jpn No.65, pp 1-8

Kondo N, Akiyama Y, Fujiwara M, Sugahara K (1980a) Sulfite oxidizing activities in plants. In: Studies on the effects of air pollutants on plants. Res. Rep. Natl. Inst. Environ. Stud. No.11, pp 137-150

Kondo N, Maruta I, Sugahara K (1980b) Effects of sulfite and pH on abscisic acid-dependent transpiration and on stomatal opening. Plant Cell Physiol 21:817-828

Koziol MJ, Jordan CF (1978) Changes in carbohydrate levels in red kidney bean (*Phaseolus vulgaris* L.) exposed to sulphur dioxide. J Exp Bot 29:1057-1043

Kropff MJ (1987) Physiological effects of sulfur dioxide: 1. The effect of sulfur dioxide on photosynthesis and stomatal regulation of *Vicia faba* L. Plant Cell Environ 10:753-760

Laisk A, Pfanz H, Heber U (1988) Sulfur dioxide fluxes into different cellular compartments of leaves photosynthesizing in a polluted atmosphere: II. Consequences of sulfur dioxide uptake as revealed by computer analysis. Planta 173:241-252

Leustek T, Saito K (1999) Sulfate transport and assimilation in plants. Plant Physiol 120:637-643

Libera W, Ziegler I, Ziegler H (1975) The action of sulfite on the HCO_3^- fixation and the fixation pattern of isolated chloroplasts and leaf tissue slices. Z Pflanzenphysiol 74:420-433

Lorenc-Plucińska G, Ziegler H (1987) The effect of sulphite on sucrose uptake and translocation in the cotyledons of castor bean (*Ricinus communis* L.). J Plant Physiol 127:97-110

Lucas PW (1990) The effects of prior exposure to sulfur dioxide and nitrogen dioxide on the water relations of timothy grass (*Phleum pratense*) under drought conditions. Environ Pollut 66:117-138

Lüttge U, Osmond CB, Ball E, Brinckmann E, Kinze G (1972) Bisulfite compounds as metabolic inhibitors: nonspecific effects on membranes. Plant Cell Physiol 13:505-514

Maas FM, De Kok LJ (1988) *In vitro* NADH oxidation as an early indicator for growth reduction in spinach exposed to H_2S in the ambient air. Plant Cell Physiol 29:523-526

Maas FM, De Kok LJ, Strik-Timmer W, Kuiper PJC (1987) Plant responses to H_2S and SO_2 fumigation. II. Differences in metabolism of H_2S and SO_2 in spinach. Physiol Plant 70:722-728

Maas FM, van Loo EN, van Hasselt PR (1988) Effect of long-term H_2S fumigation on photosynthesis in spinach. Correlation between CO_2 fixation and chlorophyll a fluorescence. Physiol Plant 72:77-83

Majerník O, Mansfield TA (1970) Direct effect of SO_2 pollution on the degree of opening of stomata. Nature 227:377-378

Malhotra SS, Hocking D (1976) Biochemical and cytological effects of sulphur dioxide on plant metabolism. New Phytol 76:227-237

Mansfield TA, Majerník O (1970) Can stomata play a part in protecting plants against air pollutants? Environ Pollut 1:149-154

Martin C, Thimann K (1972a) The role of protein synthesis in the senescence of leaves. I. The formation of protease. Plant Physiol 49:64-71

Martin C, Thimann K (1972b) The role of protein synthesis in the senescence of leaves. II. The influence of amino acids on senescence. Plant Physiol 50:432-437

Maurousset L, Bonnemain J-L (1990) Mechanism of the inhibition of phloem loading by sodium sulfite: effect of the pollutant on the transmembrane potential difference. Physiol Plant 80:233-237

Maurousset L, Raymond P, Gaudillere M, Bonnemain J-L (1992) Mechanism of the inhibition of phloem loading by sodium sulfite: effect of the pollutant on respiration, photosynthesis and energy charge in the leaf tissues. Physiol Plant 84:101-105

Mehlhorn H (1990) Ethylene-promoted ascorbate peroxidase activity protects against hydrogen peroxide, ozone and paraquat. Plant Cell Environ 13:971-976

Mehlhorn H, Wellburn AR (1987) Stress ethylene formation determines plant sensitivity to ozone. Nature 327:417-418

Mejstřík V (1980) The influence of low SO_2 concentrations on growth reduction of *Nicotiana tabacum* L. cv. Samsun and *Cucumis sativus* L. cv. Unikat. Environ Pollut Ser A 21:73-76

Menser HA, Heggestad HE (1966) Ozone and sulfur dioxide synergism: injury to tobacco plants. Science 153:424-435

Miller JE, Xerikos PB (1979) Residence time of sulfite in SO_2 'sensitive' and 'tolerant' soybean cultivars. Environ Pollut 18:259-264

Minchin PEH, Gould R (1986) Effect of SO_2 on phloem loading. Plant Sci 43:179-183

Miszalski Z, Ziegler I (1979) Increase in chloroplastic thiol groups by SO_2 and its effects on light modulation of NADP-dependent glyceraldehyde-3-phosphate dehydrogenase. Planta 145:383-387

Mukerji SK, Yang SF (1974) Phosphoenolpyruvate carboxylase from spinach leaf tissue. Plant Physiol 53:829-834

Murray F (1985) Changes in growth and quality characteristics of Lucerne (*Medicago sativa* L.) in response to sulphur dioxide exposure under field conditions. J Exp Bot 36:449-457

Murray F, Wilson S (1990) Growth responses of barley exposed to SO_2. New Phytol 114:537-541

Nakamura S (1970) Initiation of sulfite oxidation by spinach ferredoxin-NADP reductase and ferredoxin system: a model experiment on the superoxide anion radical production by metalloflavoproteins. Biochem Biophys Res Commun 41:177-183

Navari-Izzo F, Izzo R, Quartacci MF, Lorenzini G (1989) Growth and solute leakage in *Hordeum vulgaris* exposed to long-term fumigation with low concentrations of SO_2. Physiol Plant 76:445-450

Neighbour EA, Cottam DA, Mansfield TA (1988) Effects of sulphur dioxide and nitrogen dioxide on the control of water loss by birch (*Betula* spp.). New Phytol 108:149-157

Okupodu CM, Alscher RG, Grabau EA, Cramer CL (1996) Physiological, biochemical and molecular effects of sulfur dioxide. J Plant Physiol 148:309-316

Omasa K, Abo F, Natori T, Totsuka T (1980) Analysis of air pollutant sorption by plants. (3) Sorption under fumigation with NO_2, O_3 or NO_2 + O_3. In: Studies on the effects of air pollutants on plants and mechanisms of phytotoxicity. Res. Rep. Natl. Inst. Environ. Stud. Jpn No. 11, pp 213-224

Omasa K, Hashimoto Y, Aiga I (1981) A quantitative analysis of the relationships between SO_2 or NO_2 sorption and their acute effects on plant leaves using image instrumentation. Environ Control Biol 19:59-67

Omasa K, Hashimoto Y, Aiga I (1983) Observation of stomatal movements of intact plants using an image instrumentation system with a light microscope. Plant Cell Physiol 24:281-288

Omasa K, Hashimoto Y, Kramer PJ, Strain BR, Aiga I, Kondo J (1985) Direct observation of reversible and irreversible stomatal responses of attached sunflower leaves to SO_2. Plant Physiol 79:153-158

Oren A, Padan E, Malkin S (1979) Sulfide inhibition of photosystem II in Cyanobacteria (blue-green algae) and tobacco chloroplasts. Biochim Biophys Acta 546:270-279

Osmond CB, Avadhani PN (1970) Inhibition of the β-carboxylation pathway of CO_2 fixation by bisulfite compounds. Plant Physiol 45:228-230

Peiser GD, Yang SF (1979) Ethylene and ethane production from sulfur dioxide-injured plants. Plant Physiol 63:142-145

Peiser GD, Lizada MCC, Yang SF (1982) Sulfite-induced lipid peroxidation in chloroplasts as determined by ethane production. Plant Physiol 70:994-998

Pfanz H, Martinoia E, Lange O-T, Heber U (1987a) Mesophyll resistance to SO_2 fluxes into leaves. Plant Physiol. 85: 922-927

Pfanz H, Martinoia E, Lange O-T, Heber U (1987b) Flux of SO_2 into leaf cells and cellular acidification by SO_2. Plant Physiol 85:928-933

Pierre M, Queiroz O (1981) Enzymic and metabolic changes in bean leaves during continuous pollution by subnecrotic leaves of SO_2. Environ. Pollut Ser A 21:41-51

Pierre M, Queiroz O (1982) Modulation by leaf age and SO_2 concentration of the enzymic response to subnecrotic SO_2 pollution. Environ Pollut Ser A 28:209-217

Pierre M, Queiroz O (1988) Air pollution by SO_2 amplifies the effects of water stress on enzymes and total proteins of spruce needles. Physiol Plant 73:412-417

Price S, Long SP (1989) An in vivo analysis of the effect of SO_2 fumigation on photosynthesis in *Zea mays*. Physiol. Plant 76:193-200

Priebe A, Klein H, Jäger H-J (1978) Role of polyamines in SO_2-polluted pea plants. J Exp Bot 29:1045-1050

Robinson MF, Heath J, Mansfield TA (1998) Disturbances in stomatal behaviour caused by air pollutants. J Exp Bot 49:461-469

Rothermel B, Alscher R (1985) A light-enhanced metabolism of sulfite in cells of *Cucumis sativus* L. cotyledons. Planta 166:105-110

Sakaki T, Kondo N (1984) Sulfite inhibition of uptake and fixation of inorganic carbon in mesophyll protoplasts isolated from *Vicia faba* L. In: Studies on effects of air pollutant mixtures on plants, Part 1. Res. Rep. Natl. Inst. Environ. Stud. Jpn No. 65, pp 35-43

Sakaki T, Kondo N (1985) Inhibition of photosynthesis by sulfite in mesophyll protoplasts isolated from *Vicia faba* L. in relation to intracellular sulfite accumulation. Plant Cell Physiol 26:1045-1055

Schiff JA, Hodson RC (1973) The metabolism of sulfate. Annu Rev Plant Physiol 24:381-414

Sekiya J, Schmidt A, Wilson LG, Filner P (1982a) Emission of hydrogen sulfide by leaf tissue in response to L-cysteine. Plant Physiol 70:430-436

Sekiya J, Wilson LG, Filner P (1982b) Resistance to injury by sulfur dioxide. Correlation with its reduction to, and emission of, hydrogen sulfide in Cucurbitaceae. Plant Physiol 70:437-441

Shimazaki K, Sugahara K (1979a) Specific inhibition of photosystem II activity in chloroplasts by fumiation of spinach leaves with SO_2. Plant Cell Physiol 20:947-955

Shimazaki K, Sugahara K (1979b) Inhibition site of the electron transport system in lettuce chloroplasts by fumigation of leaves with SO_2. Plant Cell Physiol 21:125-135

Shimazaki K, Sakaki T, Kondo N, Sugahara K (1980) Active oxygen participation in chlorophyll destruction and lipid peroxidation in SO_2-fumigated leaves of spinach. Plant

Cell Physiol 21:1193-1204
Shimazaki K, Nakamachi K, Kondo N, Sugahara K (1984a) Sulfite inhibition of photosystem II in illuminated spinach leaves. Plant Cell Physiol 25:337-341
Shimazaki K, Ito K, Kondo N, Sugahara K (1984b) Reversible inhibition of the photosynthetic water-splitting enzyme system by SO_2-fumigation assayed chlorophyll fluorescence and EPR signal in vivo. Plant Cell Physiol 25: 795-803
Silvius JE, Ingle M, Baer CH (1975) Sulfur dioxide inhibition of photosynthesis in isolated spinach chloroplasts. Plant Physiol. 56: 434-437
Slovik S, Siegmund A, Kindermann G, Riebeling R, Balazs A (1995) Stomatal SO_2 uptake and sulfate accumulation in needles of Norway spruce stands (*Picea abies*) in Central Europe. Plant Soil 168-169:405-419
Slovik S, Hüve K, Kinderman G, Kaiser WM (1996) SO_2-dependent cation competition and compartmentalization in Norway spruce needles. Plant Cell Environ 19:813-824
Soldatini GF, Ranieri A, Lencioni L, Lorenzini G (1992) Effects of continuous SO_2 fumigation on SH-containing compounds in two wheat cultivars of different sensitivities. J Exp Bot 43:797-801
Sugahara K, Uchida S, Takimoto M (1980) Effects of sulfite ions on water-soluble chlorophyll proteins. In: Studies on the effects of air pollutants on plants and mechanisms of phytotoxicity. Res. Rep. Natl. Inst. Environ. Stud. Jpn No. 11, pp 103-112
Takahama U, Veljovic-Iovanovic S, Heber U (1992) Effects of the air pollutant SO_2 on leaves. Inhibition of sulfite oxidation in the apoplast by ascorbate and of apoplastic peroxidase by sulfite. Plant Physiol 100:261-266
Takahashi S, Yoshida Y, Tamura G (1996) Purification and characterization of ferredoxin-sulfite reductases from leek (*Allium tuberosum*) leaves. J Plant Res 109:45-52
Tanaka H, Takanashi T, Yatazawa M (1972) Experimental studies on sulphur dioxide injuries in higher plants. I. Formation of glyoxylate bisulphite in plant leaves exposed to sulphur dioxide. Water Air Soil Pollut 1:205-211
Tanaka K, Sugahara K (1980) Role of superoxide dismutase in defense against SO_2 toxicity and an increase in superoxide dismutase activity with SO_2 fumigation. Plant Cell Physiol 21:601-611
Tanaka K, Kondo N, Sugahara K (1982a) Accumulation of hydrogen peroxide in chloroplasts of SO_2-fumigated spinach leaves. Plant Cell Physiol 23:999-1007
Tanaka K, Otsubo T, Kondo N (1982b) Participation of hydrogen peroxide in the inactivation of Calvin cycle SH enzymes in SO_2-fumigated spinach leaves. Plant Cell Physiol 23:1009-1018
Tanaka K, Suda Y, Kondo N, Sugahara K (1985) O_3 tolerance and the ascorbate-dependent H_2O_2 decomposing system in chloroplasts. Plant Cell Physiol 26:1425-1431
Thomas FM, Runge M (1992) Proton neutralization in the leaves of English oak (*Quercus robur* L.) exposed to sulphur dioxide. J Exp Bot 43:803-809
Thomas MD, Hendricks RH, Collier TR, Hill GR (1943) The utilization of sulfate and sulfur dioxide for the nutrition of alfalfa. Plant Physiol 18:345-371
Thomas MD, Hill GR Jr (1935) Absorption of sulphur dioxide by alfalfa and its relation to leaf injury. Plant Physiol 10:291-307
Tschanz A, Landolt W, Bleuler P, Brunold C (1986) Effect of SO_2 on the activity of adenosine 5'-phosphosulfate sulfotransferase from spruce trees (*Picea abies*) in fumigation chambers and under field conditions. Physiol Plant 67:235-241
Veeranjaneyulu K, N'Soukpoé-Kossi CN, Leblanc RM (1994) Emission of sulfur dioxide

from sulfite-treated birch leaves. J Plant Physiol 144:420-423
Veljovic-Jovanovic S, Bilger W, Heber U (1993) Inhibition of photosynthesis, acidification and stimulation of zeaxanthin formation in leaves by sulfur dioxide and reversal of these effects. Planta 191:365-376
Weigl J, Ziegler H (1962) Die Raumliche Verteilung von ^{35}S und die Art der Markierten Verbindungen in Spinatblattern nach Begasung mit $^{35}SO_2$. Planta 58:435-447
Wilson LG, Bressan RA, Filner P (1978) Light-dependent emission of hydrogen sulfide from plants. Plant Physiol 61:184-189
Wyss H-R, Brunold C (1980) Regulation of adenosine 5'-phosphosulfate sulfotransferase by sulfur dioxide in primary leaves of beans (*Phaseolus vulgaris*). Physiol Plant 50:161-165
Yang SF (1967) Biosynthesis of ethylene. Ethylene formation from methional by horseradish peroxidase. Arch Biochem Biophys 122:481-487
Zelitch I (1957) α-Hydroxysulfonates as inhibitors of the enzymatic oxidation of glycolic and lactic acids. J Biol Chem 224:251-260
Ziegler I (1972) The effect of SO_3^- on the activity of ribulose-1,5-diphosphate carboxylase in isolated spinach chloroplasts. Planta 103:155-163
Ziegler I (1973) Effect of sulphite on phosphoenolpyruvate carboxylase and malate formation in extracts of *Zea mays*. Phytochemistry 12:1027-1030
Ziegler I (1974) Malate dehydrogenase in *Zea mays*: properties and inhibition by sulfite. Biochim Biophys Acta 364:28-37

10
Elevated Levels of Hydrogen Sulfide in the Plant Environment: Nutrient or Toxin

Luit J. De Kok, C. Elisabeth E. Stuiver, Sue Westerman, and Ineke Stulen

Laboratory of Plant Physiology, University of Groningen, P.O. Box 14, 9750 AA Haren, The Netherlands

1. Introduction

Hydrogen sulfide (H_2S) is a malodorous gas with a typical "rotten egg" odor that can be smelt at levels of 0.02 $\mu l\ l^{-1}$ and higher (for chemical and physical properties of H_2S, see the review of Beauchamp et al. 1984). Normally H_2S is only present in trace concentrations in the plant environment, but under specific conditions plants may have to cope with elevated levels of H_2S in either the pedosphere or atmosphere in both natural vegetation and agriculture. Even though sulfide is a normal intermediate in plant metabolism, the impact of H_2S on plants is paradoxical. On the one hand, it may be utilized as a sulfur nutrient, and on the other hand, above a certain threshold level it may negatively affect plant growth and functioning. In this chapter, our present knowledge on the impact of elevated levels of H_2S on plants both as nutrient and toxin is reviewed.

The predominant natural sources of H_2S in aquatic and terrestrial ecosystems are the biological decay of organic sulfur and the activity of dissimilatory sulfate-reducing bacteria (Beauchamp et al. 1984; Noggle et al. 1986; Kelly 1988; Trudinger 1986; Bates et al. 1992). Also, plants may contribute to biogenic H_2S emission (Schröder 1993). The total global annual sulfide turnover is huge, and H_2S production by sulfate-reducing bacteria alone already approaches 3 x 10^{14} mol year^{-1} (10^{10} tons year^{-1}) (Kelly 1988). However, the greater proportion of the produced sulfide either never leaves the soils and waters or is reoxidized by bacteria or chemical oxidation (Trudinger 1986; Kelly 1988).

Air Pollution and Plant Biotechnology
–Prospects for Phytomonitoring and Phytoremediation–
Edited by K. Omasa, H. Saji, S. Youssefian, and N. Kondo
© Springer -Verlag Tokyo 2002

In the pedosphere, permanent or temporary elevated H_2S levels may occur in anoxic soils, such as marshes and tideland wetlands, and in poorly drained and water-logged soils (e.g., rice paddies; Trudinger 1986; Van Diggelen et al. 1987; Bates et al. 1992; Ouattara and Jacq 1992). Little is known about in situ free sulfide concentrations, since generally much of the sulfide is precipitated by iron (Tisdale et al. 1986; Van Diggelen et al. 1987). The measured free sulfide concentrations in West European and North American salt marshes range from 0.02 to 1.4 mM in the soil moisture around the root zone (Carlson and Forrest 1982; Van Diggelen et al. 1987); in rice fields, concentrations up to 0.02 mM have been measured (Allam and Hollis 1972).

The total global natural sulfur emission into the atmosphere is estimated at 0.8 x 10^{12} mol year^{-1}, although only a minor proportion of it, about 0.5 x 10^{10} mol year^{-1}, is released as H_2S (derived from Bates et al. 1992). Not H_2S, but dimethyl sulfide (DMS) and carbonyl sulfide (COS) are the predominant naturally emitted sulfides (Beauchamp et al. 1984; Noggle et al. 1986; Kelly 1988; Bates et al. 1992). Global anthropogenic sulfur emission, mainly SO_2, largely exceeds that of natural emissions and is estimated as 25 x 10^{12} mol year^{-1} (Bates et al. 1992), but less than 1% of this is emitted as H_2S (Noggle et al. 1986). Once in the atmosphere, H_2S has a relatively short residence time (15 days) because it is rapidly oxidized by hydroxy radicals and other atmospheric oxidants (Trudinger 1986). Hence, H_2S is generally only present in trace amounts and, in rural areas, the atmospheric concentrations range from 0.02 to 0.3 nl l^{-1} (Kellogg et al. 1972; Slatt et al. 1978; Beauchamp et al. 1984).

In the atmosphere, highly elevated H_2S levels only occur locally in the vicinity of volcanoes, fumaroles, sulfur springs, and geothermal wells, where atmospheric concentrations may exceed 0.1 μl l^{-1} (Hendrickson 1979; Mudd 1979; Ernst 1997; Schulte et al. 1997). Highly elevated H_2S levels may also occur in industrialized areas, caused by the refining of oil, utilization of fossil fuels, and surface water pollution by paper mills and farina factories, or in areas with intensive bioindustry. Here, H_2S concentrations frequently exceed the odor threshold level of 0.02 μl l^{-1} (Beauchamp et al. 1984) and may cause complaints about "rotten egg" odor (Ajax and Lee 1976; Urone 1976; Hendrickson 1979).

2. Elevated H_2S and Plant Growth

There are few data available for the assessment of the global significance and economic consequences of the impact of elevated levels of H_2S on natural vegetation and agriculture. Several studies show that elevated H_2S levels in either the pedosphere or atmosphere may negatively affect plant growth and functioning, although there is a wide variation in tolerance among species.

Elevated levels of H_2S in the root environment may be phytotoxic and negatively affect plant growth (Allam and Hollis 1972; Ford 1973; Joshi et al. 1973, 1975; Joshi and Hollis 1977; Mudd 1979; Carlson and Forrest 1982; Fry et al. 1982; Tisdale et al. 1986; Van Diggelen et al. 1987; Ouattara and Jacq 1992). In submerged rice soils, its occurrence may be related to the development of several diseases in rice, such as the so-called Akiochi (autumn decline) disease in Japan, the mentek disease in Java, and the straighthead disease in the United States (Allam and Hollis 1972; Joshi et al. 1973; Mudd 1979; Tisdale et al. 1986). Levels as low as 0.002 mM sulfide may negatively affect root respiration and nutrient uptake (Allam and Hollis 1972; Joshi et al. 1973, 1975). Root injury may occur upon prolonged exposure to sulfide levels of 0.08 mM and higher (Ford 1973). In salt marshes, however, some plant species are able to cope with levels up to 5 mM sulfide (Joshi and Hollis 1977; Carlson and Forrest 1982; Van Diggelen et al. 1987), and here differences in sulfide tolerance between species likely play a primary role in their zonation (Van Diggelen et al. 1987). The pedospheric sulfide tolerance is probably determined by the in situ sulfide level in the roots, which depends on the rate of oxidation of sulfide in the rhizosphere by bacteria such as *Beggiotoa* or in the plant, and by the sulfide resistance of metabolic processes (Joshi et al. 1973, 1975; Joshi and Hollis 1977; Carlson and Forrest 1982; Fry et al. 1982; Van Diggelen et al. 1987).

Elevated levels of atmospheric H_2S also may affect plant growth and functioning. Biomass production may be reduced upon prolonged exposure to $\geq 0.03\ \mu l\ l^{-1}$ H_2S, whereas visible injury may occur upon exposure to levels of $\geq 0.3\ \mu l\ l^{-1}$ (Thompson and Kats 1978; Krause 1979; De Kok et al. 1983b, 1985, 1998; Maas et al. 1985, 1987a; De Kok 1989, 1990). Acute extremely high levels ($\geq 20\ \mu l\ l^{-1}$) can instantly result in wilting and the development of leaf necrosis (see references in Mudd 1978). In some species, exposure to levels of 0.03 - 0.1 $\mu l\ l^{-1}$ H_2S results in an enhanced biomass production, even when there is an ample sulfur supply to the roots (Thompson and Kats 1978; De Kok 1989, 1990; De Kok et al. 1983b). There is a large variation in susceptibility between plant species toward atmospheric H_2S. The basis for this variation is still unclear; it may largely be determined by the fate of the absorbed sulfide within the plant and differences in plant morphology (De Kok 1989, 1990; De Kok et al. 1998). This is illustrated by the observation that in general dicotyledons are more susceptible toward H_2S than monocotyledons. In the latter, H_2S has hardly any direct access to the vegetation point (Stulen et al. 1990, 2000). The magnitude of the in situ sulfide concentration in the cytoplasm is largely determined by the fate of the absorbed sulfide within the plant and its direct metabolism (De Kok 1989, 1990; De Kok et al. 1998).

Sulfide is a very reactive compound and, similar to cyanide, complexes with high affinity to metallo groups in proteins (Mudd 1979; Beauchamps et al. 1984). This reaction is likely the primary biochemical basis for the phytotoxicity of both pedospheric and atmospheric H_2S. In both roots and shoots, H_2S exposure may result in an inhibition of copper- or heme-containing enzymes. For instance, in rice roots the sulfide-induced inhibition of respiration can directly be explained by a reaction of sulfide with the heme-group of cytochrome *c* oxidase (Allam and Hollis 1972). Shoot respiration is not susceptible to exposure to elevated atmospheric H_2S levels (De Kok et al. 1986b; Maas et al. 1988). Nevertheless, in shoots, H_2S exposure may result in a decrease in activity of a wide group of cyanide-sensitive (likely heme-containing) NADH oxidizing enzymes (Maas and De Kok 1988). There is even a direct relation between the in vitro NADH oxidation capacity of extracts of spinach shoots and the reduction of the growth rate of spinach by H_2S, whereas in H_2S tolerant species the in vitro NADH oxidizing capacity does not decrease upon exposure (Stulen et al. 1990, 2000).

H_2S exposure may negatively affect both photosynthetic electron transport and photosynthetic carbon dioxide fixation (Oliva and Steubing 1976; Coyne and Bingham 1978; Steubing 1979; De Kok et al. 1983a; Taylor and Selvidge 1984; Maas et al. 1985, 1988). However, in general, photosynthesis is only reduced at relatively high levels of atmospheric H_2S after prolonged exposure, and is likely the consequence rather than the primary basis of the toxicity of H_2S (De Kok 1989, 1990). Similarly, H_2S exposure also may indirectly affect plant function at low temperature and result in decreased freezing tolerance of the foliage (Stuiver et al. 1992).

3. Uptake and Metabolism of H_2S

Sulfur is an essential element for plant growth. The sulfur requirement of plants depends on their developmental stage and varies considerably among species. Its content in plants ranges between 0.1 % and 1.5 % of dry weight (Duke and Reisenauer 1986). Sulfate is, under normal conditions, the major sulfur source of the plant, and is taken up by the roots and transported to the shoots. After its reduction, the sulfur is incorporated into various essential organic compounds. The main reactions in the sulfur assimilatory pathway are summarized in Fig. 1. Because sulfide is a normal intermediate in plant sulfur metabolism, one would expect that plants are able to utilize absorbed pedospheric and atmospheric H_2S as a nutrient.

Information on the uptake and utilization of pedospheric sulfide by roots is scarce. In theory, plant roots are able to take up sulfide-sulfur from marsh sediments because at pH 6.0 - 6.5 of pore water H_2S is largely undissociated (the first dissociation step with ionization of a single proton and the hydrosulfide anion has a pK_a value of 7.04; Beauchamp et al. 1984) and will easily pass through the membrane of the absorbing roots cells (Spedding et al. 1980a,b; Carlson and Forest 1982). However, it is not yet clear to what extent plant roots take up or metabolize the sulfides directly, or to what extent the sulfide is oxidized near the root surface before its uptake (Fig. 1; Joshi and Hollis 1977; Carlson and Forrest 1982; Fry et al. 1982; Van Diggelen et al. 1987). In addition, it is not yet clear to what extent the absorbed sulfur is directly metabolized in the roots. Roots contain all the necessary enzymes for sulfur reduction and assimilation, but under normal conditions the shoot appears to be the predominant site for the reduction and assimilation of sulfur in plants (Fig. 1; Brunold 1990, 1993).

Plant shoots may be both sink and source of atmospheric H_2S, with uptake and

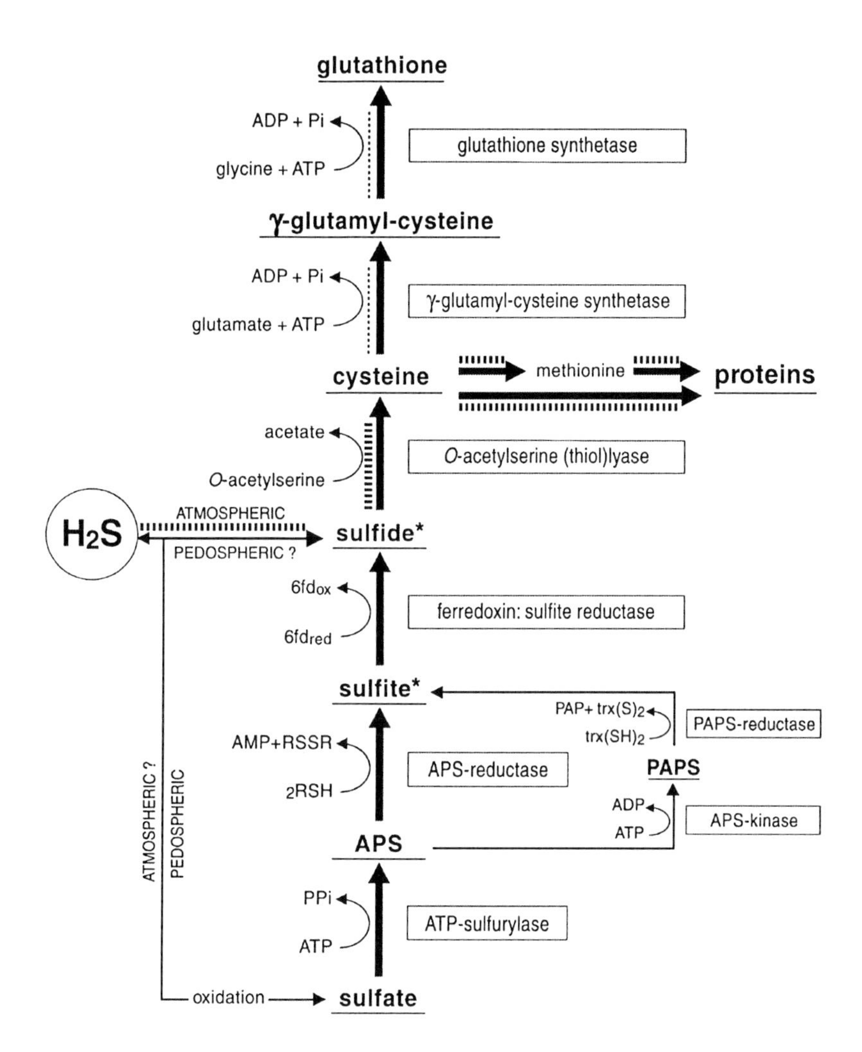

Fig. 1. Assimilation of pedospheric and atmospheric H_2S in plants. (Sulfur assimilatory pathway as adapted from Brunold 1990, 1993; Brunold and Rennenberg 1997; Gutierrez-Marcos et al. 1997; Hell 1997; Schwenn 1997) APS, adenosine 5'-phosphosulfate; fd_{red}, fd_{ox}, reduced and oxidized ferredoxin; PAPS, adenosine 3'-phosphate 5'-phosphosulfate; RSH, RSSG, thiol compound or carrier thiol (reduced and oxidized); $trx(SH)_2$, $trx(S_2)$, reduced and oxidized thioredoxin. *Sulfite and sulfide may be bound to a carrier thiol (possibly glutathione)

emission proceeding predominantly via the stomates (De Kok et al. 1989, 1991,1998; Schröder 1993). At present there are no actual data on the magnitude of the H_2S compensation point (atmospheric concentration where the uptake and emission are in equilibrium). Nevertheless, at atmospheric H_2S levels higher than 0.001 $\mu l\ l^{-1}$, plant shoots act as a sink rather than a source of H_2S (Cope and Spedding 1982; Taylor et al. 1983; De Kok et al. 1989, 1991, 1997, 1998; Poortinga and De Kok 1997; Tausz et al. 1998; Van der Kooij and De Kok 1998). H_2S emission from plants is assumed to have physiological significance as a regulatory factor in the assimilation of sulfur to maintain the sulfur pools in the plant (Schröder 1993). However, from estimates of sulfur fluxes in plants it is obvious that under normal conditions H_2S emission comprises only a negligible fraction of the total sulfur assimilation rate in plants (Stulen and De Kok 1993). More likely, H_2S emission partially reflects the in situ activity of sulfate reduction and the affinity of the cysteine synthesizing enzyme(s) for sulfide (Fig. 1). Only if plants are exposed to high sulfur burdens, such as high levels of atmospheric SO_2, the rate of H_2S emission by shoots may be substantial (Ernst 1990). Nevertheless, under normal conditions, minute rates of H_2S emission may have significance in the protection of plants against pests and environmental stress (Schnug 1997).

The uptake of atmospheric H_2S by plant foliage follows characteristic kinetics, which greatly differ from those observed for other gaseous air pollutants. The uptake is determined by physiological rather than chemicophysical factors and depends strongly on the ambient temperature, with low uptake rates at low temperatures (<10°C) and high rates at moderate temperatures (10-30°C; De Kok et al. 1991). The uptake of H_2S by shoots follows saturation kinetics with respect to the atmospheric H_2S level, whereas levels up to 1 $\mu l\ l^{-1}$ hardly affect stomatal aperture (De Kok et al. 1989, 1991, 1997, 1998; Poortinga and De Kok 1997; Tausz et al. 1998; Van der Kooij and De Kok 1998). The uptake of H_2S by the foliage at high atmospheric H_2S levels and at low temperatures is not limited by its diffusion through the stomates but by the internal resistance of absorbing cells toward H_2S. The apoplastic pH of leaf cells is, in general, lower than 6.4 and the undissociated H_2S will therefore easily pass through the plasma membrane and further diffuse into the cell (De Kok et al. 1991). The internal resistance toward H_2S is largely determined by the rate of its metabolism within the cell (Cope and Spedding 1982; De Kok et al. 1989, 1991, 1997, 1998; Buwalda et al. 1992). The kinetics of H_2S uptake by plant foliage can even be described by the Michaelis-Menten equation, and the apparent maximum uptake rate of H_2S (JH_2S_{max}) differs considerably among species (Fig. 2). The KH_2S, the concentration at which $\frac{1}{2}JH_2S_{max}$ is reached, ranges from 0.14 to 0.6 $\mu l\ l^{-1}$ H_2S.

O-Acetylserine (thiol) lyase plays an active role in the uptake of atmospheric H_2S by plants (Fig. 1; De Kok et al. 1989, 1991, 1997; 1998; Buwalda et al. 1992). At high atmospheric levels, the availability of *O*-acetylserine appears to be limiting to the uptake and metabolism of H_2S, and the maximum uptake rate can be enhanced by the direct supply of *O*-acetylserine, the substrate for *O*-acetylserine (thiol)lyase, to foliar tissue (Buwalda et al. 1992). From the observed low KH_2S values it is evident that *O*-acetylserine (thiol)lyase in situ must have a high affinity to sulfide (De Kok 1990; De Kok et al. 1989, 1991). For instance, if the plant shoot is considered as an aqueous solution in equilibrium with the determined

KH_2S concentrations (Fig. 2), then the theoretical estimated H_2S concentration in the shoot will be ≤0.1 μM (De Kok et al. 1989). Apparently the H_2S uptake kinetics reflect the resultant of the activity of *O*-acetylserine (thiol)lyase, the availability of *O*-acetylserine, and the affinity of the enzyme for H_2S. To what extent the differences in the H_2S uptake rate between species coincides with their differences in sulfur growth requirements needs further investigation (De Kok et al. 1997, 1998). The subcellular site where the absorbed H_2S is metabolized also still needs to be resolved, as *O*-acetylserine (thiol) lyase is present in both chloroplasts and cytosol (Brunold 1990, 1993; Hesse et al. 1997; Gotor et al. 1997; Hell 1997; Saito et al. 1997).

Cysteine desulfhydrase may have significance for the production and emission of H_2S by plants through catabolism of cysteine (Schröder 1993); however, in vitro it also catalyzes synthesis of cysteine in the reverse reaction, with sulfide and β-chloro-L-alanine as substrates (Poortinga et al. 1997). There is circumstantial evidence that cysteine desulfhydrase may also be involved in the assimilation of atmospheric H_2S (Schütz et al. 1991). Still, the significance of the in situ contribution of cysteine desulfhydrase to the direct metabolism of atmospheric H_2S

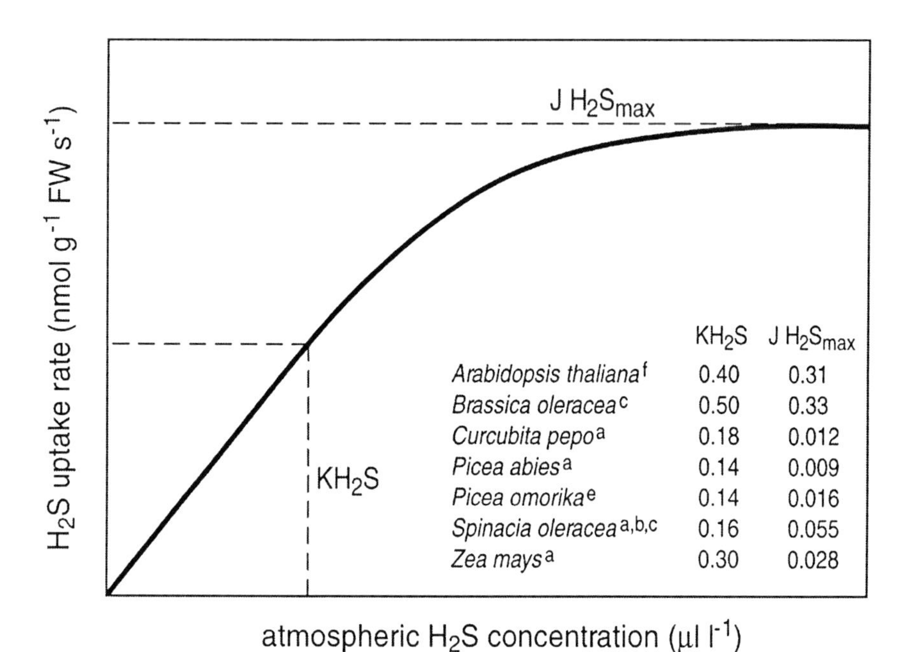

Fig. 2. Uptake kinetics of H_2S to plant shoots at 20°-25°C in the light. JH_2S_{max} represents the apparent maximum uptake rate of H_2S and KH_2S the concentration at which ½J H_2S_{max} is reached. (Data derived from De Kok et al. 1989[a], 1991[b], 1997[c]; Poortinga and De Kok 1997[d]; Tausz et al. 1998[e]; Van der Kooij and De Kok 1998[f]). FW, fresh weight

by plants requires further investigation.

If plants are able to actively metabolize atmospheric H_2S with high affinity into cysteine, a sulfur compound that is the basis of the reduced sulfur source of most organic sulfur compounds in plants (Fig. 1), one may assess the possible significance of atmospheric H_2S as a sulfur nutrient. Nevertheless, H_2S impact studies, as is discussed below, provide more information on the regulatory aspects involved in sulfur assimilation and on the interaction between atmospheric and pedospheric sulfur nutrition in plants.

4. Atmospheric H_2S, Sulfur Nutrition, and Sulfur Assimilation

4.1 H_2S as a Plant Nutrient

The significance of pedospheric H_2S in anoxic soils as a direct sulfur source for plant growth still needs to be evaluated. However, the importance of elevated levels of atmospheric H_2S in plant sulfur nutrition appears to be rather limited. Definitely, if plants are deprived of sulfur in the root environment, they are able to grow with atmospheric H_2S as the sulfur source (De Kok et al. 1997, 2000). Nevertheless, with an ample sulfur supply to the roots, plants switch only in part from sulfate taken up by the roots to atmospheric H_2S taken up by the shoot as a sulfur source for plant growth (De Kok et al. 1997, 2000; Westerman et al. 2000). Constant atmospheric H_2S levels of >0.1 $\mu l\ l^{-1}$ are necessary to meet the sulfur need of plants during the growing period (De Kok et al. 1997, 1998, 2000). Such high concentrations only occur as peak levels in heavily polluted areas or in areas with volcanic activity.

4.2 H_2S and Thiol Metabolism

The water-soluble nonprotein thiol content constitutes only a small proportion of the organic reduced sulfur in plants (about 2%; Stulen and De Kok 1993; De Kok et al. 2000). Glutathione (γ-glu-cys-gly) is generally the most abundant thiol compound present in plant tissue, and its content comprises 90% or more of the water-soluble nonprotein thiol fraction (Bergmann and Rennenberg 1993; Noctor et al. 1998). The thiol pool in the plant is very dynamic and its size may be greatly affected by environmental factors. Changes in the thiol level are, for the greater part, ascribed to changes in the glutathione level (De Kok and Stulen 1993; Noctor et al. 1998).

After H_2S exposure, part (up to 30%) of the atmospheric H_2S metabolized by plants can be identified in the water-soluble nonprotein thiols, the content of which is generally rapidly enhanced upon exposure (Maas et al. 1985, 1987a,b,c; De Kok et al. 1985, 1986a,b, 1997, 1998, 2000; De Kok 1989, 1990; Stuiver et al. 1992; Poortinga and De Kok 1995, 1997; Tausz et al. 1998). In shoots, an increase in the thiol level can already be measured after 1 or 2 hours of exposure at atmospheric levels of ≥ 0.03 μl l^{-1} H_2S, generally reaching a maximal value after 1 or 2 days of exposure, irrespective of the atmospheric H_2S concentration. In shoots, the thiol level may increase up to fivefold; however, it varies considerably between species and depends on the magnitude of the atmospheric H_2S concentration. After some delay, the thiol level in roots may also increase upon H_2S exposure, but generally to a much lesser extent (maximal twofold; Herschbach et al. 1995a,b; De Kok et al. 1997; Poortinga and De Kok 1997; Stuiver and De Kok 1997; Tausz et al. 1998; Westerman et al. 2000). In contrast to that in roots, the increase in thiol levels in shoots upon H_2S exposure is not solely due to an enhanced glutathione level, but is accompanied by a relatively higher enhanced level of cysteine (to more than 30-fold) and of γ-glutamyl-cysteine, the latter compound predominantly in the dark (Buwalda et al. 1988, 1990, 1993, 1994, De Kok et al. 1988; Poortinga and De Kok 1997; Tausz et al. 1998). After a dark-light transition, the accumulated γ-glutamyl-cysteine rapidly disappears and is converted into glutathione (Buwalda et al. 1988). Some plant families contain homologues of glutathione, such as homoglutathione (γ-glu-cys-β-ala) (Bergmann and Rennenberg 1993; Brunold and Rennenberg 1997; Noctor et al. 1997; Rennenberg 1997). Species that contain this glutathione homologue also accumulate this thiol upon H_2S exposure (Buwalda et al. 1993). In both shoots and roots, glutathione is predominantly present in its reduced form, as well as after H_2S exposure (>84 %) (De Kok et al. 1986a; Tausz et al. 1998), even though the glutathione reductase activity is not substantially affected by short-term H_2S exposure (De Kok et al. 1986a). The thiol content of water-soluble proteins is not affected by short-term H_2S exposure (De Kok et al. 1985).

The nature of the change in size and composition of the thiol pool in plant tissue upon H_2S exposure requires further examination. The chloroplast is likely to be the predominant site of sulfate reduction, and of cysteine and glutathione synthesis under normal conditions (Brunold 1990, 1993; Bergmann and Rennenberg 1993; Rennenberg 1997; Hell 1997; Schwenn 1997). However, upon H_2S exposure part of the cysteine may not be synthesized in the chloroplast but in the cytosol (see above), and may be beyond the existing systems of feedback control of assimilation. Furthermore, little is known about differences in participation between the different leaf cells in sulfate reduction and assimilation. Part of the absorbed atmospheric H_2S may be metabolized in cells, which normally do not fulfil a role in sulfur assimilation but still have the ability to synthesize cysteine. Nevertheless, it is evident that at these sites neither the necessary enzymes nor the substrates are sufficiently available for the incorporation of the surplus cysteine into glutathione. Likewise, in the dark, the availability of glycine is not sufficient for the conversion of synthesized γ-glutamyl-cysteine to glutathione (Buwalda et al. 1990), and increase in γ-glutamyl-cysteine levels in the dark can be prevented by adding glycine directly to the leaf tissue, which results in glutathione accumulation (Buwalda et al. 1990, 1993). Upregulation of glutathione synthesis in transgenic plants, by overexpression of a γ-glutamyl-cysteine synthetase gene, results in a similar "shortage" in glycine to complete the synthesis of glutathione (Noctor et al. 1997, 1998).

The question arises to what extent the increase in thiol levels upon H_2S exposure represents either a surplus of reduced organic sulfur temporarily stored in a specific subcellular compartment of cells or an altered but dynamic equilibrium in the size of the thiol pool. At present, there is no information available on the cellular or subcellular distribution of the accumulated thiols upon H_2S exposure. After cessation of the H_2S exposure, usually the accumulated thiols in shoots rapidly disappear, without the emission of H_2S (De Kok 1989, 1990). The cysteine and glutathione levels decrease simultaneously and reach comparable values to those of unexposed plants after 1 or 2 days (Buwalda et al. 1994). A dark-light transition immediately after the termination of the H_2S exposure leads to a rapid decrease in accumulated γ-glutamyl-cysteine and a simultaneous increase in the glutathione level, after which both cysteine and glutathione decrease at similar rates (Buwalda et al. 1994). Evidently, the accumulated thiols are rapidly metabolized, without specific preference, after cessation of the H_2S exposure and/or may in part be translocated to the roots (De Kok 1989, 1990).

4.3 H_2S and the Regulation of Sulfate Uptake and Reduction

Even though the size and composition of the thiol pool appear to be largely beyond direct, strict regulatory control upon H_2S exposure, their significance in plant functioning is likely to be rather insignificant (see the following section on sulfur assimilation and phytotoxicity). More interesting is the fact that thiol compounds may fulfil signal functions in the regulation of uptake and assimilation of sulfur in plants. Uptake and transport of sulfate are affected by the sulfur status; plants with a high sulfur status generally are less efficient in sulfate uptake by the roots than those with a low sulfur status (Cram 1990; Clarkson et al. 1993). Sulfate uptake is regulated by negative feedback from sulfate itself and by repression of the sulfate transporter protein, most likely by such reduced sulfur compounds as glutathione (Cram 1990; Clarkson et al. 1993; Hell 1997; Smith et al. 1997). Because atmospheric H_2S can be utilized as a sulfur source for plant growth, the question arises as to the extent of the interaction between atmospheric and pedospheric sulfur uptake and assimilation and the signals involved in the possible mutual regulation.

Indeed, exposure of plants to atmospheric H_2S results in a partial repression of the uptake of sulfate by the roots (Fig. 2; Brunold and Erismann 1974; Herschbach 1995a,b; De Kok et al. 1997, 1998, 2000; Westerman et al. 2000) and its further transport to the shoots (Herschbach et al. 1995a,b). In *Brassica oleracea* L. (curly kale) the H_2S-induced repression of the sulfate uptake by the roots is never greater than 60% and it is already maximal at an atmospheric H_2S level of about 0.2 $\mu l\ l^{-1}$

after 2 days of exposure (Westerman et al. 2000). Evidently, at a normal sulfur supply to the roots, plants switch only in part to H_2S taken up by the shoots as a sulfur source for plant growth. However, a higher repression of the sulfate uptake appears not to be necessary, since upon exposure to 0.2 $\mu l\ l^{-1}$ H_2S the proportion of the absorbed atmospheric sulfur appears to be sufficient to cover the organic sulfur need for growth of this species (De Kok et al. 2000). It is still unclear to what extent an enhanced glutathione level is the trigger of the H_2S-induced repression of sulfate uptake. The glutathione content may increase in roots upon H_2S exposure; however, the pattern of its increase appears not to be in tune with the repression of sulfate uptake (Westerman et al. 2000). Further H_2S impact studies may have great significance as tools for studying the signals involved in regulating the expression and activity of the sulfate transporter.

Some species accumulate sulfate in the shoot upon exposure to H_2S (De Kok 1989, 1990). The question arises whether this sulfate originates from oxidation of the absorbed atmospheric H_2S or is a reflection of the absence of an H_2S-induced repression of sulfate uptake by the roots and transport to the shoot. The regulation of assimilatory sulfate reduction predominantly occurs at the sites of ATP sulfurylase and adenosine 5'-phosphosulfate reductase (APS reductase), with sulfide, *O*-acetylserine, or cysteine being the most likely regulators (Brunold 1990, 1993; Hawkesford et al. 1995). High atmospheric levels of H_2S ($\geq 15\ \mu l\ l^{-1}$) repress the activities of both these enzymes (Brunold and Erismann 1975; Brunold and Schmidt 1976). At low atmospheric H_2S levels (0.2 - 0.8 $\mu l\ l^{-1}$) the activity of APS reductase in shoots of *Brassica oleracea* is rapidly decreased (up to 80%), whereas the activity of ATP sulfurylase and that of the enzymes involved in the synthesis of cysteine (serine acetyltransferase and *O*-acetylserine (thiol) lyase) are not affected (De Kok et al. 2000; Westerman et al. 2001).

4.4 H_2S and Sulfolipid Biosynthesis

Sulfolipids, predominantly sulfoquinovosyl diacylglycerol in the shoot, may constitute up to 3%-6% of the total sulfur fraction. The route of sulfolipid biosynthesis is still obscure, and especially the sulfur precursor of the sulfolipid is unknown. Adenosine 5'-phosphosulfate (APS), adenosine 3'-phosphate 5'-phosphosulfate (PAPS), sulfite (see Fig. 1) and cysteic acid, an oxidation product of cysteine, are all possible candidates (Heinz 1993). The level of sulfur nutrition and H_2S exposure have a significant impact on lipid and sulfolipid metabolism. Sulfolipids can be synthesized in plants with H_2S as the sole sulfur source; however, their levels are lower than in plants grown with sulfate in the root environment (De Kok et al. 1997). Apparently, cysteine or one of its metabolites, which are derived from H_2S, are not favorable substrates for sulfolipid biosynthesis. Alternatively, the substrates which may be formed in a different intracellular compartment (the cytosol; see earlier) than that in which the sulfolipids are synthesized (the chloroplast; Heinz 1993).

5. H_2S Metabolism Versus Toxicity

At present, it is not clear to what extent variation in susceptibility between species can be related to differences in rates of metabolism of the absorbed H_2S (De Kok et al. 1989, 1998; Stuiver et al. 1992). In the species tested so far, there is no direct relation between H_2S uptake and the growth response of plants. Because H_2S is a very reactive compound, the in situ abundance of sulfide in the plant, especially in meristematic cells, is presumably one of the most important factors determining the phytotoxicity of H_2S (De Kok 1989, 1990; De Kok et al. 1989; Stulen et al. 1990, 2000; Stuiver et al. 1992). The in situ sulfide concentration and the possible degree of intercellular penetration will depend on the fate of the sulfide at the site of its absorption within the plant relative to its reactivity with cellular compounds, its metabolism into cysteine and subsequently into other organic sulfur compounds. Particularly, the affinity and activity of the enzyme(s) involved in the fixation of atmospheric H_2S into cysteine, the rate of synthesis, and the availability of *O*-acetylserine may be of great importance for the height of in situ sulfide concentration. In this view, is interesting that transgenic tobacco plants with a fivefold increased level of *O*-acetylserine sulfhydrylase show a higher tolerance (less visible injury) to acute high H_2S levels (Youssefian et al. 1993). However, the significance of the response of these transgenic plants, and the involvement of the direct metabolism of H_2S in the protection against the toxic effects of H_2S requires evaluation at more realistic exposure levels.

It is unlikely that disturbed regulation of the size and composition of the thiol pool is directly involved in the phytotoxicity of H_2S (De Kok 1990). Even though enhanced intracellular cysteine levels are considered to be toxic (Rennenberg 1981; Filner et al. 1984), several plant species (especially monocotyledons) tolerate high levels of thiols, including cysteine, without any negative growth responses (De Kok 1989, 1990; Stulen et al. 1990, 2000). Likewise, in some species H_2S exposure results in enhanced sulfate levels, but there is no evidence for a direct relation between the level of sulfate accumulation (likely in the vacuole) and the toxicity of H_2S (De Kok 1989, 1990).

6. Concluding Remarks

In contrast to most other gaseous pollutants, there are in general few data available on the significance and the level of H_2S pollution. In addition, most countries do not yet have standards for H_2S pollution (Schulze and Stix 1990). However, in several natural and industrial areas all over the world, plants have to cope with elevated levels of H_2S in either the pedosphere or atmosphere, which may negatively affect plant growth and functioning. Furthermore, there is a wide variation in susceptibility toward H_2S among species that at present remains largely unexplained.

On the other hand, plants are able to use H_2S as a sulfur source for growth, and they even actively fix it from the atmosphere and incorporate it into cysteine, one of the key sulfur compounds in plants. This fact makes H_2S impact studies useful tools for elucidating the signals involved in the regulation of sulfate uptake, transport, and reduction. Furthermore, they are helpful in providing further insight into the interaction between shoot and roots in sulfur assimilation, the route of sulfolipid biosynthesis, and the mutual regulation of sulfur and nitrogen metabolism in plants.

Glutathione is presumed to be significant in various processes in plants, not only in sulfur metabolism but also in selenium metabolism, in the modulation of gene expression, and in protection against oxidative and environmental stress (De Kok and Stulen 1993). To obtain more insight into the involvement of glutathione in these processes, several techniques have been used to modulate glutathione levels in plants, including the overexpression of the enzymes involved in glutathione synthesis (Hell 1997; Noctor et al. 1997, 1998). It is evident that H_2S exposure results in a rapid accumulation of thiols, especially in the shoots, which in most species is predominantly glutathione. Therefore, H_2S impact studies are valuable tools for testing the role and significance of glutathione in these processes (Stuiver et al. 1992).

References

Ajax R, Lee RE Jr (1976) Non-pesticidal air pollution from agricultural processes In: Lee EE Jr (ed) Air pollution from pesticides and agricultural processes. CRC Press, Boca Raton, pp 227-223

Allam AI, Hollis JP (1972) Sulfide inhibition in rice roots. Phytopathology 62:634-639

Bates TS, Lamb, BK, Guenther A, Dignon J, Stoiber RE (1992) Sulfur emissions to the atmosphere from natural sources. J Atmos Chem 14:315-337

Beauchamp RO Jr, Bus JS, Popp JA, Boreiko CJ, Andjelkovich DA (1984) A critical review of the literature on hydrogen sulfide toxicity. CRC Crit Rev Toxicol 13:25-97

Bergmann L, Rennenberg H (1993) Glutathione metabolism in plants. In: De Kok LJ, Stulen I, Rennenberg H, Brunold C, Rauser WE (eds) Sulfur nutrition and assimilation in higher plants: regulatory agricultural and environmental aspects. SPB Academic Publishing, The Hague, pp 109-123

Brunold C (1990) Reduction of sulfate to sulfide. In: Rennenberg H, Brunold C, De Kok LJ, Stulen I (eds) Sulfur nutrition and sulfur assimilation in higher plants. SPB Academic Publishing, The Hague, pp 13-33

Brunold C (1993) Regulatory interactions between sulfate and nitrate assimilation. In: De Kok LJ, Stulen I, Rennenberg H, Brunold C, Rauser WE (eds) Sulfur nutrition and assimilation in higher plants: regulatory agricultural and environmental Aspects. SPB Academic Publishing, The Hague, pp 61-75

Brunold C, Erismann KH (1974) H_2S als Schwefelquelle bei *Lemna minor* L: Einfluss auf das Wachstum den Schwefelgehalt und die Sulfataufnahme. Experientia 30:465-467

Brunold C, Erismann KH (1975) H_2S as sulfur source in *Lemna minor* L: II Direct incorporation into cysteine and inhibition of sulfate assimilation. Experientia 31:508-

510

Brunold C, Schmidt A (1976) Regulation of adenosine 5'-phosphosulfate sulfotransferase activity by H_2S in *Lemna minor* L. Planta 133:85-88

Brunold C, Rennenberg H (1997) Regulation of sulfur metabolism in plants: first molecular approaches. Prog Bot 58:164-186

Buwalda F, De Kok LJ, Stulen I, Kuiper PJC (1988) Cysteine γ-glutamyl-cysteine and glutathione contents of spinach leaves as affected by darkness and application of excess sulfur. Physiol Plant 74:663-668

Buwalda F, Stulen, I De Kok LJ, Kuiper PJC (1990) Cysteine γ-glutamyl-cysteine and glutathione contents of spinach leaves as affected by darkness and application of excess sulfur. II. Glutathione accumulation in detached leaves exposed to H_2S in the absence of light is stimulated by the supply of glycine to the petiole. Physiol Plant 80:196-204

Buwalda F, De Kok LJ, Stulen I (1992) The flux of atmospheric H_2S to spinach leaves can be affected by the supply of *O*-acetylserine. Phyton 32(3):15-18

Buwalda F, De Kok LJ, Stulen I (1993) Effects of atmospheric H_2S on thiol composition of crop plants. J Plant Physiol 142:281-285

Buwalda F, De Kok LJ, Stulen I (1994) The pool of water-soluble non-protein thiols is not regulated within narrow limits in spinach leaves exposed to atmospheric H_2S. Plant Physiol Biochem 32:533-537

Carlson PRJr, Forrest J (1982) Uptake of dissolved sulfide by *Spartina alterniflora*: evidence from natural sulfur isotope abundance ratios. Science 216:633-635

Clarkson DT, Hawkesford MJ, Davidian J-C (1993) Membrane and long-distance transport of sulfate. In: De Kok LJ, Stulen I, Rennenberg H, Brunold C, Rauser WE (eds) Sulfur nutrition and assimilation in higher plants: regulatory agricultural and environmental aspects. SPB Academic Publishing, The Hague, pp 3-19

Cram WJ (1990) Uptake and transport of sulfate. In: Rennenberg H, Brunold C, De Kok LJ, Stulen I (eds) Sulfur nutrition and sulfur assimilation in higher plants. SPB Academic Publishing, The Hague, pp 3-11

Cope DM, Spedding DJ (1982) Hydrogen sulphide uptake by vegetation. Atmos Environ 16: 349-353

Coyne PI, Bingham GE (1978) Photosynthesis and stomatal light responses in snap beans exposed to hydrogen sulfide and ozone. J Air Pollut Control Assoc 28:1119-1123

De Kok LJ (1989) Responses of sulfur metabolism in plants to atmospheric hydrogen sulfide. Phyton 29:189-201

De Kok LJ (1990) Sulfur metabolism in plants exposed to atmospheric sulfur. In: Rennenberg H, Brunold C, De Kok LJ, Stulen I (eds) Sulfur nutrition and sulfur assimilation in higher plants. SPB Academic Publishing, The Hague, pp 111-130

De Kok LJ, Stulen I (1993) Functions of glutathione in plants under oxidative stress. In: De Kok LJ, Stulen I, Rennenberg H, Brunold C, Rauser WE (eds) Sulfur nutrition and assimilation in higher plants: regulatory agricultural and environmental aspects. SPB Academic Publishing, The Hague, pp 125-138

De Kok LJ, Thompson CR, Kuiper PJC (1983a) Sulfide-induced oxygen uptake by isolated spinach chloroplasts catalyzed by photosynthetic electron transport. Physiol Plant 59:19-22

De Kok LJ, Thompson CR, Mudd JB, Kats G (1983b) Effect of H_2S fumigation on water-soluble sulfhydryl compounds in shoots of crop plants. Z Pflanzenphysiol 111:85-89

De Kok LJ, Bosma W, Maas FM, Kuiper PJC (1985) The effect of short-term H_2S fumigation on water-soluble sulfhydryl compounds and glutathione levels in spinach.

Plant Cell Environ 8:189-194

De Kok LJ, Maas FM, Godeke J, Haaksma AB, Kuiper PJC (1986a) Glutathione a tripeptide which may function as a temporary storage of excessive reduced sulphur in H_2S fumigated spinach plants. Plant Soil 91:349-352

De Kok LJ, Stulen I, Bosma W, Hibma J (1986b) The effect of short-term H_2S fumigation on nitrate reductase activity in spinach leaves. Plant Cell Physiol 27:1249-1254

De Kok LJ, Stahl K, Rennenberg H (1989) Fluxes of atmospheric hydrogen sulfide to plant shoots. New Phytol 112:533-542

De Kok LJ, Rennenberg H, Kuiper PJC (1991) The internal resistance in spinach shoots to atmospheric H_2S deposition is determined by metabolism processes. Plant Physiol Biochem 29:463-470

De Kok LJ, Stuiver CEE, Rubinigg M, Westerman S, Grill D (1997) Impact of atmospheric sulfur deposition on sulfur metabolism in plants: H_2S as sulfur source for sulfur deprived *Brassica oleracea* L. Bot Acta 110:411-419

De Kok LJ, Stuiver CEE, Stulen I (1998) The impact of elevated levels of atmospheric H_2S on plants. In: De Kok LJ, Stulen I (eds) Responses of plant metabolism to air pollution and global change. Backhuys Publishers, Leiden, pp 51-63

De Kok LJ, Westerman S, Stuiver CEE, Stulen I (2000) Atmospheric H_2S as plant sulfur source: interaction with pedospheric sulfur nutrition - a case study with *Brassica oleracea* L. In: Brunold C, Rennenberg H, De Kok LJ, Stulen I, Davidian J-C (eds) Sulfur nutrition and sulfur assimilation in higher plants: molecular, biochemical and physiological aspects. Paul Haupt, Bern, pp 41-55

Duke SH, Reisenauer HM (1986) Roles and requirements of sulfur in plant nutrition. In: Tabatabai MA (ed) Sulfur in agriculture. American Society of Agronomy, Madison, pp 123-168

Ernst WHO (1990) Ecological aspects of sulfur metabolism. In: Rennenberg H, Brunold C, De Kok LJ, Stulen I (eds) Sulfur nutrition and sulfur assimilation in higher plants. SPB Academic Publishing, The Hague, pp 131-144

Ernst WHO (1997) Life-history syndromes and the ecology of plants from high sulphur habitats. In: Cram WJ, De Kok LJ, Stulen I, Brunold C, Rennenberg H (eds) Sulphur metabolism in higher plants: molecular ecophysiological and nutritional aspects. Backhuys Publishers, Leiden, pp 131-146

Filner P, Rennenberg H, Sekya J, Bressan RA, Wilson LG, Le Cureux L, Shimei T (1984) Biosynthesis and emission of hydrogen sulfide by higher plants. In: Koziol MJ, Whatley FR (eds) Gaseous air pollutants and plant metabolism. Butterworths, London, pp 291-312

Ford HW (1973) Levels of hydrogen sulfide toxic to citrus roots. J Am Soc Hortic Sci 98: 66-68

Fry B, Scalan, RS, Winters JK, Parker PL (1982) Sulphur uptake by salt grasses, mangroves, and seagrasses in anaerobic sediments. Geochim Cosmochim Acta 46: 1121-1124

Gotor C, Cejudo FJ, Barroso C, Vega JM (1997) Cytosolic *O*-acetylserine(thiol)lyase is highly expressed in trichomes of *Arabidopsis*. In: Cram WJ, De Kok LJ, Stulen I, Brunold C, Rennenberg H (eds) Sulphur metabolism in higher plants: molecular ecophysiological and nutritional aspects. Backhuys Publishers, Leiden, pp 221-223

Gutierrez-Marcos JF, Roberts MA, Campbell EI, Wray JL (1997) Molecular evidence supports an APS-dependent pathway of reductive sulphate assimilation in higher plants. In: Cram WJ, De Kok LJ, Stulen I, Brunold C, Rennenberg H (eds) Sulphur metabolism in higher plants: molecular ecophysiological and nutritional aspects. Backhuys Publishers, Leiden, pp 187-189

Hawkesford MJ, Schneider A, Belcher AR, Clarkson DT (1995) Regulation of enzymes involved in the sulphur-assimilatory pathway. Z Pflanzenernähr Bodenk 158:55-57

Heinz E (1993) Recent investigations on the biosynthesis of the plant sulfolipid. In: De Kok LJ, Stulen I, Rennenberg H, Brunold C, Rauser WE (eds) Sulfur nutrition and assimilation in higher plants: regulatory agricultural and environmental aspects. SPB Academic Publishing, The Hague, pp 163-178

Hell R (1997) Molecular physiology of plant metabolism. Planta 202:138-148

Hendrickson ER (1979) Hydrogen sulfide - its properties occurrences and uses. In: Hydrogen sulfide. University Park Press, Baltimore, pp 1-9

Herschbach C, De Kok LJ, Rennenberg H (1995a) Net uptake of sulfate and its transport to the shoot in spinach plants fumigated with H_2S or SO_2: does atmospheric sulfur affect the 'inter-organ' regulation of sulfur nutrition? Bot Acta 108:41-46

Herschbach C, De Kok LJ, Rennenberg H (1995b) Net uptake of sulfate and its transport to the shoot in tobacco plants fumigated with H_2S or SO_2. Plant Soil 175:75-84

Hesse H, Lipke J, Altman T, Höfgen R (1997) Expression analysis and subcellular localization of cysteine synthase isoforms from *Arabidopis thaliana*. In: Cram WJ, De Kok LJ, Stulen I, Brunold C, Rennenberg H (eds) Sulphur metabolism in higher plants: molecular ecophysiological and nutritional aspects. Backhuys Publishers, Leiden, pp 227-230

Joshi MM, Hollis JP (1977) Interaction of *Beggiatoa* and rice plant: detoxification of hydrogen sulfide in the rice rhizosphere. Science 195:179-180

Joshi MM, Ibrahim IKA, Hollis JP (1973) Oxygen release from rice seedlings. Physiol Plant 29:269-271

Joshi MM, Ibrahim IKA, Hollis JP (1975) Hydrogen sulfide: effects on physiology of rice plants and relation to straighthead desease. Phytopathology 65:1165-1170

Kellogg WW, Cadle RD, Allen ER, Lazrus AL, Martell EA (1972) The sulfur cycle. Science 175:587-596

Kelly DP (1988) Oxidation of sulphur compounds. In: Colle JA, Ferguson, SJ (eds) The nitrogen and sulphur cycles. Cambridge University Press, Cambridge, pp 65-98

Krause GHM (1979) Relative Phytotoxizität von Schwefelwasserstoff. Staub-Reinhalt Luft 39:165-167

Maas FM, De Kok LJ, Kuiper PJC (1985) The effect of H_2S fumigation on various spinach (*Spinacia oleracea* L.) cultivars. Relation between growth inhibition and accumulation of sulphur compounds in the plant. J Plant Physiol 119:219-226

Maas FM, De Kok LJ, Hoffmann I, Kuiper PJC (1987a) Plant responses to H_2S and SO_2 fumigation. I. Effects on growth, transpiration and sulfur content of spinach. Physiol Plant 70:713-721

Maas FM, De Kok LJ, Strik-Timmer W, Kuiper PJC (1987b) Plant responses to H_2S and SO_2 fumigation. II. Differences in metabolism of H_2S and SO_2 in spinach. Physiol Plant 70:722-728

Maas FM, De Kok LJ, Peters JL, Kuiper PJC (1987c) A comparative study on the effects of H_2S and SO_2 fumigation on the growth and accumulation of sulfate and sulfhydryl compounds in *Trifolium pratense* L., *Glycine max* Merr. and *Phaseolus vulgaris* L. J Exp Bot 38:1459-1469

Maas FM, van Loo EN, van Hasselt PR (1988) Effect of long-term H_2S fumigation on photosynthesis in spinach. Correlation between CO_2 fixation and chlorophyll *a* fluorescence. Physiol Plant 72:77-83

Maas FM, De Kok LJ (1988) *In vitro* NADH oxidation as an early indicator for growth reduction in spinach exposed to H_2S in the ambient air. Plant Cell Physiol 29:523-526

Mudd JB (1979) Effects on vegetation and aquatic animals In: Hydrogen sulfide. University Park Press, Baltimore, pp 67-79

Noctor G, Arisi ACM, Jouanin L, Valadier MH, Roux Y, Foyer CH (1997) Light-dependent modulation of foliar glutathione synthesis and associated amino acid metabolism in poplar overexpressing γ-glutamylcysteine synthetase. Planta 202:357-369

Noctor G, Arisi ACM, Jouanin L, Kunert KR, Rennenberg H, Foyer CH (1998) Glutathione: biosynthesis, metabolism and relationship to stress tolerance explored in transformed plants. J Exp Bot 49:623-647

Noggle JC, Meagher JF, Jones US (1986) Sulfur in the atmosphere and its effect on plant growth. In: Tabatabai MA (ed) Sulfur in agriculture. American Society of Agronomy, Madison, pp 251-278

Oliva M, Steubing L (1976) Untersuchungen über der Beeinflussung von Photosynthese, Respiration and Wasserhaushalt durch H_2S bei *Spinacia oleracea*. Angew Bot 50:1-17

Ouattara AS, Jacq VA (1992) Characterization of sulfate-reducing bacteria isolated from Senegal ricefields. FEMS Microbiol Ecol 101:217-228

Poortinga AM, De Kok LJ (1995) Utilization of H_2S by plant foliar tissue: its interaction with sulfate assimilation. Z Pflanzenernähr Boedenk 158:59-62

Poortinga AM, De Kok LJ (1997) Uptake of atmospheric H_2S by *Spinacia oleracea* L. and consequences for thiol content and composition in shoots and roots. In: Cram WJ, De Kok LJ, Stulen I, Brunold C, Rennenberg H (eds) Sulphur metabolism in higher plants: molecular cophysiological and nutritional aspects. Backhuys Publishers, Leiden, pp 285-288

Poortinga AM, Hoen G, De Kok LJ (1997) Cysteine desulfhydrase of *Spinacia oleracea* L. may catalyze synthesis of cysteine. In: Cram WJ, De Kok LJ, Stulen I, Brunold C, Rennenberg H (eds) Sulphur metabolism in higher plants: molecular ecophysiological and nutritional aspects. Backhuys Publishers, Leiden, pp 233-234

Rennenberg H (1981) Differences in the use of cysteine and glutathione as sulfur source in photoheterotrophic tobacco suspension cultures. Z Pflanzenphysiol 105:31-40

Rennenberg H (1997) Molecular approaches to glutathione biosynthesis In: Cram WJ, De Kok LJ, Stulen I, Brunold C, Rennenberg H (eds) Sulphur metabolism in higher plants: molecular ecophysiological and nutritional aspects. Backhuys Publishers, Leiden, pp 59-70

Saito K, Takahashi H, Takagi Y, Inoue K, Nojim M (1997) Molecular characterization and regulation of cysteine synthase and serine acetyltransferase from plants. In: Cram WJ, De Kok LJ, Stulen I, Brunold C, Rennenberg H (eds) Sulphur metabolism in higher plants: molecular ecophysiological and nutritional aspects. Backhuys Publishers, Leiden, pp 235-238

Schnug E (1997) Significance of sulfur for the quality of domesticated plants. In: Cram WJ, De Kok LJ, Stulen I, Brunold C, Rennenberg H (eds) Sulphur metabolism in higher plants: molecular ecophysiological and nutritional aspects. Backhuys Publishers, Leiden, pp 109-110

Schröder P (1993) Plants as sources of atmospheric sulfur. In: De Kok LJ, Stulen I, Rennenberg H, Brunold C, Rauser WE (eds) Sulfur nutrition and assimilation in higher plants: regulatory agricultural and environmental aspects. SPB Academic Publishing, The Hague, pp 253-270

Schulte M, Herschbach C, Rennenberg H (1997) Long term effects of naturally elevated CO_2, H_2S and SO_2 on sulphur allocation in *Quercus*. In: Cram WJ, De Kok LJ, Stulen I, Brunold C, Rennenberg H (eds) Sulphur metabolism in higher plants: molecular ecophysiological and nutritional aspects. Backhuys Publishers, Leiden, pp 289-291

Schulze E, Stix E (1990) Beurteilung phytotoxischer Immissionen für die noch keine Luftqualitätskriterien festgelegt sind. Angew Bot 64:225-235

Schütz B, De Kok LJ, Rennenberg H (1991) Thiol accumulation and cysteine desulfhydrase activity in H_2S-fumigated leaves and leaf homogenates of cucurbit plants. Plant Cell Physiol 32:733-736

Schwenn JD (1997) Assimilatory reduction of inorganic sulphate. In: Cram WJ, De Kok LJ, Stulen I, Brunold C, Rennenberg H (eds) Sulphur metabolism in higher plants: molecular ecophysiological and nutritional aspects. Backhuys Publishers, Leiden, pp 39-58

Slatt BJ, Natusch DFS, Prospero JM, Savoie DL (1978) Hydrogen sulfide in the atmosphere of the northern equatorial atlantic ocean and its relation to the global sulfur cycle. Atmos Environ 12:981-991

Smith FW, Hawkesford MJ, Ealing PM, Clarkson DT, Vanden Berg PJ, Belcher AR, Warrilow AGS (1997) Regulation of expression of a cDNA from barley roots encoding a high affinity sulphate transporter. Plant J 12:875-884

Spedding DJ, Ziegler I, Hampp R, Ziegler H (1980a) Effect of pH on the uptake of [^{35}S] sulfur from sulfate, sulfite and sulfide by isolated chloroplasts. Z Pflanzenphysiol 96:351-364

Spedding DJ, Ziegler I, Hampp R, Ziegler H (1980b) Effect of pH on the uptake of [^{35}S] sulfur from sulfate, sulfite and sulfide by *Chlorella vulgaris*. Z Pflanzenphysiol 97:205-214

Steubing L (1979) Wirkung von Schwefelwasserstoff auf höhere Pflanzen. Staub-Reinhalt Luft 39:161-164

Stuiver CEE, De Kok LJ (1997) Atmospheric H_2S as sulphur source for sulphur deprived *Brassica oleracea* L and *Hordeum vulgare* L. In: Cram WJ, De Kok LJ, Stulen I, Brunold C, Rennenberg H (eds) Sulphur metabolism in higher plants: molecular ecophysiological and nutritional aspects. Backhuys Publishers, Leiden, pp 293-294

Stuiver CEE, De Kok LJ, Kuiper PJC (1992) Freezing tolerance and biochemical changes in wheat shoots as affected by H_2S fumigation. Plant Physiol Biochem 30:47-55

Stulen I, De Kok LJ (1993) Whole plant regulation of sulfate uptake and metabolism - a theoretical approach and comparison with current ideas on regulation of nitrogen metabolism. In: De Kok LJ, Stulen I, Rennenberg H, Brunold C, Rauser WE (eds) Sulfur nutrition and assimilation in higher plants: regulatory agricultural and environmental aspects. SPB Academic Publishing, The Hague, pp 77-91

Stulen I, Posthumus FS, Amâncio S, De Kok LJ (1990) Why is H_2S not phytotoxic in monocots? Physiol Plant 79(2):A123

Stulen I, Posthumus FS, Amâncio S, Masselink-Beltman I, Müller M, De Kok LJ (2000) Mechanism of H_2S phytotoxicity. In: Brunold C, Rennenberg H, De Kok LJ, Stulen I, Davidian J-C (eds) Sulfur nutrition and sulfur assimilation in higher plants: molecular, biochemical and physiological aspects. Paul Haupt, Bern, pp 381-383

Tausz M, Van der Kooij TAW, Müller M, De Kok LJ, Grill D (1998) Uptake and metabolism of oxidized and reduced sulfur pollutants by spruce trees. In: De Kok LJ, Stulen I (eds) Responses of plant metabolism to air pollution and global change. Backhuys Publishers, Leiden, pp 455-458

Taylor GE, Selvidge WJ (1984) Phytotoxicity in bush bean of five sulfur-containing gases released from advanced fossil energy technologies. J Environ Qual 13:224-230

Taylor GE, McLaughlin SB, Shriner DS, Selvidge WJ (1983) The flux of sulfur containing gases to vegetation. Atmos Environ 17:789-796

Thompson CR, Kats G (1978) Effect of continuous H_2S fumigation on crop and forest plants. Environ Sci Technol 12:550-553

Tisdale SL, Reneau RB Jr, Platou JS (1986) Atlas of sulfur deficiencies. In: Tabatabai MA (ed) Sulfur in agriculture. American Society of Agronomy, Madison, pp 295-323

Trudinger PA (1986) Chemistry of the sulfur cycle. In: Tabatabai MA (ed) Sulfur in Agriculture. American Society of Agronomy, Madison, pp 1-22

Urone P (1976) The primary air pollutants - their occurrence, sources and effects. In: Stem AC (ed) Air pollution. Academic Press, New York, pp 24-75

Van der Kooij TAW, De Kok LJ (1998) Kinetics of deposition of SO_2 and H_2S to shoots of *Arabidopsis thaliana* L. In: De Kok LJ, Stulen I (eds) Responses of plant metabolism to air pollution and global change. Backhuys Publishers, Leiden, pp 481-483

Van Diggelen J, Rozema J, Broekman R (1987) Growth and mineral relations of salt-marsh species on nutrient solutions containing various sodium sulphide concentrations. In: Huiskes AHL, Blom CWPM, Rozema J (eds) Vegetation between land and sea. Junk Publishers, Dordrecht, pp 260-168

Westerman S, De Kok LJ, Stuiver CEE, Stulen I (2000) Interaction between metabolism of atmospheric H_2S in the shoot and sulfate uptake by the roots of curly kale (*Brassica oleracea*). Physiol Plant 109:443-449

Westerman S, Stulen I, Stuter M, Brunold C, De Kok LJ (2001) Atmospheric H_2S as sulfur source for Brassica oleracea: consequences for the activity of the enzymes of the assimilatory sulfate reduction pathway. Plant Physiol Biochem 39: 425-432

Youssefian S, Nakamura M, Sano H (1993) Tobacco plants transformed with the *O*-acetylserine (thiol) lyase gene of wheat are resistant to toxic levels of hydrogen sulphide gas. Plant J 4:759-769

11
Metabolism and Detoxification of Nitrogen Dioxide and Ammonia in Plants

Tadakatsu Yoneyama[1], Hak Y. Kim[2], Hiromichi Morikawa[3], and Hari S. Srivastava[4]

[1]Department of Applied Biological Chemistry, Graduate School of Agricultural and Life Sciences, The University of Tokyo, Yayoi 1-1-1, Bunkyo-ku, Tokyo 113-8657, Japan
[2]Institute of Agricultural Science and Technology, Kyungpook National University, 1370 Sankyuk-dong, Puk-ku, Taegu 702-701, Korea
[3]Department of Mathematical and Life Sciences, Graduate School of Science, Hiroshima University, Kagamiyama 1-3-1, Higashi-Hiroshima, Hiroshima 739-8526, Japan
[4]Department of Plant Science, Rohilkhand University, Bareilly 243006, India

1. Introduction

One (T.Y.) of the authors studied NO_2 uptake and metabolism in plants from 1976 to 1979 in the National Institute of Environmental Studies, Tsukuba, and most of the results were published by 1980. After about 17 years, we have been asked to write a review on NO_2 metabolism in plants. The task has become considerably easier with the joining of the coauthors (H.M., H.S.S.), who have been actively engaged in the study of plant responses to NO_2 during the past. It has been interesting to discuss what was known up to 1980 and what has become known since then about NO_2 metabolism and detoxification in plants. In addition, another nitrogen-containing but reduced form of gas, NH_3, is also included in this chapter in the hope that a comparison of the behavior and fate of oxidized and reduced gases in plants may provide further insight into their metabolic and regulatory pathways.

The combustion of fossil energy carriers (oil, gas, and coal) is connected with

Air Pollution and Plant Biotechnology
–Prospects for Phytomonitoring and Phytoremediation–
Edited by K. Omasa, H. Saji, S. Youssefian, and N. Kondo
© Springer -Verlag Tokyo 2002

the generation of trace gases such as CO_2, SO_2, and NOx (NO_2 + NO; see Nouchi, Chapter 1, this volume). Because of the relatively short half-life of NOx in the atmosphere (≈ 1 d), high concentrations are found only in the vicinity of strong sources. Plants are important sinks of NO_2. The major sources of NH_3 are agricultural lands treated with fertilizers, animal wastes, and the direct emission of NH_3 from livestock. The half-life of this gas is also short (≈ 7 h), and NH_3 can be deposited on soils and plants.

2. Absorption and Metabolism of NO_2

Unequivocal evidence of NO_2 uptake by foliar tissues was provided by ^{15}N enrichment of these tissues when they were fumigated with ^{15}N-labeled NO_2 (Durmishidze and Nutsubidze 1976; Yoneyama et al. 1978; Rogers et al. 1979; Kaji et al. 1980). The $^{15}NO_2$ uptake was highly correlated with atmospheric $^{15}NO_2$ concentrations (Rogers et al. 1979). However, such early studies were conducted at high NO_2 concentrations (ppm levels), and it is only more recently that uptake of $^{15}NO_2$ at ppb levels, which are close to the NO_2 concentrations in urban areas, have been investigated for the sunflower (Segschneider et al. 1995) and Norway spruce (Nussbaum et al. 1993). Diurnal variations in NO_2 uptake by various herbaceous plant species were also recognized before 1980 (Yoneyama et al. 1979, 1980), with the high rates of NO_2 uptake during the daytime, suggesting that stomatal resistance to NO_2 diffusion was the key factor in NO_2 uptake.

To examine the differences in NO_2 uptake in various species, analysis of NO_2 absorption by 217 plant taxa, which included 50 wild herbaceous plants collected from roadsides (42 genera, 15 families), 60 cultivated herbaceous plants (55 genera, 30 families), and 107 cultivated woody plants (74 genera, 45 families), has been conducted in fumigation chambers after feeding plants with 4 ppm ^{15}N-labeled NO_2 (Morikawa et al. 1992, 1998) (Table 1). Two parameters, NO_2-N content (NO_2-derived nitrogen content) in fumigated plant leaves (mg N g^{-1} dry wt), and NO_2-utilization index (percentage of the NO_2-derived nitrogen in the total nitrogen), were determined. The NO_2-N content differed 657 fold between the highest (*Eucalyptus viminalis*; 6.57) and the lowest (*Tillandsia ionantha* and *T. caput-medusae*; 0.01) values in the 217 taxa; 62 fold in one family (Theaceae), and 26 fold in one species (*Solidago altissima*). Nine species had NO_2-utilization indices greater than 10%, of which *Magnolia kobus*, *Eucalyptus viminalis*, *Populus nigra*, *Nicotiana tabacum*, and *Erechtites hieracifolia* had NO_2-N contents >4.9. These plants can be considered "NO_2-philic" because the NO_2-nitrogen may serve important function(s) for their growth and development as a nitrogen nutrient.

The Compositae and Myrtaceae had high values for both parameters, whereas the monocots and gymnosperms had low values. These findings, which suggest that the NO_2-nitrogen metabolic pathways differ among plant species, may well be useful for creating novel vegetation technology to reduce the atmospheric concentrations of NO_2. The proportion of plant taxa showing NO_2-absorbing activities higher than 1.0 mg NO_2-N g^{-1} dry wt was about 60% for plants from near NO_2-polluted roads and 36% for cultivated plants. The differences in stomatal aperture between the higher and lower NO_2-absorbing plants were so minute that the large differences in NO_2-absorbing activities appear to be regulated by factors other than stomatal aperture (Morikawa et al. 1993).

So far, no release of NO_2 from plant leaves has been reported (Seqschneider et al. 1995), indicating that the NO_2 compensation concentration is zero in most plants. In a few studies, however, the loss of oxidized nitrogen from the foliage of plants, well-fertilized with soil nitrogen, has been demonstrated (Weiland and Stutte 1979; Da Silva and Stutte 1981).

Until 1980, we knew that in the leaves of plants fumigated with NO_2, the NO_2-derived N was rapidly assimilated into organic forms, although a small amount of NO_2-derived N was also found in nitrite and nitrate in cell saps (Mayumi and Yamazoe 1979; Matsumaru et al. 1979; Yoneyama and Sasakawa 1979) (Table 2). At that time, we considered that the disproportionation of NO_2 in cell water as nitrate and nitrite (1:1), was probably a major process of NO_2 acquisition. However, the rapid transformation of NO_2-derived N into organic forms, the pseudo-first-order uptake involving concentration-dependent absorption (Segschneider et al. 1995), and the almost-zero concentrations of NO_2 on the cell surfaces (negligible mesophyll resistance to NO_2), suggested that other processes, which involved rapid removal of NO_2 from stomatal gas phases and rapid metabolism of dissolved inorganic N (Matsumaru et al. 1979; Kaji et al. 1980), were more important.

Ramge et al. (1993) hypothesized that the disproportionation of NO_2 in water is not the major pathway of NO_2 uptake by plant cells, but that the reaction of NO_2 with antioxidants, in particular with the reduced form of ascorbic acid, is the most plausible mechanism. The latter reaction produces dehydroascorbate and nitrite. The dehydroascorbate may be regenerated to ascorbate by dehydroascorbate reductase (Loewus and Loewus 1987), and the nitrite, which otherwise damages plant cells (Beevers and Hageman 1980), can be further reduced to ammonia (NH_3) and organic N with the high activity of nitrite reductase (NiR). In darkness, where nitrite reductase activity (NiRA) is less than in the light, nitrite accumulates in leaves fumigated with NO_2 (Yoneyama et al. 1979). Rowland-Bamford et al. (1989) reported that NO_2 uptake by leaves was not influenced by the activity of nitrate reductase (NR), as revealed by NO_2 uptake by NR-deficient mutants. Since some nitrate is produced in $^{15}NO_2$-fumigated plants (see Table 2), the dissolution of NO_2 by disproportionation may still partly operate, although nitrite formation by the ascorbate reaction would be the major pathway.

NO_2 fixed by leaves may be assimilated into amino acids by the sequential action of nitrite reductase, glutamine synthetase (GS), and glutamate synthase (GOGAT), as shown in Fig. 1A. The incorporation of ^{15}N into individual amino acids was investigated in the leaves of spinach (Yoneyama and Sasakawa 1979; Kaji et al. 1980) and sunflower (Kaji et al. 1980), and the typical results from such

Table 1. NO_2-N contents and NO_2-utilization indices of the ten highest and lowest taxa of wild herbaceous, cultivated herbaceous, and woody plants (partly from Morikawa et al. 1998)

Taxa	NO_2-N content[a]	NO_2-utilization index[b]	Family
Wild herbaceous, collected from roadside			
Erechtites hieracifolia	5.72	10.14	Compositae
Crassocephalum crepidioides	5.07	9.16	Compositae
Bidens frondosa	2.98	7.98	Compositae
Lactuca indica	2.96	7.96	Compositae
Oenothera biennis	2.77	7.38	Onagraceae
Erigeron annuus	2.73	8.41	Compositae
Artemisia princeps	2.53	6.48	Compositae
Plantago lanceolata	2.24	6	Plantaginaceae
Chenopodium album	2.24	4.68	henopodiaceae
Capsella bursa-pastoris	2.1	3.72	Cruciferae
Coix lacryma-jobi	0.63	2.47	Gramineae
Taraxacum japonicum	0.63	3.52	Compositae
Aster ageratoides	0.61	2.35	Compositae
Eragrostis ferruginea	0.59	2.2	Gramineae
Digitaria ciliaris	0.46	1.98	Gramineae
Miscanthus sinensis	0.46	1.5	Gramineae
Bromus unioloides	0.33	1.78	Gramineae
Paspalum dilatatum	0.26	0.98	Gramineae
Solidago virgaurea	0.26	1.02	Compositae
Portulaca oleracea	0.25	1.6	Portulacaceae
Cultivated herbaceous			
Nicotiana tabacum	5.72	11.44	Solanaceae
Carthamus tinctorius	3.41	5.91	Compositae
Chrysanthemum sp.	3.15	7.09	Compositae
Arabidopsis thaliana	3.03	5.21	Cruciferae
Levisticum officinale	3.02	8.89	Umbelliferae
Impatiens sp.	2.8	7.34	Balsaminaceae
Petunia hybrida	2.72	5.77	Solanaceae
Matthiola incana	2.7	5.34	Cruciferae
Cosmos bipinnatus	2.63	6.24	Compositae

Chamaemelum nobile	2.26	6.54	Compositae
Aspidistra elatior	0.14	1.2	Liliaceae
Dieffenbachia sp.	0.13	0.39	Araceae
Schlumbergera x buckleyi	0.13	0.47	Cactaceae
Tillandsia geminiflora	0.12	4.11	Bromeliaceae
Euphorbia pulcherrima	0.07	0.23	Euphorbiaceae
Saintpaulia confusa	0.06	0.27	Gesneriaceae
Cymbidium sp.	0.04	0.37	Orchidaceae
Tillandsia usneoides	0.02	0.23	Bromeliaceae
Tillandsia caput-medusae	0.01	0.18	Bromeliaceae
Tillandsia ionantha	0.01	0.16	Bromeliaceae
Woody plants			
Eucalyptus viminalis	6.57	12.54	Myrtaceae
Populus nigra	5.14	10.7	Salicaceae
Magnolia kobus	4.92	12.68	Oleaceae
Robinia pseudo-acacia	4.73	8.7	Leguminosae
Eucalyptus grandis	4.57	8.5	Myrtaceae
Eucalyptus globulus	4.08	9.36	Myrtaceae
Populus sp.	3.8	8.43	Salicaceae
Sophora japonica	3.26	7.11	Leguminosae
Prunus cerasoides	3.23	6.67	Rosaceae
Sapium sebiferum	2.85	10.1	Euphorbiaceae
Citrus tachibana	0.18	1.12	Rutaceae
Aucuba japonica	0.15	1.36	Cornaceae
Bougainvillea x buttiana	0.15	1	Nyctaginaceae
Schefflera arboricola	0.14	1.32	Araliaceae
Rhododendron simsii	0.14	0.88	Ericaceae
Juniperus procumbens	0.12	0.86	Cupressaceae
Chamaedorea microcarpa	0.1	0.63	Palmae
Thea sinensis	0.09	0.49	Theaceae
Camellia japonica cultivar	0.04	0.32	Theaceae
Codiaeum variegatum	0.04	0.15	Euphorbiaceae

[a] Content of NO_2-derived reduced nitrogen (mg N/g dry wt).
[b] Percentage of the NO_2-derived reduced nitrogen of the total reduced nitrogen (%).

Table 2. Distribution of ^{15}N in leaves exposed to ^{15}N-labeled NO_2 at 4-6 ppm for 20 min (redrawn from Kaji et al. 1980)

Fraction	Nitrogen content (μg N/ g f.w.)		^{15}N content (atom % excess)		Excess ^{15}N (μg N/ g fw)	
	Day	Night	Day	Night	Day	Night
Sunflower						
Insoluble	7193	5634	0.13	0.04	9.4 (17)[c]	2.3 (14)
Soluble	753	755	6.31	1.88	47.5 (83)	14.2 (86)
$NO_3^- + NO_2^-$	137	138	0.4	1.32	0.5 (1)	1.8 (11)
NO_2^-	0.02	0.03	94.7 [a]	94.7 [a]	0.02	0.03
NO_3^-	137	138	0.35[b]	1.28 [b]	0.48	1.77
Other	616	617	7.63[b]	2.01[b]	47.0 (82)	12.4 (75)
Total	7946	6389			56.9 (100)	16.5 (100)
Spinach						
Insoluble	3782	3668	0.09	0.04	3.4 (16)	1.5 (7)
Soluble	602	610	3.03	3.5	18.2 (84)	21.3 (93)
$NO_3^- + NO_2^-$	136	227	0.16	1.56	0.2 (1)	3.5 (15)
NO_2^-	0.03	1.5	94.7 [a]	94.7 [a]	0.03	1.4
NO_3^-	136	225	0.13 [b]	0.93[b]	0.17	2.1
Other	466	383	3.99 [b]	4.64[b]	18.6 (83)	17.8 (78)
Total	4384	4278			21.6 (100)	22.8 (100)

[a] All NO_2^- in the leaves was assumed to originate from atmospheric NO_2.
[b] Calculated by the following equation: (Excess ^{15}N/nitrogen content)×100.
[c] Numerals in parentheses are values relative to "Total" (100).

studies are shown in Table 3. The highest ^{15}N atom % was found in the amide of glutamine by 20 min fumigation with $^{15}NO_2$, as would be expected from the proposed scheme of NO_2 metabolism in plant cells shown in Fig. 1A. Thus, the ^{15}N-amino acids produced in this manner could be translocated through the phloem to other parts, including the growing leaves and roots. The translocation of ^{15}N, derived from $^{15}NO_2$, from leaves to other plant parts have been investigated in *Phaseolus vulgaris* (Rogers et al. 1979), corn (Kaji et al. 1980), sunflower (Yoneyama et al. 1980), rice (Okano et al. 1983), and Norway spruce (Nussbaum et al. 1993). In sunflower and rice, differential translocation of ^{15}N and of ^{13}C was

Table 3. The ^{15}N atom % excess in free amino acids and amides in sunflower and spinach leaves fumigated with ^{15}N-labeled NO_2 at 4-6 ppm for 20min [a] (redrawn from Kaji et al. 1980)

Amino acid and amide	Sunflower		Spinach	
	Day	Night	Day	Night
Aspartic acid	10.0 (44)	1.51 (9)	7.63 (68) [b]	3.20 (15)
Glutamic acid	11.6 (51)	1.40 (9)	8.88 (79)	3.35 (16)
Serine	6.26 (28)	0.14 (1)	2.69 (24)	0.38 (2)
Alanine	14.3 (63)	1.39 (9)	6.91 (61)	2.19 (10)
γ-Amino butyric acid	12.3 (54)	1.95 (12)	5.54 (49)	1.89 (9)
Leucine+isoleucine	-	0.40 (3)	1.79 (16)	0.19 (1)
Phenylalanine	1.30 (6)	0.17 (1)	1.11 (10)	0.07 (0)
Proline	1.25 (6)	-	0.94 (8)	0.29 (1)
Histidine	0.50 (2)	0.20 (1)	-	-
Arginine	0.79 (3)	0.09 (1)	-	0.27 (1)
Glutamine				
Amino-N	8.30 (37)	0.60 (4)	6.54 (58)	1.44 (7)
Amide-N	22.6 (100)	16.0 (100)	11.3 (100)	21.4 (100)
Aspargine				
Amino-N	0.91 (4)	-	2.73 (24)	0.17 (1)
Amide-N	2.52 (11)	-	3.56 (32)	2.85 (13)

[a] ^{15}N-labeled $^{15}NO_2$ (94.7 atom% excess ^{15}N) was exposed to plants in the daytime or nighttime.
[b] Numerals in parentheses are relative values to the ^{15}N content in the amide nitrogen of glutamine.

revealed when a single mature leaf was fed with $^{15}NO_2$ and $^{13}CO_2$ simultaneously (Yoneyama et al. 1980; Okano et al. 1983).

3. Toxicity and Detoxification of NO_2

Before 1980, we had observed that visible damage, such as wilting, of leaves fumigated with high concentrations of NO_2 was accompanied by accumulation of nitrite in the leaf tissues (Yoneyama et al. 1979). When three plant species, a high NiRA-containing spinach, a medium NiRA-containing kidney bean, and a low-NiRA-containing sunflower, were fumigated with 4 ppm NO_2, accumulation of nitrite was observed in kidney bean and sunflower leaves, with more nitrite accumulating at lower levels of irradiation. It is also noteworthy that NiRA is increased by NO_2 fumigation (Yoneyama et al. 1979; Rowland et al. 1987), and that NR activity was also found to increase in most of the species examined (Srivastava et al. 1995a). However, sometimes during our studies, we observed visible damage without an accompanying increase in nitrite or a suppression of dry matter increase in the NO_2-fumigated leaves. We inferred from these observations that the high reactivity of NO_2 may have induced membrane disruption mediated through peroxidation of the double bonds of unsaturated lipids, as reported for mammalian cells (Menzel 1976).

In vitro studies, employing electron spin resonance measurements, have also indicated an interaction of NO_2 with membrane components, such as with phospholipids, phosphatidyl ethanolamine, lecithin, and fatty acids, with the consequent production of stable free radicals (Estefan et al. 1970). In another in vitro study, addition of NO_2 to the double bonds of cyclohexene was demonstrated, and this serves a model reaction of the NO_2 with unsaturated fatty acids in membranes (Pryor and Lightsey 1981). Microscopic investigations also indicated the destruction of chloroplast membranes in zelkova tree leaves fumigated with NO_2, although no visible injury was observed (Matsushima 1979). More recently, Ramge et al. (1993) proposed a possible interaction of NO_2 with

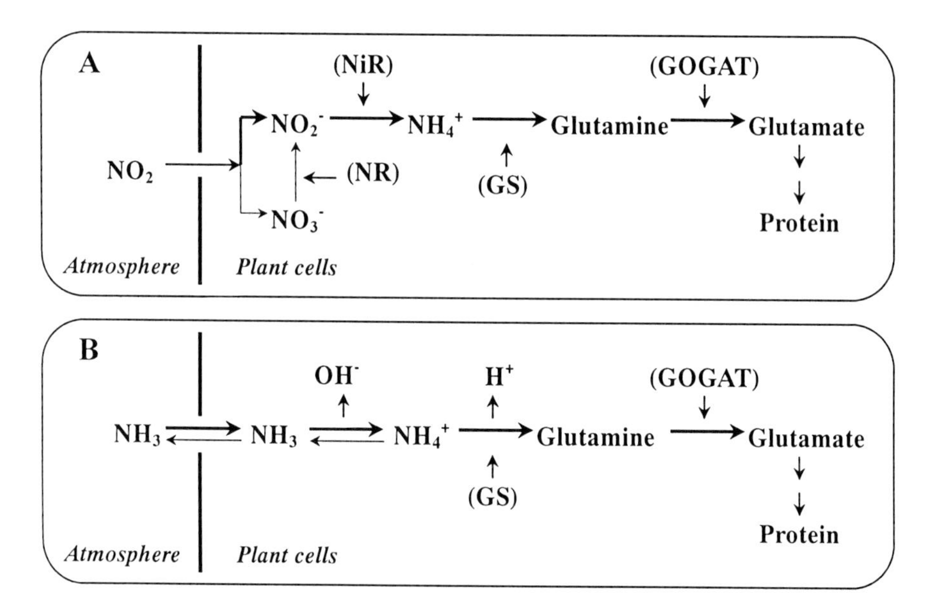

Fig. 1. Schemes of the possible metabolism of NO_2 (**A**) and NH_3 (**B**) in leaf cells. Enzymes: NR, nitrate reductase; NiR, nitrite reductase; GS, glutamine synthetase; GOGAT, glutamate synthase

the double bonds of unsaturated lipids.

With regards to situations that lead to NO_2-induced foliar damage, they also suggested that insufficient ascorbic acid, which may be important for conversion of NO_2 to nitrite in leaf apoplasts, may result in the destruction of membranes by NO_2. Another proposed mechanism of phytotoxicity may be the generation of free radicals, such as O_2^- and OH· which lead to subsequent inactivation/destruction of important biomolecules (see Aono and Polli, this volume; Srivastava 1992). NO_2 on its own has also been to shown to possess a strong radical nature (Ramge et al. 1993). Further more, Shimazaki et al. (1992) demonstrated that spinach plants, which were more tolerant than kidney beans to NO_2, also had higher levels of superoxide dismutase activity, which could metabolize O_2^- to the less toxic H_2O_2. In another study, the polyamine spermidine, which is also believed to possess free radical scavenging properties, prevented to some extent the decline in leaf dry mass in bean plants exposed to 300 nl l^{-1} NO_2 for 3d (Srivastava et al. 1995b).

4. Absorption and Metabolism of NH_3

Atmospheric NH_3 levels are the resultant balance between the release of NH_3 from the soil and plants and the removal of NH_3 through plant absorption and dissolution into rain or aerosols. In some areas, atmospheric NH_3 levels are increased tremendously by volatilization from fields grazed by cattle or treated with chemical fertilizers from animal slurry or manure (Sutton et al. 1993) and sewage sludge (Beauchamp et al. 1978).

Direct evidence for NH_3 uptake by plants was obtained from ^{15}N enrichment of plant tissues fumigated with ^{15}N-labeled NH_3 (Porter et al. 1972). The NH_3-derived N was largely assimilated in the leaves and translocated to the petioles (Grodzinski et al. 1984) and roots (Porter et al. 1972; Lockyer and Whitehead 1986). Whereas the NH_3 in nonpolluted areas may have an insignificant contribution to the plant nitrogen economy (Raven 1988), high NH_3 concentrations, of the order of 85 μg NH_3 m^{-3} as is possible in grazed lands, provided 15%-20% of the total N in Italian ryegrass canopy (Whitehead and Lockyer 1987).

Absorption of NH_3 occurs largely through the stomata (Hutchinson et al. 1972), especially under low-humidity conditions. The NH_3 dissolves in apoplastic water (pH 5-6.5) in stomata, and produces NH_4^+ and OH^-, then causing alkalization of the tissue. The NH_4^+ is then efficiently assimilated into glutamine by GS, which has a very low K_m for NH_4^+ (Miflin and Lea 1980), and H^+ is released in the process. The balance of OH^- and H^+ release in NH_3 assimilation results in the generation of 0.22 mol H^+ per NH_3 (Wollenweber and Raven 1993). Before 1980, studies using $^{15}NH_4^+$ indicated that assimilation of NH_4^+ occurred through activities of GS and GOGAT in leaf tissues (Ito et al. 1978; Ito and Kumazawa 1978). However, direct evidence for the operation of GS and GOGAT in this process was provided by Berger et al. (1986) using $^{15}NH_3$. A scheme of NH_3

metabolism in plant cells is shown in Fig. 1B. Thus, fumigation of plants with NH_3 results in an increase in organic nitrogen contents. Increases in low C/N compounds, such as in arginine, glutamine, and asparagine (Pérez-Soba et al. 1994; Lohaus and Heldt 1997), and decrease of soluble carbohydrates, such as sucrose and phosphoenolpyrunate (Lohaus and Heldt 1997), have also been observed.

An efflux of NH_3 from leaf stomata has also been reported (Farquhar et al. 1980). The NH_4^+ in leaf cells accumulates by transport from the roots through the xylem (Mattsson and Schjoerring 1996), from the photorespiratory cycle (Keys et al. 1978), and from nitrate reduction (Beevers and Hageman 1980) and amino acid deamination (Miflin and Lea 1980) in the leaves. Because the pH of chloroplasts and the cytoplasm is rather alkaline (between 7 and 8), and the pH of apoplastic water is increased (from 6.0 to 6.8) by the accumulation of NH_4^+ (Husted and Schjoerring 1995), the equilibrium between gaseous NH_3 and aqueous NH_4^+ may lead to some loss of NH_3. However, the gaseous loss of N from crops is small, and agronomically insignificant (Yoneyama 1983; Schjoerring et al. 1993). The compensation concentrations of NH_3 at the leaf surface, at which point plants show no net fluxes of NH_3, are 1 to 4 $\mu g\ m^{-3}$ (Farquhar et al. 1980). In natural ecosystems, such as in mountaneous forests where the canopy has a low NH_3 concentration, the plants can be a source of NH_3, whereas in a canopy with a high NH_3 concentration, such like as in agricultural areas or near livestock farms, the plants can serve as an NH_3 sink (Langford and Fehsenfeld 1992). Direct damage of naturally growing plants by NH_3 does not appear except at very high NH_3 concentrations (Van der Eerden 1982), although accumulation of excess H^+ in leaf cells fumigated with NH_3 may disturb the acid-base balance of the leaves (Pearson and Stewart 1993). In addition, the acidification and eutrophication of soils, where deposited NH_3 is nitrified to nitrate, may induce plant damage (Heil et al. 1988; Sutton et al. 1993).

5. Conclusions and Perspectives

Although it has been postulated that NO_2 may destroy membranes by disrupting the double bonds of lipids, there is no evidence supporting the relationship between NO_2 fumigation and membrane destruction in vivo. The involvement of ascorbic acid in the detoxification of NO_2 is also still an assumption, although several results concur with this ascorbic acid hypothesis. Recently, Ammann et al. (1995) suggested from $^{13}NO_2$-feeding experiments (*Picea abies* L.) that disproportionation of NO_2 into nitrate and nitrite in the shoots of spruce may be the major reaction rather than nitrite formation by the reaction of NO_2 and antioxidants. Thus, the mechanisms of NO_2 acquisition appear to vary according to both the plant species and the physiological conditions, such as the antioxidant contents, and may account for the great variations in NO_2 uptake reported by Morikawa et al. (1998).

Air pollution by anthropogenic NO_2 and NH_3 is still a serious problem both in urban areas and in agricultural lands (such as near cattle feedlots). Long-term measurements of the effects of these pollutants on ecosystems and the development of technical methods for their removal is essential. As plants are able to absorb and metabolize these gases, their role as filters is worth considering. However, it has to be kept in mind that the ability of plants and ecosystems to absorb and assimilate nitrogenous pollutants has limitations. Employment of technical methods to decrease the evolution of these gases is perhaps the best remedy.

Acknowledgment. We thank Drs. H. Förstel, J.K. Schjorring, and L. Van der Eerden for kindly providing reports of their studies.

References

Ammann M, Stalder M, Suter M, et al (1995) Tracing uptake and assimilation of NO_2 in spruce needles with ^{13}N. J Exp Bot 46:1685-1691

Beauchamp EG, Kidd GE, Thurtell G (1978) Ammonia volatilization from sewege sludge applied in the field. J Environ Qual 7:141-146

Beevers L, Hageman RH (1980) Nitrate and nitrite reduction. In: Miflin BJ (ed) Amino acids and derivatives. Academic Press, New York, pp 115-168

Berger MG, Klaus RE, Fock HP (1986) Assimilation of gaseous ammonia by sunflower leaves during photosynthesis. Aust J Plant Physiol 13:211-219

Da Silva PRF, Stutte CA (1981) Nitrogen volatilization from rice leaves. II. Effect of source of applied nitrogen in nutrient culture solution. Crop Sci 21:913-916

Durmishidze SV, Nutsubidze NN (1976) Asorption and conversion of nitrogen dioxide by higher plants. Dokl Biochem 227:104-107

Estefan RH, Gause EM, Rowlands JR (1970) Electron spin resonance and optical studies of the interaction between NO_2 and unsaturated lipid components. Environ Res 3:62-78

Farquhar GD, Firth PM, Wetselaar R, et al (1980) On the gaseous exchange of ammonia between leaves and the environment: determination of the ammonia compensation point. Plant Physiol 66:710-714

Grodzinski B, Jahnke S, Thompson K (1984) Translocation profiles of [^{11}C] and [^{13}N]-labelled metabolites after assimilation of $^{11}CO_2$ and [^{13}N]-labelled ammonia gas by leaves of *Helianthus annuus* L. and *Lupinus albus* L. J Exp Bot 35:678-690

Heil GW, Werger WJT, de Mol W, et al (1988) Capture of atmospheric ammonium by grassland canopies. Science 239:764-765

Husted S, Schjoerring JK (1995) Apoplastic pH and ammonium concentration in leaves of *Brassica napus* L. Plant Physiol 109:1453-1460

Hutchinson GL, Millington DW, Peters DB (1972) Atmospheric ammonia: absorption by plant leaves. Science 175:771-772.

Ito O, Kumazawa K (1978) Amino acid metabolism in plant leaf. III. The effect of light on the exchange of ^{15}N-labelled nitrogen among several amino acids in sunflower discs. Soil Sci Plant Nutr 24:327-336

Ito O, Yoneyama T, Kumazawa K (1978) Amino acid metabolism in plant leaf. IV. The

effect of light on ammonium assimilation and glutamine metabolism in the cells isolated from spinach leaves. Plant Cell Physiol 19:1109-1119
Kaji M, Yoneyama T, Totsuka T, et al (1980) Absorption of atmospheric NO_2 by plants and soils. VI. Transformation of NO_2 absorbed in the leaves and transfer of the nitrogen through the plants. Res Rep Natl Inst Environ Stud 11:51-58
Keys AJ, Bird IF, Cornelius MJ, et al (1978) Photorespiratory nitrogen cycle. Nature 275:741-743
Lea PJ, Wallsgrove RM, Miflin BJ (1978) Photorespiratory nitrogen cycle. Nature 275:741-743
Langford AO, Fehsenfeld FC (1992) Natural vegetation as a source or sink for atmospheric ammonia: a case study. Science 255:581-583
Lockyer DR, Whitehead DC (1986) The uptake of gaseous ammonia by the leaves of Italian ryegrass. J Exp Bot 37:919-927
Loewus FA, Loewus MW (1987) Biosynthesis and metabolism of ascorbic acid in plants. CRC Crit Rev Plant Sci 5:101-119
Lohaus G, Heldt HW (1997) Assimilation of gaseous ammonia and the transport of its products in barley and spinach leaves. J Exp Bot 18:1779-1786
Mattsson M, Schjoerring JK (1996) Ammonia emission from young barley plants: influence of N source, light/dark cycles and inhibition of glutamine synthetase. J Exp Bot 47:477-484
Matsumaru T, Yoneyama T, Totsuka T, et al (1979) Absorption of atmospheric NO_2 by plants and soils. (I) Quantitative estimation of absorbed NO_2 in plants by ^{15}N method. Soil Sci Plant Nutr 25:255-265
Matsushima J (1977) Comparison of microscopic structures of zelkova tree leaves treated with O_3, NO_2, and SO_2 without visible injury. Air Pollut Stud Jpn 11:360-369
Mayumi H, Yamazoe F (1979) Absorption and transformation of nitrogen dioxide in crops. Jpn J Soil Sci Plant Nutr 50:116-122
Menzel DB (1976) The role of radicals in toxicity of air pollutants. Nitrogen dioxide and ozone. In: Pryor WA (ed) Free radicals in biology, vol II. Academic Press, New York, pp181-203
Miflin BJ, Lea PJ (1980) Ammonia assimilation. In: Miflin BJ (ed) Amino acids and derivatives. Academic Press, New York, pp 169-202
Morikawa H, Higaki A, Nohno M, et al (1992) "Air-pollutant-philic plants" from nature. In: Murata N (ed) Research in photosynthesis, vol IV. Kluwer, Dordrecht, pp 79-82
Morikawa H, Kamada M, Higaki A, et al (1993) "Air-pollutant-philic plants" produced by particle bombardment.In: Proceedings JSPS-NUS seminar, Tsukuba, pp 152-158
Morikawa H, Higaki A, Nohno M, et al (1998) More than a 600-fold variation in nitrogen dioxide assimilation among 217 plant taxa. Plant Cell Environ 21:180-190
Nussbaum S, von Ballmoos P, Gfeller H, et al (1993) Incorporation of atmospheric $^{15}NO_2$-nitrogen into free amino acids by Norway spruce *Picea abies* (L.) Karst. Oecologia 94:408-414
Okano K, Tatsumi J, Yoneyama T, et al (1983) Investigation on the carbon and nitrogen transfer from a terminal leaf to the root system of rice plant by a double tracer method with ^{13}C and ^{15}N. Jpn J Crop Sci 52:331-341
Pearson J, Stewart RG (1993) Atmospheric ammonia deposition and its effects on plants. New Phytol 125:283-305
Pérez-Soba M, Stulen I, Van der Eerden LJM (1994) Effect of atmospheric ammonia on the nitrogen metabolism of Scots pine (*Pinus sylvestris*) needles. Physiol Plant 90:629-636

Porter LK, Viets FG Jr, Hutchinson GL (1972) Air containing nitrogen-15 ammonia: foliar absorption by corn seedlings. Science 135:759-761

Pryor WA, Lightsey JW (1981) Mechanisms of nitrogen dioxide reactions: Initiation of lipid peroxidation and production of nitrous acid. Science 214:435-437

Ramge P, Badeck F-W, Plöchl M, et al (1993) Apoplastic antioxidants as decisive elimination factors within the uptake process of nitrogen dioxide into leaf tissues. New Phytol 125:771-785

Raven JA (1988) Acqusition of nitrogen by the shoots of land plants: its occurrence and implications for acid-base regulation. New Phytol 109:1-20

Rogers HH, Campbell JC, Volk RJ (1979) Nitrogen-15 dioxide uptake and incorporation by *Phaseolus vulgaris* (L.). Science 206:333-335

Rowland AJ, Drew MC, Wellburn AR (1987) Foliar entry and incorporation of atmospheric nitrogen dioxide into barley plants of different nitrogen status. New Phytol 107:357-471

Rowland-Bamford AJ, Lea PJ, Wellburn AR (1989) NO_2 flux into leaves of nitrate reductase-deficient barley mutants and corresponding changes in nitrate reductase activity. Environ Exp Bot 29:439-444

Segschneider H-J, Wildt J, Förstel H (1995) Uptake of $^{15}NO_2$ by sunflower (*Helianthus annuus*) during exposures in light and darkness: quantities, relationship to stomatal aperture and incorporation into different nitrogen pools within the plant. New Phytol 131:109-119

Schjoerring JK, Kyllingsbaek A, Mortensen JV, et al (1993) Field investigation of ammonia exchange between barley plants and the atmosphere. I. Concentration profiles and flux densities of ammonia. Plant Cell Environ 16:161-167

Shimazaki K, Yu S-W, Sasaki T, et al (1992) Differences between spinach and kidney bean plants in terms of sensitivity to fumigation with NO_2. Plant Cell Physiol 33:267-273

Srivastava HS (1992) Nitrogenous pollutants in the atmosphere. Their assimilation and phytotoxicity. Curr Sci 63:310-317

Srivastava HS, Ormrod DP, Hale BA (1995a) Assimilation of nitrogen dioxide by plants and its effects on nitrogen metabolism. In: Srivastava HS, Singh RP (eds) Nitrogen nutrition in higher plants. Associated Publishing, New Delhi, pp 417-430

Srivastava HS, Ormrod DP, Hale BA (1995b) Polyamine mediated modification of bean leaf response to nitrogen dioxide. J Plant Physiol 146:313-317

Sutton MA, Piteairn CER, Fowler D (1993) The exchange of ammonia between the atmosphere and plant communities. Adv Ecol Res 24:301-393

Van der Eerden LTM (1982) Toxicity of ammonia to plants. Agric Environ 7:223-235

Weiland RT, Stutte CA (1979) Pyro-chemiluminescent differentiation of oxidized and reduced N forms evolved from plant foliage. Crop Sci 19:545-547

Whitehead DC, Lockyer DR (1987) The influence of the concentration of gaseous ammonia on its uptake by the leaves of Italian ryegrass, with and without an adequate supply of nitrogen to the roots. J Exp Bot 38:818-827

Wollenweber B, Raven JA (1993) Implications of N acquisition from atmospheric NH_3 for acid-base and cation-anion balance of *Lolium perenne*. Physiol Plant 89:519-523

Yoneyama T (1983) Distribution of nitrogen absorbed during different times of growth in the plant parts of wheat and contribution to the grain amino acids. Soil Sci Plant Nutr 29:193-207

Yoneyama T, Arai K, Totsuka T (1980) Transfer of nitrogen and carbon from a mature sunflower leaf – $^{15}NO_2$ and $^{13}CO_2$ feeding studies. Plant Cell Physiol 21:1367-1381

Yoneyama T, Sasakawa H (1979) Transformation of atmospheric NO_2 absorbed in spinach leaves. Plant Cell Physiol 20:263-266

Yoneyama T, Sasakawa H, Ishizuka S, et al (1979) Absorption of NO_2 by plants and soils (II) Nitrite accumulation, nitrite reductase activity and diurnal change of NO_2 absorption in leaves. Soil Sci Plant Nutr 25:267-275

Yoneyama T, Sasakawa H, Totsuka T, et al (1978) Response of plants to atmospheric NO_2 fumigation. (5) Measurements of $^{15}NO_2$ uptake, nitrite accumulation and nitrite reductase activity. In: Studies on evaluation and amelioration of air pollution by plants, Progress report in 1976-1977. Report of Special Research Project, NIES R-2, pp 103-111

Yoneyama T, Totsuka T, Hayakawa N, et al (1980) Absorption of atmospheric NO_2 by plants and soils. (V) Day and night NO_2-fumigation effect on the plant growth and estimation of the amount of NO_2-nitrogen absorbed by plants. In: Studies on the effects of air pollutants on plants and mechanisms of phytotoxicity. Res Rep Natl Inst Environ Stud 11:31-50

12
Plant Resistance to Ozone: the Role of Ascorbate

Jeremy Barnes[1], Youbin Zheng[2], and Tom Lyons[1]

[1]Department of Agricultural and Environmental Science, Ridley Building, Newcastle University, Newcastle upon Tyne, NE1 7RU, UK
[2]Department of Plant Agriculture, Bovey Building, Gordon Street, University of Guelph, Guelph, Ontario, Canada

1. Introduction

Ground-level concentrations of ozone (O_3) have risen rapidly over the past century (≈1.6% per annum in the Northern Hemisphere; Marenco et al. 1994), primarily due to increased emissions of nitrogen oxides and volatile hydrocarbons from all manner of domestic and industrial sources (Hough and Derwent 1990). Concentrations are expected to continue to rise for the foreseeable future; forecasts suggest that ground-level O_3 concentrations will continue to climb at a rate of between 0.3% and 1.0% per annum over the next 50 years (Chameides et al. 1994). As a consequence, present-day concentrations of O_3 commonly exceed the accepted threshold of 40 parts per billion (ppb; billion = 10^9) for damage to the most sensitive elements of natural and managed ecosystems (Fuhrer et al. 1997; Fuhrer and Achermann 1999) and the gas is now recognized to be the most prevalent and damaging air pollutant to which vegetation is exposed in many parts of Central/Southern Europe, North/South America and Asia (Yunus and Iqbal 1996). In many of these regions, there is incontrovertible evidence that O_3 levels are high enough to reduce crop yields (Heck et al. 1983; Runeckles and Chevone 1992; Turcsányi et al. 2000a), to cause shifts in the genetic composition of natural and seminatural vegetation (Davison and Barnes 1998; Barnes et al. 1999a) and predispose forest trees to damage by secondary stress factors (Sandermann et al.

Air Pollution and Plant Biotechnology
–Prospects for Phytomonitoring and Phytoremediation–
Edited by K. Omasa, H. Saji, S. Youssefian, and N. Kondo
© Springer -Verlag Tokyo 2002

1997; Chappelka and Samuelson 1998; Skärby et al. 1998; Matyssek and Innes 1999; Akimoto and Sakugawa 2000). There is, however, considerable genetic variation (both within and between species) in the resilience of plants to this powerful oxidant (Barnes et al. 1999a,b). In this chapter, we highlight what is known about the factors responsible for this variation, and examine the possibility that present-day levels of O_3 may be high enough to drive shifts in the O_3 resistance of plant populations/crop lineages in the field. In particular, we focus on recent evidence that suggests an important role for L-ascorbate (ASC) in mediating tolerance to O_3, and discuss the prospect of manipulating O_3 tolerance through this powerful antioxidant.

2. Genetic Basis of Ozone Resistance

2.1 Evolution of Resistance in the Field

There is good evidence that novel stresses can act as powerful agents driving the selection of resistant genotypes in the field (Bradshaw and McNeilly 1991); evolution of resistance to heavy metals (Bradshaw 1984), herbicides (Murphy 1983), and SO_2 (Bell et al. 1991) has been described in a number of herbaceous species. Adaptation of plant populations to O_3, on the other hand, has received rather less attention (Heagle et al. 1991; Davison and Barnes 1998; Barnes et al. 1999b). Early work conducted by Dunn (1959) in Los Angeles provided circumstantial evidence that populations of *Lupinus bicolor* had undergone selection driven by photochemical smog (of which O_3 is the major component). More recently, comparisons of the O_3 resistance of geographically distinct populations of *Populus tremuloides* L. in the United States (Berrang et al. 1986, 1989, 1991) and *P. major* in Britain (Reiling and Davison 1992) and Europe (Lyons et al. 1997), as well as open-top chamber studies conducted on populations of *Trifolium repens* L. (Heagle et al. 1991), suggest that O_3 levels experienced in the natural environment may be high enough to drive the selection of resistant genotypes. Figure 1 shows the strong correlation between O_3 resistance and O_3 climate observed in discrete populations *P. major* originating from across Europe. Indeed, fieldwork conducted by Davison and Reiling (1995), and supported by recent studies of the underlying population genetics (Wolff et al. 2000), suggests that rapid shifts in the O_3 resistance of *P. major* populations can be driven by the pollutant. It is difficult, however, to categorically prove that O_3 is the primary (or indeed only) factor involved in driving selection under such circumstances, and it is noteworthy that phytotoxic O_3 concentrations generally coincide with periods of warm, dry, sunny weather (Finlayson-Pitts and Pitts 1986). Thus, it is probably no surprise that relationships between O_3 climate and the O_3 resistance of discrete populations are often marred by correlations between O_3 resistance and several

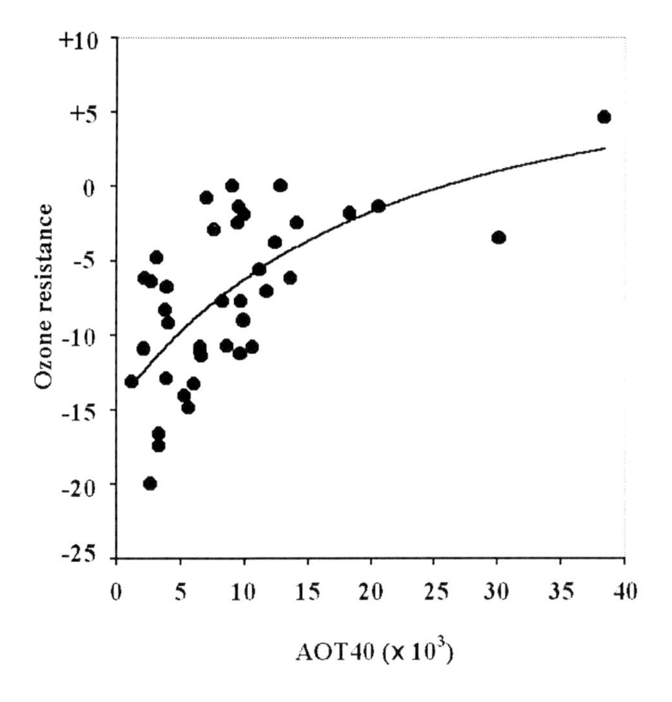

Fig. 1. Relationship between ozone resistance (% change in plant relative growth rate induced by a 2-week exposure to 70 ppb ozone 7 h d^{-1}) and ozone climate (AOT40; the accumulated hourly ozone exposure over 40 ppb during daylight hours for the consecutive period of the year when the highest ozone concentrations were experienced the year before seed collection) in 41 populations of *Plantago major*. Equation of the fitted line is $y=108.3(23/(1+0.000025\,x)^{2.049})$, $r^2 = 0.426$, $P = 0.0002$

other key environmental parameters (Davison and Barnes 1998). The available data suggest, therefore, that selection for O_3 resistance in the field may be linked with other climatic factors (Bell at al. 1991). Nevertheless, the potential of O_3 as a selective force can be demonstrated under artificial conditions. Whitfield et al. (1997), for example, showed that it was possible to increase the O_3 resistance (based upon changes in rosette diameter in plants exposed in fumigation chambers to 70 ppb O_3) of a sensitive population of *P. major* and enhance the O_3 sensitivity of a resistant one (Figure 2). Employing RAPDs, Squirrell and Wolff (unpublished) have recently been able to demonstrate that the contrasting O_3 sensitivity of these lines has a genetic basis. Also, ongoing research by Clamp and Davison (unpublished) has demonstrated similar potential for the selection of 'O_3 resistant' and 'O_3 sensitive' lines in *rc-Brassica rapa* L. (in this case, based upon

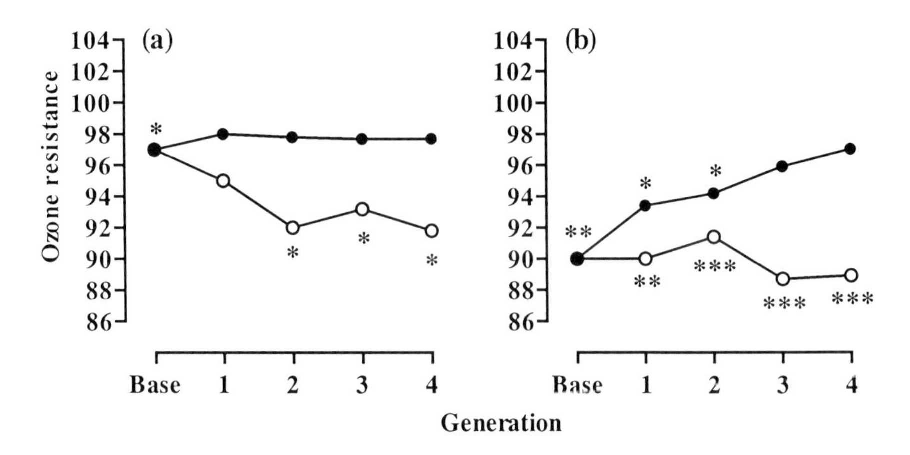

Fig. 2a,b. Changes in ozone resistance (% change in plant relative growth rate induced by a 2-week exposure to 70 ppb O_3 7 h d^{-1}) over four generations in lines selected for sensitivity (open circles) or resistance (closed circles) to ozone from two populations of *Plantago major*: (a) Lullington Heath, initially ozone resistant; (b) Bush, initially ozone sensitive. Asterisks denote significant effects of ozone in each generation (* $P< 0.05$, ** $P< 0.01$, *** $P< 0.001$) (Redrawn from Whitfield et al. 1997)

the expression of visible leaf injury under acute O_3 exposure in controlled conditions [150 ppb for 6 h]).

Most crop plants are harvested annually. Consequently, they are not subjected to the rigors of natural selection in the same way as many wild plants. However, it has been suggested that inadvertent selection may form an inevitable component of all cultivar development programs carried out in polluted areas; selection in field trials targeted at improving yields (and/or physiological traits associated with high yields) resulting in the unconscious choice of lines with modified air pollution resistance (Barnes et al. 1999b). Considering the narrow genetic base of the majority of commercial crops, once established in a cultivar or variety, such unconsciously introduced traits would be expected to be maintained from generation to generation, especially in inbreeding species such as cereals. Few studies have, however, examined whether crop cultivars bred in regions experiencing high levels of O_3 exhibit greater resistance to the pollutant (and potentially other oxidative stresses) than those bred in times prior to the rise in O_3 concentrations or in regions experiencing levels of O_3 that do not exceed the threshold for damage. There is the suggestion that cultivars of potato (*Solanum tuberosum* L.), alfalfa (*Medicago sativa* L.), and cotton (*Gossypium hirsutum* L.) bred in polluted parts of the United States are more resistant to O_3 than their counterparts bred in nonpolluted regions (Reinert et al. 1982), while common sense suggests that farmers operating in polluted regions select cultivars that show enhanced yields and reduced 'injury' regardless of cause (Gimeno et al. 1999). Inadvertent selection for O_3 resistance would, however, be expected to be restricted to those situations where selected lines outyield their more susceptible counterparts in all environments, since variation in performance would commonly lead to their rejection by breeders (Austin et al. 1980).

Regardless of whether O_3 acts as a primary selection pressure in the field, the finding that O_3-resistant genotypes exist in regions experiencing phytotoxic levels of the pollutant is of considerable significance (Barnes et al. 1999a,b). Although the 'costs' associated with O_3 resistance have not yet been studied in any detail, Davison and Reiling's (1995) observation of rapid changes in the O_3 resistance of *P. major* populations, dependent on annual variations in O_3 climate, suggest that while some plants may be better able to survive under conditions of O_3 stress, they may be at a competitive disadvantage in cleaner air, or in years when the level of O_3 does not exceed the threshold for damage; i.e., there may be hidden 'costs' associated with the evolution of resistance to this air pollutant (sensu Heagle et al. 1991), as there is to others (Bell et al. 1991).

2.2 Directed Selection of Resistance

There is a voluminous literature documenting the negative effects of present-day O_3 concentrations on the yield and quality of field-grown agricultural and horticultural crops. In the United States, research suggests that O_3, either alone or in combination with oxides of nitrogen and sulphur dioxide, is responsible for up to 90% of crop losses attributable to air pollution (Heck et al. 1982). Indeed, recent estimates indicate that a 25% reduction in ambient O_3 levels would provide benefits of $1-2 billion to the US agricultural sector alone (Murphy et al. 1999). Given the demonstration that O_3-resistance is a heritable trait (De Vos et al. 1982; Johnston et al. 1983; Mebrahtu et al. 1990), it is rather surprising that so little emphasis has been placed on the selection, and breeding, of O_3 resistant crops. One reason for this is probably the toll in terms of costs and time that need to be invested in screening and selecting plants in the field. Novel means of cutting this heavy burden would almost certainly make the breeding of O_3-resistant crops a more attractive proposition. One promising possibility is to use the impacts of O_3 on pollen as a predictor of effects on the sporophyte (Hormaza et al. 1996). This could provide a rapid and inexpensive method of promoting selection during plant breeding programs and warrants further detailed investigation.

The few plant breeding programs that have been targeted at improving O_3 resistance have met with some notable successes - modern varieties of tobacco (*Nicotiana tabacum* L.), alfalfa (*Medicago sativa* L.), sweet corn (*Zea mays* L.), snap bean (*Phaseolus vulgaris* L.), eastern white pine (*Pinus strobus* L.), and aspen (*Populus tremuloides* Michx.) have been bred that exhibit considerably greater resistance to O_3 than their predecessors (Howell et al. 1971; Aycock 1972; Cameron 1975; Campbell et al. 1977; Mebrahtu et al. 1990; Karnosky 1991). It is generally concluded that O_3 resistance is a quantitative trait (Howell et al. 1971; De Vos et al. 1982; Roose 1991), with some authors concluding that resistance involves many genes (Hucl and Beversdorf 1982) while others show that very few genes are involved (Guri 1983; Mebrahtu et al. 1990). In onion (*Allium cepa* L.), and more recently *Arabidopsis thaliana* L., single genes have been proven to play a vital role in mediating the extent of O_3-induced visible injury (Engle and Gabelman 1966; Conklin et al. 1996, 2000), and there is growing reason to believe that O_3 resistance may be controlled by a relatively small number of key genes (Davison and Barnes 1998).

3. Factors Governing Ozone Resistance

Ozone resistance is governed by a combination of factors that (i) limit the flux of the pollutant reaching its primary target (generally accepted to be the plasmalemma), (ii) control the reaction cascades that occur once the pollutant reacts with its primary target, and/or (iii) control the upregulation of metabolic processes to compensate for the damage incurred (Barnes and Wellburn 1998). Herein, we have elected to focus on the first of these possibilities, as this is an area where understanding has increased dramatically in recent years.

The leaf cuticle is effectively impermeable to O_3, so the ingress of the pollutant to the leaf interior is almost entirely controlled by stomatal conductance (a product of stomatal frequency × aperture and driven by the diffusion gradient between O_3 concentration inside and outside the leaf). Consequently, stomatal conductance plays a key role in determining the biologically effective O_3 flux impinging on the plasma membrane (J_{PLASMA}), and partial stomatal closure, a response often induced by O_3 as well as other air pollutants (Wolfenden and Mansfield 1991), constitutes an effective means of excluding the pollutant from the leaf interior. Some of the older literature suggests that the decline in stomatal conductance induced by O_3 may play a crucial role in determining O_3 resistance, but it is worth bearing in mind that changes in conductance induced by O_3 are known to result from 'damage' to either the photosynthetic machinery (Farage and Long 1995) or the stomatal apparatus (Torsethaugen et al. 1999). This suggests that by the time shifts in stomatal conductance are observed, severe damage has already occurred (i.e., O_3-induced shifts in conductance are a product of injury, rather than a primary preventative measure). Indeed, patterns of change in stomatal conductance following exposure to O_3 are often more complex than first perceived (Reiling and Davison 1995).

Commonly overlooked, is the potentially important role played by leaf anatomy in governing J_{PLASMA}, in particular the exposed mesophyll cell surface area to leaf area ratio (A_{MESO}) and the thickness of the mesophyll cell walls (l_{CW}). Values for both these parameters are inversely related to J_{PLASMA}; A_{MESO}, because the same pollutant deposition is spread across a greater internal surface area (Tingey and Taylor 1982; Taylor et al. 1988) and l_{CW}, because the thicker the mesophyll cell walls then the greater the diffusion pathlength for O_3 and the greater the probability of interaction with compounds in the cell wall that can intercept the pollutant (Plöchl et al. 2000). Once O_3 dissolves into the aqueous milieu associated with mesophyll cell walls (i.e., the pollutant passes from the gaseous to the aqueous phase), its passage to the vulnerable plasma membrane may be impeded by reaction with constituents of this fluid (Dietz 1997). The majority of research has focused on the role played by ascorbate (ASC) in screening the plasmalemma from O_3-induced oxidative stress. This powerful antioxidant is found in relatively high concentration in the leaf apoplast and has an extremely high rate of reaction with O_3 and several of its primary reaction products (Lyons et al. 1999a). Little is known about the involvement of additional extracellular antioxidants in controlling the titre of reactive oxygen species (ROS) in the leaf apoplast under the influence of O_3. This represents a serious gap in current knowledge.

3.1 The Role of Ascorbate in Mediating Ozone Resistance

The phytotoxicity of O_3 stems from its strong oxidizing properties. Moreover, once the pollutant has entered the leaf interior, additional ROS are generated through chemical reaction (Grimes et al. 1983; Kanofsky and Sima, 1991, 1995; Byvoet et al. 1995) and/or endogenous production (Schraudner et al. 1998; Pellinen et al. 1999; Rao and Davis 1999). Consequently, the antioxidant systems that have evolved to control concentrations of ROS in all aerobic organisms are considered to play a vital role in combating the oxidative stress imposed by ozone, as with other oxidative stresses (Runeckles and Chevone 1992; Smirnoff and Pallanca 1996; Heath and Taylor 1997; Sharma and Davis 1997; Davey et al. 2000). The effective control of ROS is expedited through enzyme-metabolite couplings that attain sufficient rates of removal of ROS to prevent damage to key biomolecules (Smirnoff 1996; Noctor and Foyer 1998; Smirnoff 2000). These systems are especially important in leaf cells, since these are probably exposed to higher concentrations of ROS than many other types of living cell (Noctor et al. 2000).

Ascorbate is believed to play a key role in defending leaf cells from various physiological stresses that result in the accelerated production of ROS, including O_3 (Figure 3). The compound was first reported to afford effective protection against O_3 during the early 1960s (Freebairn 1960; Freebairn and Taylor 1960; Menser 1964), with more-sophisticated experiments conducted since confirming earlier findings (Mächler et al. 1995; Moldau et al. 1998; Zheng et al. 2000). Figure 4 illustrates how the manipulation of leaf ASC content (via foliar application or feeding plants the immediate biosynthetic precursor, L-galactono-1,4-lactone) can protect against the detrimental effects of O_3 (in this case, assessed through the depression in leaf photosynthetic capacity induced by O_3). In addition, variation in O_3 resistance both within and between species has been shown to

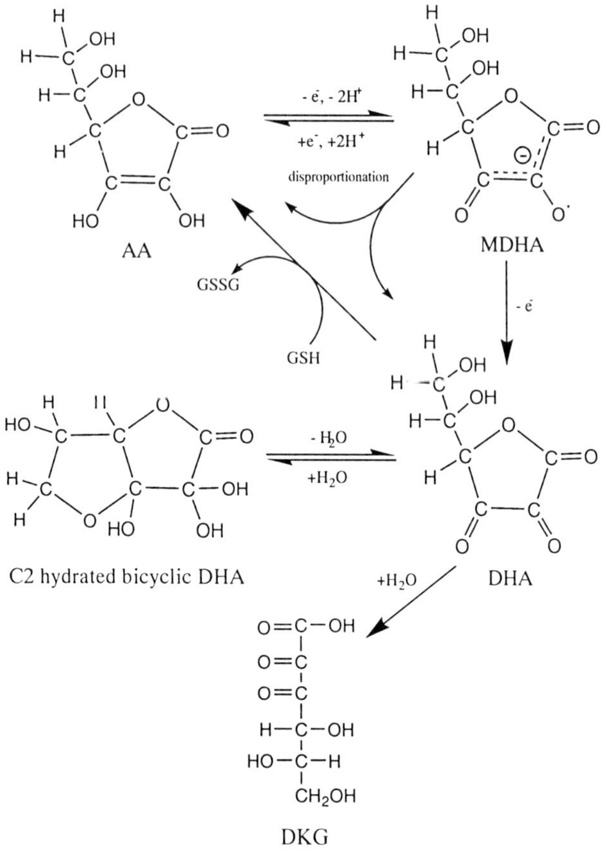

Fig. 3. L-Ascorbic acid (*AA*) is oxidized via two consecutive reversible 1-e$^-$ steps. The first product, the free radical monodehydro-L-ascorbic acid (*MDHA*), may spontaneously disproportionate to AA and dehydro-L-ascorbic acid (*DHA*), be reduced to AA enzymatically at the expense of reductant (NADH), or be further oxidized to DHA. In aqueous solution, DHA exists mainly in a hydrated form. In vivo, it is rapidly reduced back to AA by both enzymatic and nonenzymatic reactions – as a consequence glutathione (*GSH*) is oxidized (*GSSG*) - or it may irreversibly hydrolyze to form 2,3-diketo-L-gulonic acid (*DKG*). (Redrawn from Nishikawa and Kurata 2000)

correlate with ASC content (Lee et al. 1984; Bilodeau and Chevrier 1998; Lyons et al. 1999b; Zheng et al. 2000). Indeed, recent selection studies on *rc-B. rapa* conducted by Clamp and Davison (unpublished) indicate that O_3 resistance cosegregates with foliar ascorbate levels in this species – in much the same way as observed by Conklin and coworkers (Conklin et al. 1996, 2000) during work on

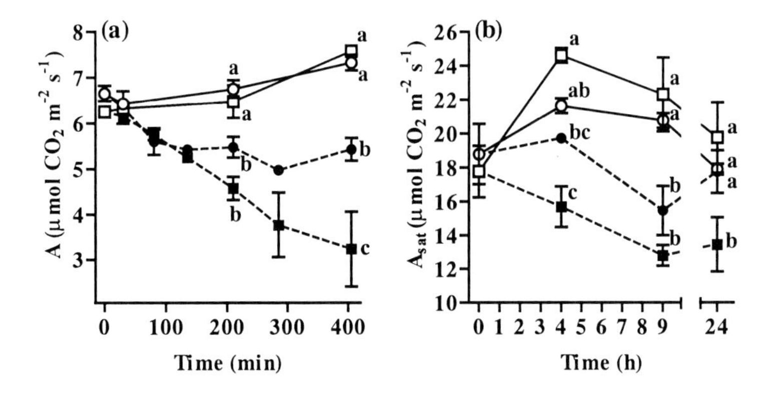

Fig. 4a,b. The impact of ozone exposure on (a) the rate of CO_2 assimilation (*A*) in *Plantago major*. Prior to fumigation, leaves were sprayed with 10 mM $NaHCO_3$ (squares) or 10 mM sodium-ascorbate (circles). Plants were exposed to charcoal/Purafil®-filtered air (CFA; open symbols) or CFA plus 400 nmol·mol^{-1} ozone for 7 h (CFA+O_3; closed symbols). Points represent the mean ± SE (n = 4). Letters denote significant difference (P< 0.05) between treatments at individual points in time (redrawn from Zheng et al. 2000). (b) The light-saturated rate of CO_2 assimilation (A_{sat}) in *Raphanus sativus*. Prior to fumigation, hydroponically-cultivated plants were fed nutrient solution (squares) or nutrient solution plus 50 mM L-galactono-1,4-lactone (circles). Plants were exposed to charcoal/Purafil®-filtered air (CFA; open symbols) or CFA plus 180 nmol·mol^{-1} ozone for 9 h (CFA+O_3; closed symbols). Points represent the mean ± SE of four independent observations. Letters denote significant difference (P< 0.05) between treatments at individual points in time (redrawn from Maddison et al. 2001). Parallel measurements revealed no effects of the ASC manipulation treatments on stomatal conductance (i.e., O_3 fluxes to the leaf interior were equivalent in control and ASC-manipulated plants)

a range of ascorbate-deficient *Arabidopsis thaliana* mutants.

The importance of compartment-specific scavenging of ROS has become increasingly recognized. In the case of O_3, much recent attention has focused on the potential for O_3 scavenging by constituents of the leaf apoplast. A positive relationship between O_3 resistance (assessed in terms of the depression in growth induced by O_3) and the ASC content of the leaf apoplast has been shown to exist in *P. major* and *R. sativus* (Figure 5). Theoretical models (based upon O_3 detoxification in the leaf apoplast through direct reaction with ASC), plus a growing number of experimental studies (Chameides 1989; Luwe et al. 1993; Polle et al. 1995; Moldau et al. 1998; Plöchl et al. 2000; Turcsányi et al. 2000b), suggest that the pool of ASC located in the leaf apoplast may be sufficient to afford a significant degree of protection against O_3 (Figure 6), a scenario first proposed by Castillo and Greppin (1988). However, many important questions remain

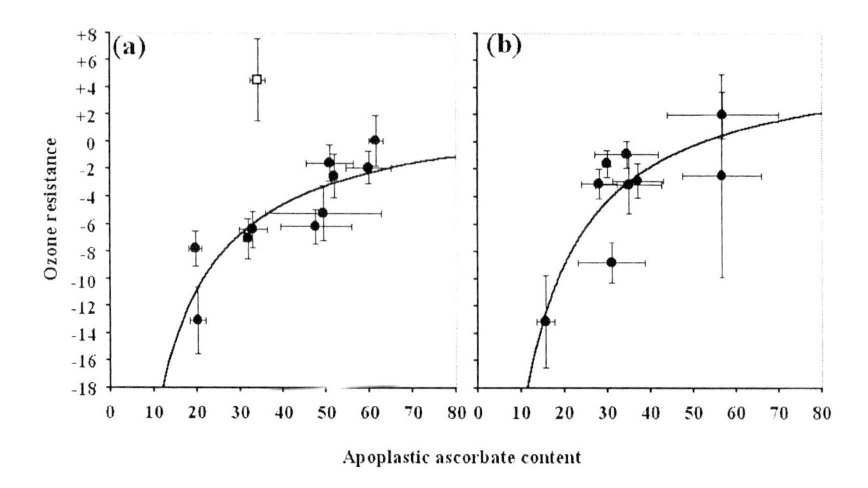

Fig. 5a,b. Relationship between ascorbate content of the leaf apoplast (nmol g^{-1} FW) and ozone resistance (% change in plant relative growth rate induced by exposure to 70 ppb ozone 7 h d^{-1}) in (a) populations of *Plantago major*; $y = 102.7\,x\,/(3.1 + x)$, $r^2 = 0.731$, $P = 0.002$ and (b) varieties of *Raphanus sativus*; $y = 106.5\,x\,/(3.4 + x)$, $r^2 = 0.714$, $P = 0.004$. Data represent means (n = 10-15) ± SE. (Replotted from Barnes et al. 1999b)

unanswered, and there is growing recognition that additional apoplast constituents are likely to contribute to O_3 interception, possibly to varying degrees in different species (Jakob and Heber 1998; Lyons et al. 1999a).

3.2 Modification of Ozone Resistance by Manipulation of Ascorbate Metabolism

Attempts to enhance O_3 resistance through the altered expression of key enzymes linked with cellular antioxidative defences have met with mixed success, dependent upon the gene selected for manipulation, the success in achieving the desired goal, and the subcellular compartment in which the gene product is expressed (see reviews by Foyer et al. 1994; Harris and Bailey-Serres 1994; Örvar and Ellis 1997; Lea et al. 1998; Srivastava 1999; see the chapter by M. Aono, this volume). Since the extracellular targeting sequences are unknown, it has not yet proved possible to investigate whether the manipulation of antioxidant enzyme activities in the apoplast could alter O_3 resistance. In contrast, the recent resolution of the biosynthetic pathway for ASC by Wheeler et al. (1998) and Conklin et al. (1999) should pave the way for the manipulation of extracellular ASC content via transgenic technology. Several enzymes involved in ASC biosynthesis have already been characterized (e.g., GDP-mannose

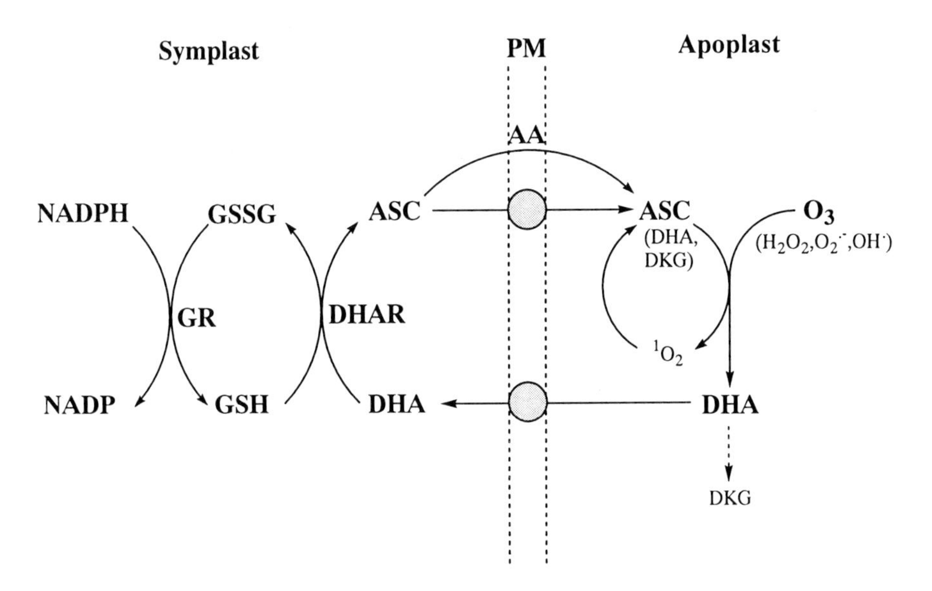

Fig. 6. Possible mechanism whereby ozone (*O_3*) and its reactive derivatives hydrogen peroxide (*H_2O_2*), superoxide radical (*$O_2^{\cdot-}$*), and hydroxyl radical (*$OH^{\cdot}$*) may be detoxified by direct reaction with ascorbate (*ASC*) in the leaf apoplast. Supply of ASC to the apoplast is driven by diffusion, facilitated by a transporter in the plasmalemma (*PM*) (Horemans et al. 2000), and/or simple diffusion of the neutral species ascorbic acid (*AA*) (Plöchl et al. 2000). Oxidation produces singlet O_2 (*$^1O_2$*), which may react with further ASC (Kanofsky and Sima 1995); the oxidized form of ASC, dehydroascorbic acid (*DHA*), must return to the symplast for reduction, or hydrolyze to diketogulonic acid (*DKG*). It is possible that DHA and DKG may function as antioxidants (Deutsch 1998). Regeneration of DHA in the symplast is mediated by DHA reductase (*DHAR*), utilizing reduced glutathione (*GSH*). The oxidized glutathione (*GSSG*) is in turn reduced by GSSG reductase (*GR*), which uses NADPH as reductant

pyrophosphorylase [Keller et al. 1999], galactonolactone-1,4-lactone dehydrogenase [Østergaard et al. 1997; Imai et al. 1998], galactose dehydrogenase [Smirnoff 2000]), and several other key enzymes are in various stages of cloning and purification (Smirnoff 2000). The potential afforded by these studies is exemplified by work on potato carrying antisense GDP-mannose pyrophosphorylase (Keller et al. 1999) and an *Arabidopsis* mutant (*vtc1*) that shows decreased activity of the same enzyme (Conklin et al. 1999). Both sets of plants exhibit much-reduced accumulation of ASC compared to the wild-type. Alternative avenues aimed at the manipulation of foliar ASC content are also being pursued e.g. a gene encoding gulonolactone oxidase, the final enzyme in the ASC biosynthetic pathway in animals (Nishikimi and Yagi 1996), has been expressed in potato (Imai, Vanacker and Foyer, unpublished), tobacco, and lettuce (Jain and Nessler 2000) with the result that transgenic plants accumulated up to seven times more ASC than controls. The genetic manipulation of enzymes involved in ASC-turnover may also prove a useful tool in the future (Davey et al. 2000). Of particular interest, in terms of the manipulation of O_3 resistance through the level of ASC in the leaf apoplast, is ASC oxidase. The characterization and manipulation of this poorly understood enzyme may afford one means to alter the size of the ASC pool in the leaf apoplast and with it, potentially, ozone resistance.

4. Conclusions

There is unequivocal evidence that present-day concentrations of O_3 are responsible for substantial losses in crop yield and shifts in the diversity of wild plant populations across the globe. It is therefore imperative that research is directed toward a better understanding of the mechanisms underlying the phytotoxicity of this ubiquitous air pollutant. This knowledge is the key to designing future crops with improved resistance to O_3 and possibly other physiological stresses that cause the accelerated production of ROS.

It appears that O_3-resistant genotypes may be naturally (or in the case of crops, inadvertently) favored in areas experiencing phytotoxic levels of the pollutant. Further work should be undertaken to assess the ecophysiological significance of these evolutionary shifts in population genetics and a thorough cost/benefit analysis is required. The immediate prospect of developing crops with enhanced levels of ASC using transformation techniques is exciting. Genetically manipulated crops of this kind might show enhanced resistance not only to ozone, but also to a variety of additional stresses that induce similar oxidative insult, while exhibiting substantially improved nutritive value (Davey et al. 2000).

Acknowledgments. The authors thank Mr. K. Taylor, Mr. A. White, and Mr. P. Green for technical assistance during the course of some of the experimental studies presented. JB and TL are indebted to the Royal Society and Newcastle University's Management and Information Systems [MAIS] Project for providing the funding required to pursue their research interests. YZ was funded by awards made to JB by Newcastle University Research Committee, The Swales Foundation, and the U.K. Overseas Development Agency (now the UK Department for International Development).

References

Akimoto H, Sakugawa H (2000) Oxidants/acidic species and forest decline in East Asia. Japan Science and Technology Corporation-Core Research for Evolutionary Science and Technology, Higashi-Hiroshima

Austin RN, Bingham J, Blackwell RD, et al (1980) Genetic improvements in winter wheat yield since 1900 and associated physiological changes. J Agric Sci 94:675-689

Aycock MK (1972) Combining ability estimates for weather fleck in *Nicotiana tabacum* L. Crop Sci 12:672-689

Barnes JD, Davison AW, Balaguer L, et al (1999a) Resistance to air pollutants: from cell to community. In: Pugnaire FI, Valladares F (eds) Handbook of functional plant ecology. Dekker, New York, pp 735-770

Barnes J, Bender J, Lyons T, et al (1999b) Natural and man-made selection for air pollution resistance. J Exp Bot 50:1423-1435

Barnes JD, Wellburn AR (1998) Air pollutant combinations. In: De Kok LJ, Stulen I (eds) Responses of plant metabolism to air pollution and global change. Backhuys, Lieden, pp 147-164

Bell JNB, Ashmore MR, Wilson GB (1991) Ecological genetics and chemical modifications of the atmosphere. In: Taylor GE, Pitelka LF, Clegg MT (eds) Ecological genetics and air pollution. Springer-Verlag, New York, pp 33-59

Berrang P, Karnosky DF, Mickler RA, et al (1986) Natural selection for ozone tolerance in *Populus tremuloides*. Can J For Res 16:1214-1216

Berrang P, Karnosky DF, Bennett JP (1989) Natural selection for ozone tolerance in *Populus tremuloides*: field verification. Can J For Res 19:519-522

Berrang P, Karnosky DF, Bennett JP (1991) Natural selection for ozone tolerance in *Populus tremuloides*: an evaluation of nationwide trends. Can J For Res 21:1091-1097

Bilodeau C, Chevrier N (1998) Endogenous ascorbate level modulates ozone tolerance in *Euglena gracilis* cells. Plant Physiol Biochem 36:695-702

Bradshaw AD (1984) Adaptation of plants to soils containing toxic metals – a test for conceit. In: Evered D, Collins GM (eds) Origins and development of adaptation. Pitman, London, pp 4-19

Bradshaw AD, McNeilly T (1991) Evolution in relation to environmental stress. In: Taylor GE, Pitelka LF, Clegg MT (eds) Ecological genetics and air pollution. Springer-Verlag, New York, pp 11-31

Byvoet P, Balis JU, Shelley SA, et al (1995) Detection of hydroxyl radicals upon interaction of ozone with aqueous media or extracellular surfactant – the role of trace iron. Arch Biochem Biophys 319:464-469

Cameron JW (1975) Inheritance in sweet corn for resistance to acute ozone injury. J Am Soc Hortic Sci 100:577-579

Campbell TA, Devine TE, Howell PK (1977) Diallel analysis of resistance to air pollutants in alfalfa. Crop Sci 17:664-665

Castillo FJ, Greppin H (1988) Extracellular ascorbic acid and enzyme activities related to ascorbic acid metabolism in *Sedum album* L. leaves after ozone exposure. Environ Exp Bot 28:231-238

Chameides WL (1989) The chemistry of ozone deposition to plant leaves: role of ascorbic acid. Environ Sci Technol 23:595-600

Chameides WL, Kasibhatla PS, Yienger J, et al (1994) Growth of continental-scale metro-agro-plexes, regional ozone pollution, and world food production. Science 264:74-77

Chappelka AH, Samuelson LJ (1998) Ambient ozone effects on forest trees of the eastern United States: a review. New Phytol 139:109-122

Conklin PL, Williams EH, Last RL (1996) Environmental stress sensitivity of an ascorbate-deficient *Arabidopsis* mutant. Proc Natl Acad Sci USA 93:9970-9974

Conklin PL, Norris SN, Wheeler GL, et al (1999) Genetic evidence for the role of GDP-mannose in plant ascorbic acid (vitamin C) biosynthesis. Proc Natl Acad Sci USA 96:4198-4203

Conklin PL, Saracco SA, Norris SR, et al (2000) Identification of ascorbic acid-deficient *Arabidopsis thaliana* mutants. Genetics 154:847-856

Davey MW, Van Montagu M, Inzé D, et al (2000) Plant L-ascorbic acid: chemistry, function, metabolism, bioavailability and effects of processing. J Sci Food Agric 80:825-860

Davison AW, Barnes JD (1998) Impacts of ozone on wild species. New Phytol 139:135-151

Davison AW, Reiling K (1995) A rapid change in the ozone resistance of *Plantago major* after summers with high ozone concentrations. New Phytol 131:337-344

Deitz K-J (1997) Functions and responses of the leaf apoplast under stress. Prog Bot 58:221-254

Deutsch JC (1998) Ascorbic acid oxidation by hydrogen peroxide. Anal Biochem 255:1-7

De Vos NE, Hill RR, Pell EJ, et al (1982) Quantitative inheritance of ozone resistance in potato. Crop Sci 22:992-995

Dunn DB (1959) Some effects of air pollutants on *Lupinus* in the Los Angeles area. Ecology 40:621-625

Engle RL, Gabelman WH (1966) Inheritance and mechanism for resistance to ozone damage in onion. Proc Am Soc Hortic Sci 89:423-430

Farage PK, Long SP (1995) An *in vivo* analysis of photosynthesis during short-term O_3 exposure in three contrasting species. Photosynth Res 43:11-18

Finlayson-Pitts BJ, Pitts JN (1986) Atmospheric chemistry: fundamentals and experimental techniques. Wiley, New York

Foyer CH, Descourvières P, Kunert KJ (1994) Protection against oxygen radicals: an important defence mechanism studied in transgenic plants. Plant Cell Environ 17:507-523

Freebairn HT (1960) The prevention of air pollution damage to plants by use of vitamin C sprays. J Air Pollut Control Assoc 10:314-317

Freebairn HT, Taylor OC (1960) Prevention of plant damage from airborne oxidising agents. Proc Am Soc Hortic Sci 76:693-699

Fuhrer J, Achermann B (1999) Critical levels for ozone - level II. Environmental Documentation No. 115, Swiss Agency for Environment, Forest and Landscape, Bern

Fuhrer J, Skarby L, Ashmore MR (1997) Critical levels for ozone effects on vegetation in Europe. Environ Pollut 97:91-106

Gimeno BS, Bermejo V, Reinert RA, et al (1999) Adverse effects of ambient ozone on watermelon yield and physiology at a rural site in Eastern Spain. New Phytol 144:245-260

Grimes HD, Perkins KK, Boss WF (1983) Ozone degrades into hydroxyl radicals under physiological conditions. Plant Physiol 72:1016-1020

Guri A (1983) Attempts to elucidate the genetic control of ozone sensitivity in seedlings of *Phaseolus vulgaris* L. Can J Plant Sci 63:727-731

Harris MJ, Bailey-Serres JN (1994) Ozone effects on gene expression and molecular approaches to breeding for air pollution resistance. In: Basra AS (ed) Stress-induced gene expression in plants. Harwood, Singapore, pp 185-207

Heagle AS, McLaughlin MR, Miller JE, et al (1991) Adaptation of a white clover population to ozone stress. New Phytol 119:61-68

Heath RL, Taylor GE (1997) Physiological processes and plant responses to ozone exposure. In: Sandermann H, Wellburn AR, Heath RL (eds) Forest decline and ozone: a comparison of controlled chamber and field experiments. Ecological Studies vol. 127. Springer-Verlag, Berlin, pp 317-368

Heck WW, Taylor OC, Adams R, et al (1982) Assessment of crop loss from ozone. J Air Pollut Control Assoc 32:353-361

Heck WW, Adams RM, Cure WW, et al (1983) A reassessment of crop loss from ozone. Environ Sci Technol 12:572A-581A

Horemans N, Foyer CH, Asard H (2000) Transport and action of ascorbate at the plant plasma membrane. Trends Plant Sci 5:263-267

Hormaza JI, Pinney K, Polito VS (1996) Correlation in the tolerance to ozone between sporophytes and male gametophytes of several fruit and nut tree species (Rosaceae). Sex Plant Reprod 9:44-48

Hough AM, Derwent RG (1990) Changes in the global concentration of tropospheric ozone due to human activities. Nature 344:645-648

Howell RK, Devine TE, Hanson CH (1971) Resistance of selected alfalfa strains to ozone. Crop Sci 11:114-115

Hucl P, Beversdorf WD (1982) The inheritance of ozone insensitivity in selected *Phaseolus vulgaris* L. populations. Can J Plant Sci 62:861-865

Imai T, Karita S, Shatori G, et al (1998) L-Galactono-γ-lactone dehydrogenase from sweet potato: purification and cDNA sequence analysis. Plant Cell Physiol 39:1350-1358

Jain AK, Nessler CL (2000) Metabolic engineering of an alternative pathway for ascorbic acid biosynthesis in plants. Mol Breed 6:73-78

Jakob A, Heber U (1998) Apoplastic ascorbate does not prevent the oxidation of fluorescent amphiphilic dyes by ambient and elevated concentrations of ozone in leaves. Plant Cell Physiol 39:313-322

Johnston WJ, Haaland RL, Dickens R (1983) Inheritance of ozone resistance in tall fescue. Crop Sci 23:235-236

Kanofsky JR, Sima S (1991) Singlet oxygen production from the reactions of ozone with biological molecules. J Biol Chem 266:9039-9042

Kanofsky JR, Sima S (1995) Singlet oxygen generation from the reaction of ozone with plant leaves. J Biol Chem 270:7850-7852

Karnosky DF (1991) Ecological genetics and changes in atmospheric chemistry: the application of knowledge. In: Taylor GE, Pitelka LF, Clegg MT (eds) Ecological genetics and air pollution. Springer-Verlag, New York, pp 321-336

Keller R, Springer F, Renz A, et al (1999) Antisense inhibition of the GDP-mannose pyrophosphorylase reduces the ascorbate content in transgenic plants leading to developmental changes during senescence. Plant J 19:131-141

Lea PJ, Wellburn FAM, Wellburn AR, et al (1998) Use of transgenic plants in the assessment of physiological responses to atmospheric pollutants. In: de Kok LJ, Stulen I (eds) Responses of plant metabolism to air pollution and global change. Backhuys, Leiden, pp 241-250

Lee EH, Jersey JA, Gifford C, et al (1984) Differential ozone tolerance in soybean and snap beans: analysis of ascorbic acid in O_3-suseptible and O_3-resistant cultivars by high-performance liquid chromatography. Environ Exp Bot 24:331-341

Luwe MWF, Takahama U, Heber U (1993) Role of ascorbate in detoxifying ozone in the apoplast of spinach (*Spinacia oleracea* L.) leaves. Plant Physiol 101:969-976

Lyons TM, Barnes JD, Davison AW (1997) Relationships between ozone resistance and climate in European populations of *Plantago major*. New Phytol 136:503-510

Lyons T, Plöchl M, Turcsányi E, et al (1999a) Extracellular antioxidants: a protective screen against ozone? In: Agrawal SB, Agrawal M (eds) Environmental pollution and plant responses. CRC Press/Lewis Publishers, Boca Raton, pp 183-201

Lyons T, Ollerenshaw JH, Barnes JD (1999b) Impacts of ozone on *Plantago major*: apoplastic and symplastic antioxidant status. New Phytol 141:253-263

Mächler F, Wasescha MR, Krieg F, et al (1995) Damage by ozone and protection by ascorbic acid in barley leaves. J Plant Physiol 147:469-473

Maddison J, Lyons T, Plöchl M, Barnes J.D. Hydroponically-cultivated radish fed L-galactono-1, 4-lactone exhibit increased ozone tolerance. Planta. In Press

Marenco A, Gouget H, Nedelec P, et al (1994) Evidence of a long-term increase in tropospheric ozone from Pic Du Midi data series – consequences - positive radiative forcing. J Geophys Res Atmos 99:16617-16632

Matyssek R, Innes JL (1999) Ozone – a risk factor for trees and forests in Europe? Water Air Soil Pollut 116:199-226

Mebrahtu T, Mersie W, Rangappa M (1990) Inheritance of ambient ozone insensitivity in common bean (*Phaseolus vulgaris* L.). Environ Pollut 67:79-89

Menser HA (1964) Response of plants to air pollutants. III. A relation between ascorbic acid levels and ozone-susceptibility of light-preconditioned tobacco plants. Plant Physiol 39:564-567

Moldau H, Bichelle I, Hüve K (1998) Dark-induced ascorbate deficiency in leaf cell walls increases plasmalemma injury under ozone. Planta 207:60-66

Murphy JJ, Delucchi MA, McCubbin DR, Kim HJ (1999) The cost of crop damage caused by ozone air pollution from motor vehicles. J Environ Manage 55:273-289

Murphy KJ (1983) Herbicide resistance. Biologist 30:211-218

Noctor G, Foyer CH (1998) Ascorbate and glutathione: keeping active oxygen under control. Annu Rev Plant Physiol Plant Mol Biol 49:249-279

Noctor G, Veljovic-Jovanovic S, Foyer CH (2000) Peroxide processing in photosynthesis: antioxidant coupling and redox signalling. Phil Trans R Soc Lond B 355:1465-1475

Nishikawa Y, Kurata T (2000) Interconversion between dehydro-L-ascorbic acid and L-ascorbic acid. Biosci Biotech Biochem 64:476-483

Nishikimi M, Yagi K (1996) Biochemistry and molecular biology of ascorbic acid biosynthesis. In: Harris RJ (ed) Ascorbic acid: biochemistry and biomedical cell biology. Plenum Press, New York, pp 17-39

Østergaard J, Persiau G, Davey MW, et al (1997) Isolation and cDNA cloning for L-galactono-γ-lactone dehydrogenase, an enzyme involved in the biosynthesis of ascorbic acid in plants. J Biol Chem 272:30009-30016

Örvar BL, Ellis BE (1997) Transgenic tobacco plants expressing antisense RNA for cytosolic ascorbate peroxidase show increased susceptibility to ozone injury. Plant J 11:1297-1305

Pellinen R, Palva T, Kangasjärvi J (1999) Subcellular localization of ozone-induced hydrogen peroxide production in birch (*Betula pendula*) leaf cells. Plant J 20:349-356

Plöchl M, Lyons T, Ollerenshaw J, et al (2000) Simulating ozone detoxification in the leaf apoplast through the direct reaction with ascorbate. Planta 210:454-467

Polle A, Weiser G, Havranek WM (1995) Quantification of ozone influx and apoplastic ascorbate content in needles of Norway spruce trees (*Picea abies* L., Karst.) at high altitude. Plant Cell Environ 18:681-688

Rao MV, Davis KR (1999) Ozone-induced cell death occurs via two distinct mechanisms in *Arabidopsis*: the role of salicylic acid. Plant J 17:603-614

Reiling K, Davison AW (1992) Spatial variation in ozone resistance of British populations of *Plantago major* L. New Phytol 122:699-708

Reiling K, Davison AW (1995) Effects of ozone on stomatal conductance and photosynthesis in populations of *Plantago major* L. New Phytol 129:587-594

Reinert RA, Heggestad HE, Heck WW (1982) Response and genetic modification of plants for tolerance to air pollutants. In: Christiansen MN, Lewis CF (eds) Breeding plants for less favourable environments. Wiley, Chichester, pp 259-292

Roose ML (1991) Genetics of response to atmospheric pollutants. In: Taylor GE, Pitelka LF, Clegg MT (eds) Ecological genetics and air pollution. Springer-Verlag, New York, pp 111-126

Runeckles VC, Chevone BI (1992) Crop responses to ozone. In: Lefohn AS (ed). Surface-level ozone exposures and their effects on vegetation. Lewis Publishers, Chelsea, pp 189-270

Sandermann H, Wellburn AR, Heath RL (1997) Forest decline and ozone: a comparison of controlled chamber and field experiments. Ecological Studies vol 127. Springer-Verlag, Berlin

Schraudner M, Moeder W, Wiese C, et al (1998) Ozone-induced oxidative burst in the ozone biomonitor plant, tobacco Bel W3. Plant J 16:235-245

Sharma YK, Davis KR (1997) The effects of ozone on antioxidant responses in plants. Free Radic Biol Med 23:480-488

Skärby L, Ro-Poulsen H, Wellburn FAM, et al (1998) Impacts of ozone on forests: a European perspective. New Phytol 139:109-122

Smirnoff N (1996) The function and metabolism of ascorbic acid in plants. Ann Bot 78:661-669

Smirnoff N (2000) Ascorbic acid: metabolism and functions of a multi-facetted molecule. Curr Opin Plant Biol 3:229-235

Smirnoff N, Pallanca JE (1996) Ascorbate metabolism in relation to oxidative stress. Biochem Soc Trans 24:472-478

Srivastava HS (1999) Biochemical defence mechanisms of plants to increased levels of ozone and other atmospheric pollutants. Curr Sci 76:525-533

Taylor GE, Hanson PJ, Baldocchi DD (1988) Pollutant deposition to individual leaves and plant canopies: sites of regulation and relationship to injury. In: Heck WW, Taylor OC, Tingey DT (eds) Assessment of crop loss from air pollutants. Elsevier, London, pp 227-257

Tingey DT, Taylor GE (1982) Variation in plant response to ozone: a conceptual model of physiological events. In: Unsworth MH, Ormrod DP (eds) Effects of gaseous air pollutants in agriculture and horticulture. Butterworth Scientific, London, pp 113-138

Torsethaugen G, Pell EJ, Assmann SM (1999) Ozone inhibits guard cell K^+ channels implicated in stomatal opening. Proc Natl Acad Sci USA 96:13577-13582

Turcsányi E, Cardoso-Vilhena J, Daymond J, et al (2000a). Impacts of tropospheric ozone: past, present and likely future. In: Singh SN (ed) Trace gas emissions and plants. Kluwer, Dordrecht, pp 249-272

Turcsányi E, Lyons T, Plöchl M, et al (2000b) Does ascorbate in the mesophyll cell walls form the first line of defence against ozone? Testing the concept using broad bean (*Vicia faba* L.). J Exp Bot 51:901-910

Wheeler GL, Jones MA, Smirnoff N (1998) The biosynthetic pathway of vitamin C in higher plants. Nature 393:365-369

Whitfield CP, Davison AW, Ashenden TW (1997) Artificial selection and heritability of ozone resistance in two populations of *Plantago major*. New Phytol 137:645-655

Wolfenden J, Mansfield TA (1991) Physiological disturbances in plants caused by air pollutants. Proc R Soc Edin 97B:117-138

Wolff K, Morgan-Richards M, Davison AW (2000) Patterns of molecular genetic variation in *Plantago major* and *P. intermedia* in relation to ozone resistance. New Phytol 145:501-509

Yunus M, Iqbal M (1996) Global status of air pollution: an overview. Wiley, Chichester

Zheng Y, Lyons T, Ollerenshaw JH, et al (2000) Ascorbate in the leaf apoplast is a factor mediating ozone resistance in *Plantago major*. Plant Physiol Biochem 38:403-411

13
Detoxification of Active Oxygen Species and Tolerance in Plants Exposed to Air Pollutants and CO_2

Shigeto Morita[1,2] and Kunisuke Tanaka[1,2]

[1]Faculty of Agriculture, Kyoto Prefectural University, Shimogamo-Hangicho 1-5, Sakyo-ku, Kyoto 606-8522, Japan
[2]Basic Research Division, Kyoto Prefectural Institute of Agricultural Biotechnology, Ohji 74, Kitainayazuma, Seika, Soraku, Kyoto 619-0244, Japan

1. Introduction

With the increasing the combustion of fossil fuels, the ambient levels of air pollutants and carbon dioxide (CO_2) have been steadily increasing. Air pollutants, ozone, and sulfur dioxide (SO_2) are known to have damaging effects on plants. Ozone and SO_2 cause visible foliar damages at high concentrations and inhibition of photosynthesis at lower concentrations (Schraudner et al. 1996; Tanaka 1994). Because the damages caused by these pollutants are dependent on oxygen, it has been proposed that active oxygen species (AOS) are involved in the toxicity of the pollutants (Tanaka 1994). Elevations in CO_2 level are also of concern because they have various effects on plants directly and can also affect vegetation through climate changes. CO_2 has been shown to affect the growth, photosynthesis, and yield of plants (Murray 1997) and even to affect plant stress sensitivity by influencing their antioxidative defense systems (see Section 4 of this chapter).

In this chapter, the responses of antioxidative defense systems to ozone, SO_2, and CO_2 in plants are reviewed, and the correlation between the AOS scavenging systems and plant tolerance to air pollutants is discussed.

Air Pollution and Plant Biotechnology
–Prospects for Phytomonitoring and Phytoremediation–
Edited by K. Omasa, H. Saji, S. Youssefian, and N. Kondo
© Springer -Verlag Tokyo 2002

1.1 Active Oxygen Species and Antioxidative Defense Systems

Oxygen is essential for the survival of aerobic organisms, but it can be converted to AOS during cellular metabolism. AOS such as superoxide, hydrogen peroxide (H_2O_2), and hydroxyl radical have deleterious effects because they react with cellular components, which results in the oxidative breakdown of lipids, DNA, and proteins. All aerobic organisms have antioxidative defense systems that scavenge AOS for protection from oxidative injuries. If the scavenging of AOS is overwhelmed by AOS production, cells are faced with the threat of toxicity of AOS. Such stresses caused by AOS are known as oxidative stresses.

Antioxidative systems consist of low molecular weight antioxidants (ascorbate and glutathione, etc.) and enzymes involved in the AOS scavenging reactions (Bowler et al. 1992; Asada 1997; Noctor and Foyer 1998; Fig. 1). Ascorbate (AsA) and glutathione (GSH) can react directly with AOS to scavenge them. On the other hand, the enzymatic scavenging reaction of superoxide is catalyzed by superoxide dismutase (SOD). In higher plants, scavenging of H_2O_2 is catalyzed by ascorbate peroxidase (APX) and catalase (CAT). APX uses AsA as a specific substrate to reduce H_2O_2 to water, resulting in the production of monodehydroascorbate, which is then converted to dehydroascorbate. These molecules are reduced back to AsA by the reaction of monodehydroascorbate

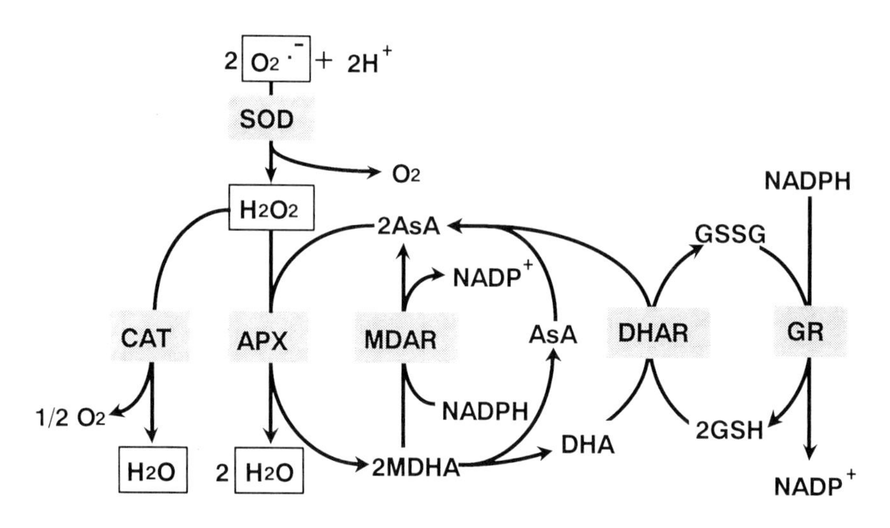

Fig. 1. Active oxygen scavenging pathway in higher plants. *SOD*, superoxide dismutase; *CAT*, catalase; *APX*, ascorbate peroxidase; *MDAR*, monodehydroascorbate reductase; *DHAR*, dehydroascorbate reductase; *GR*, glutathione reductase; *AsA*, ascorbate; *MDHA*, monodehydroascorbate; *DHA*, dehydroascorbate; *GSH*, reduced glutathione; *GSSG*, oxidized glutathione

reductase, dehydroascorbate reductase, and glutathione reductase (GR). The series of reactions of APX to GR is known as the AsA-GSH cycle. The oxidized products of AsA (monodehydroascorbate and dehydroascorbate) and GSH (oxidized glutathione; GSSG) by nonenzymatic scavenging of AOS are also regenerated by AsA-GSH cycle enzymes.

All these enzymes of the AOS scavenging system have multiple isozymes, and the isozymes of each enzyme are differentially localized within cells, with the exception of CAT, which is localized only in peroxisomes. Therefore, each of the intracellular compartments (chloroplasts, mitochondria, peroxisomes, and cytosol) contains each set of the isozymes of the SOD and AsA-GSH cycle enzymes (Bowler et al. 1992; Asada 1997; Jimenez et al. 1997).

In addition to the enzymes described above, plants contain glutathione peroxidase (GPX) and glutathione *S*-transferase (GST), which are thought to be involved in the reduction of lipid peroxides, the oxidation products of membrane lipids (Beeor-Tzahar et al. 1995; Marrs 1996). There are also numerous peroxidases other than APX and GPX, although their substrates and localizations are not clear. Some of these peroxidases may be involved in the scavenging of H_2O_2.

The expression of the enzymes of the AOS scavenging system is controlled to minimize the AOS toxicity. For example, many environmental stresses cause enhanced AOS production within plant cells. To cope with oxidative stress, plants induce the enzymes of the AOS scavenging systems under various environmental stresses such as drought, high light, cold, and heat as well as air pollutants (Noctor and Foyer 1998).

2. Response of Antioxidative Systems to Ozone

2.1 Oxidative Stress Induced by Ozone

The involvement of AOS in ozone toxicity is demonstrated by the observation that ozone fumigation causes the formation of malondialdehydes, which are oxidation products of lipids (Sakaki et al. 1983). This result indicates that ozone-fumigated plants are faced with oxidative stress. Previously, it was thought that ozone entered leaves through the stomata, and then was dissolved in the apoplastic fluid and converted into such AOS as superoxide and hydroxyl radical (Grimes et al. 1983; Mehlhorn et al. 1990). The toxic effects of AOS thus produced diminish membrane integrity, leading to further cytoplasmic damage.

However, it has been revealed that plants produce AOS actively in response to ozone fumigation (Schraudner et al. 1998). The transient AOS production during ozone fumigation occurs in both ozone-sensitive and ozone-tolerant cultivars of tobacco. In contrast, additional AOS production "after" ozone exposure is observed only in sensitive cultivars. AOS production takes place in apoplastic fluid in both the first and the second phase. The second accumulation of AOS in sensitive cultivars is localized in spotlike sites, which suggests a possible correlation between localized AOS production and subsequent localized cell death.

Ozone-induced active AOS accumulation has also been observed by Pellinen et al. (1999) in birch and by Rao and Davis (1999) in *Arabidopsis*. In birch leaves, high levels of H_2O_2 are accumulated on the plasma membrane and in cell walls, in addition to the lower levels accumulated in mitochondria and peroxisomes. It has also been demonstrated that both NAD(P)H oxidase, which produces superoxide, and apoplast-localized peroxidases are involved in the ozone-induced H_2O_2 accumulation (Pellinen et al. 1999). The involvement of NAD(P)H oxidase in the accumulation of H_2O_2 and cell death caused by ozone has also been demonstrated in *Arabidopsis* (Rao and Davis 1999).

The ozone-induced active AOS production and subsequent cell death are quite similar to the pathogen-induced AOS production (the so-called oxidative burst) and hypersensitive response (HR) that are seen in pathogen defense response. Ozone fumigation also elicits responses similar to the pathogen defense response, such as the induction of phytoalexin synthesis, and the induction of GST, phenylalanine ammonia lyase and pathogenesis-related (PR) genes (Sandermann et al. 1998). The resemblance of ozone-induced response to pathogen defense response suggests that ozone-induced lesions, rather than being simply the result of oxidative injuries, might constitute an HR-like cell death that is activated by ozone fumigation.

2.2 Response of Antioxidative Systems and Tolerance to Ozone

As mentioned in the previous section, ozone fumigation causes AOS production and oxidative stress within plant cells, which leads to toxic effects. In response to ozone fumigation, plants induce antioxidative defense systems, including low molecular weight antioxidants and AOS scavenging enzymes. In this section, the responses of antioxidative systems to ozone are discussed.

2.2.1 Antioxidants

AsA and GSH are efficient AOS scavenging agents utilized in both enzymatic and nonenzymatic reactions (Nocter and Foyer 1998). These agents are oxidized in such reactions, resulting in a decrease in the reduced form of antioxidants. Ozone fumigation has been associated with decreases in the reduced forms of AsA and GSH in several species (Sen Gupta et al. 1991; Badiani et al. 1993; Pasqualini et al. 1999; Tausz et al. 1999), indicating that AsA and GSH function in the amelioration of ozone-induced oxidative stress.

Wellburn and Wellburn (1997) investigated changes in the levels of antioxidants following ozone fumigation in ozone-sensitive and ozone-tolerant cultivars of six species. The results were highly complex and difficult to interpret. In tobacco and plantain, AsA and GSH were increased both in sensitive and tolerant cultivars, but in radish, a decrease in GSH and an increase in AsA were observed in the tolerant cultivar, whereas a decrease in AsA level was observed in the sensitive cultivar. The levels of AsA and GSH are determined by the rate of oxidation by AOS, the rate of regeneration of oxidized form, and de novo synthesis. Thus, while ozone-induced oxidative stress enhances the oxidation of antioxidants, it is probable that the changes in the rate of regeneration and de novo synthesis following ozone fumigation differ among species.

Although the response of antioxidants observed by Wellburn and Wellburn (1997) did not appear to be correlated with ozone tolerance, correlations between the levels of AsA or GSH and ozone tolerance have been demonstrated. In spinach, ozone-tolerant cultivars contain higher levels of AsA and GSH than less-tolerant cultivars under normal ambient conditions (Tanaka et al. 1985). Similar results have been reported in tobacco (Pasqualini et al. 1999). The ozone-tolerant tobacco cultivar has a higher AsA level and higher ratio of GSH/GSSG than the ozone-sensitive cultivar. Also, an *Arabidopsis* mutant with reduced AsA level (*vtc1*) is more sensitive to ozone, SO_2, and UV-B than the wild type (Conklin et al. 1996). These observations suggest that, under unstressed conditions, the basal levels of AsA and GSH seem to be correlated with ozone tolerance.

2.2.2 AOS Scavenging Enzymes

Like other environmental stresses, ozone fumigation induces AOS scavenging enzymes. Numerous studies have demonstrated ozone-induced elevation of the activities of SOD (Sen Gupta et al. 1991; Pitcher et al. 1992; Rao et al. 1996), APX (Tanaka et al. 1985; Badiani et al. 1993; Kubo et al. 1995; Rao et al. 1996), GR (Tanaka et al. 1988a; Edwards et al. 1994; Rao et al. 1996), and peroxidase (Kubo et al. 1995; Rao et al. 1996; Schraudner et al. 1998). In addition, ozone-induced increases in mRNA levels of antioxidative genes, including GPX, GST, and the copper chaperone gene, which is involved in detoxification of superoxide, have been observed (see Table 1).

Although ozone induces the enzymes of AOS scavenging systems, individual genes encoding different isozymes show different responses to ozone. For example, one of the three catalase genes, *Cat1*, is suppressed, whereas the other isogenes, *Cat2* and *Cat3*, are induced by ozone in *Nicotiana plumbaginifolia* (Willekens et al. 1994). Multiple genes of SOD isoforms also show different responses to ozone. Cytosolic CuZn-SOD is induced, mitochondrial Mn-SOD is not affected, and chloroplastic Fe-SOD and some chloroplastic CuZn-SODs (e.g., *Arabidopsis CSD2*) are suppressed (Willekens et al. 1994; Conklin and Last 1995; Kliebenstein et al. 1998).

It should be noted that in many cases reported to date, the induction of antioxidative genes encoding cytosolic enzymes is prominent (e.g., SOD, APX, GPX, GST, and peroxidase; see Table 1). As mentioned in the previous section,

Table 1. The responses of antioxidative defense genes to ozone

Genes encoding	Plant species (cultivar)	Reference
Genes induced by ozone		
Cytosolic CuZn-SOD	Tobacco (PBD6)	Willekens et al. 1994
	Arabidopsis	Sharma and Davis 1994
	Arabidopsis	Conklin and Last 1995
	Arabidopsis	Kliebenstein et al. 1998
Chloroplastic CuZn-SOD	*Arabidopsis*	Rao and Davis 1999
Cytosolic APX	Tobacco (PBD6)	Willekens et al. 1994
	Tobacco (Bel W3)	Orvar et al. 1997
	Arabidopsis	Kubo et al. 1995
	Arabidopsis	Conklin and Last 1995
	Arabidopsis	Rao and Davis 1999
CAT (*Cat2*, *Cat3*)	*Nicotiana plumbaginifolia*	Willekens et al. 1994
CAT (SA-binding CAT)	Tobacco (Bel W3)	Orvar et al. 1997
Peroxidase	*Arabidopsis*	Sharma and Davis 1994
GPX	*Nicotiana plumbaginifolia*	Willekens et al. 1994
	Tobacco (Bel B and Bel W3)	Schraudner et al. 1998
Chloroplastic GPX	*Arabidopsis*	Rao and Davis 1999
GST	*Arabidopsis*	Sharma and Davis 1994
	Arabidopsis	Conklin and Last 1995
	Arabidopsis	Sharma et al. 1996
	Arabidopsis	Rao and Davis 1999
Copper chaperone	*Arabidopsis*	Himelblau et al. 1998
Genes repressed by ozone		
Chloroplastic CuZn-SOD	*Arabidopsis*	Kliebenstein et al. 1998
Chloroplastic Fe-SOD	*Nicotiana plumbaginifolia*	Willekens et al. 1994
	Arabidopsis	Conklin and Last 1995
	Arabidopsis	Kliebenstein et al. 1998
CAT (*Cat1*)	*Nicotiana plumbaginifolia*	Willekens et al. 1994
GR	*Arabidopsis*	Conklin and Last 1995
Chloroplastic GR	*Arabidopsis*	Rao and Davis 1999
GPX	*Arabidopsis*	Rao and Davis 1999

The tobacco cultivars, PBD6 and Bel W3, are ozone sensitive, and Bel B is ozone tolerant.

H_2O_2 production by ozone fumigation occurs in apoplasts, cytoplasm, mitochondria, and peroxisomes (Pellinen et al. 1999). It has been shown that "cytosolic" SOD is localized in apoplasts (Ogawa et al. 1996), and cytosolic APX activity also exists in apoplasts as well as cytoplasm (Vanacker et al. 1998). Therefore, the induction of cytosolic isoforms of AOS scavenging enzymes is reasonable for detoxification of AOS produced by ozone fumigation. In addition, the induction of catalase observed in *Nicotiana* (Willekens et al. 1994) contributes to the amelioration of ozone-induced oxidative stress by scavenging H_2O_2 generated in peroxisomes.

In contrast to cytosolic enzymes, some genes of chloroplastic isozymes of CuZn-SOD, Fe-SOD, and GR are suppressed by ozone fumigation (Willekens et al. 1994; Conklin and Last 1995; Kliebenstein et al. 1998; Rao and Davis 1999). The downregulation of these genes apparently opposes the defense against ozone-induced oxidative stress. Although ozone-induced production of H_2O_2 is not observed in chloroplasts (Pellinen et al. 1999), the importance of AOS scavenging in chloroplasts is demonstrated in transgenic plants with elevated SOD activity in chloroplasts and enhanced tolerance to ozone (Van Camp et al. 1994). Madamanchi et al. (1994) have suggested that translational or posttranslational regulation of chloroplastic CuZn-SOD exists in the SO_2-fumigated pea. Similar posttranscriptional controls may be involved in the regulation of chloroplastic SOD genes in response to ozone.

2.2.3 Possible Signal Transducer of Ozone

When exposed to ozone, plants increase their production of ethylene and salicylic acid, both of which are thought to transduce the ozone signal (Sandermann et al. 1998). As mentioned in the previous section, ozone induces the genes involved in pathogen defense response as well as antioxidative genes (Sandermann et al. 1998). Salicylic acid is a well-known signaling molecule of pathogen defense response (Lamb and Dixon 1997). The induction of some of the defense genes (GST, chloroplastic GPX, and *PR-1*) by ozone is mediated by salicylic acid, whereas the inductions of cytosolic APX and PAL by ozone are independent of salicylic acid (Sharma et al. 1996; Rao and Davis 1999). Pretreatment with ethylene or methyl jasmonate can protect plants from subsequent ozone injuries, indicating that ethylene and jasmonate are signal transducers for activation of defense responses against ozone (Mehlhorn 1990; Orvar et al. 1997). Ethylene and methyl jasmonate both induce the cytosolic APX gene (Mehlhorn 1990; Orvar et al. 1997), suggesting that these signal molecules might be involved in the response of cytosolic APX to ozone.

Despite its deleterious effects, H_2O_2 is known to act as a signal transducer during pathogen defense (HR and systemic acquired resistance) and cold and heat acclimation (Lamb and Dixon 1997; Prasad et al. 1994; Dat et al. 1998). In these cases, H_2O_2 is involved in the upregulation of GST, GPX (Levine et al. 1994), CAT, and peroxidase (Prasad et al. 1994). The resemblance between the oxidative burst and the induction of defense genes caused by pathogen infection and those caused by ozone fumigation (Sandermann et al. 1998) tempts us to speculate that H_2O_2 may be involved in ozone signaling as well as pathogen defense. H_2O_2 has been shown to induce both cytosolic SOD and cytosolic APX (Pastori and Trippi 1992; Morita et al. 1999). Therefore, some of the antioxidative genes might be regulated by H_2O_2 under ozone fumigation.

2.3 Antioxidative Enzymes and Ozone Tolerance

The involvement of antioxidative enzymes in tolerance to ozone-induced oxidative damage has been clearly demonstrated in antisense transgenic plants. Transgenic plants with reduced activities of cytosolic APX or CAT showed increased sensitivity to ozone (Orvar and Ellis 1997; Willekens et al. 1997). These observations indicate that the scavenging of AOS is essential for defense against ozone toxicity.

Many transgenic plants with increased AOS scavenging enzymes have been produced to examine whether the overexpression of antioxidative genes can elevate stress tolerance (for review, see Allen et al. 1997). Some of these plants (Van Camp et al. 1994; Broadbent et al. 1995) were revealed to be more tolerant to ozone compared wild type plants (see the chapter by Aono, this volume).

The targets of overexpression of AOS scavenging enzymes have been mainly chloroplastic isoforms (or overexpression of the enzymes targeted to chloroplasts) because chloroplasts are thought to be the largest source of AOS generation. However, Pitcher and Zilinskas (1996) examined ozone tolerance in transgenic plants overexpressing cytosolic SOD and found that elevated activity of cytosolic SOD can protect plants from ozone injuries. This result is consistent with the observation that production of AOS induced by ozone fumigation occurs in apoplast (Schraudner et al. 1998; Pellinen et al. 1999). This observation demonstrates the indispensable role of cytosolic SOD in the defense against ozone toxicity, although the mechanisms of the protective effects of overexpression of cytosolic SOD remain unclear. It is probable that increased SOD activity protects cells from oxidative injuries and results in the decreased foliar necrosis. In addition, it can be postulated that overexpression of cytosolic SOD alters the levels of AOS, thereby modifying the signaling pathway. If H_2O_2 participates in the signaling of defense responses under ozone fumigation as well as during pathogen infection, then the elevation of H_2O_2 level caused by SOD overexpression could enhance the activation of defense responses. Alternatively, a decrease in superoxide level could affect ozone sensitivity. Rao and Davis (1999) suggest that there are two types of ozone-induced cell death, one that is likely to be an HR-like, programmed cell death, and the other a necrotic cell death caused by oxidative injuries. Superoxide can act as a signal for initiating programmed cell death (Jabs et al. 1996). A decrease in superoxide level could thus suppress HR-like death, resulting in a decrease of foliar injuries.

Because the toxic effects of H_2O_2 and the role of H_2O_2 in the defense system are not fully understood, the effect of overexpression of APX cannot presently be determined. It is probable that sensitivity to ozone (and also to pathogens) varies among individual transgenic plants depending on their APX activity. With respect to ozone injuries, transgenic plants with increased APX activity in chloroplasts are not significantly different from nontransgenic plants (Torsethaugen et al. 1997). In this case, the increased APX activity might diminish the H_2O_2 signal and fail to activate the defense response effectively.

The existence of multiple isoforms of AOS scavenging enzymes and differential expression of individual isoforms are probably the result of the dual function of AOS, with toxic reagents causing oxidative damages and signal transducers regulating defense responses. For the fine control of AOS levels at specific sites, AOS scavenging enzymes are regulated in a complex manner. The transgenic plants containing antioxidative genes reported to date have transgenes driven by constitutive promoters. To effectively enhance the tolerance to ozone or pathogens by a transgenic approach, it will be necessary to control the expression of transgenes using inducible promoters.

3. Response of Antioxidative Systems and Tolerance to SO_2

Several studies have demonstrated the involvement of AOS in SO_2 toxicity and a correlation between antioxidative systems and tolerance to SO_2. Tanaka and Sugahara (1980) reported induction of SOD activity in poplar plants in response to a low concentration of SO_2 (0.1 ppm), and plants with higher SOD activity were more tolerant to subsequent exposure to 2.0 ppm SO_2 than control plants. In addition, paraquat-tolerant tobacco plants generated from paraquat-tolerant calli and having high SOD activity showed less damage than control plants following fumigation with SO_2 (Tanaka et al. 1988b). Madamanchi and Alscher (1991) demonstrated that the antioxidative systems of two pea cultivars having different SO_2 sensitivity showed different responses when exposed to SO_2. SOD activity was elevated by SO_2 in the SO_2-tolerant cultivar but not in the sensitive cultivar. Also, the SO_2-induced increases in GSH content and the GR activity were faster and more pronounced in the tolerant cultivar than in the sensitive cultivar.

The intracellular sites of SO_2-induced AOS production in plants are obscure. Several antioxidative genes are induced by SO_2, but the genes responsive to SO_2 are not always chloroplastic isoforms (see Table 2), and they vary among plant species and experimental conditions. In an SO_2-tolerant pea cultivar, SO_2 exposure causes elevation of activities and protein levels of cytosolic and chloroplastic CuZn-SOD (Madamanchi et al. 1994). However, elevation of the mRNA level of chloroplastic CuZn-SOD was not observed during these periods, suggesting that some translational or posttranslational mechanisms might be involved in the response of chloroplastic CuZn-SOD. In *Nicotiana plumbaginifolia*, SO_2 causes induction of *Cat2* and GPX, and repression of *Cat1* and chloroplastic Fe-SOD genes (Willekens et al. 1994). In *Arabidopsis*, on the other hand, activities of APX and peroxidase are increased by SO_2, and the mRNA level of cytosolic APX is elevated by SO_2 (Kubo et al. 1995). The induction of APX, CAT, and peroxidase suggests that H_2O_2 scavenging is important in the

Table 2. The responses of antioxidative defense genes to SO_2

Genes encoding	Plant species (cultivar)	Reference
Genes induced by SO_2		
Cytosolic APX	*Arabidopsis*	Kubo et al. 1995
CAT (*Cat2*)	*Nicotiana plumbaginifolia*	Willekens et al. 1994
GPX	*Nicotiana plumbaginifolia*	Willekens et al. 1994
Genes repressed by SO_2		
Chloroplastic CuZn-SOD	Pea (Nugget)	Rao and Davis 1999 [a]
Chloroplastic Fe-SOD	*Nicotiana plumbaginifolia*	Willekens et al. 1994
CAT (*Cat1*)	*Nicotiana plumbaginifolia*	Willekens et al. 1994

[a] The protein level of chloroplastic CuZn-SOD was elevated by SO_2.
Pea cultivar, Nugget is SO_2 sensitive.

defense against SO_2. Although the response of antioxidative genes differs widely between *Nicotiana* and *Arabidopsis*, SO_2 and ozone have similar effects on the gene expression of AOS scavenging genes in these species. The genes induced by SO_2 exposure are also induced by ozone fumigation (Willekens et al. 1994; Kubo et al. 1995). Because SO_2 and ozone both induce enhanced synthesis of phytoalexins and polyamines in addition to gene expression of antioxidative enzymes, it is suggested that these air pollutants activate the same defense system, possibly by the same signaling pathway (Willekens et al. 1994). The molecular mechanism of signaling of SO_2 remains poorly understood. Whether active AOS production is caused by SO_2, as it is by ozone, and whether H_2O_2, salicylic acid, or ethylene are also involved in the SO_2 signaling are intriguing questions.

Transgenic tobacco plants that overexpress GR in chloroplasts have been reported to show enhanced tolerance to SO_2 compared with wild-type plants (Aono et al. 1993). However, the effects of overexpression of other AOS scavenging enzymes on SO_2 tolerance remain to be investigated in future studies.

4. Response of Antioxidative Systems to CO_2

The world's atmospheric level of CO_2 has been steadily increasing as a result of fossil fuel consumption and tropical deforestation. It is estimated that CO_2 concentration will increase to twice the present ambient level during the twenty-first century (Roeckner 1992). Elevated CO_2 levels have been implicated in global climate change, and particularly in global warming.

CO_2 is an essential component of photosynthesis. In chloroplasts, electrons excited by photochemical reaction cause reduction of $NADP^+$, with the resulting NADPH being utilized for CO_2 fixation. When the supply of CO_2 is limited, the electron sink ($NADP^+$) is saturated, and the excess electrons cause an overreduction of the photosynthetic electron transport pathway (Arisi et al. 1998). In such cases, the reduction of molecular oxygen to form superoxide radicals (Mehler reaction) can constitute a sink of excess electrons (Foyer 1997; Asada 1999). Under severe stress conditions, the scavenging of H_2O_2 generated from superoxide does not proceed efficiently due to the inactivation of chloroplastic APX, and thus photooxidative damages occur (Shikanai et al. 1998). Limited CO_2 availability is likely to occur under drought conditions due to stomatal closure. Thus, a shortage of CO_2 under illuminated leaves is deleterious.

In contrast, elevation of the CO_2 level increases both the growth and biomass of plants. However, the notion that high levels of CO_2 are beneficial for higher yield is highly oversimplified. An elevation in CO_2 can increase the grain yield of wheat, rice, and soybean, but it simultaneously reduces the N content in the grains, impairing nutritional quality (Murray 1997). Moreover, it is expected that plants encounter high CO_2 in conjunction with other environmental stresses, such as increased UV-B, air pollutants, high temperature, and drought. Under stress conditions, the positive effects of high CO_2 on the growth of plants can be decreased in some species. For example, high CO_2-induced increases in seed yield and biomass are decreased by UV-B radiation in wheat and rice (Teramura et al. 1990; Ziska and Teramura 1992). In other species, however, the positive effects of high CO_2 can counteract the negative effects of ozone and SO_2. The ozone-induced inhibition of shoot growth is suppressed under high CO_2 concentration in wheat (Rao et al. 1995). Also, it has been shown that SO_2 fumigation causes a reduction in photosynthetic rate, but the negative effect of SO_2 can be compensated by exposure to a combination of high CO_2 and SO_2 (Sandhu et al. 1992).

The effects of high CO_2 on antioxidative systems have been investigated by Schwanz et al. (1996). Under high CO_2, the activities of SOD, APX, and CAT are decreased. The downregulation of these enzymes suggests that, under a high CO_2 condition, plants suffer less from oxidative stress than when grown under an ambient CO_2 level. It is probable that high CO_2 enhances the utilization of NADPH for carbon fixation, which leads to the increased availability of $NADP^+$ for the electron sink. In these situations, the production of superoxide might be decreased as a result of the reduced electron flux toward the Mehler reaction.

High CO_2 also affects stress sensitivity to drought and ozone fumigation. Under a combination of high CO_2 and drought, the activities of SOD and CAT in oak and the activities of SOD and peroxidase in pine are increased compared with those in the plants treated with drought stress alone (Schwanz et al. 1996). Rao et al. (1995) reported that high CO_2 can ameliorate ozone-induced oxidative damage in foliar proteins. They also observed that the activities of SOD, APX, GR and peroxidase were higher in plants under high CO_2 and ozone than in those under normal CO_2 and high ozone. The mechanism by which high CO_2 increases the activities of antioxidative enzymes is unclear. It should be noted that the AOS scavenging enzymes are not induced by high CO_2 alone (Rao et al. 1995; Schwanz et al. 1996), which indicates that induction of antioxidative enzymes by high CO_2 occurs only under stressed conditions.

It is postulated that oxidative damage by other environmental stresses may be ameliolated under high CO_2 condition. The effects of CO_2 in stressed plants, especially at high temperature, require further investigation.

Acknowledgments. The authors are grateful for financial support from the Ministry of Education, Culture, Sports, Science and Technology (Grants-in-Aid for Scientific Research 10460149 and 11740448), the Japan Society for the Promotion of Science (the "Research for the Future" Program), and the Ministry of Agriculture, Forestry and Fisheries of Japan (Rice Genome Project MP-2106).

References

Allen RD, Webb RP, Schake SA (1997) Use of transgenic plants to study antioxidant defenses. Free Radical Biol Med 23:473-479

Aono M, Kubo A, Saji H, Tanaka K, Kondo N (1993) Enhanced tolerance to photooxidative stress of transgenic *Nicotiana tabacum* with high chloroplastic glutathione reductase activity. Plant Cell Physiol 34:129-135

Arisi A-CM, Cornic G, Jouanin L, Foyer CH (1998) Overexpression of iron superoxide dismutase in transformed poplar modifies the regulation of photosynthesis at low CO_2 partial pressures or following exposure to the prooxidant herbicide methyl viologen. Plant Physiol 117:565-574

Asada K (1997) The role of ascorbate peroxidase and monodehydroascorbate reductase in H_2O_2 scavenging in plants. In: Scandalios JG (ed) Oxidative stress and the molecular biology of antioxidant defenses. Cold Spring Harbor Laboratory Press, New York, pp 715-735

Asada K (1999) The water-water cycle in chloroplasts: scavenging of active oxygens and dissipation of excess photons. Annu Rev Plant Physiol Plant Mol Biol 50:601-639

Badiani M, Schenone G, Paolacci AR, Fumagalli I (1993) Daily fluctuations of antioxidants in bean (*Phaseolus vulgaris* L.) leaves as affected by the presence of ambient air pollutants. Plant Cell Physiol 34:271-279

Beeor-Tzahar T, Ben-Hayyim G, Holland D, Faltin Z, Eshdat Y (1995) A stress-associated citrus protein is a distinct plant phospholipid hydroperoxide glutathione peroxidase. FEBS Lett 366:151-155

Bowler C, Van Montagu M, Inze D (1992) Superoxide dismutase and stress tolerance. Annu Rev Plant Physiol Plant Mol Biol 43:83-116

Broadbent P, Creissen GP, Kular B, Wellburn AR, Mullineaux PM (1995) Oxidative stress responses in transgenic tobacco containing altered levels of glutathione reductase activity. Plant J 8:247-255

Conklin PL, Last RL (1995) Differential accumulation of antioxidant mRNAs in *Arabidopsis thaliana* exposed to ozone. Plant Physiol 109:203-212

Conklin PL, Williams EH, Last RL (1996) Environmental stress sensitivity of an ascorbic

acid-deficient *Arabidopsis* mutant. Proc Natl Acad Sci USA 93:9970-9974
Dat JF, Lopez-Delgado H, Foyer CH, Scott IM (1998) Parallel changes in H_2O_2 and catalase during thermotolerance induced by salicylic acid or heat acclimation in mustard seedlings. Plant Physiol 116:1351-1357
Edwards EA, Enard C, Creissen GP, Mullineaux PM (1994) Synthesis and properties of glutathione reductase in stressed peas. Planta 192:137-143
Foyer CH (1997) Oxygen metabolism and electron transport in photosynthesis. In: Scandalios JG (ed) Oxidative stress and the molecular biology of antioxidant defenses. Cold Spring Harbor Laboratory Press, New York, pp 587-621
Grimes HD, Perkins KK, Boss WF (1983) Ozone degrades into hydroxyl radical under physiological conditions. Plant Physiol 72:1016-1020
Himelblau E, Mira H, Lin S-J, Culotta VC, Penarrubia L, Amasino RM (1998) Identification of a functional homolog of the yeast copper homeostasis gene *ATX1* from *Arabidopsis*. Plant Physiol 117:1227-1234
Jabs T, Dietrich RA, Dangl JL (1996) Initiation of runaway cell death in an *Arabidopsis* mutant by extracellular superoxide. Science 273:1853-1856
Jimenez A, Hernandez JA, del Rio LA, Sevilla F (1997) Evidence for the presence of the ascorbate-glutathione cycle in mitochondria and peroxisomes of pea (*Pisum sativum* L.) leaves. Plant Physiol 114:275-284
Kliebenstein DJ, Monde R-A, Last RL (1998) Superoxide dismutase in *Arabidopsis*: an eclectic enzyme family with disparate regulation and protein localization. Plant Physiol 118:637-650
Kubo A, Saji H, Tanaka K, Kondo N (1995) Expression of *arabidopsis* cytosolic ascorbate peroxidase gene in response to ozone or sulfur dioxide. Plant Mol Biol 29: 479-489
Lamb C, Dixon RA (1997) The oxidative burst in plant disease resistance. Annu Rev Plant Physiol Plant Mol Biol 48:251-275
Levine A, Tenhaken R, Dixon R, Lamb C (1994) H_2O_2 from the oxidative burst orchestrates the plant hypersensitive disease resistance response. Cell 79:583-593
Madamanchi NR, Alscher RG (1991) Metabolic bases for differences in sensitivity of two pea cultivars to sulfur dioxide. Plant Physiol 97:88-93
Madamanchi NR, Donahue JL, Cramer CL, Alscher RG, Pedersen K (1994) Differential response of Cu,Zn superoxide dismutases in two pea cultivars during a short-term exposure to sulfur dioxide. Plant Mol Biol 26:95-103
Marrs KA (1996) The functions and regulation of glutathione S-transferase. Annu Rev Plant Physiol Plant Mol Biol 47:127-158
Mehlhorn H (1990) Ethylene-promoted ascorbate peroxidase activity protects against hydrogen peroxide, ozone and paraquat. Plant Cell Environ 13: 971-976
Mehlhorn H, Tabner BJ, Wellburn AR (1990) Electron spin resonance evidence for the formation of free radicals in plants exposed to ozone. Physiol Plant 79:377-383
Morita S, Kaminaka H, Masumura T, Tanaka K (1999) Induction of rice cytosolic ascorbate peroxidase mRNA by oxidative stress; the involvement of hydrogen peroxide in oxidative stress signalling. Plant Cell Physiol 40:417-422
Murray DR (1997) Carbon dioxide and plant responses. Research Studies Press, Taunton, Somerset, England
Noctor G, Foyer CH (1998) Ascorbate and glutathione: keeping active oxygen under control. Annu Rev Plant Physiol Plant Mol Biol 49:249-279
Ogawa K, Kanematsu S, Asada K (1996) Intra- and extra-cellular localization of "cytosolic" CuZn-superoxide dismutase in spinach leaf and hypocotyl. Plant Cell

Physiol 37:790-799

Orvar BL, Ellis BE (1997) Transgenic tobacco plants expressing antisense RNA for cytosolic ascorbate peroxidase show increased susceptibility to ozone injury. Plant J 11:1297-1305

Orvar BL, McPherson J, Ellis BE (1997) Pre-activating wounding response in tobacco prior to high-level ozone exposure prevents necrotic injury. Plant J 11: 203-212

Pasqualini S, Batini P, Ederli L, Antonielli M (1999) Reponses of the xanthophyll cycle pool and ascorbate-glutathione cycle to ozone stress in two tobacco cultivars. Free Radical Res 31:s67-s73

Pastori GM, Trippi VS (1992) Oxidative stress induces high rate of glutathione reductase synthesis in a drought-resistant maize strain. Plant Cell Physiol 33:957-961

Pellinen R, Palva T, Kangasjarvi J (1999) Subcellular localization of ozone-induced hydrogen peroxide production in birch (*Betula pendula*) leaf cells. Plant J 20:349-356

Pitcher LH, Zilinskas BA (1996) Overexpression of copper/zinc superoxide dismutase in the cytosol of transgenic tobacco confers partial resistance to ozone-induced foliar necrosis. Plant Physiol 110:583-588

Pitcher LH, Brennan E, Zilinskas BA (1992) The antiozonant ethylenediurea does not act via superoxide dismutase induction in bean. Plant Physiol 99:1388-1392

Prasad TK, Anderson MD, Martin BA, Stewart CR (1994) Evidence for chilling-induced oxidative stress in maize seedlings and a regulatory role for hydrogen peroxide. Plant Cell 6:65-74

Rao MV, Davis KR (1999) Ozone-induced cell death occurs via two distinct mechanisms in *Arabidopsis*: the role of salicylic acid. Plant J 17:603-614

Rao MV, Hale BA, Ormrod DP (1995) Amelioration of ozone-induced oxidative damage in wheat plants grown under high carbon dioxide. Plant Physiol 109:421-432

Rao MV, Paliyath G, Ormrod DP (1996) Ultraviolet-B- and ozone-induced biochemical changes in antioxidant enzymes of *Arabidopsis thaliana*. Plant Physiol 110:125-136

Roeckner E (1992) Past, present and future levels of greenhouse gases in the atmosphere and model predictions of related climatic changes. J Exp Bot 43:1097-1109

Sakaki T, Kondo N, Sugahara K (1983) Breakdown of photosynthetic pigments and lipids in spinach leaves with ozone fumigation: role of active oxygens. Physiol Plant 59:28-34

Sandermann H Jr, Ernst D, Heller W, Langebartels C (1998) Ozone: an abiotic elicitor of plant defence reactions. Trends Plant Sci 3:47-50

Sandhu R, Li Y, Gupta G (1992) Sulphur dioxide and carbon dioxide induced changes in soybean physiology. Plant Sci 83:31-34

Schraudner M, Langebartels C, Sandermann H Jr (1996) Plant defence systems and ozone. Biochem Soc Trans 24:456-461

Schraudner M, Moeder W, Wiese C, Van Camp W, Inze D, Langebartels C, Sandermann H Jr (1998) Ozone-induced oxidative burst in the ozone biomonitor plant, tobacco Bel W3. Plant J 16:235-245

Schwanz P, Picon C, Vivin P, Dreyer E, Guehl J-M, Polle A (1996) Response of antioxidative systems to drought stress in penduncenculate oak and maritime pine as modulated by elevated CO_2. Plant Physiol 110:393-402

Sen Gupta A, Alscher RG, McCune D (1991) Response of photosynthesis and cellular antioxidants to ozone in *Populus* leaves. Plant Physiol 96:650-655

Sharma YK, Davis KR (1994) Ozone-induced expression of stress-related genes in *Arabidopsis thaliana*. Plant Physiol 105:1089-1096

Sharma YK, Leon J, Raskin I, Davis KR (1996) Ozone-induced responses in *Arabidopsis thaliana*: the role of salicylic acid in the accumulation of defense-related transcripts and induced resistance. Proc Natl Acad Sci USA 93:5099-5104

Shikanai T, Takeda T, Yamauchi H, Sano S, Tomizawa K, Yokota A, Shigeoka S (1998) Inhibition of ascorbate peroxidase under oxidative stress in tobacco having bacterial catalase in chloroplasts. FEBS Lett 428:47-51

Tanaka K (1994) Tolerance to herbicides and air pollutants. In: Foyer CH, Mullineaux PM (eds) Causes of photooxidative stress and amelioration of defense systems in plants. CRC Press, Boca Raton, pp 365-378

Tanaka K, Sugahara K (1980) Role of superoxide dismutase in defense against SO_2 toxicity and an increase in superoxide dismutase activity with SO_2 fumigation. Plant Cell Physiol 21:601-611

Tanaka K, Suda Y, Kondo N, Sugahara K (1985) O_3 tolerance and the ascorbate-dependent H_2O_2 decomposing system in chloroplasts. Plant Cell Physiol 26:1425-1431

Tanaka K, Saji H, Kondo N (1988a) Immunological properties of spinach glutathione reductase and inductive biosynthesis of the enzyme with ozone. Plant Cell Physiol 29:637-642

Tanaka K, Furusawa I, Kondo N, Tanaka K (1988b) SO_2 tolerance of tobacco plants regenerated from paraquat-tolerant callus. Plant Cell Physiol 29:743-746

Tausz M, Bytnerowicz A, Arbaugh MJ, Weidner W, Grill D (1999) Antioxidants and protective pigments of *Pinus ponderosa* needles at gradients of natural stresses and ozone in the San Bernardino Mountains in California. Free Radical Res 31:s113-s120

Teramura AH, Sullivan JH, Ziska LH (1990) Interaction of elevated ultraviolet-B radiation and CO_2 on productivity and photosynthetic characteristics in wheat, rice and soybean. Plant Physiol 94:470-475

Torsethaugen G, Pitcher LH, Zilinskas BA, Pell EJ (1997) Overproduction of ascorbate peroxidase in the tobacco chloroplast does not provide protection against ozone. Plant Physiol 114:529-537

Van Camp W, Willekens H, Bowler C, Van Montagu M, Inze D, Reupold-Popp P, Sandermann H Jr, Langebartels C (1994) Elevated levels of superoxide dismutase protect transgenic plants against ozone damage. Biotechnology 12:165-168

Vanacker H, Carver TL, Foyer CH (1998) Pathogen-induced changes in the antioxidant status of the apoplast in barley leaves. Plant Physiol 117:1103-1114

Wellburn AR, Wellburn FAM (1997) Air pollution and free radical protection responses of plants. In: Scandalios JG (ed) Oxidative stress and the molecular biology of antioxidant defenses. Cold Spring Harbor Laboratory Press, New York, pp 861-876

Willekens H, Van Camp W, Van Montagu M, Inze D, Langebartels C, Sandermann H Jr (1994) Ozone, sulfur dioxide, and ultraviolet B have similar effects on mRNA accumulation of antioxidant genes in *Nicotiana plumbaginifolia* L. Plant Physiol 106:1007-1014

Willekens H, Chamnongpol S, Davey M, Schraudner M, Langebartels C, Van Montagu M, Inze D, Van Camp W (1997) Catalase is a sink for H_2O_2 and is indispensable for stress defence in C_3 plants. EMBO J 16:4806-4816

Ziska LH, Teramura AH (1992) CO_2 enhancement of growth and photosynthesis in rice (*Oryza sativa*). Plant Physiol 99:473-481

14
Countermeasures with Fertilization to Reduce Oxidant-Induced Injury to Plants

Haruko Kuno and Kazushi Arai

Environmental Ecology Division, Tokyo Metropolitan Forestry Experiment Station, Hirai 2753-1, Hinode, Nishitama-gun, Tokyo 190-0182, Japan

1. Introduction

Visible injury to field and garden plants, as a result of photochemical smog in Los Angeles, CA, USA, was reported in the 1940s. In Japan, visible injuries to taro (*Colocasia esulenta* Schott), tobacco (*Nicotiana rustica* L.), and pak-choi (*Brassica campestris* L.) seedlings and to welsh onion (*Allium fistulosum* L.), komatsuna (*Brassica campestris* L.), and spinach (*Spinaca oleracea* L.) leaves due to ozone (O_3) and peroxyacetyl nitrate (PAN) have also been identified (Matsuoka et al. 1971; Sawada et al. 1972; Fukuda 1973; Matsumoto et al. 1977) in the 1970s. In addition, visible injuries from photochemical oxidants and their resultant effects on the growth and yields of a variety of agricultural crops, such as rice (*Oryza sativa* L), and soybean [*Glycine max* (L.) Merr. et al.], have been observed (Matsuoka et al. 1976; Takasaki et al. 1976; Saio et al. 1978). Recently, areas with high levels of photochemical oxidants have been found to be extending from coastal urban areas to inland areas (Ohara et al. 1995), and the development of severe foliar damage from O_3 and PAN, is being identified in areas where only limited plant injuries were reported several years ago (Air Pollution Section, Center of Promoting Countermeasure against Pollution, the Kanto Area 1999). Various forms of foliar damage to agricultural crops caused by air pollutants have also been observed in Taiwan (Sun 1994a, b), China, and other countries (Khan et al 1996) in Asia. To overcome the damaging effects of air pollutants on

Air Pollution and Plant Biotechnology
–Prospects for Phytomonitoring and Phytoremediation–
Edited by K. Omasa, H. Saji, S. Youssefian, and N. Kondo
© Springer -Verlag Tokyo 2002

agricultural crops and trees, various studies have been conducted including the selection of plant varieties resistant to ambient oxidants or O_3 (Kuno 1989b; Tanaka et al. 1990; Izuta et al. 1999), the creation of transgenic plants (Aono et al. 1995), the use of fertilizers for cultural management (Montes et al. 1982; Fangmeier et al. 1996; Pell et al. 1990), and the supply of chemicals (Gatta et al. 1997; Godzik and Manning 1998). In this review, changes in visible injuries of plants resulting from exposure to ambient oxidants or O_3 as a result of fertilization are discussed. In addition, practical examples of fertilization to avoid visible injuries of vegetable crops conducted in Japan are described.

2. Effects of Nutritional Components on Ozone-Induced Visible Injury

Foliar damage to field and garden crops by photochemical oxidants or O_3 has been identified in many countries. The effects of soil moisture content (Flagler et al. 1987), temperature (Ormrod et al. 1973; Flagler et al. 1987), light (Moldau et al. 1998), plant variety (Kuno 1989a), and growth periods (Kuno 1989a) on the extent of the injuries are well documented. The physiological relationships between such damage and the kind and amounts of various nutritional components, however, are inadequately understood (Flagler et al. 1987; Heagle et al. 1993; Fangmeier et al. 1997). In the following sections, the effects of the major fertilizer components, nitrogen (N), phosphate (P), potassium (K), and calcium (Ca) on visible foliar damage are discussed in view of the results of several studies summarized in Table 1.

2.1 Nitrogen

Numerous studies have been conducted on the relationship between N application and the susceptibility of tobacco to O_3. Kitamura and Monden (1973) cultivated the tobacco variety, Hicks 2, with various amounts of N in the field, and found that the area and number of injured leaves decreased with increasing N applications. A negative correlation was thus found between the extent of injury and protein N content, with the susceptibility of tobacco to O_3 being closely related to the decline in N content of leaf proteins. Foliar damage to the middle and lower positions of spinach by photochemical oxidants was observed most around harvest time. Increased sensitivity of spinach to O_3 was found to be correlated with lower protein and chlorophyll contents and SOD activities of leaves located in the middle and lower positions of the plant compared with leaves in the upper positions (Kuno 1989a). Spinach plants, cultivated in pots with various fertilizers, were exposed to O_3 during harvest time, and the extent of visible injury was determined (Fig. 1). Severe O_3-induced damage was observed with application of the standard amount of N, but a reduction in O_3-induced damage was found in treatments with 2- or

4fold the standard N amounts. In treatments with no N application, almost no injury was found because of the abnormal growth of the plants (Fig. 1). In accordance with the increased amounts of applied N, the total foliar contents of N, protein, and chlorophyll also increased, but there was no clear physiological relationship between these increases and the O_3-induced injuries.

Sugawara (1957) found that higher ascorbic acid contents were observed on doubling the N amount. Ascorbic acid, which is involved in resistance to O_3 (Tanaka et al. 1985), was measured in the middle and lower leaf positions of spinach before O_3 exposure (Kuno 1988). The high ascorbic acid content in the spinach leaves at zero-N application was associated with indicated poor growth and almost no injury, while application of four times the standard N amount

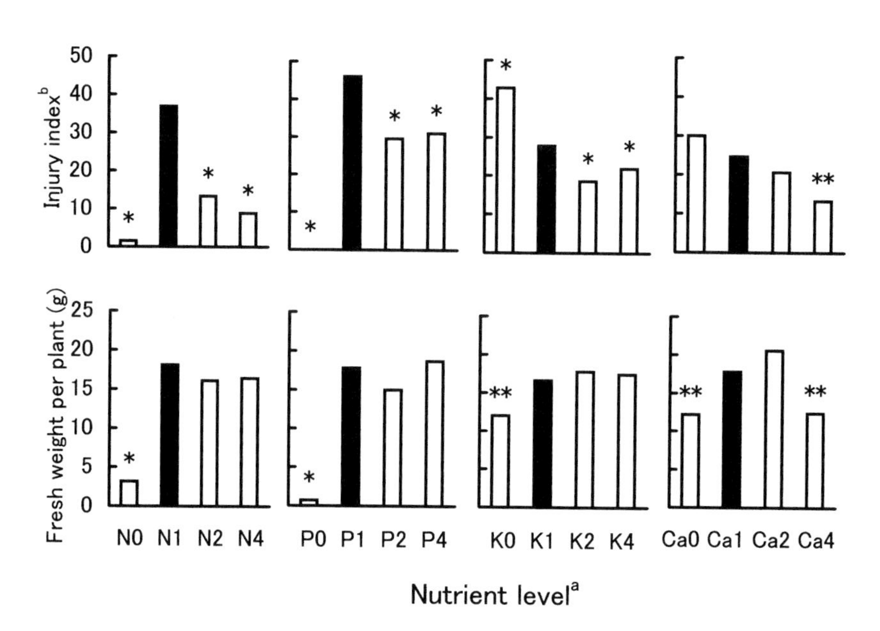

Fig.1. Visible injury index and fresh weight per plant of spinach exposed to ozone, using N, P, K, and Ca nutrient levels (Kuno 1988). Standard dosages of fertilizer used as controls were N 0.5 g, P_2O_5 0.75 g, K_2O 0.5 g, MgO 0.5 g, and CaO 2.75 g per 2.4-l pot. [a]Nutrient levels: N0 : N1 : N2 : N4 = 0.0 : 0.5 : 1.0 : 2.0 g N/pot; P0 : P1 : P2 : P4 = 0 : 0.75 : 1.5 : 3.0 g P_2O_5/pot; K0 : K1 : K2 : K4 = 0.0 : 0.5 : 1.0 : 2.0 g K_2O/pot; Ca0 : Ca1 : Ca2 : Ca4 = 0.0 : 2.75 : 5.5 : 11.0 g CaO/pot

[b]Injury index= $\left(\sum_{n=0}^{5} nl\right) \times \frac{100}{5L}$

where n, Grades 0 to 5 (5 = very severe injury), and l, Numbers of leaves injured or not injured by oxidants. L , Numbers of total leaves. *P<0.05; **P<0.01 versus control

Table 1. Nutrient effects on visible injury of plants to ozone or ambient oxidants

N nutrition

Plant	Cultivar	N application rate and growing medium [a]	Exposure conditions (Ozone conc. period)	Foliar injury	Reference[b]
Tobacco	Hicks 2	N 3, 7, 11, 15 kg/10 ares. Field	Ox[c] Okayama, Japan	Visible injury decreased with increasing N.	1
Spinach	Shinryoku	N 1.0, 2.0 g per 1/5,000 ares Wougnel pot. volcanic ash soil	O_3 0.20 ppm 3 h	Visible injury decreased with increasing N.	2
Spinach	F_1 Shinryoku	N 0.5, 1.0, 2.0 g per 2.4-l pot. kuroboku soil + peat + sand	O_3 0.11 ppm 4 h	Visible injury decreased with increasing N.	3
Radish		N 0.3, 0.6, 1.2 g per 1/5,000 ares Wougnel pot. volcanic ash soil + peat	Ox Max. 0.16 ppm in Tokyo	Visible injury decreased with increasing N.	4
Soybean	Enrei	N 0, 0.3, 0.6, 0.9 g per 1/2,000 ares Wougnel pot.	Ox Gunma, Japan	Visible injury decreased with increasing N.	5
Morning glory	Scarlet Ohara	N 0.15, 0.3, 0.6, 1.2 g per 1/5,000 ares Wougnel pot. volcanic ash soil + peat	Ox Max. 0.19 ppm in Tokyo	Visible injury decreased with increasing N.	4
Rice	Yamabiko	N 0, 0.5, 1.0, 2.0 g per 1/10,000 ares Wougnel pot. soil + peat	O_3 0.3 ppm 2 h	Visible injury increased with increasing N.	6
Rice	Yamabiko	N 0, 0.5, 1.0 g per 1/2,000 ares Wougnel pot. soil + peat	O_3 0.3 ppm 3 h	Visible injury increased with increasing N.	6
Rice	Jinheung	N 0, 5, 10, 20 kg/10 ares, 1/2,000 ares Wougnel pot. soil	O_3 0.3 ppm 3 h	Visible injury increased with increasing N.	7
Rice	Nongback	N 0, 5, 10, 20 kg/10 ares, 1/2,000 ares Wougnel pot. soil	O_3 0.3 ppm 3 h	Visible injury increased with increasing N.	7
Rice	Tongil	N 0, 5, 10, 20 kg/10 ares, 1/2,000 ares Wougnel pot. soil	O_3 0.3 ppm 3 h	Visible injury increased with increasing N.	7
Rice	Milyang no.23	N 0, 5, 10, 20 kg/10 ares, 1/2,000 ares Wougnel pot. soil	O_3 0.3 ppm 3 h	Visible injury increased with increasing N.	7
Rice	Nihonmasari	N 8, 16, 24 kg/10 ares. Field	Ox Gunma, Japan	Visible injury increased with increasing N.	5
Birch		N 74, 150 kg/ha/y, 7.5-l pot. Soil	O_3 AA*1.7 in Kuopio	Yellow leaves % decreased with increasing N.	8

P nutrition

Plant	Cultivar	P application rate and growing medium [a]	Exposure conditions (Ozone conc. period)	Foliar injury	Reference[b]
Tobacco	Matsukawa	P_2O_5 0, 2, 10 g per 1/5,000 ares Wougnel pot	Ox Okayama, Japan	Visible injury decreased with increasing P.	9
Tobacco	Matsukawa	P_2O_5 0, 2, 6, 18 g per 1/5,000 ares Wougnel pot	Ox Okayama, Japan	Visible injury decreased with increasing P.	9
Spinach	F_1 Shinryoku	P_2O_5 0.75, 1.5, 3.0 g per 2.4-l pot. kuroboku soil + peat + sand	O_3 0.12 ppm 4 h	Visible injury decreased with increasing P.	3
Radish		P_2O_5 0.0, 0.6, 1.2, 2.4 g per 1/5,000 ares Wougnel pot. volcanic ash soil + peat	Ox Max. 0.16 ppm in Tokyo	Visible injury decreased with increasing P.	4
Soybean	Enrei	Phosphate absorption coefficient 0%, 1%, 3%, 9%, 1/2000 ares Wougnel pot. soil	Ox Gunma, Japan	Visible injury decreased with increasing P.	5
Morning glory	Scarlet Ohara	P_2O_5 0.0, 0.6, 1.2, 2.4 g per 1/5,000 ares Wougnel pot. volcanic ash soil + peat	Ox Max. 0.19 ppm in Tokyo	Visible injury decreased with increasing P.	4
Rice	Yamabiko	P_2O_5 0, 1.0 g per 1/2,000 ares Wougnel pot	O_3 0.3 ppm 3 h	Visible injury decreased with non-P.	6

K nutrition

Plant	Cultivar	K application rate and growing medium [a]	Exposure conditions (Ozone conc.		period)	Foliar injury	Reference[b]
Spinach	Shinryoku	K_2O 2.0, 3.0 g per 1/5,000 ares Wougnel pot. valcanic ash soil	O_3	0.20 ppm	3 h	Visible injury decreased with increasing K.	2
Spinach	F_1 Shinryoku	K_2O 0, 0.5, 1.0, 2.0 g per 2.4-l pot. kuroboku soil + peat + sand	O_3	0.11 ppm	5 h	Visible injury decreased with increasing K.	3
Onion		K 0, 8 me/l per 1/2,000 ares Wougnel pot	O_3	0.2 ppm	5 h	Visible injury increased with non-K.	10
Radish		K_2O 0.0, 0.3, 0.6, 1.2 g per 1/5,000 ares Wougnel pot. volcanic ash soil + peat	Ox	Max. 0.16 ppm in Tokyo		Visible injury decreased with increasing K.	4
Pinto bean	Pinto	K 105, 710 meq/l. Hoagland nutrient solution	O_3	8 Level		Visible injury increased with decreasing K.	11
Soybean	Dare	K 105, 710 meq/l. Hoagland nutrient solution	O_3	8 Level		Visible injury increased with decreasing K.	11
Tomato		K 0, 4 me/l per 1/2,000 ares Wougnel pot	O_3	0.15 ppm	3 h	Visible injury increased with non-K.	10
Morning glory	Scarlet Ohara	K_2O 0.0, 0.3, 0.6, 1.2 g per 1/5,000 ares Wougnel pot. volcanic ash soil + peat	Ox	Max. 0.19 ppm in Tokyo		Visible injury decreased with increasing K.	4
Rice	Yamabiko	K_2O 0, 0.5 g per 1/5,000 ares Wougnel pot. soil + peat	O_3	0.3 ppm	2.5 h	Visible injury increased with non-K.	6
Rice	Yamabiko	K_2O 0, 0.5 g per 1/2,000 ares Wougnel pot. soil + peat	O_3	0.3 ppm	3 h	Visible injury increased with non-K.	6
Rice		K_2O 0, 5.6 kg/10 ares. Field	Ox	Max. 0.07 ppm in Aichi		No influence with non-K.	12
Rice	Nihonmasari	K_2O 0, 8, 16, 24 kg/10 ares. Field	Ox	Gunma, Japan		Visible injury decreased with increasing K.	5
Norway spruce		K_2SO_4 -K, +K 3 plants/40-l pot. soil	O_3	0.08 ppm	5 months	Visible injury increased with increasing K.	13

Ca nutrition

Plant	Cultivar	Ca application rate and growing medium [a]	Exposure conditions (Ozone conc.		period)	Foliar injury	Reference[b]
Onion, Tomato		$CaCO_3$ 0, 6 g per 1.2 kg(dry weight) soil in pot	O_3	0.2 ppm	5 h	Visible injury decreased with increasing Ca.	14
Chard		$CaCO_3$ 0, 6 g per 1.2 kg(dry weight) soil in pot	O_3	0.2 ppm	5 h	Visible injury increased with increasing Ca.	14
Spinach	F_1 Shinryoku	CaO 0,2.75,5.5,11.0 g per 2.4-l pot. kuroboku soil + peat + sand	O_3	0.11 ppm	4 h	Visible injury decreased with increasing Ca.	3
Spinach	Hoyo	$CaCO_3$ 0, 5.0 g per 2.5 kg soil in 1/5,000 ares Wougnel pot	Ox	Tokyo, Japan		Visible injury decreased with increasing Ca.	15

[a] Normal growth is shown, expect abnomal growth section.

[b] 1, Kitamura et al. (1973); 2, Shinozaki et al. (1975); 3, Kuno (1988); 4, Kuno and Terakado (1983); 5, Kotoyori et al. (1983); 6, Baba et al. (1979); 7, Jeong et al. (1981); 8, Pääkkönen et al. (1995); 9, Uno et al. (1973); 10, Asakawa et al. (1978); 11, Dunning et al. (1974); 12, Imamura et al. (1980); 13, Lippert et al. (1997); 14, Hyogo-ken Agricultural Experimental Station (1977); 15, Mayumi (1983).

[c] Ox is ambient oxidants.

Table 2. Activities of enzymes involved in the scavenging of reactive oxygen species in leaves of spinach grown with different levels of N and K fertilizers in pots (Kuno 1988)

Enzyme	N0	N1	N2	N4
SOD (units/cm^2) (%)	11.9 (53)	22.6 (100)	19.4 (86)	18.1 (80)
Asc. per. (μmol/cm^2/min) (%)	0.568 (73)	0.778 (100)	0.825 (106)	0.723 (93)
GR (μmol/cm^2/min) (%)	3.0 (20)	14.7 (100)	21.1 (144)	27.3 (186)
Cat. (μmol/cm^2/min) (%)	4.6 (26)	17.5 (100)	23.3 (133)	23.1 (132)
G. per. (units/cm^2) (%)	0.150 (49)	0.307 (100)	0.227 (74)	0.231 (75)
Protein (mg/dm^2) (%)	23.1 (48)	48.2 (100)	55.1 (114)	49.7 (103)

Enzyme	K0	K1	K2	K4
SOD (units/cm^2) (%)	14.7 (112)	13.1 (100)	15.3 (117)	18.9 (144)
Asc. per. (μmol/cm^2/min) (%)	0.473 (110)	0.430 (100)	0.498 (116)	0.550 (128)
GR (μmol/cm^2/min) (%)	27.5 (89)	30.9 (100)	46.2 (150)	41.6 (135)
Cat. (μmol/cm^2/min) (%)	11.8 (101)	11.7 (100)	14.7 (126)	14.0 (120)
G. per. (units/cm^2) (%)	0.104 (61)	0.170 (100)	0.166 (98)	0.133 (78)
Protein (mg/dm^2) (%)	47.1 (102)	46.0 (100)	47.0 (102)	53.2 (116)

SOD, Superoxide dismutase; Asc. per., Ascorbate peroxidase; GR, Glutathione reductase; Cat., Catalase; G. per., Guaiacol peroxidase
Spinach used was the O_3-sensitive F_1 Shinryoku cultivar.
Each value represents the mean of the middle leaf discs of 5 plants.
For N and K nutrient levels refer to legend in Fig. 1.
(%), Percent of activity of enzymes in leaf discs of plants grown under no (N0 or K0) or higher (N2 or K2 and N4 or K4 with 1.0 and 2.0 g N or K_2O/pot, respectively) N or K levels compared to plants grown with normal N or K levels (N1 or K1, 0.5 g N or K_2O/pot).

Table 3. Ozone injury in four rice varieties cultivated with different N levels (Jeong et al. 1981)

Type	Variety	Ozone injury (%)		
		Nitrogen		
		5 kg/10 ares	10 kg/10 ares	20 kg/10 ares
Japonica type	Jinheung	7.5	43.5	45.5
Japonica type	Nongback	3.7	36.0	40.2
Japonica×Indica type	Tongil	2.1	8.2	9.1
Japonica×Indica type	Milyang No.23	1.2	10.5	24.1

Note: Rice plants were exposed to 0.3 ppm ozone for 3 h at the 7- to 8-leaf stage

resulted in a dramatic decrease in injuries. In contrast, other reports (Meguro et al. 1991; Takebe et al. 1995) suggested an increase in the total ascorbic acid content in spinach with decreasing amounts of applied N, at ranges from 0 to 30 gN/m^2, in the field. From this study, a relationship between the increased ascorbic acid content and a decrease in O_3-induced injury by N application was suggested. The relationships between the activity of various enzymes involved in the scavenging of reactive oxygen species (ROS) (Tanaka et al. 1988, 1990) and the amounts of applied N (N0, N1, N2, and N4) are shown in Table 2. Here, higher activities of GR and Cat are observed with two- to fourfold the amounts of applied N (Kuno 1988). It is assumed that N applications, above the standard amount, increase the activity of the scavenging enzymes and hence enhance the tolerance.

In addition, reduced foliar damage, from O_3 or oxidants, was also observed by increased N applications to other plants, including radish (*Raphanus sativus* L.) (Kuno and Terakado 1983), soybean (Kotoyori et al. 1983; Flagler et al. 1987), morning glory (*Pharbitis nil* Choisy) (Kuno and Terakado 1983), and birch (*Betula pendula* Roth.) (Pääkkönen and Holopainen 1995). However, there are also numerous reports on increased visible injury to the leaves of rice plants, by O_3, with increasing amounts of applied N (Table 1) (Baba and Teraoka 1979; Jeong and Ota 1981; Kotoyori et al. 1983). Visible foliar injuries developed during the growing season from transplantation to the formation of young spikes. Four rice varieties, with different sensitivities to O_3, showed an increased tendency for visible injury with elevated N amounts, although there were large differences in the extent of injury between these varieties. Specifically, the Japonica varieties Jinheung and Nangback showed extremely high sensitivities to O_3, whereas there was little effect of N treatment on the Japonica × Indica varieties Tongil and Milyang no. 23 (Table 3) (Jeong and Ota 1981).

Abscisic acid (ABA) is well known as a hormone that controls the opening and closure of stomata in plants. In addition, increased resistance to foliar damage by air pollutants was found to be associated with higher endogenous ABA contents (Kondo and Sugahara 1978). The O_3-resistant variety of rice, Touitsu, was found to contain more ABA than the variety Nihonbare, which is susceptible to O_3

Table 4. ABA content in four rice varieties cultivated with different nitrogen levels (Jeong et al. 1981)

		ABA content (μg/kg FW)			
		Nitrogen			
Type	Variety	0 kg/10 ares	5 kg/10 ares	10 kg/10 ares	20 kg/10 ares
Japonica type	Jinheung	6.48	6.23	1.85	2.25
Japonica type	Nongback	6.89	6.04	2.00	2.89
Japonica×Indica type	Tongil	7.58	7.41	4.63	3.70
Japonica×Indica type	Milyang No.23	7.39	7.06	5.67	2.22

ABA, abscisic acid.
Note: Data were taken at the flowering stage

(Jeong et al. 1980). Furthermore, the amount of endogenous ABA in rice leaves was found to decrease with surplus N applications with O_3-susceptible Japonica varieties showing a greater decrease in ABA content than O_3-resistant Japonica × Indica varieties (Table 4) (Jeong and Ota 1981). In addition, a negative correlation was found between the ABA content and endogenous ethylene, which is thought to augment O_3-induced injuries (Jeong et al. 1981). Finally, a clear positive correlation was observed between N application rates and opening of stomata (Baba and Teraoka 1979; Ishihara et al. 1978). Based on these findings, the visible injuries induced by O_3 and enhanced by increased amounts of N applied to rice could arise from the following phenomena: a decrease in endogenous foliar ABA due to a large amount of N fertilizer application results in (1) enlarged opening of stomata and in turn increased uptake of O_3 into leaves, and (2) increased levels of endogenous ethylene, which promotes development of foliar injury.

However, although this severe of events may explain the increased sensitivity of rice plants to O_3 with increasing N applications, it is also clear that alternative events may take place in other plants such as spinach, in which N application reduces their sensitivity to O_3.

2.2 Phosphorus

Foliar damage caused by oxidants was reduced by increased rates of P application to pot cultures of tobacco. Some fertilizers and configurations such as superphosphate or triphosphorate, that is, water-soluble phosphates, showed greater alleviation of the injuries than fused phosphate (Uno and Miyake 1973). In spinach (Kuno 1988), radish, and morning glory (Kuno and Terakado 1983), foliar injuries decreased with increasing P application. When phosphate having 1%, 3%, or 9% of phosphate absorption coefficients in soil was applied, soybean showed increased growth and reduced visible injuries because of oxidants with the higher P application (Kotoyori et al. 1983). In rice plants, there were no consistent tendencies of the effects of P application on pollutant-induced foliar damage.

2.3 Potassium

The relationship between K application and O_3- or oxidant-induced injuries on spinach (Kuno 1988), radish, and morning glory (Kuno and Terakado 1983) has been studied. Severe foliar injuries were observed under conditions of shortage or low K concentrations, suggesting that such injuries were reduced by heavy K applications. Pinto bean (*Phaseolus vulgaris* L.) and soybean showed more severe O_3-induced injury to their leaves at K concentrations lower than standard concentrations in water culture (Dunning et al. 1974), and Welsh onion and tomato showed severe injuries due to K shortage (Asakawa et al. 1978). The K accumulated by roots was found to activate RNases (Sasakawa et al. 1973), and also to activate the synthesis and activity of enzymes involved in carbohydrate and nitrogen metabolism. K was clearly associated with the synthesis of glutathione, a tripeptide necessary for the detoxification of ROS (Evans and Sorger 1966). Without K application to spinach, a low foliar content of total glutathione was observed, whereas increased K application tended to increase the total glutathione level (Kuno 1988). In addition, as shown in Table 2, GR activity decreased with a zero application of K (K0), but increased with twofold (K2) or fourfold (K4) the standard (K1) level of K application. In addition, SOD activities tended to increase at the K4 treatment level. Based on these results, the increased foliar K content following abundant K application to spinach plants is assumed to be an important factor for enhancing the tolerance of spinach to O_3-induced injury.

Similarly, in rice plants, the development of visible injuries was slightly reduced with increasing K applications (Kotoyori et al. 1983). In addition, reduced K application to rice plants in pot cultures was found to result in increased O_3-induced injuries (Baba and Teraoka 1979; Imamura et al. 1980), and in an associated increase in stomatal opening (Baba and Teraoka 1979). In contrast, an experimental study using Norway spruce (*Picea abies* (L.) Karst.) demonstrated enhanced development of chlorotic mottles after O_3 treatment with K application (Lippert et al. 1997).

2.4 Calcium

The application of calcium is quite common for acidic soils in Japan, with dicotyledonous plants generally being high accumulators of calcium, whereas monocotyledons such as rice are low calcium accumulators. In vegetables, calcium nourishment is becoming an important problem, especially because many vegetables are caliciphiles. The importance of calcium was also studied in the relationship between oxidant-induced injury and the amount of applied calcium. When sufficient amounts of calcium were applied to spinach, decreased injuries were found (see Fig. 1), but surplus calcium applications, fourfold that of standard amounts, slightly inhibited growth (Kuno 1988). In experimental sections of N fertilizer application, supplementary calcium carbonate application to spinach was found to reduce the extent of oxidant-induced injury (Mayumi and Yamazoe 1983).

2.5 Mixtures of Nutrient Components

Heagle (1979) reported that supplementing four kinds of soil with a mixed N-P-K (6-25-15) fertilizer reduced the extent of foliar damage to four soybean varieties following O_3 treatment. Furthermore, using open-top chambers, Heagle et al. (1983) examined the effect of N-P-K fertilizer applications on the soybean variety Davis for 2 years exposure to different O_3 concentrations in a nonfiltered air chamber. Increased injuries were observed with elevated O_3 concentrations, but although increased fertilizer application reduced foliar injuries in 1977, no clear effect of fertilization was observed in 1978. In a rice paddy field, application of 75 or 225 kg of compost, containing all four elements ($N:P_2O_5:K_2O:Ca$ =10:8.6:5.6:60 kg/10 ares), resulted in a marginal increase in growth but had little effect on the extent of oxidant-induced visible injury (Imamura et al. 1980). With regards to several tree species, foliar visible injury has been demonstrated (Günthardt-Goerg et al. 1997). With the exception of willow (*Salix nigra* Marsh.) (Greitner and Winner 1989), well-fertilized birch (Landolt et al. 1997) and hybrid poplar (*Populus maximowiczii* × *P.trichocarpa*) (Harkov and Brennan 1980) showed reduced O_3-induced damage compared with trees that were not wellfertilized.

As mentioned previously, it is evident that several of the nutrient elements have clear effects on the extent of foliar damage in plants by oxidants or O_3. Indeed, N and K supplementation are found to have particular effects on stomatal opening and on the amounts of ABA, ascorbic acid, glutathione, and ethylene as well as on the activities of various ROS-scavenging enzymes. However, the effects of other nutrient elements on these factors have not yet been analyzed and should be the focus of further studies.

3. Countermeasures for Reducing Damage Caused by Oxidants to Spinach: Methods for Fertilizer Application

Plant foliar damage by photochemical oxidants may also create serious economic problems, especially for leafy vegetables that, like spinach, are highly susceptible to oxidants and which are sold directly at the market. To achieve effective countermeasures against such reductions, extensive studies on the effects of oxidants on spinach under fertilization have been conducted. In the following sections, countermeasures to protect spinach from ambient oxidant levels are discussed in relation to different fertilizer regimens.

Spinach, a major product of suburban areas, is susceptible to ambient levels of oxidants between spring and autumn, and the development of visible damage can result in severe economic losses. Characteristic injuries to spinach result from O_3 and PAN, and differences in susceptibility to these oxidants depend mostly on plant growth stage and variety. By using spinach varieties resistant to these oxidants, visible injuries could be decreased or avoided (Kuno 1989a). However, as these resistant varieties were European in origin and have different growth characteristics as well as different taste, they have not been popular in Japan. Spinach now being sold in Japan is mainly F_1 varieties, whose level of O_3 resistance is intermediate between that of the susceptible Asian varieties and the resistant European varieties (Kuno 1989b). Therefore, improved cultivation techniques are required to decrease the levels of oxidant injury to the plants. To address the question as to whether fertilizer could decrease such damage, a study using five cultivars of spinach (on andosol) at Tokyo Metropolitan Agriculture Experiment Station was undertaken (Kuno 1988). In the standard section of the experimental field, 16 kg/10 ares of mixed fertilizer ($N:P_2O_5:K_2O=8:8:8$) was applied, while 32 kg/10 ares and 64 kg/10 ares of K_2O were additionally applied to the two- and four-times sections, respectively, in seeding. The effects of ambient oxidants (90 ppb daily maximum O_3 concentration) on spinach grown in the different fertilized sections were analyzed. A general decrease in injuries on five varieties grown on the two-times sections was observed compared with those grown on the standard section, while a further decrease in injuries was found for two susceptible varieties of spinach grown on the four-times sections (Kuno 1988).

Kuno (1988) examined a method for supplementary application of small quantities of fertilizer at the most effective application time. In the field, after applying 16 kg/l0 ares mixed fertilizer ($N:P_2O_5:K_2O=8:8:8$) in seeding, three kinds of fertilizers, that is, N at 8 kg/10 ares, K_2O at 8 kg/10 ares, and N plus K_2O, each at 8 kg/10 ares, were additionally applied 1 week before harvest. Visible injuries by ambient oxidants (100 ppb daily maximum O_3 concentration) were then surveyed (Fig. 2). Injuries to the variety Ujo were reduced by 41% under additional N application, by 55% in the additional K section, and by 68% in the additional N plus K section, in comparison with the control section that had no additional fertilizer. Injuries on the variety F_1 Shinryoku were reduced by 79% in additional N plus K section in comparison with the control section, but no significant effects on damage were observed in the additional N section and the additional K section. For both varieties, the additional application of N plus K, showed the best choice for the reduction of injury.

A confirmatory experiment on the reduction of injuries, using additional fertilizer applications at the field cultivated by farmers, was conducted using the spinach hybrid varieties, Parade or Lead, in the suburbs of Tokyo. Additional application of fertilizer resulted in a 50% reduction in visible injuries due to ambient oxidants (110 ppb daily maximum O_3 concentration) of hybrid varieties, Parade, compared with the control section. In a subsequent experiment, fertilizer supplementation resulted in a 40% reduction in injuries to the spinach hybrid Lead after exposure to oxidants (80 ppb daily maximum O_3 concentration). Therefore, injuries could clearly be reduced by additional applications of fertilizer in upland

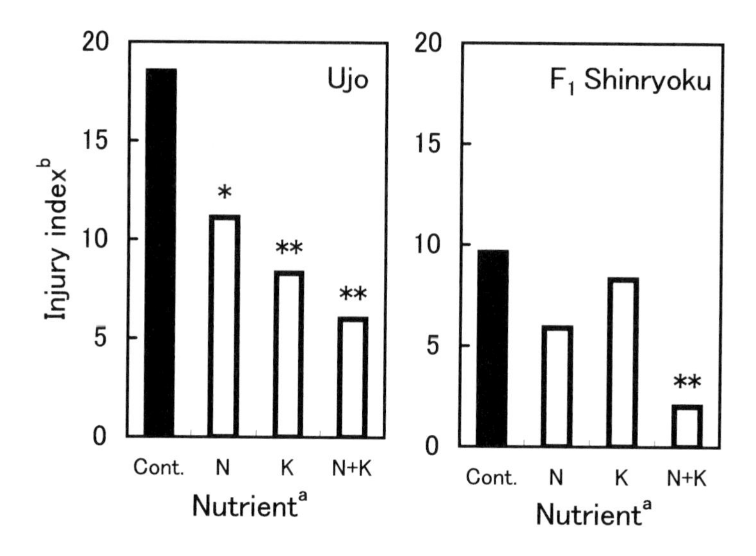

Fig. 2. Reductions in oxidant-induced foliar damage of spinach cultivars, Ujo and Shinryoku, by additional field application of fertilizer at Tachikawa, Tokyo, in 1985 (Kuno 1988). [a]Nutrient : C, Control; N, additional N application (8 kg N/10 ares); K, additional K application (8 kg K_2O/10 ares); N+K, additional N and K application (8 kg N+8 kg K_2O/10 ares) [b]Injury index is as shown in Fig. 1. $^{*}P<0.05$; $^{**}P<0.01$ versus control

fields. In addition to the additional application of fertilizers, the selection of resistant varieties could lead to reduce foliar damage by oxidants. Kuno (1989b) reported that hybrids between European individuals resistant to O_3 (used as the female parent) and Asian varieties (used as the male parent) had good taste and were comparatively O_3 resistant.

An optimal process for fertilizer use against oxidant damage could consist of the following regimen: the standard amount of fertilizer is applied to hybrid varieties of spinach and, when true leaves begin to rapidly grow, one-half of the standard amount of fertilizer, mixed with N and K, is then applied.

Acknowledgments. The authors sincerely acknowledge the support of Professor Emeritus Hiroshi Hirata, Tokyo University of Agriculture and Technology, and of Professor Emeritus Toshiichi Okita, Obirin University. Furthermore, the authors gratefully acknowledge Dr. Kenji Omasa and Dr. Shohab Youssefian, the editors, for their kind suggestions.

References

Air Pollution Section, Center of Promoting Countermeasure against Pollution, the Kanto Area (1999) Total report in 1973-1998, of the survey on plant effects of photochemical smog (in Japanese).

Aono M, Saji H, Sakamoto A, et al (1995) Paraquat tolerance of transgenic *Nicotiana tabacum* with enhanced activities of glutathione reductase and superoxide dismutase. Plant Cell Physiol 36:1687-1691

Asakawa F, Maekawa M, Tanaka H, et al (1978) Relationship between inorganic nutrition and ozone injury in plants. 2. Effects of potassium deficiency on the susceptibility of welsh onion and tomato plants to ozone (in Japanese). Kinki-Tyugoku Agric Res 56:82-85

Baba I, Teraoka S (1979) Physiological studies on the mechanism of the occurrence of air pollution damage in crop plants. V. Stomatal aperture and the Eh value of tissue fluids of shoots in relation to the occurrence of leaf injuries caused by ozone fumigation (in Japanese). Nogaku Kenkyu 57:163-188

Dunning JA, Heck WW, Tingey DT (1974) Foliar sensitivity of pinto bean and soybean to ozone as affected by temperature, potassium nutrition and ozone dose. Water Air Soil Pollut 3:305-313

Evans HJ, Sorger GJ (1966) Role of mineral elements with emphasis on the univalent cations. Annu Rev Plant Physiol 17:47-76

Fangmeier A, Grüters U, Hertstein U, et al (1996) Effects of elevated CO_2, nitrogen supply and tropospheric ozone on spring wheat. I. Growth and yield. Environ Pollut 91:381-390

Fangmeier A, Grüters U, Högy P, et al (1997) Effects of elevated CO_2, nitrogen supply and tropospheric ozone on spring wheat. II. Nutrients (N, P, K, S, Ca, Mg, Fe, Mn, Zn). Environ Pollut 96:43-59

Flagler RB, Patterson RP, Heagle AS, et al (1987) Ozone and soil moisture deficit effects on nitrogen metabolism of soybean. Crop Sci 27: 1177-1184

Fukuda M (1973) Studies on weather fleck on tobacco leaves (in Japanese). Bull Okayama Tobacco Exp Sta 33:1

Gatta L, Mancino L, Federico R (1997) Translocation and persistence of EDU (ethylenediurea) in plants: the relationship with its role in ozone damage. Environ Pollut 96:445-448

Godzik B, Manning WJ (1998) Relative effectiveness of ethylenediurea, and constitutent amounts of urea and phenylurea in ethylenediurea, in prevention of ozone injury to tobacco. Environ Pollut 103:1-6

Greitner CS, Winner WE (1989) Nutrient effects on responses of willow and alder to ozone. In: Olson RK, Lefon AS (eds) Transactions, effects of air pollution on western forests. Air & Waste Management Association, Anaheim, CA, pp 493-511

Günthardt-Goerg MS, McQuattie CJ, Scheidegger C, et al (1997) Ozone-induced cytochemical and ultrastructural changes in leaf mesophyll cell walls. Can J For Res 27:453-463

Harkov R, Brennan E (1980) The influence of soil fertility and water stress on the ozone response of hybrid poplar trees. Phytopathology 70:991-994

Heagle AS (1979) Effects of growth media, fertiliser rate and hour and season of exposure on sensitivity of four soybean cultivars to ozone. Environ Pollut 18:313-322

Heagle AS, Letchworth MB, Mitchell CA (1983) Effects of growth medium and fertilizer rate on the yield response of soybeans exposed to chronic doses of ozone. Phytopathology 73:134-139

Heagle AS, Miller JE, Sherrill DE, et al (1993) Effects of ozone and carbon dioxide mixtures on two clones of white clover. New Phytol 123:751-762

Hyogo-ken Agricultural Experimental Station (1977) Influence of inorganic components in soil to susceptibilities of agricultural crops against ozone (in Japanese). In: Saitama-ken Agricultural Experimental Station et al. (eds) Studies on visible injuries of agricultural crops by photochemical smog, pp222-227

Imamura S, Okino H, Ido Y, et al (1980) Visible injuries on rice leaf by photochemical oxidants (in Japanese with English summary). Res Bull Aichi Agric Res Cent 12:407-413

Ishihara K, Ebara H, Hirasawa T, et al (1978) The relationship between environmental factors and behavior of stomata in the rice plants. VII. The relation between nitrogen content in leaf blades and stomatal aperture (in Japanese with English summary). Jpn J Crop Sci 47:664-673

Izuta T, Takahashi K, Matsumura H, et al (1999) Cultivar difference of *Brassica campestris* L. in the sensitivity to O_3 based on the dry weight growth. J Jpn Soc Atmos Environ 34:137-146

Jeong Y, Ota Y (1981) Physiological studies on photochemical oxidant injury in rice plants. IV. Effect of nitrogen application on endogenous abscisic acid (ABA) production and ozone injury of rice plants (in Japanese with English summary). Jpn J Crop Sci 50:570-574

Jeong Y, Nakamura H, Ota Y (1980) Physiological studies on photochemical oxidant injury in rice plants. I. Varietal difference of abscisic acid content and its relation to the resistance to ozone (in Japanese with English summary). Jpn J Crop Sci 49:456-460

Jeong Y, Nakamura H, Ota Y (1981) Physiological studies on photochemical oxidant injury in rice plants. II. Effect of abscisic acid (ABA) on ozone injury and ethylene production in rice plants (in Japanese with English summary). Jpn J Crop Sci 50:560-565

Khan MR, Khan MW, Khan AA (1996) Evaluation of the sensitivity of some vegetable crops to ozone. Ann Appl Biol 128:94-95

Kitamura T, Monden E (1973) Studies on weather fleck on tobacco leaves. IX. Effect of nitrogen application rate on fleck injury (type II and III) to flue-cured tobacco (in Japanese with English summary). Bull Okayama Tobacco Exp Sta 33:71-77

Kondo N, Sugahara K (1978) Changes in transpiration rate of SO_2-resistant and -sensitive plants with SO_2 fumigation and the participation of abscisic acid. Plant Cell Physiol 19:365-373

Kotoyori T, Yamada K, Miyahara K (1983) Relationships among various types of fertilizer, application of soil improvement material and degree of injuries caused by the oxidants. In: Gunma-ken Agricultural Experimental Station et al. (eds) Analysis and countermeasures of agricultural crop injuries against photochemical smog, pp 258-259 (in Japanese)

Kuno H (1988) Fertilizer application system to reduce spinach leaf injury. In: Effects of photochemical oxidants on the horticultural plants (in Japanese with English summary). Bull Tokyo Agric Exp Sta 21:97-128

Kuno H (1989a) Charactestics of spinach leaf injury by photochemical oxidants and mechanisms of ozone resistance in spinach cultivars. I. Relationship of growth stages to leaf injury (in Japanese with English summary). J Jpn Soc Air Pollut 23:180-187

Kuno H (1989b) Characteristics of spinach leaf injury of photochemical oxidants and mechanisms of ozone resistance in spinach cultivars. II. Resistance to oxidants in spinach cultivars (in Japanese with English summary). J Jpn Soc Air Pollut 24:188-195

Kuno H, Terakado K (1983) Studies on indicator plants for photochemical oxidant. IV. Effect of nitrogen, phosphorus and potassium nutrition on the response of morning glory and radish to oxidants in the atmosphere (in Japanese with English summary). Bull Tokyo Agric Exp Sta 16:187-193

Landolt W, Günthardt-Goerg MS, Pfenninger I, et al (1997) Effect of fertilization on ozone-induced changes in the metabolism of birch (*Betula pendula*) leaves. New Phytol 137:389-397

Lippert M, Steiner K, Pfirrmann T, et al (1997) Assessing the impact of elevated O_3 and CO_2 on gas exchange characteristics of differently K supplied clonal Norway spruce trees during exposure and the following season. Trees 11:306-315

Matsumoto S, Akiyama M, Takagi N, et al (1977) Studies on the crop injury induced by air pollution in Okayama prefecture. 2. On the oxidant injury to vegetable (in Japanese). Bull Okayama Agric Exp Sta 2:11-19

Matsuoka Y, Takasaki T, Udagawa R (1971) Relationship between injury of agricultural products and oxidants in Chiba (in Japanese). J Jpn Soc Air Pollut 6:140

Matsuoka Y, Takasaki T, Morikawa M, et al (1976) Studies on the visible injury to rice plants caused by photochemical oxidants. I. Identification of the leaf injury caused by photochemical oxidants (in Japanese with English summary). Jpn J Crop Sci 45:124-234

Mayumi H, Yamazoe F (1983) Effects of photochemical air pollution on plants (in Japanese with English summary). Bull Nat Inst Agric Sci Ser B35:1-71

Meguro T, Yoshida K, Yamada J, et al (1991) Guide line of internal quality (nitrate and vitamin C) to spinach in the summer season (in Japanese). Jpn J Soil Sci Plant Nutr 62:435-438

Moldau H, Bichele I, Hüve K (1998) Dark-induced ascorbate deficiency in leaf cell walls increases plasmalemma injury under ozone. Planta 207:60-66

Montes RA, Blum U, Heagle AS (1982) The effects of ozone and nitrogen fertilizer on tall fescue, ladino clover, and a fescue-clover mixture. I. Growth, regrowth, and forage production. Can J Bot 60:2745-2752

Ohara T, Wakamatsu S, Uno I, et al (1995) An analysis of annual trends of photochemical oxidants in the Kanto and Kansai areas. J Jpn Soc Air Pollut 30:137-148

Ormrod DP, Adedipe NO, Hofstra G (1973) Ozone effects on growth of radish plants as influenced by nitrogen and phosphorus nutrition and by temperature. Plant Soil 39:437-439

Pääkkönen E, Holopainen T (1995) Influence of nitrogen supply on the response of clones of birch (*Betula pendula* Roth.) to ozone. New Phytol 129:595-603

Pell EJ, Winner WE, Vinten-Johansen C, et al (1990) Response of radish to multiple stress. New Phytol 115:439-446

Saio K, Matsuyama K, Enoki M (1978) On the analysis of photochemical oxidants injury to several plants by utilization of filtered air chamber (in Japanese with English summary). Bull Osaka Agric Res Cent 15:77-86

Sasakawa H, Yamamoto Y, Yatazawa M (1973) Studies on increases in UV absorption of rice callus cells by K^+ (in Japanese). Jpn J Soil Sci Plant Nutr 44:138-143

Sawada T, Komeiji T, Nouchi I, et al (1972) Symptoms and distribution of plant damage which was thought to be due to photochemical smog in Tokyo (in Japanese with English summary). J Jpn Soc Air Pollut 7:232

Sugawara Y (1957) Studies on vitamin C in crops of field and garden crops (in Japanese). Yoken-do, pp 35-37

Sun E (1994a) Ozone injury to leafy sweet potato and spinach in northern Taiwan. Bot Bull Acad Sin 35:165-170

Sun E (1994b) Air pollution injuries to vegetation in Taiwan. Plant Dis 78:436-440

Takasaki T, Morikawa M, Matsuoka Y, et al (1976) The recognition of the visible injuries on rice plant leaves by photochemical oxidants. On the discrimination from *Akagare* disease (in Japanese with English summary). Bull Chiba Agric Exp Sta 17:160-168

Takebe M, Ishihara T, Matsuno K, et al (1995) Effect of nitrogen application on the contents of sugar, ascorbic acid, nitrate and oxalic acid in spinach (*Spinacia oleracea* L.) and Komatsuna (*Brassica campestris* L.) (in Japanese with English summary). Jpn J Soil Sci Plant Nutr 66:238-246

Tanaka K, Suda Y, Kondo N, et al (1985) O_3 tolerance and the ascorbate-dependent H_2O_2 decomposing system in chloroplasts. Plant Cell Physiol 26:1425-1431

Tanaka K, Saji H, Kondo N (1988) Immunological properties of spinach glutathione reductase and inductive biosynthesis of the enzyme with ozone. Plant Cell Physiol 29:637-642

Tanaka K, Machida T, Sugimoto T (1990) Ozone tolerance and glutathione reductase in tobacco cultivars. Agric Biol Chem 54:1061-1062

Uno Y, Miyake Y (1973) Studies on weather fleck on tobacco leaves. XI. Influence of soil conditions on the development of fleck (Type I) (in Japanese with English summary). Bull Okayama Tobacco Exp Sta 33:91-101

III. Image Diagnosis of Plant Response and Gas Exchange

15
Image Instrumentation of Chlorophyll *a* Fluorescence for Diagnosing Photosynthetic Injury

Kenji Omasa and Kotaro Takayama

Department of Biological and Environmental Engineering, Graduate School of Agricultural and Life Sciences, The University of Tokyo, Yayoi 1-1-1, Bunkyo-ku, Tokyo 113-8657, Japan

1. Introduction

Environmental stress factors such as air pollutants, agricultural chemicals, water deficit, chilling, and UV light can affect the health of plants (Larcher 1995; Kramer and Boyer 1995; Yunus and Iqbal 1996; De Kok and Stulen 1998). The first symptoms to appear are decreases in photosynthesis and growth. Later symptoms include visible injury of leaves and withering. In recent years forest decline, which may be due to environmental stress factors including meteorological changes, has been widely reported (Schulze et al. 1989; Larcher 1995; Sandermann et al. 1997).

Recent advances in imaging of physiological functions of intact plants to diagnose early abnormal symptoms are remarkable (Omasa and Aiga 1987; Omasa 1990, 2000; Hashimoto et al. 1990; Häder 1992, 2000; Lichtenthaler 1996). For example, thermal imaging (Omasa and Aiga 1987; Omasa and Croxdale 1992; Omasa 1994, this volume; Jones 1999), which gives information on stomatal response and gas exchange, including transpiration, photosynthesis, and absorption of air pollutants, has been applied to remote sensing of outdoor trees and to spatial analysis of physiological functions of leaves in indoor experiments. Multispectral image analyses of leaf reflection (Omasa and Aiga 1987; Omasa 1990, 2000;

Air Pollution and Plant Biotechnology
–Prospects for Phytomonitoring and Phytoremediation–
Edited by K. Omasa, H. Saji, S. Youssefian, and N. Kondo
© Springer -Verlag Tokyo 2002

Wessman 1990; Chappelle et al. 1992; Renxz 1999), and steady-state fluorescence (Lichtenthaler 1996; Kim et al. 1996, 1997, this volume; Saito et al. 1997) have been used for early detection of alteration and bleaching of plant pigments related to photosynthetic injuries. Laser-induced fluorescence (LIF) may be effective for obtaining specific information on changes in cell walls bound by phenolics, compounds in vacuoles, and other fluorophores in living mesophylls in addition to the foregoing information.

Meanwhile, originally developed by Omasa et al. (1987) and Daley et al. (1989), the techniques of image analysis of chlorophyll (Chl) *a* fluorescence of plant leaves in situ has been widely used as a sensitive and nondestructive way to assess the functional state of the photosynthetic apparatus. These techniques are used for early detection of changes in patchy stomatal response and photosynthetic activity caused by abiotic stress factors such as air pollutants, low concentrations of O_2, water deficit, UV light, chilling, and agricultural chemicals (Omasa et al. 1987, 1991, 2001; Daley et al. 1989; Omasa and Shimazaki 1990; Siebke and Weis 1995a; Rolfe and Scholes 1995; Takayama et al. 2000) and biotic stress factors (Balachandran et al. 1994; Osmond et al. 1998). These techniques can also be used to analyze the development of the photosynthetic apparatus of attached leaves and cultured tissues (Croxdale and Omasa 1990a, 1990b; Omasa 1992). Recently, field-portable imaging systems (Daley 1995; Osmond et al. 1998; this volume) and the LIF imaging system (Omasa 1988, 1998) for remote measurement of Chl fluorescence induction have been developed.

In the following section, examples of the diagnosis of photosynthetic dysfunction of plants caused by environmental stress factors such as air pollutants and agricultural chemicals using techniques of image analysis of Chl *a* fluorescence are reviewed.

2. Chlorophyll *a* Fluorescence

2.1 Chlorophyll Fluorescence Induction

Figure 1 shows an excitation and emission matrix of steady-state fluorescence of a healthy cucumber (*Cucumis sativus* L. cv. Sharp7) leaf. When the leaf is irradiated with visible rays of light (about 380 to 600 nm), fluorescence with very strong intensity in the spectral range of about 660 to 770 nm is emitted from Chl *a*. Because Chl *a* fluorescence is the reemission of light energy trapped by antenna chlorophyll and is not used in the photochemical reaction, fluorescence intensity depend on the magnitude of the photochemical reaction (Papageorgiou 1975; Lichtenthaler 1988; Krause and Weis 1991; Govindjee 1995). Rapid changes in the intensity of Chl *a* fluorescence (peak wavelength, 683 nm) emitted from photosystem II (PSII) antenna chlorophyll during a dark-light transition (Chl

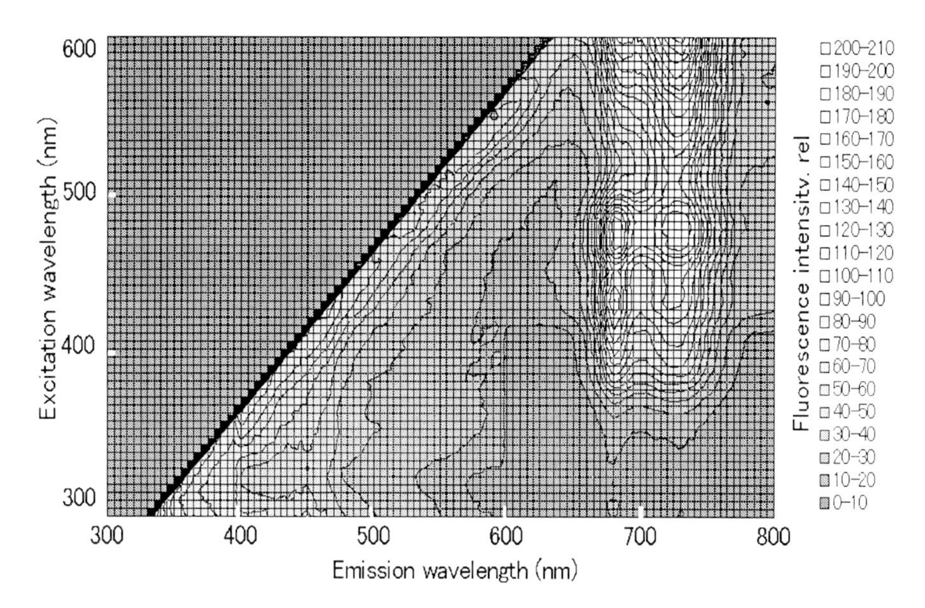

Fig. 1. Matrix of excitation and emission of steady-state fluorescence of a healthy cucumber leaf measured by a fluorescence spectrophotometer (Hitachi F-4500)

fluorescence induction, CFI) reflect the various reactions of photosynthesis, especially those of the photosynthetic electron transport system. This phenomenon is wellknown as the Kautsky effect (Kautsky and Hirsch 1931).

Figure 2 shows CFI transients of a healthy cucumber (*Cucumis sativus* L. cv. Natsusairaku) leaf under different intensities of actinic blue-green light. These measurements clearly revealed the typical CFI transients with inflection points labeled as O, I, D, P, S, M, and T (Papageorgiou and Govindjee 1968; Munday and Govindjee 1969a, 1969b) under light intensities from 50 to 200 μmol photons m^{-2} s^{-1}. The fluorescence increases from O (origin, not shown) to I (inflection or intermediary peak), decreases to D (dip), and increases to P (peak); this is called the fast phase. The fluorescence decreases from P to S (quasi-steady state), and declines to T (terminal steady state) via M (a maximum) (Govindjee 1995); this is called the slow phase.

The fast phase takes a maximum of a few seconds after the start of irradiation with actinic light, although the appearance is faster with increased light intensity. It is closely correlated with the redox reactions of Q_A (Krause and Weis 1991). Because Q_A is the primary electron acceptor of PSII, fluorescence intensity is low when Q_A is oxidized, and high when Q_A is reduced, in the fast phase of CFI. The transient OI represents the photoreduction of Q_A by the PSII reaction center; ID represents rapid oxidation of Q_A by plastoquinone (PQ) pool and photosystem I

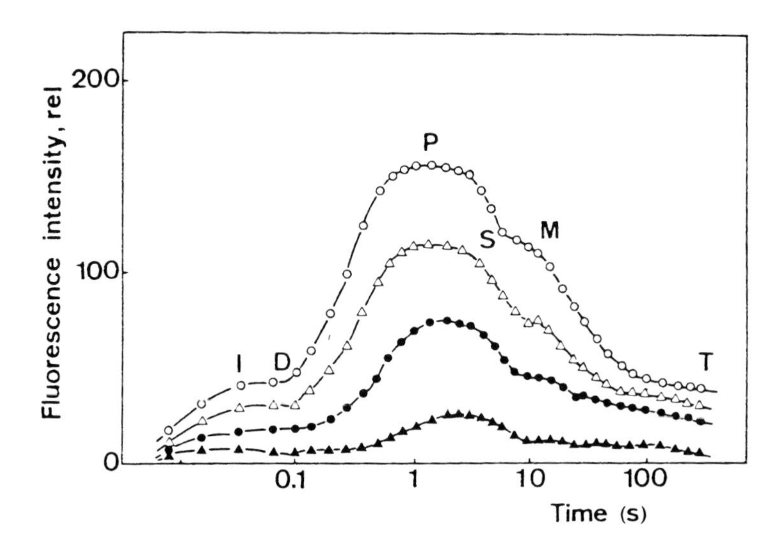

Fig. 2. Chlorophyll fluorescence induction (CFI) transients of a small area (about 1 mm^2) of a healthy cucumber leaf measured under different intensities of actinic blue-green light (Omasa et al. 1987). Intensities of actinic light (○, 200; △, 150; ●,100; ▲,50 μmol photons m^{-2} s^{-1}) were changed by xenon lamps with blue glass filters (380 to 620 nm). Fluorescence was measured by a CCD camera system with an interference filter (683 nm; half-band width, 10 nm) (see Fig.4)

(PSI); and DP indicates the photoreduction of Q_A by the PSII reaction linked to the water-splitting enzyme system. The transient from P to T in the slow phase requires a couple of minutes. The change includes two components, photochemical quenching and nonphotochemical quenching. It may also reflect an interaction between electron transport and carbon fixation.

The CFI transient represents a complex polyphasic process in which the details depend on experimental conditions. We should note that the CFI transient is influenced by (1) PSII cooperativity, (2) PSII heterogeneity, (3) size of the PQ pool and rate of its reoxidation, (4) rate of electron transport beyond PSI including carbon metabolism, and (5) rate of electron donation to P^+_{680} (Krause and Weis 1991).

2.2 Chlorophyll Fluorescence Quenching

Under continuous light, fluorescence intensity shows a lower value than its maximum. "Chl fluorescence quenching" denotes all processes that lower the fluorescence intensity. The quenching consists of photochemical and nonphotochemical components. Photochemical quenching is closely correlated with the oxidation state of Q_A. Nonphotochemical quenching is an index of the ability of the photosynthetic apparatus to generate a high *trans*-thylakoid pH gradient, to sustain electron transport, and to waste excess excitation energy as heat. By means of a saturation light pulse, it is possible to quantitatively estimate both photochemical and nonphotochemical quenching (Quick and Horton 1984; Schreiber et al. 1986; Krause and Weis 1991; Siebke and Weis 1995a, 1995b; Govindjee 1995). This fluorescence analysis is called the saturation pulse method.

Figure 3 shows a schematic diagram for measuring both CFI images and pulse-saturated fluorescence images. After 1 h dark, the leaf is irradiated with the saturation light pulse, and F_m image is measured. After the leaf is kept in the dark for 30 min, CFI images (fluorescence intensities: F_I at I, F_D at D, F_P at P, F_S at S, F_M at M, and F_T at T) are measured under actinic light. After confirmation of steady-state fluorescence under actinic light, F and F_m' images are measured just before and during irradiation of the saturation light pulse, respectively.

As quenching parameters, q_P, q_N, NPQ (nonphotochemical quenching), and Φ_{PSII} (quantum yield of PSII electron transport) have been widely used (Schreiber et al. 1986; Bilger and Björkman 1990; Kooten and Snel 1990; Krause and Weis

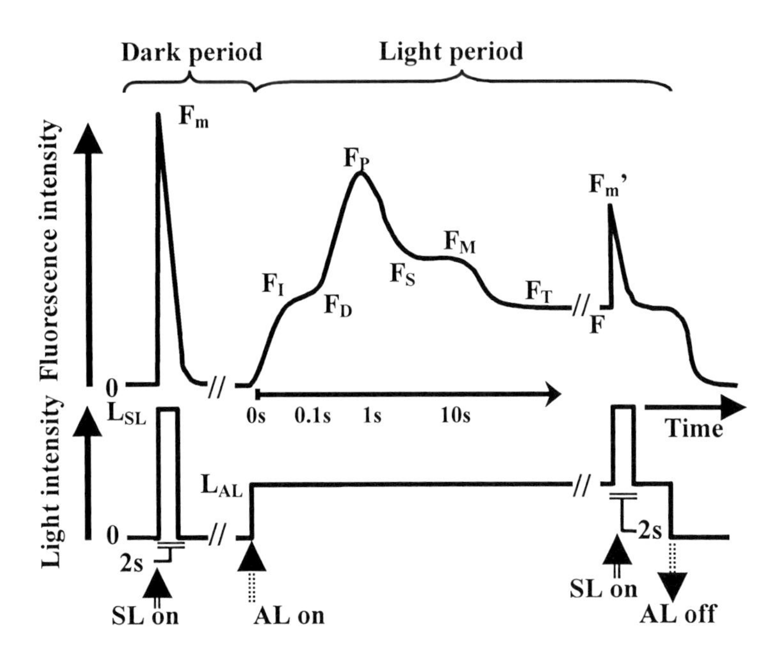

Fig. 3. Schematic diagram for measuring both CFI images and pulse-saturated fluorescence images

1991; Genty and Meyer 1995; Rolfe and Scholes 1995; Siebke and Weis 1995a; Osmond et al. 1998). These parameters are calculated from fluorescence yield images obtained from fluorescence intensity images. Fluorescence yield is the quantum yield of Chl *a* fluorescence. The absolute fluorescence yield is obtained from the total number of photons emitted divided by the total number of photons absorbed (Govindjee 1995). When actinic light and saturation light pulse are absorbed in an essentially identical manner by Chl, relative fluorescence yield images (Φ_{Fm}, Φ_F, and $\Phi_{Fm'}$) are calculated by

$$\Phi_{Fm} \cong F_m/L_{SL} \tag{1}$$

$$\Phi_F \cong F/L_{AL} \tag{2}$$

$$\Phi_{Fm'} \cong F_m'/L_{SL} \tag{3}$$

where F_m is the maximal fluorescence intensity (dark), F is the steady-state fluorescence intensity under actinic light, F_m' is the maximal fluorescence intensity under actinic light, L_{SL} is the intensity of saturation light pulse, and L_{AL} is the intensity of actinic light (Schreiber 1986; Rolfe and Scholes 1995).

Photochemical quenching is caused by the oxidized state of the primary acceptor (Q_A) of PSII. The coefficient for photochemical quenching, q_P, represents the proportion of excitons captured by open traps and being converted to chemical energy in the PSII reaction center (Krause and Weis 1991). q_P is calculated by

$$q_P = (\Phi_{Fm'} - \Phi_F)/(\Phi_{Fm'} - \Phi_{F0'})$$

$$\cong (\Phi_{Fm'} - \Phi_F)/(\Phi_{Fm'} - 0.2 \times \Phi_{Fm}) \tag{4}$$

where Φ_{Fm} is the maximal fluorescence yield (dark), that is, fluorescence yield with all PSII reaction centers closed ($q_P=0$) and all nonphotochemical quenching processes are at a minimum ($q_N=0$), $\Phi_{Fm'}$ is the maximal fluorescence yield (light), that is, fluorescence yield with all PSII reaction centers closed in any light-adapted state ($q_P=0$ and $q_N \geqq 0$), and $\Phi_{F0'}$ is the minimal fluorescence yield (light), that is, fluorescence yield with all PSII reaction centers open in any light adapted state ($q_P=1$ and $q_N \geqq 0$). $\Phi_{F0'}$ excited by the measuring beam of the PAM fluorometer is too weak to measure with the ordinary video system. Therefore, $\Phi_{F0'}$ is nearly constant and approximately $0.2 \times \Phi_{Fm}$ (Schreiber et al. 1986; Daley et al. 1989; Osmond et al. 1998).

Nonphotochemical quenching is caused in vivo by several mechanisms. However, most of nonphotochemical quenching has been found to be correlated to the energization of the thylakoids because of the build-up of a transmembrane pH gradient. The coefficient for nonphotochemical quenching, q_N, represents the proportion of the nonphotochemical quenching in all fluorescence quenching. q_N

is computed by

$$q_N=1-(\Phi_{Fm'}-\Phi_{F0'})/(\Phi_{Fm}-\Phi_{F0})$$

$$\cong(\Phi_{Fm}-\Phi_{Fm'})/(0.8\times\Phi_{Fm}) \qquad (5)$$

where Φ_{F0} is the minimal fluorescence yield (dark), that is, fluorescence yield with all PSII reaction centers open while the photosynthetic membrane is in the nonenergized state (dark- or low-light-adapted $q_P=1$ and $q_N=0$), nearly constant, and approximately $0.2 \times \Phi_{Fm}$ as Φ_{F0} is equal to $\Phi_{F0'}$ (Schreiber et al. 1986; Daley et al. 1989; Osmond et al. 1998).

Knowledge of $\Phi_{F0'}$ is indispensable in order to calculate q_P and q_N. However, the determination of $\Phi_{F0'}$ may be problematic, particularly under field conditions. Hence, NPQ as a nonphotochemical quenching parameter which does not require the determination of $\Phi_{F0'}$ was developed. NPQ is computed by

$$NPQ=(\Phi_{Fm}-\Phi_{Fm'})/\Phi_{Fm'} \qquad (6)$$

Furthermore, quantum yield of PSII electron transport, Φ_{PSII}, as a parameter for the assessment of the proportion of light energy absorbed by PSII and used in photosynthetic electron transport was developed. Φ_{PSII} can be also calculated by the following expression without determining $\Phi_{F0'}$.

$$\Phi_{PSII}=(\Phi_{Fm'}-\Phi_F)/\Phi_{Fm'}$$

$$=\Delta\Phi_F/\Phi_{Fm'} \qquad (7)$$

The product of Φ_{PSII} and irradiance is linearly related to CO_2 assimilation rate under suitable measuring conditions. Φ_{PSII} is used as an indicator of the rate of photosynthesis (Genty et al. 1989).

3. Image Instrumentation System

Figure 4 shows a diagram of the laser-induced fluorescence (LIF) image instrumentation system for analyzing Chl fluorescence and photographs of scanning argon laser. This system can be used to measure CFI images by stimulating the photosynthetic apparatus with actinic light from a scanning argon laser (457.9, 514.5 nm) in addition to halogen or xenon lamps with blue glass filters (380 to 620 nm). In the scanning laser system, a beam from an argon laser (800 mW) is scanned horizontally by a polygon laser scanner (4 in Fig. 4) and vertically by a galvanometer scanner (6). The scan rate is controlled by synchronization with the signal from a camera controller or a computer system. The maximum scanning rate is 30 s^{-1}. Another galvanometer scanner (3) and a

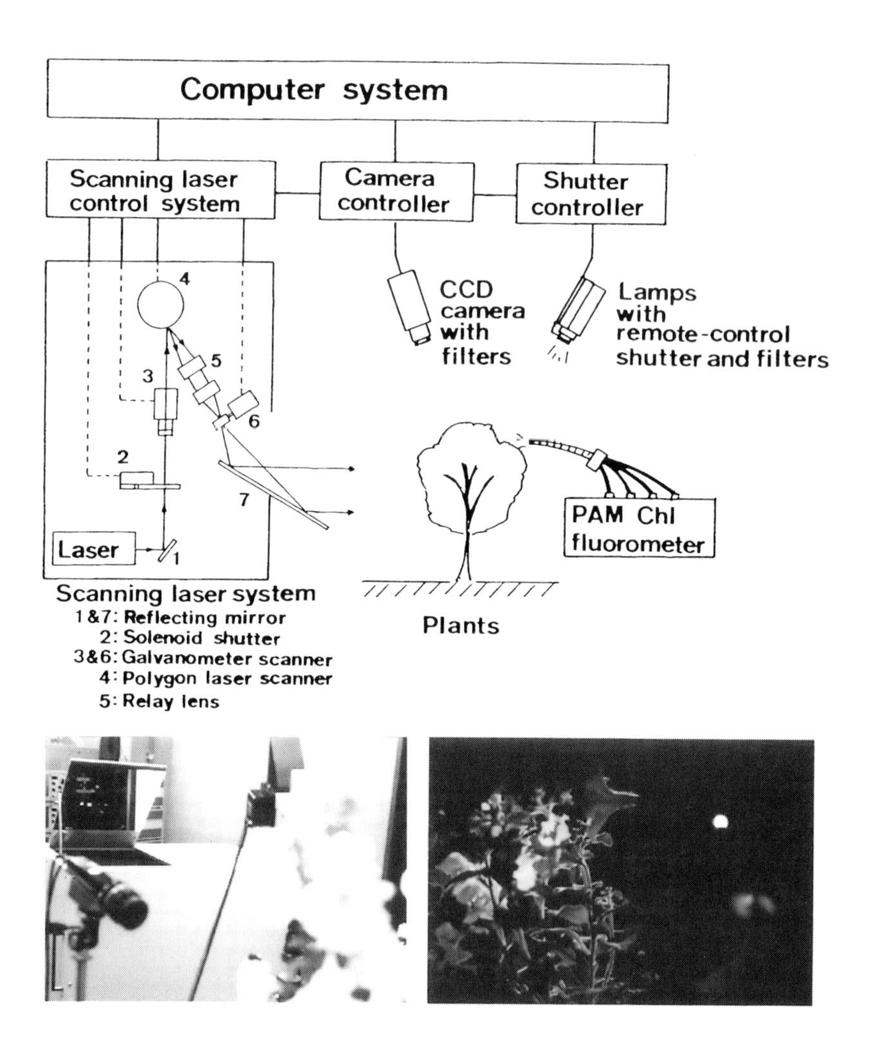

Fig. 4. Diagram and photographs of laser-induced fluorescence (LIF) image instrumentation system for analyzing CFI transients (Omasa 1988, 1998)

solenoid shutter (2) are used to remove the retrace line in the vertical scan. The left-hand photograph in Fig. 4 shows the scanning laser and the xenon lamp projector, and the right-hand photograph shows a plant irradiated by the scanning laser. The scanning laser enables measurement of plants from a greater distance than with the xenon lamp.

Figure 5 shows a comparison between CFI transients of a small area (about 1

mm^2) of a healthy sunflower (*Helianthus annuus* L. cv. Russian Mammoth) leaf measured in situ with the argon laser (LIF) and CFI transients measured with blue-green actinic light from the xenon lamp projectors at three different intensities of actinic light. Before the CFI measurement, the plant leaf was placed in the dark and allowed to adapt for 20 min. The fluorescence was continuously measured by a highly sensitive CCD (charge-coupled device) camera with uniform sensitivity, afterimage suppression through an interference filter (683 nm; half-band width, 10 nm) and a red-pass filter (>650 nm). A series of images was recorded on a computer or on a digital video recorder and then analyzed by the computer. These CFI transients clearly displayed the typical CFI pattern (see Fig. 2; Papageorgiou 1975; Govindjee 1995). The fluorescence intensity of each transient level and the amplitudes of DP, PS, and MT increased as the light intensity increased, and were almost identical in spite of the difference in light source. Figure 6 shows CFI images induced by the scanning laser taken from a petunia (*Petunia hybrida* L. cv Tytan White) plant. These results confirm that the scanning laser enables clear CFI imaging from a distance.

By the procedure shown in Fig. 3, the system can also be used to analyze Chl fluorescence quenching images as well as CFI images by irradiating with a saturation blue-green light pulse (4000 μmol photons m^{-2} s^{-1}) in addition to the

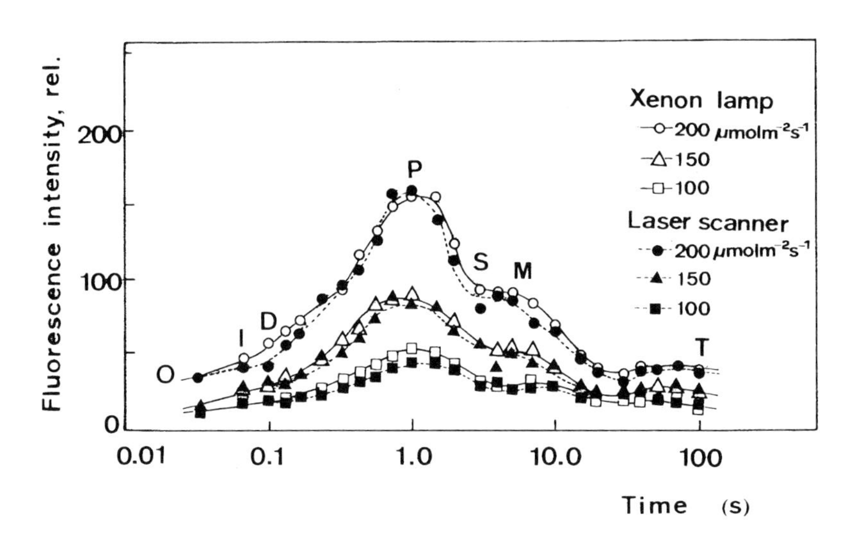

Fig. 5. Comparison between CFI transients of a small area (1 mm^2) of a healthy sunflower leaf measured in situ using the scanning laser (LIF) and CFI transients measured using xenon lamps at different light intensities (Omasa 1988, 1998). The rate of laser scanning was 30 s^{-1}

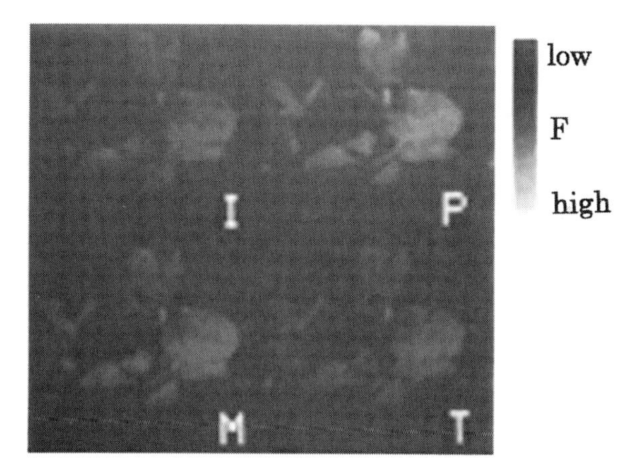

Fig. 6. CFI images at characteristic transient levels (I, P, M, and T) of a petunia plant measured with the scanning laser (Omasa 1988, 1998)

blue-green actinic light (300 μmol photons m^{-2} s^{-1}) from halogen lamps, although it is suitable only for small areas.

4. Diagnosis of Environmental Stresses

4.1 Effects of Air Pollutants

Air pollution affects the health of plants growing in urban areas and forests (Yunus and Iqbal 1996; Sandermann et al. 1997; De Kok and Stulen 1998). Decreases in photosynthesis and growth appear initially and are followed by visible injury of leaves and withering. We used CFI analysis for early detection of the effects of major air pollutants such as SO_2 and oxidants (Omasa et al. 1987; Omasa and Shimazaki. 1990).

Figure 7 shows CFI transients and intensity images at characteristic transient levels (I, P, M, and T) of an attached sunflower (*Helianthus annus* L. cv. Russian Mammoth) leaf just after fumigation with 1.5 μl l^{-1} SO_2 for 30 min. During the fumigation, half of the leaf blade was covered with aluminum foil to shield the leaf from SO_2. This procedure allows comparison between a fumigated area (F) and an unfumigated area (UF) of the same leaf. In the unfumigated area, CFI clearly showed the typical IDPSMT transients (Papageorgiou 1975; Govindjee 1995).

Because the CFI transients observed on dark-light transition of the leaf reflect the partial reactions of photosynthesis, we can detect alterations in the photosynthetic apparatus caused by SO_2 from the changes in CFI transients (Fig. 7A). As shown in Fig. 7B, the image intensity in the fumigated areas differed strikingly from those in the unfumigated areas. Fluorescence intensity in the fumigated area was higher at I, markedly lower at P, and higher at T. The amplitude of fluorescence

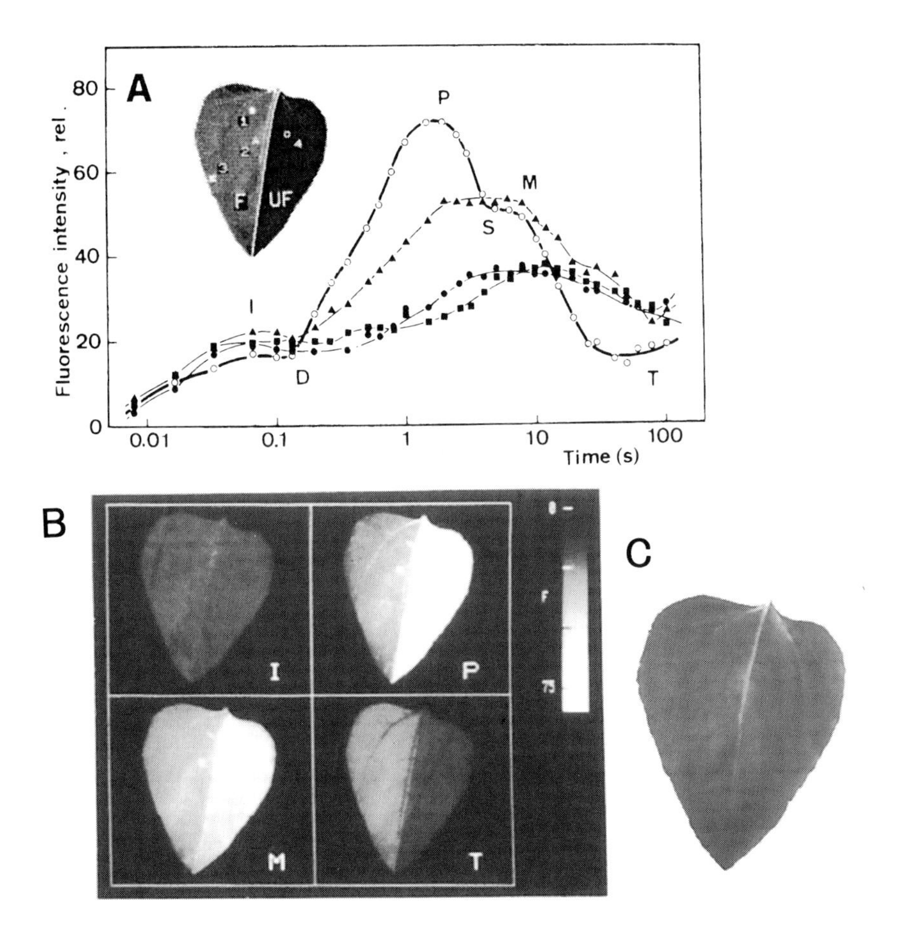

Fig. 7. CFI transients (**A**) and intensity images (**B**) at characteristic transient levels (I, P, M, and T) of an attached sunflower leaf just after 1.5 μl l^{-1} SO_2 fumigation for 30 min (Omasa et al. 1987). Fumigated area (F): ●, interveinal site 1; ▲, site 2, near a large vein; ■, site 3, near a veinlet. Unfumigated area (UF): ○, interveinal site 4. There was no visible injury (**C**)

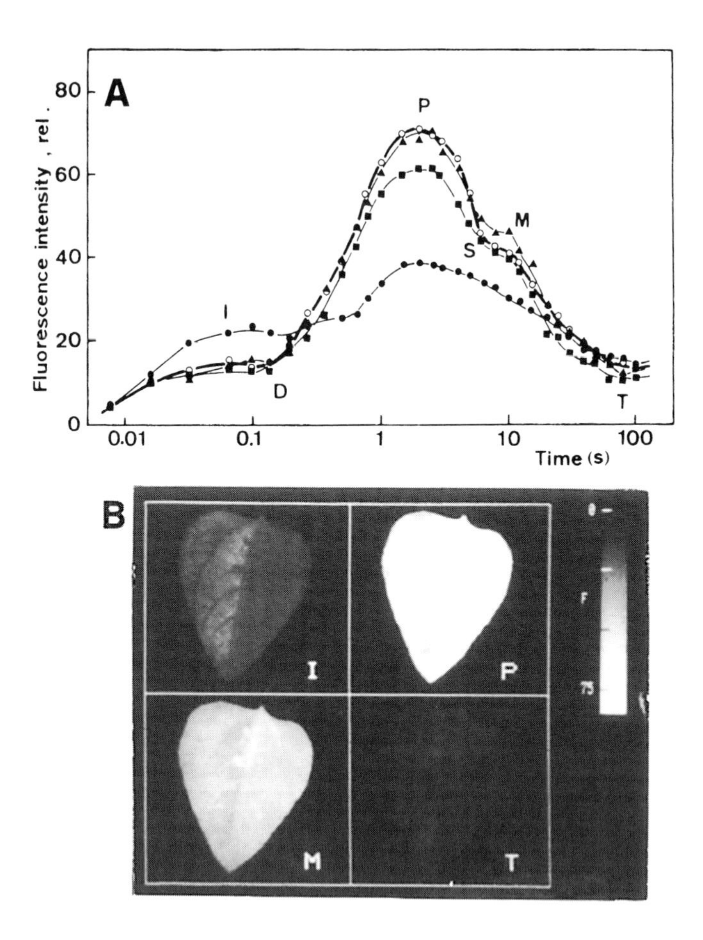

Fig. 8. Recovery of CFI transients (**A**) and intensity images (**B**) of the sunflower leaf kept in clean air for 6 h from disturbances by the SO_2 fumigation shown in Fig.7 (Omasa et al. 1987). Symbols are as in Fig. 7

transients DP, PS, and MT, indicating photosynthetic activity, was reduced in the fumigated areas of the leaf. The changes in intensity and amplitude varied with the location on the leaf surface: the effects of SO_2 were more severe between veins and near veinlets than near large veins. Contrary to the perturbation in the photosynthetic apparatus shown, there was no visible injury in the whole surface of leaf at the end of SO_2 treatment (Fig. 7C) and 2 days later.

The significance of the changes in CFI induced by SO_2 fumigation is as follows

(Omasa et al. 1987). Because fluorescence intensity in the early induction phenomena is regulated by the redox state of Q_A (Papageorgiou 1975; Krause and Weis 1991; Govindjee 1995), the elevated I level suggests that Q_A was brought to the reduced state by SO_2 fumigation. Because the DP rise in CFI reflects photoreduction of Q_A through the reductant from H_2O, a diminished rise of DP was consistent with inactivation of the water-splitting enzyme system. As PS decline involves energy-dependent quenching, its suppression suggested the depression of formation of the *trans*-thylakoid proton gradient was probably caused by the inactivation of the water-splitting enzyme system. However, the possibility that the PS decline was affected by the inhibition of electron flow from Q_A to PSI cannot be excluded because PS decline partly reflects the oxidation of Q_A by PSI. Suppression of MT decline was probably caused by the inhibition of the *trans*-thylakoid proton gradient formation in addition to unidentified reactions in chloroplasts. Although the extent of the SO_2 effect on CFI differed from area to area on a single leaf, the mode of SO_2 action was essentially the same. The fumigated plants were then placed in clean air under darkness for 6 h. Monitoring of CFI showed plant recovery (Fig. 8). Near the large veins, where the change in CFI was relatively small, the CFI recovered completely, as shown by its IDPSMT transient. Near the veinlets, CFI recovered almost completely. However, the fluorescence emitted from the interveinal area was still affected: P reappeared but its intensity was still low, and I was elevated. The elevated I level suggests that the PSII reaction centers in the chloroplasts were irreversibly injured. In contrast, the elevated T level in the quasi-stationary state became normal in this area. Despite these results, however, no injury was visible at the end of the SO_2 treatment nor at 2 days after the treatment. Thus, CFI imaging can reveal photosynthetic damage before symptoms are visible.

On the other hand, CFI responses to peroxyacetyl nitrate (PAN) were different from those to SO_2. Petunia (*Petunia hybrida* L. cv. Tytan white) plants were fumigated with 0.06 $\mu l\ l^{-1}$ PAN for 3 h. The CFI transients and fluorescence intensity images observed within 1 h after the fumigation were not affected, indicating that photosynthetic activity was not yet damaged. No visible injury was observed at this time. After the measurement, half of the leaf blade was covered with aluminum foil to prevent light illumination and the plants were allowed to stand under light (130 μmol photons $m^{-2}\ s^{-1}$) for 12 h. Consequently, both CFI change and visible injury progressed according to the time elapse (Fig. 9). Figure 10 shows a photograph and CFI images at characteristic transient levels (I, P, M, and T) after the 12 h light treatment. The fluorescence intensities were depressed markedly at visibly injured sites in the irradiated area (a), but were virtually unaffected in the unirradiated area (b). This phenomenon indicates that light played an important role in the phytotoxicity of PAN and that photosynthetic injury depends on bleaching of plant pigments, that is, destruction of cells.

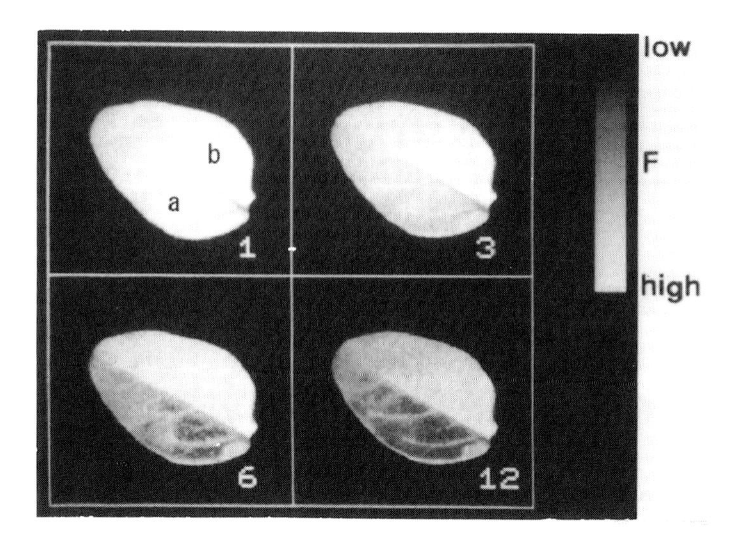

Fig. 9. Changes in intensity image at peak P of CFI of an attached petunia leaf after 0.06 μl l^{-1} peroxyacetyl nitrate (PAN) fumigation for 3 h. Numerals (1, 3, 6, 12) show the elapse of time (h) after the end of fumigation. A, area irradiated under 130 μmol photons m^{-2} s^{-1} after the fumigation; B, unirradiated area

4.2 Effects of Herbicides

Many herbicides inhibit photosynthetic functions of plants. Heterogeneous distribution of photosynthetic inhibition in leaves caused by herbicides has been widely investigated by Chl fluorescence analysis (Daley et al. 1989; Genty and Meyer 1995; Rolfe and Scholes 1995; Takayama et al. 2000; Omasa et al. 2001).

Figure 11 shows CFI and quenching images obtained by a sequence, shown in Fig.3, of kidney bean (*Phaseolus vulgaris* L. cv. Shinedogawa) leaves after treatment with 1/1000 diluted solution of a herbicide, Nekosogi-ace (Rainbow Chemicals), including 6.0% 3-(3,4-dichlorophenyl)-1,1-dimethylurea (DCMU), 3.0% 2,6-dichlorothiobenzamide (DCBN), and 1.0% 3-(5-*tert*-butylisoxazol-3-yl)-1,1-dimethylurea (isouron) from the leafstalk for 20 min. Although the fluorescence intensity was relatively uniform all over the leaf surface before treatment, it was remarkably increased at mesophyll cells around the major veins in several images of CFI and pulse-saturated fluorescence after the treatment. Because DCMU blocks electron transfer from Q_A to PQ, probably by binding to the Q_B site of the D1 protein, the fluorescence rise from O to P in CFI transients is

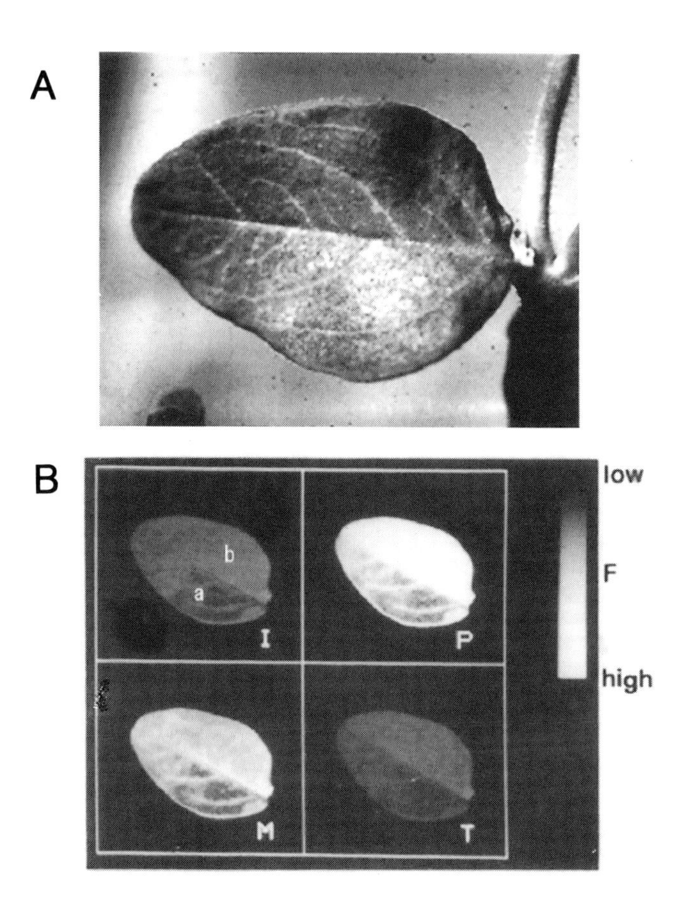

Fig. 10. A photograph (**A**) and CFI images (**B**) at characteristic transient levels (I, P, M, and T) of the attached petunia leaf at 12 h after the end of PAN fumigation shown in Fig.9 (Omasa and Shimazaki 1990). Symbols are as in Fig. 9. Visible injury was observed in irradiated area only (**A**)

much faster in the presence of DCMU than in the absence of the inhibitor (Krause and Weis 1991). Therefore, we could diagnose the fluorescence rise at sites around major veins observed in Fig. 11A as photosynthetic inhibition caused by DCMU in Nekosogi-ace transported with the transpiration stream from the leafstalk. Isouron also might cause fluorescence rise because it blocks the electron transfer. There was no visible injury on the leaf during the measurement (C).

Quenching parameters such as q_P, q_N, NPQ, and Φ_{PSII} calculated by Eqs. 4 to 7 are known as indicators of photochemical and nonphotochemical quenching. Values of nonphotochemical quenching parameters, q_N and NPQ, decreased at

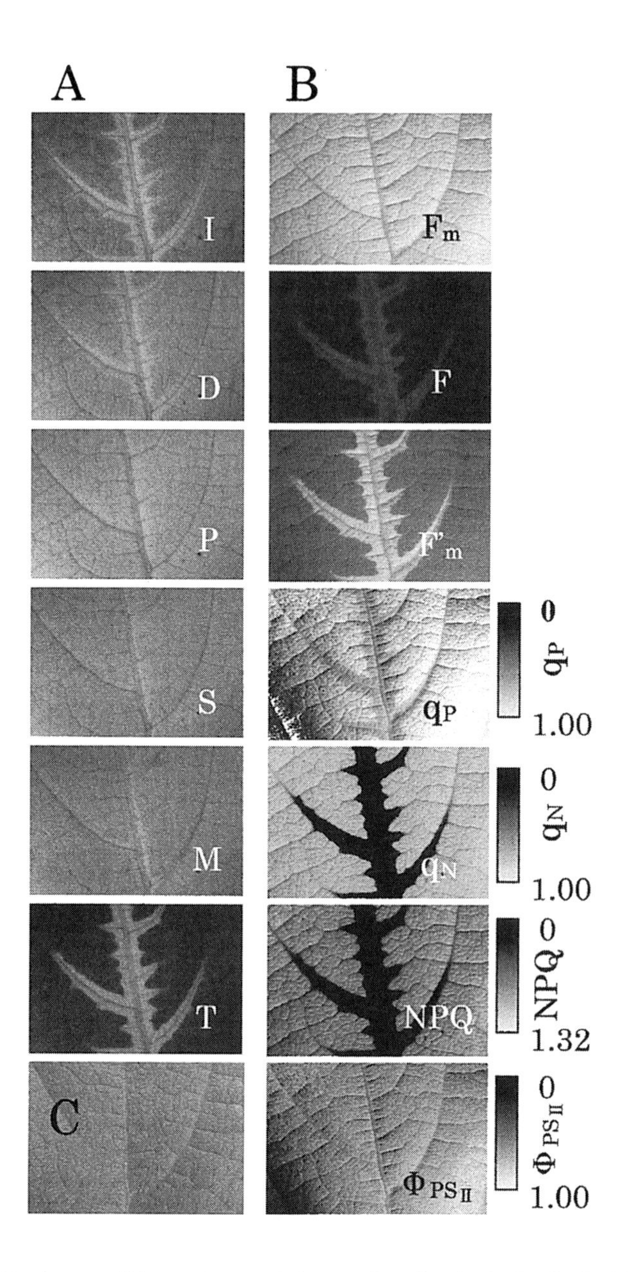

Fig. 11. CFI images (**A**) at characteristic transient levels (I, D, P, S, M, and T) and quenching images (**B**) obtained by a sequence shown in Fig. 3 of a kidney bean leaf after treatment with 1/1000 diluted solution of a herbicide, Nekosogi-ace (Rainbow Chemicals), including 6.0% DCMU, 3.0% DCBN, and 1.0% isouron, from the leafstalk for 20 min (Omasa et al. 2001). There was no visible injury just after the treatment (**C**)

about the same sites around the major veins as inhibition sites in CFI images. This result indicates decreased electron transport from PSII caused by DCMU and consequently inhibition of *trans*-thylakoid proton gradient formation and decrease in CO_2 assimilation (Schreiber et al. 1986; Krause and Weis 1991; Siebke and Weis 1995a; Omasa et al. 2001). However, DCMU inhibition was not clearly detected in the images of q_P and Φ_{PSII}. This result shows that q_P and Φ_{PSII} calculated by Eqs. 4 and 7 are limited as indicators of photochemical inhibition and CO_2 assimilation rate. Consequently, simultaneous use of images of CFI and quenching parameters enables detailed diagnosis of injuries of the photosynthetic system, especially in photochemical inhibition.

Finally, photosynthetic inhibition of intact plants caused by different herbicide treatment was compared. Figure 12 shows F_m, F_m', and NPQ images of an attached cucumber (*Cucumis sativus* L. cv. Sharp7) leaf before (A), and after (B) the application of 1/1000 diluted solution of Nekosogi-ace in soil. After 2 days of the herbicide treatment, mesophyll sites around the major veins showed a remarkable increase in F_m' as compared with noninjured sites. On the other hand, no injuries were detected in the F_m image. The NPQ value showed a decrease in the sites located near the vein.

Figure 13 shows F_m, F_m', and NPQ images of an attached cucumber leaf taken before (A), and after (B) treatment of a foliar application-type herbicide, Roundup

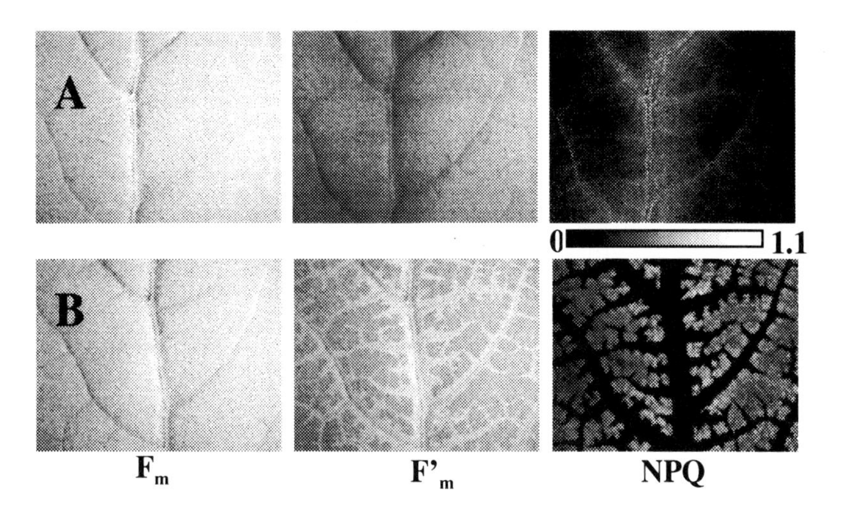

Fig. 12. F_m, F_m', and NPQ images of an attached cucumber leaf before (**A**) and 2 days after (**B**) application of 1/1000 diluted solution of Nekosogi-ace in soil (Takayama et al. 2000). There was no visible injury at 2 days after the treatment

(Monsanto), including 1.0% *N*-(phosphonomethyl)glycine (glyphosate). At 24 h after the treatment, mesophyll sites around the major veins showed a remarkable decrease in F_m and F_m' as compared with noninjured sites, an early symptom of dysfunction in photosynthetic apparatus in these sites. In the NPQ image, the sites around the major veins showed a decrease in NPQ. However, the sites surrounding the low NPQ sites and branches of veins showed a local rise in NPQ.

Photosynthetic inhibition caused by the two types of herbicides was estimated using NPQ images. The manner of appearance of the symptoms differed for the two herbicides, as shown by Figs. 12 and 13. This difference was attributed to the difference in the components of the two herbicides and the methods of application. It was also proven that the inhibition spreads from the sites along the veins for both methods of application. Low values of NPQ in the sites around the major veins predicted low assimilation rates (Daley et al. 1989; Siebke and Weis 1995a; Omasa et al. 2001) and the inhibition of the ability of chloroplasts to generate a *trans*-thylakoid pH gradient, to sustain electron transport, and to dissipate excess excitation energy as heat in these sites (Osmond et al. 1998). However, we cannot explain why the NPQ values increased at the sites surrounding the low-NPQ sites and branches of veins, as shown in Fig. 13.

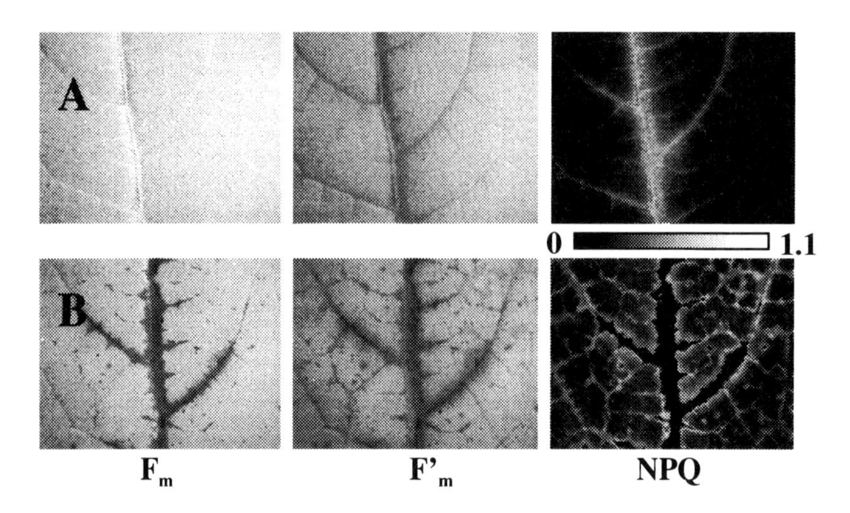

Fig. 13. F_m, F_m', and NPQ images of an attached cucumber leaf before (**A**) and 24 h after (**B**) treatment of a foliar application-type herbicide, Roundup (Monsanto), including 1.0% glyphosate (Takayama et al. 2000). There was no visible injury at 24 h after the treatment

5. Conclusions

We have introduced an image instrumentation system for diagnosing the effects of environmental stress factors such as air pollutants and agricultural chemicals (herbicides) on photosynthetic activity and sites of inhibition in the photosynthetic apparatus of attached leaves. By use of the scanning laser in addition to halogen or xenon lamps, this system enabled diagnosis of photosynthetic functions of plants at a distance from the instrumentation system. In this system, two methods of Chl *a* fluorescence imaging analysis were introduced: CFI analysis and the saturation pulse method. The saturation pulse method can analyze the photosynthetic injuries more quantitatively than by CFI analysis. However, the saturation pulse method requires an evenly distributed high level of photosynthetic active radiation (PAR), which makes it difficult to apply to a large leaf area. In contrast, CFI analysis does not require such a high level of PAR for measurements and is easier to use for examining a large leaf area. Simultaneous use of images of CFI and quenching parameters enables detailed diagnosis of injuries of the photosynthetic system, especially in photochemical inhibition. Although CFI analysis and the saturation pulse method are effective as nondestructive assays of photosynthetic injuries, they are affected by the plant materials and measurement conditions, including pigment contents, leaf age, and surrounding environment. Therefore, it is necessary to use them for accurate assays with other methods such as measurement of CO_2 uptake by assimilation chamber and biochemical analysis.

References

Balachandran S, Osmond CB, Daley PF (1994) Diagnosis of the earliest strain-specific interactions between tobacco mosaic virus and chloroplasts of tobacco leaves *in vivo* by means of chlorophyll fluorescence imaging. Plant Physiol 104:1059-1065

Bilger W, Björkman O (1990) Role of the xanthophyll cycle in photoprotection elucidated by measurements of light-induced absorbance changes, fluorescence and photosynthesis in leaves of Hedera canariensis. Photosynth Res 25:173-185

Chappelle EW, Kim MS, McMurtrey JE III (1992) Ratio analysis of reflectance spectra (RARS): an algorithm for the remote estimation of the concentrations of chlorophyll *a*, chlorophyll *b*, and carotenoids in soybean leaves. Remote Sens Environ 39:239-247

Croxdale JG, Omasa K (1990a) Chlorophyll *a* fluorescence and carbon assimilation in developing leaves of light-grown cucumber. Plant Physiol 93:1078-1082

Croxdale JG, Omasa K (1990b) Patterns of chlorophyll fluorescence kinetics in relation to growth and expansion in cucumber leaves. Plant Physiol 93:1083-1088

Daley PF (1995) Chlorophyll fluorescence analysis and imaging in plant stress and disease. Can J Plant Physiol 17:167-173

Daley PF, Raschke K, Ball JT, et al (1989) Topography of photosynthetic activity of leaves obtained from video images of chlorophyll fluorescence. Plant Physiol 90:1233-1238

De Kok, Stulen I (1998) Responses of plant metabolism to air pollution and global change. Backhuys

Genty B, Meyer S (1995) Quantitative mapping of leaf photosynthesis using chlorophyll fluorescence imaging. Aust J Physiol 22:277–284

Genty B, Briantais JM, Baker NR (1989) The relationship between the quantum yield of photosynthetic electron transport and quenching of chlorophyll fluorescence. Biochim Biophys Acta 990:87-92

Govindjee (1995) Sixty-three years since Kautsky: Chlorophyll *a* fluorescence. Aust J Plant Physiol 22:131-160

Häder DP (eds) (1992) Image analysis in biology. CRC Press

Häder DP (eds) (2000) Image analysis: Methods and applications, 2nd edn. CRC Press

Hashimoto Y, Kramer PJ, Nonami H, et al (1990) Measurement techniques in plant science. Academic Press

Jones HG (1999) Use of thermography for quantitative studies of spatial and temporal variation of stomatal conductance over leaf surface. Plant Cell Environ 22:1043-1055

Kautsky H, Hirsch A (1931) Neue Versuche zur Kohlensäureassimilation. Naturwissenschaftren 19:964

Kim MS, Krizek DT, Daughtry CST, et al (1996) Fluorescence imaging system: application for the assessment of vegetation stresses. SPIE 2959:4-13

Kim MS, Mulchi CL, Daughtry CST, et al (1997) Fluorescence images of soybean leaves grown under increased O_3 and CO_2. SPIE 3059:22-31

Kooten OV, Snel JFH (1990) The use of chlorophyll fluorescence nomenclature in plant stress physiology. Photosynth Res 25:147-150

Kramer PJ, Boyer JS (1995) Water relations of plants and soils. Academic Press

Krause GH, Weis E (1991) Chlorophyll fluorescence and photosynthesis: the basics. Annu Rev Plant Physiol Plant Mol Biol 42:313-349

Larcher W (1995) Physiological plant ecology. Springer

Lichtenthaler HK (1988) Applications of chlorophyll fluorescence. Kluwer

Lichtenthaler HK (1996) Vegetation stress. J Plant Physiol 148:599-644

Munday JCM Jr, Govindjee (1969a) Light-induced changes in the fluorescence yield of chlorophyll *a in vivo*. Ⅲ. The dip and the peak in the fluorescence transient of *Chlorella pyrenoidosa*. Biophysical J 9:1-21

Munday JCM Jr, Govindjee (1969b) Light-induced changes in the fluorescence yield of chlorophyll *a in vivo*. Ⅳ. The effect of pre-illumination on the fluorescence transient of *Chlorella pyrenoidosa*. Biophys J 9:22-35

Omasa K (1988) Image instrumentation of plants using a laser scanner (in Japanese). Proc Annu Meet Jpn Soc Environ Control Biol 1988:14-15

Omasa K (1990) Image instrumentation methods of plant analysis. In: Linskens HF, Jackson JF (eds) Modern methods of plant analysis, vol 11. Springer, New Series, pp 203-243

Omasa K (1992) Image diagnosis of photosynthesis in cultured tissues. Acta Hortic 319:653-658

Omasa K (1994) Diagnosis of trees by portable thermographic system. In: Kuttler W, Jochimsen M (eds) Immissionsökologische Forschung im Wandel der Zeit. Essener Ökologische Schriften, pp 141-152

Omasa K (1998) Image instrumentation of chlorophyll *a* fluorescence. SPIE 3382:91-99

Omasa K (2000) Phytobiological IT in agricultural engineering. Proc XIV Memorial CIGR World Congress 2000:125-132

Omasa K, Aiga I (1987) Environmental measurement: image instrumentation for evaluating pollution effects on plants. In: Singh MG (ed) Systems and control encyclopedia.

Pergamon Press, pp 1516-1522

Omasa K, Shimazaki K (1990) Image analysis of chlorophyll fluorescence in leaves. In: Hashimoto Y, Kramer PJ, Nonami H, et al (eds) Measurement technique in plant science. Academic Press, pp 387-401

Omasa K, Croxdale JG (1992) Image analysis of stomatal movements and gas exchange. In: Häder DP (ed) Image analysis in biology. CRC Press, pp 171-193

Omasa K, Shimazaki K, Aiga I, et al (1987) Image analysis of chlorophyll fluorescence transients for diagnosing the photosynthetic system of attached leaves. Plant Physiol 84:748-752

Omasa K, Maruyama S, Matthews MA, et al (1991) Image diagnosis of photosynthesis in water-deficit plants. IFAC Workshop Series 1991, No.1. Pergamon Press, pp 383-388

Omasa K, Takayama K, Goto, E (2001) Image diagnosis of photosynthetic injuries induced by herbicide in plants: comparison of the induction method with the saturation pulse method for chlorophyll a fluorescence analysis (in Japanese with English summary). J Soc High Tech Agric 13:29-37

Osmond CB, Daley PF, Badger MR, et al (1998) Chlorophyll fluorescence quenching during photosynthetic induction in leaves of *Abutilon striatum* dicks. Infected with abutilon mosaic virus, observed with a field-portable imaging system. Bot Acta 111:390-397

Papageorgiou G (1975) Chlorophyll fluorescence: an intrinsic probe of photosynthesis. In: Govindjee (ed) Bioenergetics of photosynthesis. Academic Press, pp 319-371

Papageorgiou G, Govindjee (1968) Light-induced changes in the fluorescence yield of chlorophyll *a in vivo*. Ⅱ. *Chlorella pyrenoidosa*. Biophys J 8:1316-1328

Quick WP, Horton P (1984) Studies on the induction of chlorophyll fluorescence in barley protoplasts. Ⅱ. Resolution of fluorescence quenching by redox state and the *trans*-thylakoid pH gradient. Proc R Soc Lond B 220:371-382

Rencz AN (ed) (1999) Manual of remote sensing, 3rd edn., vol 3. Wiley, New York, p707

Rolfe SA, Scholes JD (1995) Quantitative imaging of chlorophyll fluorescence. New Phytol 131:69-79

Saito Y, Takahashi K, Nomura E, et al (1997) Visualization of laser-induced fluorescence of plants influenced by environmental stress with a microfluorescence imaging system and a fluorescence imaging lidar system. SPIE 3059:190-198

Sandermann H, Wellburn AR, Heath RL (1997) Forest decline and ozone. Springer

Schreiber U (1986) Detection of rapid induction kinetics with a new type of high-frequency modulated chlorophyll fluorometer. Photosynth Res 9:261-272

Schreiber U, Schliwa U, Bilger W (1986) Continuous recording of photochemical and non-photochemical chlorophyll fluorescence quenching with a new type of modulation fluorometer. Photosynth Res 10:51-62

Schulze ED, Lange OL, Oren R (eds) (1989) Forest decline and air pollution. Springer-Verlag

Siebke K, Weis E (1995a) Assimilation images of leaves of *Glechoma hederacea*: analysis of non–synchronous stomata related oscillations. Planta 196:155-165

Siebke K, Weis E (1995b) Imaging of chlorophyll *a* fluorescence in leaves: topography of photosynthetic oscillations in leaves of *Glechoma hederacea*. Photosynth Res 45:225-237

Takayama K, Goto E, Omasa K (2000) Diagnosis of photosynthetic injury caused by agricultural chemicals using chlorophyll fluorescence imaging. Proc XIV Memorial CIGR World Congress 2000:1436-1441

Wessman CA (1990) Evaluation of canopy biochemistry. In: Hobbs RJ, Mooney HA (Eds) Remote sensing of biosphere functioning. Springer, pp 135-156

Yunus M, Iqbal M (1996) Plant response to air pollution. Wiley

16 Field-Portable Imaging System for Measurement of Chlorophyll Fluorescence Quenching

Barry Osmond* and Yong-Mok Park**

Photobioenergetics Group, Research School of Biological Sciences, Institute of Advanced Studies, Australian National University, Box 3252 Weston Creek ACT 2611, Australia
*Present Address: Biosphere 2 Center, Columbia University, 32540 S Biosphere Road, PO Box 689, Oracle AZ 85623, USA
**Present Address: Department of Life Science, College of Natural Science and Engineering, Chongju University, Chongju, 360-764, Korea

1. Introduction

Chlorophyll fluorescence is a powerful tool for noninvasive evaluation of photosynthesis, especially during induction experiments (Govindjee 1995). At ambient temperatures a small amount (about 1%) of photosynthetically active radiation (400-700nm) absorbed by chlorophyll is re-emitted as fluorescence in the near infra-red (>690nm). The fluorescence arises principally from PSII, the primary charge separating and O_2 evolving site in photosynthesis, and the quenching of this fluorescence is a reliable indicator of photochemical (useful) and nonphotochemical (wasteful) fates of the absorbed photons (Krause and Weis 1991). The functioning of the multi-component PSII is sensitive to many environmental factors, and as a consequence, fluorescence monitoring has found widespread application in stress physiology and has transformed the assessment of photosynthesis in the field (Osmond et al. 1999).

Digital imaging of chlorophyll fluorescence quenching in leaves, pioneered by Omasa et al. (1987) and Daley et al. (1989), is particularly useful for determining the uniformity of photosynthetic responses in tissues exposed to biotic and abiotic

Air Pollution and Plant Biotechnology
–Prospects for Phytomonitoring and Phytoremediation–
Edited by K. Omasa, H. Saji, S. Youssefian, and N. Kondo
© Springer -Verlag Tokyo 2002

stress (Buschmann and Lichtenthaler 1998; Chaerle and Van Der Streaten 2000). Patchy stomatal responses are common following abiotic stress (Terashima 1992). Imaging techniques have been used to demonstrate short and long term patchiness (Genty and Meyer 1995; Siebke and Weis 1995a,b; Lichtenthaler et al., 1996; Lichtenthaler and Miehe, 1997; Osmond et al. 1999a; Mott and Buckley 2000). They are especially useful for evaluation of biotic stress following infection with viral and fungal pathogens (Balachandran et al. 1994; Peterson and Aylor 1995; Rolfe and Scholes 1995; Scholes and Rolfe 1996; Lohaus et al. 2000; Chaerle and Van Der Straeten 2000).

Just as portable CO_2 exchange systems and portable chlorophyll fluorescence monitoring systems have stimulated research in plant environmental biology over the last two decades, portable fluorescence imaging systems have the potential to greatly expand our understanding of photosynthetic responses to environment. The truism that "seeing is believing" is particularly apt, and in the near future, inexpensive chlorophyll fluorescence imaging devices (Lootens and Vandecasteele 2000) will make it possible for all to grasp the fundamentals of light energy conversion in the photosynthetic apparatus. The importance of such a dramatic advance in the public understanding of plant functions, especially of the critical contribution of terrestrial plants to global scale processes, should not be underestimated. At the same time in research, because leaf photosynthesis is routinely estimated on an area basis, assessment of the uniformity or otherwise of photosynthetic activity must now become a fundamental step in any measurement system (e.g. Balachandran et al. 1997), especially in the field. This brief paper reports some lessons learned from several years experience of a prototype, almost portable, chlorophyll fluorescence imaging system (Osmond et al. 1998).

2. A Prototype (Almost) Portable System

The field-portable chlorophyll fluorescence imaging system described by Daley (1995) used state of the art components at the time of its construction (1990-92). Subsequently, minor modifications have been made to LEDs, light source control and image capture components. In the original configuration, leaf chlorophyll fluorescence was excited by 663 nm light from a ring illuminator containing 12 ultra-bright red LEDs (190HRP; 13 candela; 10 mm 4% view angle, from AND Inc, Burlingame, CA) powered by a Hansatech LC1 (Hansatech Instruments, Kings Lynne, Norfolk) light source controller from 12V batteries. As these LEDs are no longer available, and because all red LEDs tested subsequently have been found to have significant secondary emissions in the near infra-red, 24 ultra-bright blue LEDs (NSPB500S; 3 candela; 5mm 15% view angle, from Nichia Corp) are now used. The blue LEDs have no difficulty exciting DCMU enhanced fluorescence from cells adjacent to veins in a *Clematis* leaf (Figure 1) when this leaf is placed under a 1-2 mm thick *Kalanchoë* leaf (Osmond, unpublished) i.e., self absorption of fluorescence is unlikely to limit the use of the device.

The LEDs are focussed on a 5-12 mm diameter disc of tissue held in a leaf clip on attached or detached leaves, and fluorescence from the tissue observed with a video camera through a 690 nm cut-off filter. The illuminator was fitted to a 50 mm macrolens (Canon EF1: 2.5) on a Canon L1 color video camera and recorder. All commercially available video cameras are fitted with the near IR-filter over the CCD detector and this must removed to obtain sufficient sensitivity to Chl fluorescence beyond 690 nm. The illuminator was controlled by a program in Visual Basic using a PCM-D24CTR3 card (Computer Boards Inc Mansfield MA) in a laptop computer that was also used to capture images via an inexpensive frame-grabber ('Snappy": Play Inc., Rancho Cordova CA.). Computer-controlled,

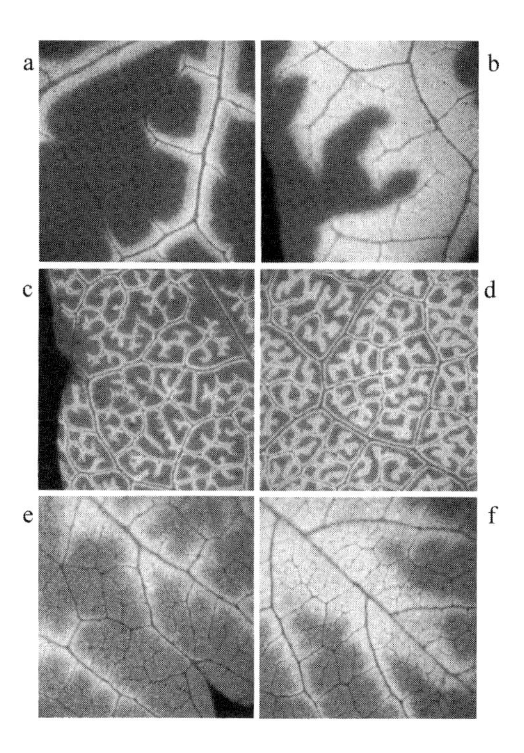

Fig. 1. Quenching maps (after 120 s) from Kautsky transients and transients with saturating flashes obtained from leaves after absorption of inhibitors of photosynthesis. *Clematis* leaves in 1 μM DCMU for 30 min (a) and 3 h (b); in 50 μM nigericin for 1 h (c) and 3 h (d); and *Potentilla* leaves in 10 mM DTT 30 min at the tip of the leaf (e) and after 1 h at the base of a leaf. Experiments were done in the Stocker Haus, Technical University of Darmstadt

saturating flashes were achieved by applying maximum power for 2 s, with image capture synchronized 200 ms into the flash. Other details of the system and subsequent image processing have been presented elsewhere (Osmond et al. 1998).

The multi-component system weighs some 20 kg, about half of which is batteries. Nevertheless, the system is field portable in airplanes and motor vehicles and has been used in laboratories on four continents. It is cumbersome to use in the field, especially with respect to fragile leaves that tend to be detached by movement of the rest of the plant in windy conditions. In hot, bright conditions the system must be protected with a mylar-coated thermal blanket, and a shade box is needed to view the computer screen. In principle, the leaf clips used to dark adapt areas of the leaf for 30 min are not more difficult to use than with the Hansatech PEA fluorometer. However, longer enclosures (several minutes) after dark adaptation are sometimes needed to image quenching patterns and NPQ. The slightest leaf movement during this time means that quenching calculations are compromised.

The system lacks automatic exposure adjustment that is rapid enough to accurately capture steady state fluorescence in strong light (Bro et al. 1996), and so cannot be used for photosystem II efficiency mapping. It is limited to analysis of the slow phases of Kautsky transients at constant excitation light and to mapping NPQ from images captured in saturating flashes (2s) applied on top of a lower intensity actinic light. The High 8 tape recorder of the video camera is particularly useful in the field, in that static images can be extracted from the record at any time subsequently. Indeed, manual operation of the apparatus in the Kautsky mode is very straightforward, but non-photochemical quenching images can also be captured manually using the x10 power switch of the Hansatech controller. Data loss due to computer malfunction can be avoided by routine use of the tape recorder as a back up system.

3. Some Applications and Results

Three applications will be described: inhibitor studies in the laboratory, water stress responses in the field and in the laboratory, and imaging of biotic stress in greenhouse grown plants.

3.1 Inhibitors of Photosynthesis

Inhibitors of photosynthesis that prevent the normal quenching of chlorophyll fluorescence are commonly applied to demonstrate the performance of imaging systems (Daley et al. 1989; Genty and Meyer 1995). Figure 1 shows the performance of the portable system tested with three inhibitors that were absorbed via the petioles of leaves of *Clematis sp* and *Potentilla reptans* standing in weak light on the laboratory bench for different periods of time. Two types of quenching were assessed. In panels (a)-(d) Kautsky transients were followed in images taken at 0 and 120 s after illumination with 230 μmol photons m^{-2} s^{-1} blue light. In panels (e) and (f) saturating flash images (950 μmol photons m^{-2} s^{-1}) were captured at 0 and 120 s from leaves illuminated with actinic light at 230 μmol photons m^{-2} s^{-1} to estimate non-photochemical quenching. The PSII binding herbicide DCMU moves rapidly along the veins preventing quenching as it diffuses into mesophyll cells, but after 30 min it has not entered all intervein regions that still show normal quenching. After 3 h all regions have been inhibited and no quenching is detectable. Nigericin, a high molecular weight inhibitor that discharges the trans-thylakoid ΔpH, moves only very slowly from the veins into mesophyll cells. The inhibitor of violaxanthin de-epoxidase, DTT, is very rapidly transferred through the veins to all cells where completely prevents the development of non-photochemical quenching as assessed in saturating flash experiments.

3.2 Water Stress

Water stress often causes wilting, especially in the leaves of meadow herbs and grasses exposed to bright light after long periods of dull weather. Figure 2 shows the typical non-uniform chlorophyll fluorescence pattern (patchiness) observed in the first 30-60 s of Kautsky transients in wilting leaves (Omasa et al. 1991; Osmond et al. 1998). Fluorescence quenching was followed continuously on tape in 240 μmol photons m^{-2} s^{-1} red light after 30 min dark adaptation and images were later captured at intervals from the tape for processing in the laboratory. The leaves of *Sanguisorba* examined in the field had folded along the mid-vein. Tissues along one or both sides of the mid-vein retained high fluorescence, similar to the patchiness found in water stressed *Potentilla reptans* (Osmond et al. 1999). It is tempting to suggest that the patchy, slower quenching of chlorophyll fluorescence in the early phases of the transient reflects water stress effects on primary photosynthetic processes, and that the highly fluorescent tissues have lower turgor, causing one side of the leaf to collapse on the other.

Slowly appearing, persistent patchiness can be observed late in the Kautsky transients of leaves exposed to more severe water stress. Detached leaves of grapevines were cut under water and allowed to dark-adapt overnight in vials of water. Controls showed rapid and uniform quenching in Kautsky experiments with 680 μmol photons m^{-2} s^{-1} of blue light (Figure 2). The leaves were then placed in a pressure bomb and water was expressed until pre-selected leaf water potentials were attained. These leaves were then imaged using the same protocols as for the controls. We found that quenching was rapid and uniform up to water potentials of -14 bars but that above -20 bars, slower quenching and pronounced patchiness was always observed. Similar results were observed in other species (Park, unpublished) and in some species, patchiness appeared at water potentials well above -20 bars.

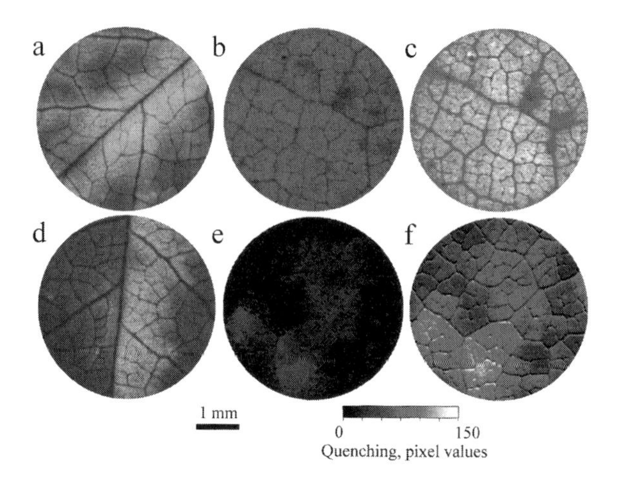

Fig. 2. Patchy fluorescence in attached, folded leaves of water stressed *Sanguisorba* observed in-situ in the meadows of the Botanic Garden, Darmstadt, imaged (a) 25 and (d) 40s into a Kautsky transient in red light. Maps showing uniform quenching in grapevine leaves at full turgor calculated after (b) 15 and (c) 60s illumination in blue light (note small lesions due to infection). Leaves examined in the same way after treatment in a pressure bomb to –32 bars show slower quenching and strong patchiness after (e) 15 and (f) 60s

These experiments highlight one of the large unknowns of leaf water relations, namely the relationships between vascular architecture and water supply to different areas of the leaf. Because the effects of wilting on the patchiness of chlorophyll fluorescence quenching are commonly seen closest to major veins, we conclude that water delivery patterns are likely to be complex. Perhaps tracking the delivery of inhibitors under controlled conditions (Figure 1) or surgical interventions (Omasa et al. 1991) could advance our understanding of water supply networks in leaves. The pressure bomb method promises to bring a more quantitative approach to the analysis of non-uniform quenching following water stress.

3.3 Biotic Stress

The instrument described above was intended for use with virus infected plants (Daley 1995). Following the successful early detection of photosynthetic lesions arising from tobacco mosaic virus infection using laboratory based chlorophyll fluorescence imaging equipment (Osmond et al. 1990; Balachandran et al. 1994), these methods have been widely applied to biotic stress (Peterson and Aylor 1995; Scholes and Rolfe 1996; Técsi et al. 1996). The portable system has been used in several laboratories, giving insights into the properties of symptomatic and asymptomatic tissues in *Abutilon striatum* infected with the geminivirus Abutilon mosaic virus infected (Osmond et al. 1998) and has questioned the carbohydrate accumulation hypothesis for symptom development (Lohaus et al. 2000). The portability of the system was exploited in the course of a one day visit to the laboratory of Professor Ichiro Terashima, Osaka University, where Dr Sachiko Funayama was researching the photosynthetic effects of another geminivirus (Tobacco leaf curl virus) in wild populations of *Eupatorium makinoi*. Figure 3 shows the relatively uniform initial maximum fluorescence in young *Eupatorium* leaves that displayed the characteristic vein clearing symptoms. The non-photochemical quenching mapping capability of the system makes it is clear that the chlorophyll deficient cells near the veins have low photosynthetic activity, as established by the physiological and biochemical analyses of Funayama et al. (1997).

4. Further Developments

The Daley instrument has proved very versatile and the scale of observation useful in many applications, not the least of which has been the development of photoinhibitory printing (Ning et al. 1995; Osmond et al. 1999b). Following the *Chromatium* printing of Englemann (1883; related by Schlegel 1999), the starch printing of Molisch (1914) and the chlorophyll printing of Ackyoyd and Harvey (related by Kemp 2000), the visualization of PSII photoinhibition by excess light in shade leaves through a photographic negative, continues to bring art to science. Using a combination of the Daley instrument and controlled confocal microscopy Osmond et al. (1999b) achieved visualization of photoinhibition at the level of individual grana in chloroplasts. The photoinhibition prints in Figure 4 demonstrate one of the limitations of the Daley design; its fixed focus. A 35 mm negative of four gurus of chlorophyll fluorescence was used to print an attractive photoinhibitory group portrait on an ivy leaf using the flexible, laboratory based imaging system of Siebke and Weis (1995a). When this image is printed and examined in the Daley instrument, the fixed focus confines one to vignettes (Osmond et al. 1999b).

The next generation of simple portable video imaging systems will be much lighter and comparatively inexpensive (cf. Lootens and Vandercasteele 2000). Small video cameras in which the infra-red filter can be readily removed and which can be fitted with small close-up lenses are available for about 1% of the cost and at less than 10 % of the weight of the camera used above (e.g. Vicam, Vista Imaging Inc CA, also known in Europe as Webcam). Driven from the USB port of lightweight computers, and supplied with their own image capture software, these cameras will overcome many of the limitations to portability. If

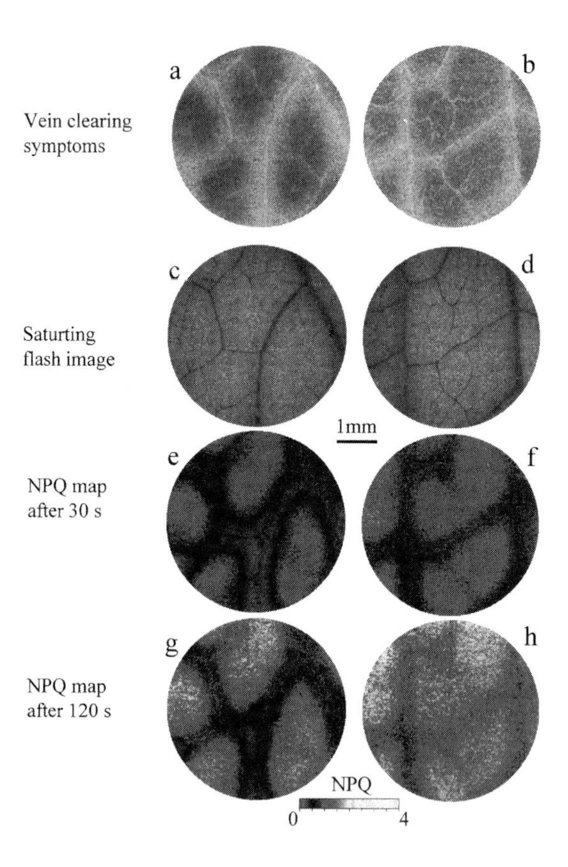

Fig. 3. The vein clearing symptoms in geminivirus infected leaves of *E makinoi* (panels a, b) cannot be distinguished from adjacent green tissue in images of maximum chlorophyll fluorescence (panels c, d) but show severely retarded non-photochemical quenching after 60 and 180s (panels e, f and g, h)

suitable dark adaptation and leaf holding devices can be constructed, the limitations of fixed focus will be overcome.

This optimistic outlook presupposes that versatile LED illuminators with uniform light fields can be developed. Ultra-bright blue LEDs now available

Fig. 4. Photoinhibitory printing of an entire 35 mm negative on a deeply shaded ivy leaf for 1 h in full sunlight (left, image courtesy of Dr Katharina Siebke). Vignettes from the same negative printed on a *Cissus* leaf (right) that just fit within the image area of the fixed focus portable chlorophyll fluorescence imaging system used here. The gurus of chlorophyll fluorescence analysis are, left to right Govindjee, Ulrich Schreiber, Olle Björkman, and Paul Falkowski

provide more efficient excitation of PSII and lower reflectance that can be removed cleanly by cut-off filters. Simple, lightweight parabolic reflectors, studded with these LEDs and driven by 9V batteries, can be envisaged. Another option would seem to be illumination from below with purpose built blue LED arrays in contact with the leaf, along the lines of the illuminators used in Licor photosynthesis measurement systems. The leaf clip configuration seems likely to remain the most effective means of dark adaptation. Access to individual leaves will be easier, and the likelihood that leaves are damaged by manipulation of equipment will be reduced. We can now look forward to an explosion in the use of chlorophyll fluorescence imaging, from classroom demonstrations using such inexpensive devices to an Imaging PAM, a portable, pulse modulated research instrument to be marketed by H Walz Effeltrich in 2001-2002 (U Schreiber, personal communication).

Acknowledgements. We are grateful to Paul Daley for constructing the system described here with funding from NSF grant BSR891451 and to Murray Badger for devising the control software. The Alexander von Humboldt Stiftung supported research by CBO that permitted evaluation of the instrument. Some of the research was carried out with support to Y-MP from the Overseas Research Program of Chongju University.

References

Balachandran S, Osmond CB, Daley PF (1994) Diagnosis of the earliest strain-specific interactions between tobacco mosaic virus and chloroplasts of tobacco leaves *in vivo* by means of chlorophyll fluorescence imaging. Plant Physiol 104: 1059-1065

Balachandran S, Hurry VM, Kelley SE, Osmond CB, Robinson SA, Rohozinski J, Seaton GGR, Sims DA (1997) Concepts of biotic stress. Some insights into the stress physiology of virus-infected plants, from the perspective of photosynthesis. Physiol Plant 100: 203-213

Bro E, Meyer S, Genty B (1996) Heterogeneity of leaf CO_2 assimilation during photosynthetic induction. Plant Cell Environ 19: 1349-1358

Buschmann C, Lichtenthaler HK (1998) Principles and characteristics of multi-colour fluorescence imaging of plants. J Plant Physiol 152: 297-314

Chaerle L, Van Der Straeten D (2000) Imaging techniques and early detection of plant stress. TIPS 5: 495-501

Daley PF (1995) Chlorophyll fluorescence analysis and imaging in plant stress and disease. Canad J Plant Pathol 17: 167-173

Daley PF, Raschke K, Ball, JT, Berry JA (1989) Topography of photosynthetic activity in leaves obtained from video images of chlorophyll fluorescence. Plant Physiol 90: 1233-1238

Funayama S, Sonoike K, Terashima I (1997) Photosynthetic properties of *Eupatorium makinoi* infected by a geminivirus. Photosynth Res 53; 253-261

Genty B, Meyer S (1995) Quantitative mapping of leaf photosynthesis using chlorophyll fluorescence imaging. Aust J Plant Physiol 22: 277-284

Govindjee (1995) Sixty three years since Kautsky: chlorophyll *a* fluorescence. Aust J Plant Physiol 22: 131-160

Kemp M (2000) Science in culture. Nature 403: 364

Krause GH, Weis E (1991) Chlorophyll fluorescence and photosynthesis: the basics. Annu Rev Plant Physiol Plant Mol Biol.42: 313-349

Lichtenthaler HK, Miehe JA (1997) Fluorescence imaging as a diagnostic tool for plant stress. TIPS 2: 316-320

Lohaus G, Heldt H, Osmond CB (2000) Infection with phloem limited *Abutilon mosaic virus* causes localised carbohydrate accumulation in leaves of *Abutilon striatum*: relationships to symptom development and effects on chlorophyll fluorescence quenching during photosynthetic induction. Plant Biol 2: 161-167

Lootens P, Vandercasteele P (2000) A cheap chlorophyll *a* fluorescence imaging system. Photosynthetica 38: 53-56

Mott KA, Buckley TN (2000) patchy stomatal conductance: emergent behaviour of stomata. TIPS 5: 258-262

Molisch H (1914) Über die Herstellung von Photographien im Laubblatte. *Sitzungsberichte der kaiserlichen Akademie der Wissenschaften, Wein*. Reprinted in Molisch, H (1922)

Populäre biologische Vorträge. Verlag Gustav Fischer, Jena, pp 243-246

Ning L, Edwards GE, Strobel A, Daley LS, Callis JB (1995) Imaging fluorometer to detect pathological and physiological change in plants. Appl Spectroscopy 49: 1381-1389

Omasa K, Shimazaki K-I, Aiga I, Larcher W, Onoe M (1987) Image analysis of chlorophyll fluorescence transients for diagnosing the photosynthetic system of attached leaves. Plant Physiol 84: 748-752

Omasa K, Murayama S, Matthews MA, Boyer JS (1991) Image diagnosis of photosynthesis in water-deficit plants. In: Hashimoto Y, Day W, Eds., Mathematical and Control Applications in Agriculture and Horticulture. Pergamon Press, Oxford, pp 383-388

Osmond CB, Berry JA, Balachandran S, Büchen-Osmond C, Daley PF Hodgson RC (1990) Potential consequences of virus infection for shade-sun acclimation in leaves. Bot Acta 103: 226-229

Osmond CB, Daley PF, Badger MR, Lüttge U (1998) Chlorophyll fluorescence quenching during photosynthetic induction in leaves of *Abutilon striatum* Dicks. infected with Abutilon mosaic virus, observed with a field-portable imaging system. Bot Acta 111: 390-397

Osmond CB, Kramer D, Lüttge U (1999a) Reversible, water stress-induced non-uniform chlorophyll fluorescence quenching in wilting leaves of *Potentilla reptans* may not be due to patchy stomatal responses. Plant Biol 1: 618-624

Osmond B, Schwartz O and Gunning B (1999b) Photoinhibitory printing on leaves, visualised by chlorophyll fluorescence imaging and confocal microscopy, is due to diminished fluorescence from grana. Aust J Plant Physiol 26; 717-724

Peterson RB, Aylor DE (1995) Chlorophyll fluorescence induction in leaves of *Phaseolus vulgaris* infected with bean rust (*Uromyces appendiculatus*). Plant Physiol 108: 163-171

Rolfe SA, and Scholes JD (1995) Quantitative imaging of chlorophyll fluorescence. New Phytol 131: 69-79

Schlegel HG (1999) Geschichte der Mikrobiologie. Acta Historica Leopoldina 28: 77-79

Scholes JD, Rolfe SA (1996) Photosynthesis in localised regions of oat leaves infected with crown rust (*Puccinia coronata*): quantitative imaging of chlorophyll fluorescence. Planta 199: 573-582

Siebke K, Weis, E (1995a) Assimilation images of leaves of *Glechoma hederacea*: analysis of non-synchronous stomata related oscillations. Planta 196: 155-165

Siebke K, Weis E (1995b) Imaging of chlorophyll-*a*-fluorescence in leaves: topography of photosynthetic oscillations in leaves of *Glechoma hederacea*. Photosynth Res 45: 225-237

Técsi LI, Smith AM, Manle AJ, Leegood RC (1996) A spatial analysis of physiological changes associated with infection of cotyledons of marrow plants with cucumber mosaic virus. Plant Physiol 111: 975-985

Terashima I (1992) Anatomy of non-uniform leaf photosynthesis. Photosynth Res 31: 195-212

17
Assessment of Environmental Plant Stresses Using Multispectral Steady-State Fluorescence Imagery

Moon S. Kim[1], Charles L. Mulchi[2], James E. McMurtrey[3], Craig S. T. Daughtry[3], and Emmett W. Chappelle[4]

[1]Instrumentation and Sensing Laboratory, USDA Agricultural Research Service, Beltsville, MD 20705, USA
[2]Department of Natural Resource Sciences, University of Maryland, College Park, MD 20742, USA
[3]Hydrology and Remote Sensing Laboratory, USDA Agricultural Research Service, Beltsville, MD 20705, USA
[4]Laboratory for Terrestrial Physics, NASA/GSFC, Greenbelt, MD 20771, USA

1. Introduction

Early recognition of environmental stress factors that may ultimately result in the loss of productivity is essential for economical assessments of agricultural and forestry practices (Moran et al. 1997). Vast resources and management approaches are implemented to increase plant productivity. Rapid noninvasive assessment methods that can detect deleterious environmental effects on crops during early stages of growth in a timely manner would be of great value. A relatively new active sensing technique available for vegetative monitoring is steady-state fluorescence.

Initially, the use of fluorescence measurements on higher plants (and algae) was confined to elucidating the fundamental mechanisms of photosynthesis. Kautsky and a co-worker in the early 1930s were the first to begin extensive investigations into transient characteristics of chlorophyll fluorescence from dark-adapted photosynthetic materials, which thus became known as the “Kautsky effect” (see

Air Pollution and Plant Biotechnology
–Prospects for Phytomonitoring and Phytoremediation–
Edited by K. Omasa, H. Saji, S. Youssefian, and N. Kondo
© Springer -Verlag Tokyo 2002

Omasa and Takayama, this volume). Studies have progressed to further enhance the elucidation of the fundamental photochemistry of the photosynthetic apparatus in higher plants (and photosynthetic bacteria) using chlorophyll (Chl) *a* fluorescence from either in vivo leaves, tissue homogenate, or isolated chloroplast.

In the mid-1980s, Chappelle and his colleagues (1984, 1985) observed a broad fluorescence emission in the blue-green region of the spectrum from intact plant leaves under UV-A (337 nm) excitation with an emission maximum centered at 440 nm (F440) and a shoulder peak at 525 nm (F525), respectively. They were the first to extensively use these additional fluorescence emission bands along with Chl fluorescence emissions at the red and far-red regions of the spectrum with emission maxima at 685 (F685) and 730 nm (F730). They suggested that such multispectral fluorescence emission bands were essential in providing the additional information required for relating fluorescence measurements to physiological status of plants. The magnitudes of the fluorescence bands have been found to change under several stress conditions or under growth environmental changes including those of both natural and anthropogenic origins; thus, these changes in spectral attributes may be utilized to assess physiological conditions of plants (Chappelle et al. 1984; Lang et al.1996; Lichtenthaler et al. 1993; McMurtrey et al. 1994; Kim et al. 1996). It has been also demonstrated that plant types may be differentiated by virtue of having distinct ranges of intensities at F440, F525, F685, and F730 bands (Chappelle et al. 1984, 1985). A detailed review of the vegetative steady-state fluorescence is provided in the next section.

2. Steady-State Fluorescence Characteristics of Vegetation

2.1 Red (F685) and Far-Red (F735) Fluorescence

Because of the wide use of fluorescence techniques for fundamental photosynthesis studies, a wealth of information on the characteristics of Chl fluorescence bands is available. In a multipigment system such as photosynthetic apparatus, resonance energy transfer is the first process that occurs after light absorption. Consequently, the energy transferred to photosystem (PS) II, but not utilized in photosynthesis, is dissipated as fluorescence at 685- and 735-nm regions of the spectrum or as heat (Lichtenthaler and Rinderle 1988). Therefore, in general, an inverse relationship exists between Chl fluorescence and photosynthesis provided that the Chl concentration remains constant.

Several studies have demonstrated that the ratio of the two steady-state Chl fluorescence peaks, namely F685/F735, can be used to assess relative Chl concentrations and photosynthetic efficiency (Lichtenthaler et al. 1993, 1996). These observations were based on two key premises. First, F685 undergoes partial reabsorption by itself because of a strong absorbance of Chl *a* in the red region of the spectrum (Hagg et al. 1992; Stober and Lichtenthaler 1993). Therefore, the F685/F735 ratio strongly depends on the Chl *a* concentration and decreases for plants with higher Chl *a* concentrations as reabsorption by Chl *a* at 685 nm becomes greater in comparison to that of 735 nm. Second, the F685/F735 ratio was also shown to be sensitive to the photosynthetic activity of plant leaves. In plants treated with DCMU, a higher F685/F735 ratio was observed with respect to that of control leaves (Lichtenthaler and Rinderle 1988). These observations indicated the dependency of the ratio to photosynthetic electron transfer efficiency from PS II to PS I. Environmental perturbations that affect the photosynthetic activity of plants may be detected by this ratio. However, Chl fluorescence changes occur when stress conditions damage the photosynthetic apparatus. This also occurs during temporary photoinhibition and serves as a protective or adaptive mechanism.

A study by Stober and Lichtenthaler (1993) showed decreased F685/F730 during the greening of etiolated leaves. D'Ambrosio et al. (1992) studied the F685/F730 ratio and Chl contents on nine tree species during the autumnal breakdown of Chl and showed increasing F685/F730 as leaves become senescent. They thus concluded that the F685/F735 ratio could be used as a remote indicator of Chl content. The data presented by these investigations showed that the ratio and Chl contents had an inverse exponential (curvilinear) relationship that reached a plateau when the total leaf Chl contents reached approximately 15 to 20 $\mu g/cm^2$. These observations imply that the ratio may be a valid indicator of Chl contents only to a certain concentration range. For instance, those conditions that resulted in significant variations in photosynthetic pigment contents on plants include nutrient deficiencies such as N (Heisel et al. 1996; McMurtrey et al. 1994).

In general, the Chl fluorescence ratio, F685/F730, increases synergistically in response to both plant stresses that exacerbate the photosynthetic mechanisms or result in decreased Chl concentrations (e.g., either breakdown or less synthesis). The occurrence of F685/F730 characteristics suggested the potential exploitation of fluorescence techniques in the detection of physiological changes in plants induced by a number of abiotic and biotic factors that affect plant physiology. However, it was recognized that Chl fluorescence measurement alone may limit the use in relating such measurements to specific stress factors responsible for physiological changes in plants. This limitation may be more evident with such stress conditions accompanied by insignificant effects on the photosynthetic activities or those stresses causing minimal changes (i.e., previsual) in photosynthetic pigment content.

2.2 Blue (F440) and Green (F525) Fluorescence

The use of blue-green fluorescence in vegetative monitoring is relatively new with respect to Chl fluorescence. Uncertainties exist in terms of the identity of major constituents and their precise location within the leaf, as well as other contributing factors or mechanisms governing blue-green fluorescence emission characteristics from intact leaves. It may be difficult to characterize factors responsible for blue-green fluorescence of in vivo plant leaves when considering the heterogeneity caused by complex morphology and the web of biochemical/biophysical processes involved. Several investigations have attempted to characterize the compounds and contributing factors responsible for blue-green fluorescence from intact leaves.

It has been known that a number of compounds present in plants, such as polyphenolics, are fluorescent under UV excitation. Harris and Hartley (1976), based on a UV fluorescence microscopy study, illustrated that cell walls were highly fluorescent and suggested cell wall-bound phenolics such as ferulic acids to be blue fluorescent compounds. Goulas et al. (1990) also suggested that free and esterified ferulic acids and *p*-coumaric acid as sources for blue fluorescence. Chappelle and Williams (1987) speculated that in vivo fluorescence in the blue band of the spectrum may emanate from NADPH, tannic acid, lignin, vitamin K_1, and plastoquinone. Moreover, flavins and β-carotene were suggested to be responsible for green fluorescence emission. Morales et al. (1994) also showed that flavins such as flavin adenine dinucleotide (FAD) and flavin mononucleotide (FMN) were responsible for in vivo green band emission.

To further assist in the identification of the compounds responsible for blue-green fluorescence emissions, fluorescence spectra of pure plant constituents as well as solvent extract were acquired (e.g., Chappelle et al. 1991; Lang et al. 1991). Lang et al. (1991) suggested cell wall-bound phenolics and compounds in the vacuoles such as chlorogenic acid, caffeic acid, sinapic acid, coumarins (aesculetin, scopoletin), and catechin, and stilbenes (*t*-stilbene, rhaponticin) as possible candidates for blue fluorescence. They also suggested alkaloid berberine and quercetin as compounds responsible for green fluorescence and stated that riboflavin may contribute little to green fluorescence.

Plant cell structural compounds such as lignin and phenolics are relatively inert chemically. The changes in the magnitude of blue-green fluorescence in response to certain environmental stress conditions indicate that the fluorescence changes may be caused by chemically dynamic (e.g., redox changes due to stress condition) molecules that are involved in photosynthesis. Chappelle et al. (1991) showed that the blue fluorescent nature of water-soluble extract from clover leaves was identified as due to NADPH that plays a vital role in photosynthesis as one of the primary intermediates during the photosynthetic process, accepting electrons from PS I. They further showed that the fluorescence ratio F440/F600 of greenhouse-grown soybean leaves showed a strong positive linear relationship with the rate of photosynthesis. Cerovic et al. (1993) showed that approximately 20% of the mesophyll blue fluorescence emanated from components located in the chloroplast of spinach (*Spinacia oleracea* L. var. Wobli). In addition, a light-induced change in blue fluorescence in the chloroplast and mesophyll cells was observed. This fluorescence change was suggested to be due to the reduction of NADP in PS I. The foregoing observations suggested that blue fluorescence emission changes may be more closely related to photosynthesis than was previously thought. However, Stober and Lichtenthaler (1993) reported that no light-induced changes in blue-green fluorescence were observed during the induction kinetics of dark-adapted soybean and wheat leaves.

Although these compounds are considered to be responsible for the blue-green fluorescence emission from intact leaves, photosynthetic pigments in the mesophyll layers, which absorb strongly in these regions of the spectrum, may selectively reabsorb the blue-green fluorescence emitted by other compounds (Lang et al. 1992; Stober and Lichtenthaler 1992); this was based on observing a decrease in the blue-green fluorescence emissions during the greening of etiolated wheat leaves. Stober and Lichtenthaler (1992) reported an increasing blue to green ratio (F440/F530) as etiolated leaves turned green. They stated that blue fluorescence was reabsorbed by all the photosynthetic pigments and that the green band was only partially reabsorbed by the accessory pigments. Morales et al. (1994) observed increased blue-green fluorescence in mesophylls of iron deficient sugar beet (*Beta vulgaris* L.) leaves. The iron deficiency treatments decreased photosynthetic pigments per unit area, and the authors suggested that most of the blue-green fluorescence increases were caused by a reduction of the screening of UV light by photosynthetic pigments. In addition, selective reabsorption of blue fluorescence by Chl molecules led to changes in the shape of the emission spectra.

Lang et al. (1991) observed significantly higher fluorescence intensities from the upper sides in comparison to the lower sides of bifacial C3 soybean leaves throughout the blue, green, red, and far-red regions of the spectrum. The contrast was speculated to be caused by anatomic differences that resulted in differential reabsorption by photosynthetic pigments. A subsequent study by Lang et al. (1992) showed significantly higher green fluorescence intensities from epidermal-stripped tobacco leaves than those of intact leaves, and thus they postulated that green fluorescence emanated solely from the mesophyll layers. In contrast, because no significant blue fluorescence changes were observed between the intact leaves and leaves stripped of the epidermal layers, blue fluorescence was suggested to emanate from the cell wall in both the epidermal and the mesophyll layers. Moreover, because of the strong absorption and reabsorption by photosynthetic pigments in the blue region of the spectrum in the mesophyll layer, the majority of the blue fluorescence has been thought to emanate from the epidermal layers of intact green leaves. For instance, Stober et al. (1994) showed fourfold higher blue fluorescence from wheat leaves treated with a bleaching herbicide norfluorazone (removal of photosynthetic pigments) compared to normal green wheat leaves. On the basis of this observation, they suggested that the major part of the blue fluorescence emanated from the cell walls of the epidermal layer of intact plants.

Fluorescence changes due to physiological perturbations in a complex matrix such as a plant leaf may depend on several diverse biochemical and biophysical mechanisms. In general, the fluorescence magnitude changes in the blue-green regions of the spectrum on intact leaves followed by exposures to stress conditions have been mainly associated with changes in concentrations of fluorophores or photosynthetic pigment contents. These variation may also depend on a number of additional factors such as viscosity of the cytosolic solution, hydration state of membrane, membrane integrity, pH, temperature, oxidation and reduction states of the fluorophores, and membrane topology (Chappelle et al. 1991; Kim et al. 1997). Improved elucidation of mechanisms and factors governing blue-green fluorescence changes should enhance the usefulness of fluorescence techniques for noninvasive detection of the physiological state of plants.

3. Multispectral Steady-State Fluorescence Techniques

Advances in low-light imaging technology have provided opportunities to capture fluorescence images from various materials. Applications of imaging technology in vegetative fluorescence studies have been reported in several recent studies (Cecchi et al. 1998; Daughtry et al. 1997; Heisel et al. 1996; Kim et al. 1996, 1997, 1998, 2001; Lang et al. 1996; Lichtenthaler et al. 1996; Omasa et al. 1987, Omasa 1998). Fluorescence emissions from plants have been shown highly variable because of morphological and physiological variations that can be amplified by environmental stress effects. The imaging technique provides information on the spatial variability of fluorescence patterns across the samples that cannot be readily obtained from nonimaging systems that only provide an integrated value of an area. In the following sections, multispectral fluorescence images of plant leaves acquired with a newly developed multispectral fluorescence imaging system are presented. Previous studies are detailed below to show the versatility of multispectral fluorescence imaging techniques for assessment of environmental plant stresses. The experiments include soybeans exposed (long term) to moderately elevated O_3 and CO_2, and soybean isolines containing varying concentration of flavonoids, UV-B protective pigments.

3.1 Effects of Air Pollutants and Varying UV-B Absorbing Compound on Soybean Plant by Fluorescence Imagery

The levels of both tropospheric ozone (O_3) and carbon dioxide (CO_2) have increased since the turn of the last century. Increases in O_3 and CO_2 are consequences of both natural and anthropogenic pollution. Tropospheric O_3 is a secondary pollutant formed through a series of photochemical reactions from nitrogen oxides (NO_x) and hydrocarbons released during the combustion process (Logan 1985). Steady increases in atmospheric CO_2 have been attributed to the burning of fossil fuels and deforestation. The levels of both tropospheric O_3 and CO_2 in the next few decades are projected to increase at rates faster than those during the past 100 years (Krupa and Kickert 1989; Hough and Derwent 1990).

Plants possess natural cellular defense mechanisms to tolerate the presence of low to moderate concentrations of O_3 (Logan 1985). When the operational tolerance capacity of plants is exceeded because of chronic O_3 exposure, this has many adverse effects on vegetation. Deleterious effects on physiological responses of plants as well as significant reductions in the overall growth and productivity have been commonly observed (Krupa and Kicker 1989; Mulchi et al. 1992, 1995; Rudorff et al. 1996). In view of the likelihood of atmospheric CO_2 increases in the near future, an overwhelming number of studies have documented the beneficial effects of a CO_2-enriched environment on physiological processes and productivity of plants (Kimball and Idso 1983; Krupa and Kickert 1989; Mulchi et al. 1992, 1995; Rudorff et al. 1996). In general, C3 plants grown in elevated CO_2 environments exhibited increased rates of photosynthesis, decreased transpiration rates due to decreased stomatal conductance, and increased biomass and yield. Moreover, CO_2-enriched environments mitigate the presence of perturbing environmental factors that exacerbate physiological responses of plants such as concomitant exposure to elevated tropospheric O_3 (Mulchi et al. 1992, 1995; Rudorff et al. 1996) and drought stress (Allen et al. 1998).

Flavonoids, produced almost exclusively by higher plants and accumulating in all plant parts, constitute one of the largest classes of plant phenolics. Various chemically heterogenous plant flavonoids are synthesized from phenylalanine produced in the shikimic acid pathway. The basic molecular structure is based on the skeleton of two aromatic benzene rings joined by a three-carbon bridge. The oxidation state of the three-carbon bridge is primarily responsible for the different classification of flavonoids. They exist as glycosides in vivo and are found in cell sap or vacuoles. In soybean and other legumes, accumulation of leaf flavonols is usually restricted to the epidermis (Cosio and McClure, 1984). Flavones and flavonols have been suggested to protect plants cells from excessive UV radiation. Elevated levels of flavonoid synthesis associated with enhanced UV-B radiation have been well documented.

We have conducted air quality field experiments for several years to study the effects of chronic exposure to combinations of altered levels of tropospheric CO_2 and O_3 on crop species. Our primary objective was to characterize fluorescence responses of soybean leaves grown under chronically elevated CO_2, elevated O_3, and both CO_2 and O_3 elevated environments with the use of a multispectral fluorescence imaging system (FIS). In addition, fluorescence images of four soybean plant cultivars with different concentrations of flavonols (kaempferol glycosides) were evaluated to study relationships between varying concentrations of flavonol glycosides with fluorescence emission intensities. This investigation utilized genetically mutated flavonol isolines grown in a constant environment, thus limiting factors known to affect fluorescence emission characteristics other than flavonol concentrations.

3.2 Multispectral Fluorescence Imaging System

A laboratory-based multispectral FIS developed in our laboratories was used to acquire steady state fluorescence images in the blue (F450), green (F550), red (F680), and far-red (F740) regions of the spectrum. The major components of FIS include a continuous wave (CW) excitation light source; nonfluorescent sample

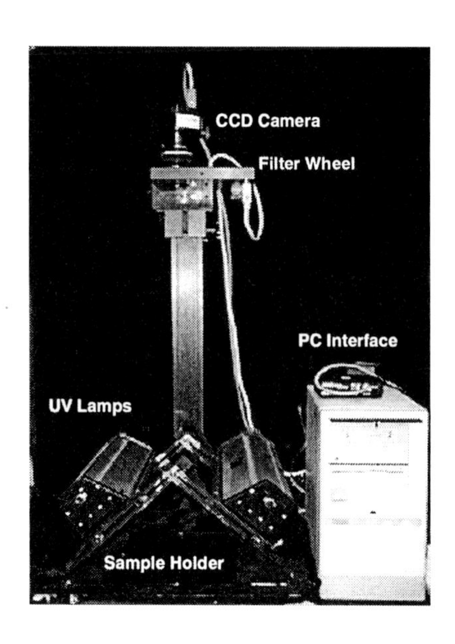

Fig. 1. Laboratory-based multispectral fluorescence imaging system (FIS)

holder; interference filters; a digital camera; and a computer interface for instrument control and data collection (Fig. 1).

Two 15-w longwave UV-A fluorescent lamps (XX-15A; Spectronics, NY, USA), which provided near uniform illumination, were used as the excitation source. The lamps were arranged at an angle, one on each side toward a central target area 0.2 m above the sample surface, to provide nearly uniform illumination. The radiation from the UV lamps was filtered with Schott UG-1 glass to block the transmission of radiation greater than 400 nm; this prevented any reflected excitation light from being detected at any of the fluorescence bands. The UV excitation intensity at the target area was 0.33 mW/cm^2 with emission maximum at 360 nm. Although the UG-1 filter transmits in the far-red region (700 nm $< \lambda <$ 730 nm) of the spectrum, the UV lamps did not emit radiation in this wavelength region.

A front-illuminated, thermoelectrically cooled charge-coupled device (CCD) camera (Lynxx-2; Spectra Source Instruments, Westlake Village, CA, USA), capable of capturing a spatial resolution of 196 H 165 (31,680 pixels) was used. It has a low noise range (quantum efficiency, > 20%) from 400 to 800 nm with a maximal quantum efficiency of 55% at the 650- to 750-nm region. Noise from the CCD is strongly dependent on temperature, and thus it was cooled via Peltier devices. A Nikon lens (*f*-1/3.5, 20 mm) mounted to the CCD camera head was coupled to an AB300 automated filter wheel (CVI Laser, NM, USA) that held up to five circular filters (5-cm diameter). Four bandpass interference filters were placed in the filter wheel holders.

Figure 2 illustrates the transmittance characteristics of the interference filters where the blue filter has a maximum transmittance at 450 nm with 25-nm FWHM, green at 550 nm with 25-nm FWHM, red at 680 nm with 10-nm FWHM, and far-red at 740 nm with 10-nm FWHM. For the blue-green region of the spectrum, broader bandpass filters were used to increase the sensitivity of the imaging system because the quantum efficiency (QE) of the CCD camera in this region was not as high as that of the red far-red region of the spectrum. The QE are approximately 25% and 55% for blue/green and red/far-red regions, respectively.

The CCD camera exposure time intervals span from 0.01 to 4000 s and are controlled by the data acquisition software provided by the manufacturer. Fluorescence images of a target area, 12 H 12 cm (and adjustable) were captured via a 12-bit PC analog to digital (A/D) board and stored to image files. When acquiring fluorescence images from plant leaves, a range of exposure time intervals, approximately 20 to 30 s, was used to take advantage of the 12-bit range of the A/D data conversion.

With the use of a flat field fluorescent target, the system was calibrated for heterogeneity in CCD responses for individual pixels and heterogeneity in excitation intensity on the target area. Although the CW UV lamps provided near-uniform illumination on the target area, this correction ensured better pixel-by-pixel comparisons within the image. Also, dark current responses of the CCD pixels were subtracted from each image. The CCD camera and the filter wheel

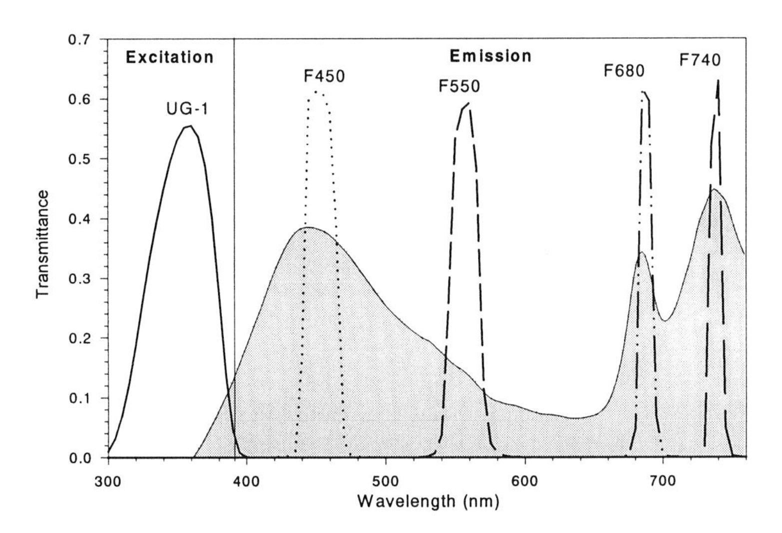

Fig. 2. Transmittance characteristics of filters used in the FIS: Schott glass UG-1 (*solid*) bandpass filter and F450 blue (*dotted*), F550 green (*short dashed*), F680 red (*dashed-dot-dot*), and F740 far-red (*long dashed*) interference filters. Transmittance characteristics of filters are shown overlaid on top of a typical fluorescence emission spectrum of a soybean leaf with UV-A excitation (*shaded area*)

were operated and controlled by the PC. The images were processed and analyzed with software developed in-house by which ratios of multiband images, gray scale stretch, and descriptive statistics on a selected polygonal area on the leaves were typically performed.

3.3 Air Quality Treatments

The field Open-top chamber (OTC) system developed in the early 1970s (Heagle et al. 1973) provides various controlled mixtures of gaseous environments for plant growth. For this investigation, we used field OTC systems to simulate four altered combinations of tropospheric CO_2 and O_3 environments. The experimental field site is located at the South Farm, USDA/BARC, Beltsville, MD, USA. The simulated gaseous environments included (1) charcoal-filtered (CF) ambient air as the control chamber; (2) elevated CO_2 that consisted of CF ambient air plus 150 ± 10 μl/l CO_2; (3) elevated O_3 that consisted of nonfiltered (NF) ambient air and addition of 35 ± 5 nl/l O_3; and (4) high CO_2 and O_3 chamber that consisted of NF ambient and additions of 150 ± 10 μl/l CO_2 and 35 ± 5 nl/l O_3. These treatments are designated as CF, CF+CO_2, NF+O_3, and NF+CO_2+O_3, respectively. Monitored air qualities for 1997 and 1998 growing seasons are summarized in Table 1.

The daily maximum O_3 concentrations for those elevated O_3 treated chambers were monitored to prevent the chamber air levels from exceeding the then current U.S. secondary standard for O_3 (120 nl/l). The elevated CO_2 concentration selected has been projected from the current rate of increase to occur by the middle of the this century (Krupa and Kickert 1989).

Two soybean [*Glycine max* (L). Merr] cultivars, ‘Forrest’ and ‘Essex’, were germinated and grown in trays containing a peat-vermiculite mix. After 2 weeks, seedlings were selected for uniformity and the most uniform plants were transplanted into the open-top chambers in rows 0.6 m apart with plants spaced 10 cm apart. The air quality treatments were applied immediately after the transplanting until harvest. Soil moisture contents were monitored with soil moisture probes to maintain moisture levels at 0 to -0.05 mPa to prevent drought stress. Plants were irrigated as needed.

The treatments were arranged using a randomized complete block design with four replicates of four air quality treatments in well-watered plots. Thus, a total of 16 OTCs (arranged in 4 rows H 4 columns) were used for these studies. Two cultivars were planted within each chamber using one-half of the chambers for each cultivar. The air quality treatments were randomly assigned to the chambers within a block.

Plant leaves from the uppermost fully expanded trifolia were used. Steady-state multispectral fluorescence images were acquired on excised leaves within 30 min of the excision, where stems well below the trifolia were cut submerged in water to maintain the water column. None of the sample leaves displayed any visual stress symptoms (e.g., discoloration, necrosis, or chlorosis). The main interests in this

Table 1. Observed 7 h/day (9:00 AM - 4:00 PM, EST) monthly and seasonal mean values for ambient and open-top chamber CO_2 and O_3 concentrations (1997 and 1998 averaged)

	Air quality treatment	July[a]	August	September	October	**Seasonal**
CO_2 (:l/l)	Ambient	342.7	344.1	352.3	362.4	**348.5**
	CF	341.7	338.7	345.6	358.7	**340.2**
	CF + CO_2	457.7	480.3	494.1	513.1	**476.4**
	NF + CO_2 + O_3	460.3	480.9	489.8	513.3	**475.7**
	NF + O_3	342.8	339.8	346.8	359.5	**341.2**
O_3 (nl/l)	Ambient	51.4	46.2	35.4	22.3	**40.3**
	CF	27.1	23.1	20.2	16.4	**22.0**
	CF + CO_2	24.9	21.0	18.0	15.0	**20.0**
	NF + CO_2 + O_3	75.8	72.8	73.2	59.1	**71.1**
	NF + O_3	75.8	73.6	71.8	58.7	**70.7**

[a]Monthly mean values were calculated from July 15 through July 31.

chapter are the fluorescence responses and chronic air pollutant exposure effects; hence, fluorescence measurements made during the pod-filling period are discussed.

3.4 Soybean Flavonol Isolines

For the investigation of fluorescence characteristics of plants having different concentrations of flavonols (kaempferol glycosides), leaves from soybean, including four F6-derived kaempferol glycoside isolines, OX922, OX941, OX942, and OX944, were used. These isolines differ mainly in concentrations of kaempferol glycosides in that OX922 and OX941 contain K3 through K6 and K9 while these are absent in OX942 and OX944. Isolines OX922 and OX941 contain anthocyanidins resulting in purple flowers while OX942 and OX944 have white flowers due to the absence of anthocyanidins.

Plants were grown in a greenhouse with a day temperature range of 18° - 30°C, a night temperature range of 15° - 20°C, and relative humidity of 50% - 90%. The maximum daytime photosynthetically active radiation (PAR) intensities were in the range of 800 - 1000μmol m^{-2} s^{-1}, measured at plant canopy height at midday.

The growth medium was a fertilized Jiffy mix, a commercial peat-vermiculite mixture used routinely to grow healthy plants for subsequent stress tests. Seeds were germinated and plant seedlings were grown in a tray containing peat-vermiculite mix. After 7 days, seedlings were selected for uniformity and transplanted to 15-cm pots (five plants per pot and four pots per isoline), also containing peat-vermiculite mix. The stand was thinned to two plants per pot 1 week after transplanting and grown for 2 more weeks. Sample leaves were selected from mid- to uppermost fully expanded trifolia before flowering. The experiment was conducted using a randomized complete block design.

4. Effects of Moderately Elevated O_3 and CO_2

Fluorescence images of soybean leaves exposed to various levels of O_3 and CO_2 environments were acquired during pod-filling period. Gray-scale fluorescence

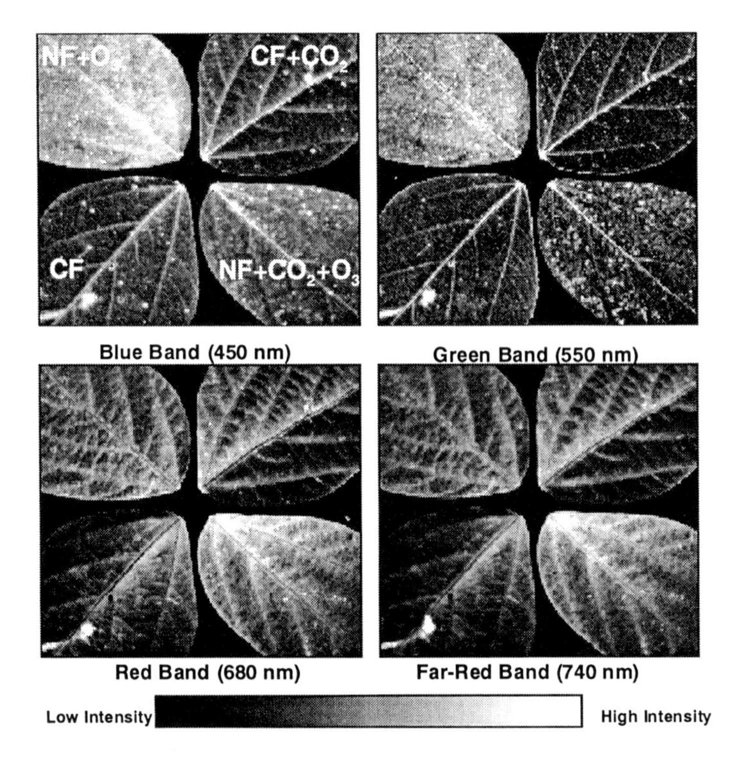

Fig. 3. Gray-scale fluorescence images of soybean (cultivar 'Essex') leaves grown in open-top chambers with four different air quality treatments; charcoal filtered (*CF*) control: elevated O_3 ($NF+O_3$); elevated CO_2 ($CF+CO_2$), and concomitantly elevated CO_2 and O_3 ($NF+CO_2+O_3$) environments

Table 2. Mean fluorescence intensities[a] acquired with fluorescence imaging system of soybean cultivars 'Forrest' and 'Essex' during pod-filling period (1997 and 1998)

	Air quality treatment	F450	F550	F680	F740
Essex[b]	CF	57.9 ab[d]	44.3 ab	64.2 a	89.2 b
	CF + CO_2	54.0 a	43.5 a	57.4 a	78.5 a
	NF + O_3	65.0 c	47.9 c	65.2 a	86.4 b
	NF + CO_2 + O_3	61.1 bc	47.0 bc	63.5 a	87.1 b
Forrest[c]	CF	56.0 b	39.3 a	77.3 a	104.4 b
	CF + CO_2	50.9 a	38.2 a	69.6 a	90.3 a
	NF + O_3	63.2 c	46.6 b	77.5 a	96.1 ab
	NF + CO_2 + O_3	55.7 b	40.8 a	74.1 a	98.6 ab

[a]Fluorescence means for each treatment were based on 32 leaf images as a result of combining 2-day measurements, 16 samples from each day.
[b]Ozone-tolerant cultivar.
[c]Ozone-sensitive cultivar.
[d]For each soybean cultivar, within-column means with same letter are not significantly different at ∀ = 0.05 according to pairwise contrasts.

images of soybean leaves (cv. 'Forrest') acquired with the FIS at F450, F550, F680, and F740 bands are shown in Fig. 3. Spatial variations of fluorescence emissions on the major portions of leaves that were difficult to characterize using integrated point-source measurements (e.g., spectrofluorometer) are readily visible. Digital values were stretched to span the whole gray scale to maximize the contrast. The darker areas represent the lowest fluorescence intensities and the white colors represent the highest intensities.

Treatment mean values for these bands are presented in Table 2. Each treatment mean value is an average of 32 samples (leaf means) in which individual leaf mean was calculated from the entire leaf portion captured with the imaging system (approximately 5000 pixels); however, the brightest-large white spots in the lower left corner of CF leaf were caused by physical damage and were excluded when the mean value were calculated for that leaf sample.

The ANOVA result in Table 2 shows that the most significant differences are observed at F450, followed by F550. In the F450 and F550 bands, NF+O_3 treatment means were significantly higher than those of other treatments. There were no significant differences between CF and NF+CO_2+O_3 treatments within cultivars. Means for CF+CO_2 was significantly lower at F450 for both cultivars compared to CF treatments. None of the treatment means was significantly different at F680. In the F740 band, CF+CO_2 treatments for both cultivars had significantly lower fluorescence intensities.

Fluorescence emission from the major vascular bundles (veins) was, in general, higher than those in the interveinal areas (see Fig. 3). This result can be attributed to the combined effects of stronger reabsorption of blue-green fluorescence by the photosynthetic pigments more abundant in the interveinal areas and higher lignin concentrations in the veins (Lichtenthaler et al. 1996). The most pronounced differences among the treatments are noticed in F450 and F550 images of Essex leaves shown in Fig. 3. Similar differences were observed for Forrest. The soybeans leaves grown under NF+O_3 had more heterogeneous variations in fluorescence intensities across the leaves, which are seen as localized irregular white spots in Fig. 3. This effect is more clearly illustrated in Fig. 4, where pixel intensities along the dotted lines across the leaves for CF, NF+O_3, and NF+CO_2+O_3 treatments for Forrest leaves at F450 are shown on the graphs. Fluorescence intensities in the interveinal regions across the leaves from CF and CF+CO_2 treatments (not seen in Fig. 4) are more uniform than those grown under NF+O_3 and NF+CO_2+O_3 environments. Plants exposed to high O_3 in the presence

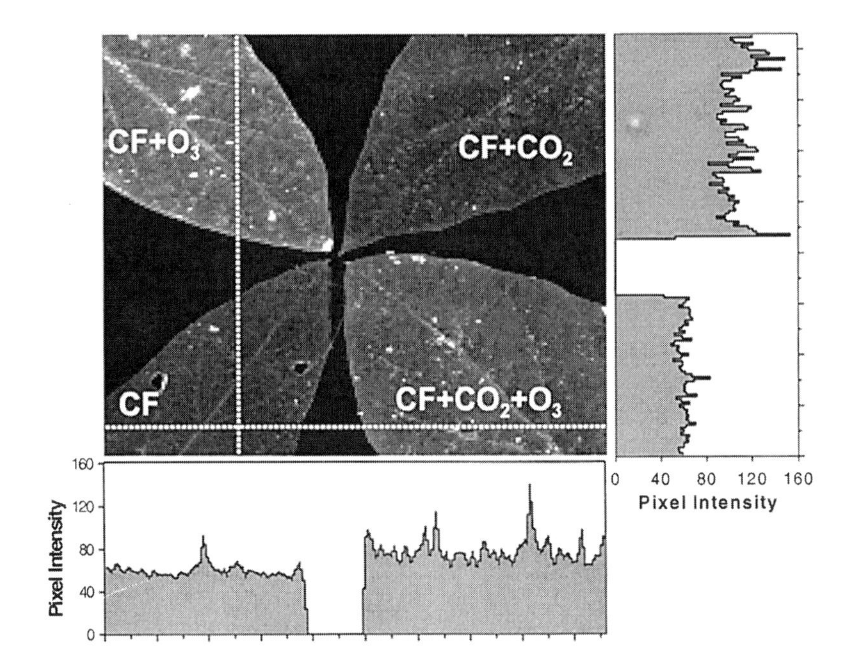

Fig. 4. Blue (F450) fluorescence image of soybean leaves (Forrest) grown under four combinations of air quality treatments. Pixel intensities of cross sections (*dotted lines*) illustrate spatial variations across leaf surface among the treatments

of elevated CO_2 generally showed a heterogeneous fluorescence response, although this was less apparent than those found in NF+O_3 treatment. It should be pointed out that not all the leaf samples grown in the elevated O_3 environments, especially those grown under NF+CO_2+O_3 treatment, displayed the heterogeneous fluorescence characteristics across the leaves seen in Fig. 4.

The major veins shown in the F680 and F740 images in Fig. 3 appear as dark strips indicating relatively low fluorescence intensity compared to the adjacent areas. The lack of chloroplasts in the veins would yield lower chlorophyll fluorescence. In contrast, fluorescence emissions in the areas adjacent to the veins are, in general, slightly higher than in the interveinal regions. An earlier study with the use of FIS at these chlorophyll *a* fluorescence bands showed highly irregular patterns of fluorescence magnitude variations in the interveinal regions across the soybean leaves when the photosynthetic apparatus attained damages (Kim et al. 1996, 2001). The results seen in the F680 and F740 band images in this study do not appear to exhibit the manifestation of damage affecting the photosynthetic apparatus.

Although the means were not significantly different among the air quality treatments at F680, the mean trends were similar to those at F450 and F550 for both Essex and Forrest. The fluorescence intensities at F680 and F740 between Forrest and Essex were notably different. Forrest had higher fluorescence emission than Essex. Mean intensities for the CF+CO_2 treatment within the cultivar were also markedly lower. If chlorophyll *a* concentrations were not the limiting factor, the lowest fluorescence intensity at F680 would indicate more efficient or active photochemistry in the photosynthetic apparatus.

Many factors including hydration status of cell walls in membrane layers can affect the blue-green fluorescence emission. Higher fluorescence emissions have been observed as the epidermal membrane becomes dehydrated (Kim et al. 1997). Damage resulting from exceeding the capacity of plants to repair from oxidative stress (i.e., chronic exposures to elevated O_3) occurs on the plasma or cell membranes and causes fluxes of various components across membranes such as amino acids, ions, and water (Logan 1985). Hence, membrane degradation caused by the chronic exposure to moderately elevated O_3 may have contributed to localized fluorescence increases at F450 and F550. These changes were more evident in the images of sample leaves grown in NF+ O_3 environment and somewhat less extensively in NF+CO_2+O_3 environment. Ultimately, accumulative effects of membrane degradation were manifested as early senescence, commonly observed in plants grown under elevated levels of O_3. Elevated CO_2 as a single agent enhances plant water relation due to partial stomatal closure (Rudorff et al. 1996). Significantly lower F450 for CF+CO_2 treated leaves may stem from the improved water relationship in which cell walls and air spaces in mesophyll layers would exist in well-hydrated states compared to leaves grown in the CF ambient environment.

Photosynthesis is affected when the physiological and biochemical processes of plants are disturbed. Thus, when plants are subjected to those stress conditions that directly exacerbate the photosynthetic processes, these are manifested as increases in the chlorophyll fluorescence emissions at F685 and F730. Conversely, lower chlorophyll fluorescence emissions at F680 (and F740), e.g., leaves grown under CF+CO_2 environment, indicated more photosynthetic efficiency and activity. Environment conditions that affect only certain biochemical and biophysical reactions and do not initially impact the integrity of photosynthetic apparatus may be observed as changes in the blue-green region of the spectrum. Moderately elevated O_3 environment may possibly be one of those conditions.

A previous physiology study conducted on soybean plants using the same experiment setup showed that negative physiological effects of elevated O_3 were partially ameliorated by a CO_2-enriched environment (Mulchi et al. 1992). Mean intensities at F450 and F550 for NF+O_3 were significantly different when compared to those under CF conditions for both cultivars. However, the mean fluorescence responses at these bands for soybean plant leaves grown under NF+CO_2+O_3 environment were not significantly affected. This observation indicated that factors responsible for the increases in the F450 and F550 emissions were mitigated by the presence of elevated CO_2. These findings are consistent with the observation described by Mulchi et al. (1992, 1995) and further demonstrate the versatility of fluorescence as a noninvasive method for the assessment of physiological responses of plants.

5. Effects of Varying Content of Flavonols

Gray-scale fluorescence images of soybean leaves from isolines OX922, OX941, OX942, and OX944 (Fig. 5A) were acquired with the FIS at the F450, F550, F680, and F740 bands. Digital values were also stretched to optimize the contrast. A total of 64 leaf images, 16 for each flavonol isoline, was acquired.

Each mean value is an average of 16 leaf samples (leaf mean values) where each leaf value was calculated from the entire portion of the leaf that was captured with the FIS (approximately 5000 pixels values). The histograms of the pixel values versus the frequency of the pixels for F450 and F680 are presented in Fig. 5B. Distinct distributions of the fluorescence emissions from the soybeans depicted as two peaks clearly reflect the mean differences. Each peak within the histogram is a convoluted distribution of fluorescence emissions of two soybean flavonol isolines where the lower and the higher distribution are due to OX921 and OX922 and to OX942 and OX944, respectively. Mean relative fluorescence intensity values are presented in Table 3. Fluorescence emissions in all four bands for the isolines containing kaempferol glycosides OX922 and OX941 were significantly lower than OX942 and OX944.

Fluorescence intensities of the major vascular bundles (veins) in the F450 and F550 were higher than those in the interveinal areas. Photosynthetic pigment concentrations (data not shown) between these isolines were not significantly

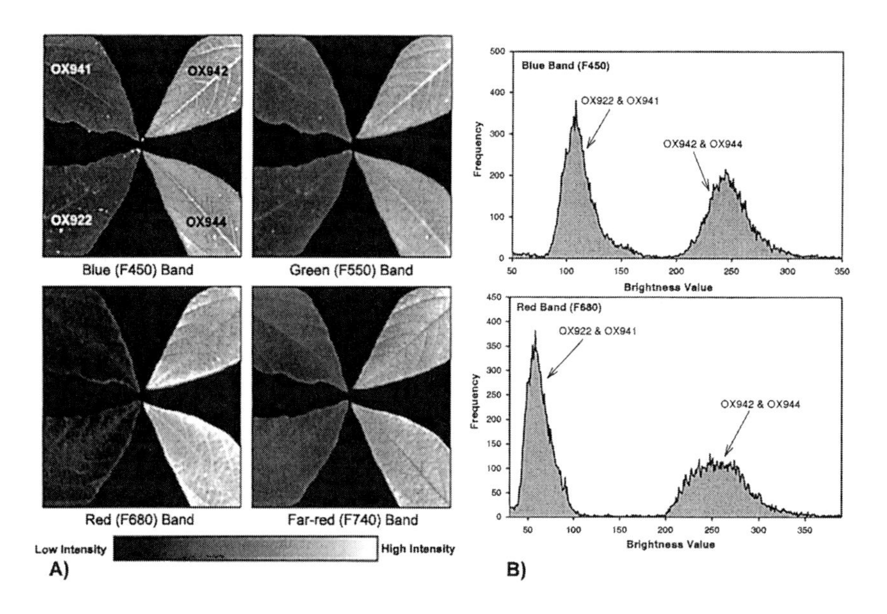

Fig. 5. **A** Gray-scale fluorescence images of the soybean flavonol isolines. **B** Histograms of the blue and red fluorescence images of the soybean isolines. The vertical axis is the frequency of the pixels with each brightness value

different (except for OX941, which had slightly lower concentrations). In addition, the intensities of the red chlorophyll band emissions were higher for the soybeans having higher rates of photosynthesis. These results imply that fluorescence emission intensities in all fluorescence emission bands are mainly affected by the presence and absence of kaempferol glycosides. Furthermore, these results suggest that the excitation energy absorbed by flavonols is not utilized in fluorescence excitation of other compounds responsible for fluorescence emissions in the visible and near infrared regions of the solar spectrum. Several studies also indicated that UV screening and nonblue fluorescent compounds such as flavonols in the epidermis may attenuate the fraction of excitation beam reaching the mesophyll layer of the intact plants (Stober and Lictenthaler 1993; Lang et al. 1996). Cerovic et al. (1993) indicated that up to 75% of the UV excitation light may be absorbed by the epidermis before reaching the mesophyll layer, based on a comparison of F680 intensities between intact leaves and leaves with the epidermal layers removed.

Conventionally, fluorescence changes in the blue-green portions of the spectrum have been mainly associated proportionally with changes in concentrations of fluorescing compounds. Above studies also direct our attention to factors that may be associated with in vivo fluorescence emissions of green plants. The anatomic

Table 3. Mean fluorescence intensities[a] of soybean flavonol isolines acquired with fluorescence imaging system

Isolines	F450	F550	F680	F740
OX922	134.4 **a**[b]	77.2 **a**	83.1 **a**	106.6 **a**
OX941	133.8 **a**	80.6 **a**	83.5 **a**	104.5 **a**
OX942	217.5 **b**	120.2 **b**	250.9 **b**	256.3 **b**
OX944	218.6 **b**	125.2 **b**	243.6 **b**	248.8 **b**

[a]Fluorescence means for each isoline were based on 16 leaf images.
[b]Within-column means with same letter are not significantly different at ∀ = 0.05 according to pairwise contrasts.

characteristics of a plant leaf are highly specialized for light energy absorption. Chloroplast rearrangement within the mesophyll cells can change the amount of light absorbed by the leaf. In addition, variations in concentrations of UV screening pigments can significantly reduce in vivo fluorescence emissions. Thus, prior light conditions may affect fluorescence emission characteristics.

6. Concluding Remarks

A multispectral fluorescence imaging system using a stable UV (360 nm) excitation light source is a viable technique to capture fluorescence images of leaves in the blue, green, red, and far-red regions of the spectrum centered at 450, 550, 680, and 740 nm, respectively. Several experiments were presented to demonstrate the versatility of the FIS. Fluorescence changes resulting from biochemical and physiological changes in plants were qualitatively and quantitatively characterized with the added benefit of allowing spatial characterizations to be made. Imaging a major portion of leaf with a sensitive spectral technique such as fluorescence may provide better means to assess the effects of stress conditions, especially those previsual conditions.

Studies of in vivo plant fluorescence consider interactions of complex biochemistry and physiology affecting the emission characteristics of fluorescence. Environmental factors associated with plant stresses, such as elevated ambient O_3 and contents of UV absorbing compounds, significantly altered the fluorescence emission characteristics of soybean leaves. Fluorescence emission characteristics of plant leaves may also depend on other variables such as viscosity of the solvent (fluidity of membrane), temperature, oxidation and reduction states of compounds, and membrane topology (plant anatomy). Further research is needed to enhance the elucidation of the mechanisms and factors contributing to intact plant fluorescence changes in response to environmental perturbations.

References

Allen LH Jr, Valle RR, Jones JW, et al. (1998) Soybean leaf water potential responses to carbon dioxide and drought. Agron J 90:375 - 383

Cecchi G, Kim MS, Bazzani M, et al. (1998) Fluorescence responses of Mediterranean sea grass *Posidonia oceanica*: Summer 1997 ATOM - LIFT Campaign. SPIE 3382:126 - 132

Cerovic ZG. Bergher M, Goulas Y, et al. (1993) Simultaneous measurement of changes in red and blue fluorescence in illuminated isolated chloroplasts and leaf pieces: the contribution of NADPH to the blue fluorescence signal. Photosynth Res 36:193 - 204

Chappelle EW, Williams DL (1987) Laser induced fluorescence (LIF) from plant foliage. IEEE Trans Geosci Remote Sens GE-25:726 - 736

Chappelle EW, Wood FM, McMurtrey JE, et al. (1984) Laser induced fluorescence of green plants: 1. A technique for the remote detection of plant stress and species differentiation. Appl Opt 23:134 - 138

Chappelle EW, Wood FM, Newcomb WW, et al. (1985) Laser induced fluorescence of green plants: 3. LIF spectral signatures of five major plant types. Appl Opt 24: 74 - 80

Chappelle EW, McMurtrey JE, Kim MS (1991) Identification of the pigment responsible for the blue fluorescence band in laser induced fluorescence (LIF) spectra of green plants, and the potential use of this band in remotely estimating rates of photosynthesis. Remote Sens Environ 36:213 - 218

Cosio EG, McClure JW (1984) Kaempferol glycosides and enzymes of flavonol biosynthesis in leaves of a soybean strain with low photosynthetic rates. Plant Physiol 74:877 - 881

D'Ambrosio N, Szabo K, Lichtenthaler HK (1992) Increase of the chlorophyll fluorescence ratio F690/F735 during the autumnal chlorophyll breakdown. Radia Environ Biophys 31:51 - 62

Daughtry CST, McMurtrey JE, Kim MS, et al. (1997) Estimating crop residue cover by blue fluorescence imaging. Remote Sens Environ 60:14 - 21

Goulas YI, Moya I, Schmuck G (1990) Time resolved spectroscopy of the blue fluorescence of spinach leaves. Photosynth Res 25:299 - 307

Hagg C, Stober F, Lichtenthaler HK (1992) Pigment content, chlorophyll fluorescence and photosynthetic activity of spruce clones under normal and limited nutrition. Photosynthetica 27(3):385 - 400

Harris PJ, Hartley RD (1976) Detection of bound ferulic acid in cell walls of the graminae by ultraviolet fluorescence microscopy. Nature 259:508 - 510

Heagle AS, Body DE, Heck WW (1973) An open-top field chamber to assess the impact of air pollution on plants. J Environ Qual 2(3):365 - 368

Heisel F, Sowinska M, Miehe JA, et al. (1996) Detection of nutrient deficiencies of maize by laser induced fluorescence imaging J Plant Physiol 148:622 - 631

Hough A, Derwent RG (1990) Changes in the global concentration of tropospheric ozone

due to human activities Nature 344:645 - 648
Kim MS, Krizek D, Chappelle EW, et al. (1996), Fluorescence imaging system (FIS): assessment of vegetation stress. SPIE 2959:4 - 13
Kim MS, Mulchi CL, Daughtry CST, et al. (1997) Fluorescence images of soybean leaves grown under increased ozone and carbon dioxide. SPIE 3059:22 – 31
Kim MS, Lee EH, Mulchi CL, et al. (1998) Fluorescence imaging of soybean flavonol isolines. SPIE 3382:170 - 178
Kim MS, McMurtrey JE, Mulchi CL, et al. (2001) Steady-state multispectral fluorescence imaging system for plant leaves. Appl Opt 40:157 – 166
Kimball BA, Idso SB (1983) Increasing atmospheric CO_2: effects on crop yield, water use and climate. Agri Water Manage 7:55 - 72
Krupa SV, Kickert RN (1989) The greenhouse effect: impact of ultraviolet-B(UV-B) radiation, carbon dioxide (CO_2), and ozone (O_3) on vegetation. Environ Pollut 61:263 - 393
Lang M, Stober F, Lichtenthaler HK (1991) Fluorescence emission spectra of plant leaves and plant constituents. Radiat Environ Biophys 30:333 - 347
Lang M, Stiffel P, Braunova Z, et al. (1992) Investigation of the blue-green fluorescence emission of plant leaves. Bot Acta 105:395 - 468
Lang M, Lichtenthaler HK, Sowinska M, et al. (1996) Fluorescence imaging of water and temperature stress in plant leaves. J Plant Physiol 148:613 - 621
Lichtenthaler HK, Rinderle U (1988) Role of chlorophyll fluorescence in the detection of stress conditions of plants. CRC Crit Rev Anal Chem 19:29 - 85
Lichtenthaler HK, Stober F, Lang M (1993) Laser-induced fluorescence emission signatures and spectral fluorescence ratios of terrestrial vegetation. Proc Int Geosci Remote Sens Symp (IGARSS)'93:1317 - 1320
Lichtenthaler HK, Lang M, Sowinska M, et al. (1996) Detection of vegetation stress via a new high resolution fluorescence imaging system. J Plant Physiol 148:599 - 612
Logan JA (1985) Tropospheric ozone: seasonal behavior, trends, and anthropogenic influence. J Geophys Res 90(D6)10:463 - 482
McMurtrey JE, Chappelle EW, Kim MS, et al. (1994) Distinguishing nitrogen fertilization levels in field corn (*Zea mays* L.) with actively induced fluorescence and passive reflectance measurements. Remote Sens Environ 47:36 - 44
Morales F, Cerovic ZG, Moya I (1994) Characterization of blue-green fluorescence in the mesophyll of sugar beet leaves affected by iron deficiency. Plant Physiol 106:127 - 133
Moran MS, Inoue Y, Barnes EM (1997) Opportunities and limitations for image-based remote sensing in precision crop management. Remote Sens Environ 61:319 - 346
Mulchi CL, Slaughter L, Saleem M, et al. (1992) Growth and physiological characteristics of soybean in open-top chambers in response to ozone and increased atmospheric CO_2. Agri Ecosyst Environ 38:107 - 118
Mulchi CL, Rudorff B, Lee EH, et al. (1995) Morphological responses among crop species to full-season exposures to enhanced concentrations of atmospheric CO_2 and O_3. Water Air Soil Pollut 85:1379 - 1386
Omasa (1998) Image instrumentation of chlorophyll a fluorescence. SPIE 3382:91 - 99
Omasa K, Shimazaki KI, Aiga I, et al. (1987) Image analysis of chlorophyll fluorescence transients for diagnosing the photosynthetic system of attached leaves. Plant Physiol 84:748 - 752
Rudorff B, Mulchi CL, Daughtry CST, et al. (1996) Growth, radiation use efficiency, and canopy reflectance of wheat and corn grown under elevated ozone and carbon dioxide

atmospheres. Remote Sens Environ 55:163 - 173
Stober F, Lichtenthaler HK (1992) Changes of the laser-induced blue, green and red fluorescence signatures during greening of etiolated leaves of wheat. J Plant Physiol 140:673 - 680
Stober F, Lichtenthaler HK (1993) Studies on the localization and spectral characteristics of the fluorescence emission of differently pigmented wheat leaves. Bot Acta 106:365 - 370
Stober F, Lang M, Lichtenthaler HK (1994) Studies on the blue, green, red fluorescence signature of green etiolated and white leaves. Remote Sens Environ 47:65 – 71

18
Diagnosis of Stomatal Response and Gas Exchange of Trees by Thermal Remote Sensing

Kenji Omasa

Department of Biological and Environment Engineering, Graduate School of Agricultural and Life Sciences, The University of Tokyo, Yayoi 1-1-1, Bunkyo-ku, Tokyo 113-8657, Japan

1. Introduction

Air pollution and meteorological changes have influenced the health of trees growing in urban and forest areas (Guderian 1985; Sakai and Larcher 1987; Schulze et al. 1989; Omasa et al. 1996; Sandermann et al. 1997; Paoletti 1998; Waring and Running 1998). Acid rain and acid fog are among the factors considered to have such an effect (Cowling 1989; Schulze et al. 1989). The first abnormal symptoms to appear are stomatal closure and decrease in photosynthesis and growth. When the injury is severe, the symptoms extend to visible injury of leaves and withering. Withered branches are often observed in polluted urban areas. In recent years a forest decline, which may possibly be due to acid rain and acid fog as well as other air pollutants and meteorological changes, has been reported in Europe, North America, and East Asia (Guderian 1985; Cowling 1989; Schulze et al. 1989; Environment Agency of Japan 1991-1993, Sandermann et al. 1997; Paoletti 1998).

The development of remote sensing from satellites and airplanes has proved important for monitoring the effects (Colwell 1983; Rencz 1999). Color infrared photographs and multispectral data taken from an airplane have often been used to estimate the visible injury of trees and forest decline. Global changes in forests have been observed by satellite remote sensing such as LANDSAT/TM,

Air Pollution and Plant Biotechnology
–Prospects for Phytomonitoring and Phytoremediation–
Edited by K. Omasa, H. Saji, S. Youssefian, and N. Kondo
© Springer -Verlag Tokyo 2002

SPOT/HRV, NOAA/AVHRR, and EOS/MODIS. For example, the LANDSAT/TM provides several spectral images with a high resolution of 30 m in the visible to near-infrared range and 120 m in the thermal infrared range. Therefore, these images have been used for analyses of decline and evapotranspiration in urban woods and local areas of forests (Colwell 1983; Nemani and Running 1989; Hobbs and Mooney 1990). A recent technical trend in remote sensing from airplanes and satellites is hyperspectral observation capable of resolving from several tens to several hundreds of spectral bands (Hobbs and Mooney 1990; Rencz 1999; Omasa, 2000). Hyperspectral analysis in the visible to near-infrared region may be able to provide more phytobiological information on changes in contents of water and biochemical components in living plants and soils, productivity and stresses of individual plants and vegetation, and classification of plant species.

Meanwhile, portable thermal cameras (thermographic system) have often been used to remotely measure changes in temperature of plants and canopy as a surrogate for stomatal conductance (=1/stomatal resistance) and photosynthesis rate (Schurer 1975; Omasa et al. 1979, 1990, 1993; Omasa 1994; Horler et al. 1980; Hashimoto et al. 1984, 1990; Inoue et al. 1990; Taconet et al. 1995; Jones 1999). In the latter half of the 1970s, thermal camera joined with a computer was developed for image analysis of leaf temperature (Omasa et al. 1979; Hashimoto et al. 1984). Consequently, Omasa et al. (1981a-c; Omasa and Croxdale 1992) quantitatively evaluated spatial distributions of stomatal resistance (=1/stomatal conductance), transpiration rate, and absorption rate of air pollutants all over the attached leaf from leaf temperature. Recently, such quantitative study has been noticed in thermal image sensing although it is difficult to analyze quantitatively energy balance over the leaf surface (Jones 1999). The microthermogram provided information on responses of stomata at sites between veins of rice plants (Omasa 1996).

It is also very difficult to spatially evaluate stomatal resistance and transpiration rate of plants growing in the field. However, the thermal image can provide information for early detection of plant stresses, because stomatal closure occurs before the appearance of visible injury, and for screening of plants with high growth and high air pollutant absorption under steady-state thermal environments (Omasa and Aiga 1987; Omasa 1990a, 1994). Helicopter-borne remote sensing by a thermal camera was effective for early detection of environmental stress of woody canopy (Omasa et al. 1993; Omasa 1994).

2. Information Obtained from Leaf Temperature

Water evaporates from mesophyll and epidermal cell walls in the substomatal cavity and diffuses into the atmosphere through the stomata and boundary layers of leaves and trees. Carbon dioxide (CO_2), for photosynthesis, and air pollutants enter the leaf in the opposite direction to that of the water vapor (Monteith 1973;

Omasa 1979; Jones 1992). A simple resistance model for heat and gas exchange between a tree and free air is shown in Fig.1. Resistance in transfer between the gas-liquid interface in the substomatal cavity and air on the leaf is represented as the bulk stomatal resistance. Resistance between the leaf boundary layer and free air is expressed as the aerodynamic resistance. Although the bulk stomatal resistance indicates stomatal opening, it also depends on the number of stomata, their size and structure. The aerodynamic resistance varies with wind velocity, atmospheric stability, leaf shape, and the spatial structure of trees.

When the leaf surface is not wet with rain or dew, the transpiration rate, W, which is the flux in diffusion of water vapor from the leaf to free air, is given by

$$W=\{X_s(T_l)-\Phi X_s(T_a)\}/(r_{aw}+r_{sw}) \quad (1)$$

where T_l is the leaf temperature, T_a is the air temperature, $X_s(T)$ is the saturated water vapor diffusion at temperature T, Φ is the relative humidity, r_{aw} is the

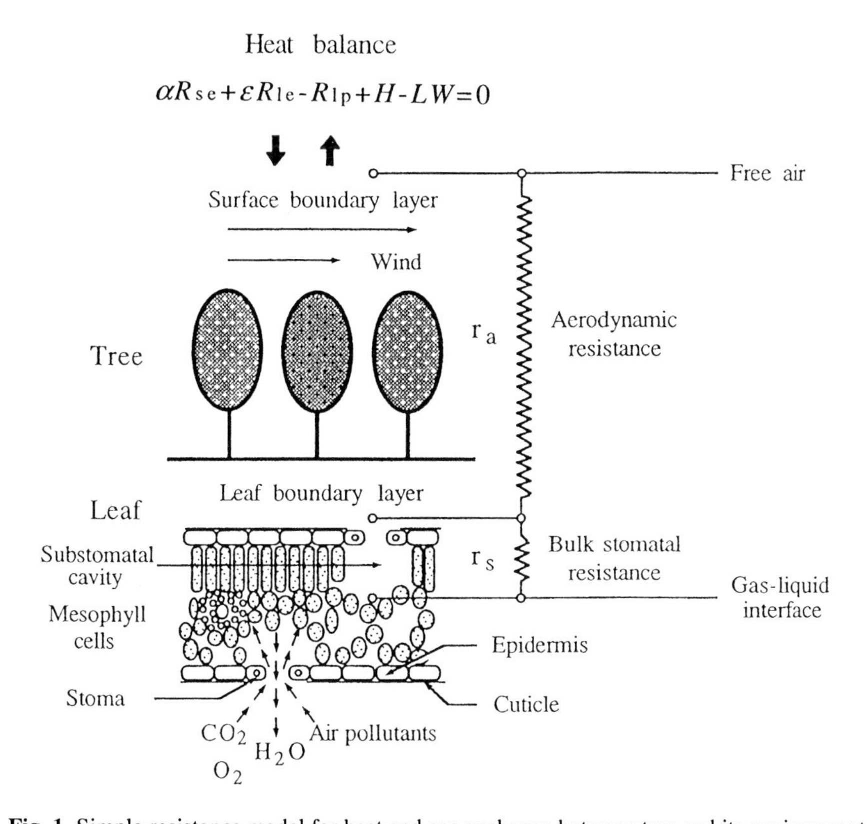

Fig. 1. Simple resistance model for heat and gas exchange between tree and its environment

aerodynamic resistance to water vapor diffusion, and r_{sw} is the bulk stomatal resistance to water vapor diffusion.

The absorption rate of gases such as CO_2 and air pollutants, Q is

$$Q=(C_a-C_l)/(r_{ag}+r_{sg}) \tag{2}$$

where C_a is the gas concentration of free air, C_l is the gas concentration at the gas-liquid interface in the stomatal cavity, r_{ag} is the aerodynamic resistance to gas diffusion, and r_{sg} is the bulk stomatal resistance to gas diffusion. The C_l of major air pollutants such as SO_2, NO_2, O_3, and formaldehyde in the healthy leaves can be assumed nearly to equal 0 $\mu l\ l^{-1}$ (ppmv) because the metabolic rate in the tissues is very rapid (see Omasa et al., this volume). However, the C_l of CO_2 dependent on photosynthesis and respiration and of most organic air pollutants varies according to species and growth conditions (Jones 1992; see Omasa et al., this volume).

On the other hand, the heat balance at the leaf surfaces is given by

$$\alpha R_{se}+\varepsilon R_{le}-R_{lp}+H-LW=0 \tag{3}$$

where R_{se} is the short-wavelength radiation ($\leq$3 μm) from the environment, α is the absorption coefficient of short-wavelength radiation of the leaf, R_{le} is the long-wavelength radiation (>3 μm) from the environment, ε is the emissivity of long-wavelength radiation of the leaf, R_{lp} is the long-wavelength radiation from the leaf surface, H is the sensible heat transfer by convection, and L is the latent heat of vaporization.

According to Planck's law, R_{lp} in Eq. 3 is

$$R_{lp}=\varepsilon\sigma T_l^4 \tag{4}$$

and H is

$$H=\rho c_p (T_a-T_l)/r_{ak} \tag{5}$$

where σ is the Stefan-Boltzmann constant, ρc_p is the volumetric heat capacity of the air, and r_{ak} is the aerodynamic resistance to heat transfer.

Substituting Eqs. 4 and 5 in Eq. 3 gives the following equation for transpiration rate, W:

$$W=\{\alpha R_{se}+\varepsilon (R_{le}-\sigma T_l^4)+ \rho c_p (T_a-T_l)/r_{ak}\}/L \tag{6}$$

In the range of growth temperature of trees, R_{lp} is approximated by

$$R_{lp}=\varepsilon\sigma (A_0 T_l+B_0) \tag{7}$$

and W is expressed as a simplified equation

$$W=AT_l+B \tag{8}$$

where

$$A=-(\varepsilon\sigma A_0+\rho c_p/r_{ak})/L \tag{9}$$

$$B=(\alpha R_{se}+\varepsilon (R_{le}-\sigma B_0)+\rho c_p T_a/r_{ak})/L \tag{10}$$

and A_0 is 1.06 x 10^8 K^3 and B_0 is -2.37 x 10^{10} K^4 in the range of 293.15 to 303.15 K (20° to 30°C).

Values for the micrometeorological parameters of outdoor trees in Eqs. 9 and 10 change with time and situation. When there is cloud and a breeze, thermal conditions such as air temperature, humidity, radiation and air current are maintained relatively constant and uniform. The poor effect of direct solar radiation and the shade of trees on the parameters also decreases under a cloudy sky. Therefore, A and B in Eqs. 9 and 10 are assumed to be constant values under such thermal conditions, and W in Eq. 8 is expressed as a linear function of T_l. Because the increase in T_l means a decrease of W in the equation, the leaf temperature may be used as an indicator of tree health and activity.

The relationship between aerodynamic resistances in heat and mass transfer is approximated by

$$r_{ak}=r_{aw}=r_{ag} \tag{11}$$

except under the conditions of stable atmosphere in the night. The bulk stomatal resistance of gas is related to that of water vapor by

$$r_{sg}=(D_w/D_g)r_{sw} \tag{12}$$

where D_w is air-water vapor diffusivity and D_g is air-gas vapor diffusivity. The bulk stomatal resistance, r_{sw}, is transformed into the following equation by substituting Eqs. 8 and 11 in Eq. 1.

$$r_{sw}=\{X_s(T_l)-\Phi X_s(T_a)\}/(AT_l+B)-r_{ak} \tag{13}$$

Equation 13 shows that the leaf temperature T_l gives information about r_{sw}, an indicator of stomatal opening, under constant thermal conditions. The stomatal conductance, which is often obtained for the measurement using a diffusion porometer, is given by $1/r_{ws}$.

Equation 2 is transformed into the following equation by substituting Eqs. 11 and 12:

$$Q=(C_a-C_l)/\{r_{ak}+(D_w/D_g)r_{sw}\} \tag{14}$$

When the gas concentration C_l at the gas-liquid interface in the stomatal cavity is known, information about the absorption rate Q of gases is also obtained from T_l.

3. Image Instrumentation of Leaf Temperature

3.1 Portable Thermal Camera

Portable thermal cameras of both optical-mechanical scanning and electric scanning types (focal plane array sensor) are on the market. The thermal camera of the optical-mechanical scanning type with an InSb (3 to 5 μm) or HgCdTe (8 to 13 μm) detector cooled by liquid nitrogen (77 K) or Stirling cryocooler has been used for a long time. The spectral range of 3 to 5 μm is not suitable for measuring the leaf temperature of trees outdoors owing to the direct effect of the sun's radiation. Consequently, a thermal camera with HgCdTe detector of 8 to 13 μm range has been generally used for field observation. Figure 2A shows a portable thermal camera (JEOL, JTG-5200) with an optical-mechanical scanning type of mirror vibration and a HgCdTe detector (8 to 13 μm, cooled by liquid nitrogen). This camera needs a flame time of more than 0.1 s. The signals detected by the camera head are converted into 16-bit digital signals (512Hx480V) and analyzed by an image processor with camera control functions. A series of thermal images is measured continuously by the system and stored in a builtin hard disk and MO disks. In outdoor situations, it is possible to carry out simple analyses of the image data using the image processor. This system gives a sensitivity (black body at 30°C) of 0.05°C and a horizontal resolution of 420 lines. The temperature-resolving power is improved to about 0.01°C by averaging the images. Measurement accuracy such as uniformity and repeatability of temperature is below 1% or 0.5°C, whichever is greater.

The recent advance in the development of thermal array sensors is remarkable. A smaller-sized, convenient thermal camera with a flame rate of 30 Hz, similar to a CCD video camera, is realized by development of new focal plane array (FPA) technology such as a 320x240 uncooled FPA (8 to 14 μm) with a sensitivity of 0.1°C and a measurement accuracy of ±2% or 2°C, whichever is greater (Avio, TVS-610) (Fig.2B). However, the measurement accuracy of an uncooled FPA has been inferior to that of the optical-mechanical scanning type until now.

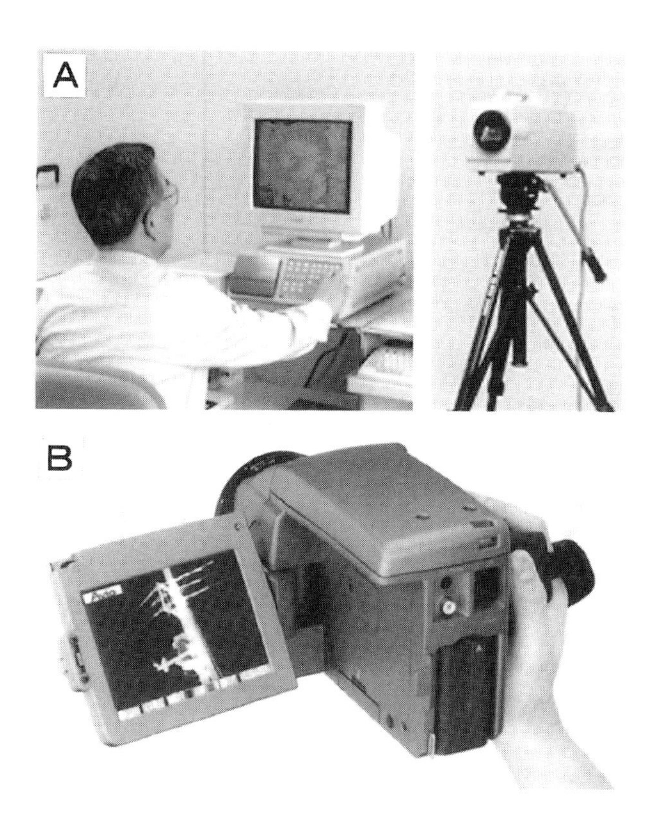

Fig. 2A, B. Portable thermal cameras of optical-mechanical scanning type (**A**, JEOL, JTG-5200) and uncooled FPA type (**B**, Avio, TVS-610)

3.2 Accuracy in Measuring Leaf Temperature

For a perfectly diffuse and opaque leaf surface, thermal radiation $R_p(T, T_s)$ from the surface of temperature T in the spectral sensitivity range of the thermal camera is given by

$$R_p(T,T_s)=\varepsilon_p R_b(T)+(1-\varepsilon_p)R_e(T_s) \tag{15}$$

where ε_p is the emissivity of the leaf surface in the spectral range, $R_b(T)$ is the spectral radiation from the black body of temperature T, and $R_e(T_s)$ is the spectral radiation from the environment of temperature T_s to the leaf surface. The emissivity ε_p of the leaf in the spectral range 8 to 13 μm is 0.95 to 0.99 (Gates et al. 1965; Fuchs and Tanner 1966; Omasa et al. 1979). The measured temperature is affected by the emissivity and radiation from the environment. Therefore, it is necessary to use the thermal camera with functions to correct for these factors to obtain an exact measurement of the leaf temperature. The influence of changes in functions, such as the radiation-electricity conversion of the detector, its amplification, and the transmission and reflection of lens, filter, etc., is corrected continuously by monitoring a builtin black body source in the camera head. It is possible to measure the leaf temperature within an accuracy of 0.1 K (Omasa et al. 1979).

The error in temperature measured with the thermal camera depends on the spatial distribution of temperature (Fig. 3). In Fig. 3, the number of slits indicates the frequency of switchover between high and low temperatures on the horizontal line of the image. When the distribution of temperature is zigzag, the error reaches its maximum. The frequency of switchover is about 50 times at an error of 5% (0.25°C) and about 65 times at an error of 10% (0.5°C).

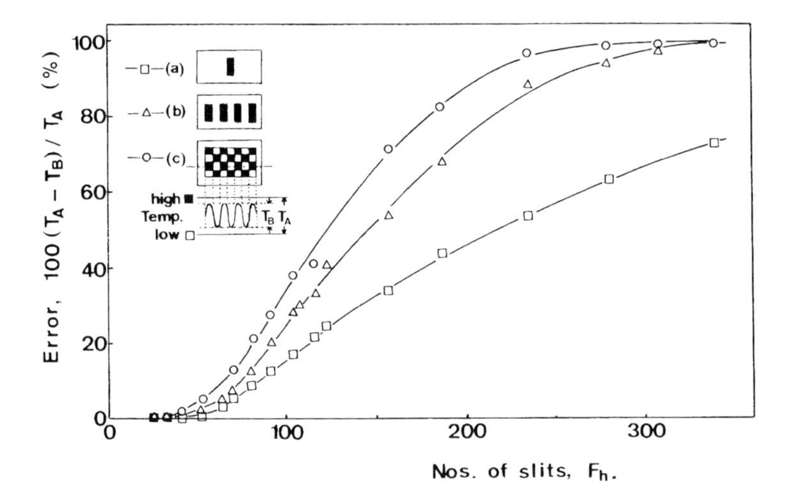

Fig. 3. Relationship between the error in temperature measured with the thermal camera (JEOL, JTG-5200) and the change in spatial distribution of temperature (from Omasa et al. 1993). The black and white patterns made by slits (*a*) to (*c*) show the difference in spatial distribution of temperature where the black area is high temperature and the white area is low temperature, T_A (=5°C) is the difference in real temperature between black and white areas, and T_B is that in measured temperature. The number of slits means frequency of switchover at high and low temperatures on a horizontal line of the image

4. Diagnosis of Trees by Leaf Temperature Image

4.1 Diagnosis of Effects of O_3 Exposure on Trees

Abiotic and biotic stresses such as air pollutants, water deficit, high and low temperature, and virus infection cause spatially heterogeneous impairment of the attached leaves (Omasa et al. 1981b, c, 1987; Hashimoto et al. 1984; Daley et al. 1989; Omasa 1990a, b; Omasa and Croxdale 1992; Osmond et al. 1998). Such heterogeneous impairment is indicated in stomatal response and photosynthetic activity. As described in Section 2, the leaf temperature provides information about stomatal response, transpiration and absorption of air pollutants and CO_2 under constant thermal conditions. The portable thermal camera can be used, therefore, to show evidence of spatially different responses of stomata in attached leaves to various stresses. In addition, it is possible to calculate images of stomatal resistance, transpiration rate, and absorption rate of NO_2, SO_2, and O_3 from the leaf temperature image measured under controlled thermal conditions in the growth chamber (Omasa et al. 1981a-c; Omasa and Croxdale 1992).

Figure 4 shows the effects of 0.1 $\mu l\ l^{-1}$ O_3 exposure during 3 weeks on Chinese laurestine and Japanese red cedar. Ozone is a major component of photochemical oxidants produced in the urban atmosphere by a series of photochemical reactions involving nitrogen oxides and gaseous hydrocarbons. Entry of O_3 into the leaf tissues through the stomata caused necrotic visible injury and defoliation to the Chinese laurestine. The Japanese red cedar showed a decrease in growth rate, but there was no visible injury. The leaf temperature in both trees increased with exposure time because of stomatal closure and death of the leaves (only Chinese laurestine). This temperature rise in the Japanese red cedar appeared especially in the upper part of the tree. It is difficult to measure spatial differences in the response of stomata of trees with attached needle leaves to environmental stimuli using ordinary porometers. However, use of the portable thermal camera makes it possible to obtain such spatial information easily.

4.2 Diagnosis of Street Trees

Street trees in urban areas grow under severe environmental conditions. The trees are exposed continuously to harmful gas from car exhausts and other air pollutants. Buildings interrupt sunlight falling on the trees, and street lamps illuminate them at night. Because the paved roads obstruct rainwater permeating into the soil together with a supply of nutrients from dead leaves, the soil water content, soil nutrients, and humidity decrease in urban areas. The effect first shows as a decrease in transpiration and photosynthesis. Although porometers and micrometeorological methods are used for measuring rates of transpiration and photosynthesis, these cannot provide the spatial distribution of these processes in

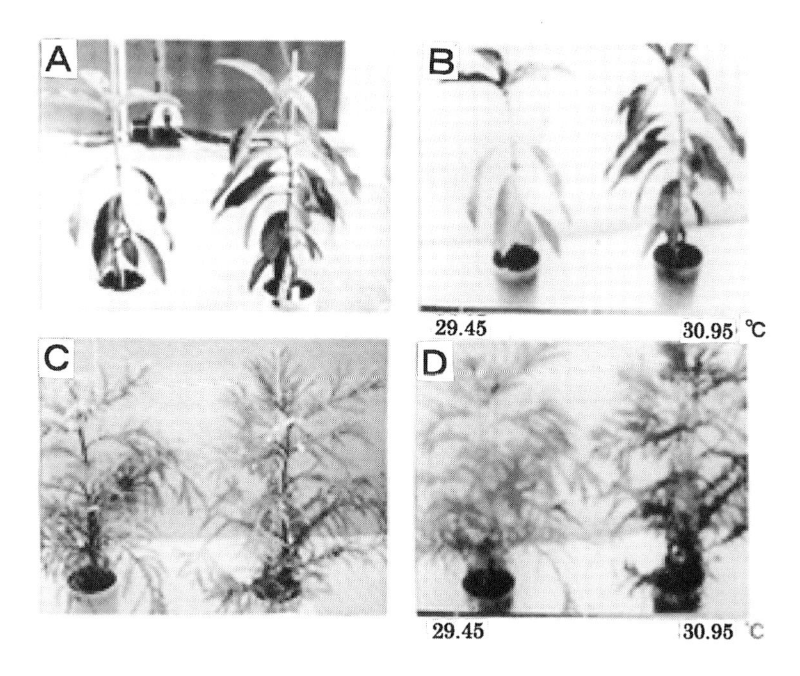

Fig. 4A-D. Effects of O_3 exposure on trees (from Shimizu et al. 1993). **A** Photograph of Chinese laurestine (*Viburnum odoratissimum* Ker-Gawler var. awabuki (K.Koch) Zabel), **B** Thermal image of the Chinese laurestine, **C** Photograph of Japanese red cedar (*Cryptomeria japonica* (L.fil.) D.Don), **D** Thermal image of the Japanese red cedar. The injured tree on the left side in the photograph and in the thermal image was exposed to 0.1 $\mu l\ l^{-1}$ O_3 for 3 weeks under 25°C, 70%RH, and 200-400 μmol photons $m^{-2}\ s^{-1}$. The healthy tree on the right side was grown under the same conditions without O_3 exposure. The *gray scale* on the underside in the thermal image represents the temperature scale. The temperature shown by *white* is higher than that shown by the *black*

leaves and branches.

Figure 5 shows a photograph and a thermal image of zelkova trees growing in an urban street. The thermal image was measured under cloudy and breezy conditions. The leaf temperature of the tree on the left (*a* in Fig.5A and 5B) was higher than on the right (*b*) and other trees; this indicates stomatal closure and decrease in transpiration and photosynthesis in the tree on the left. These phenomena might be caused by volatile matter from a gasoline service station on the left-hand side. The tree on the right (*b*) was healthy owing to a sufficient supply of light, nutrients, and water from vacant land on the right-hand side. Although growth of the tree on the left (*a*) was poorer than that of the tree on the right, leaf injury was not visible. The combined use of the thermal camera and the porometer thus makes it possible to diagnose the health of trees precisely.

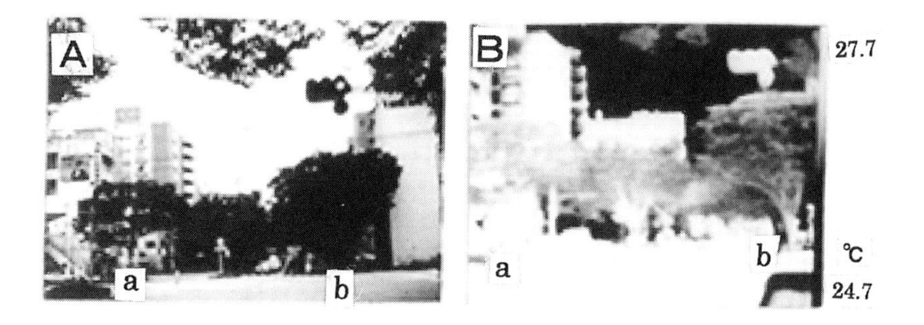

Fig. 5A, B. Photograph (**A**) and thermal image (**B**) of zelkova (*Zelkova serrata* (Thunb.) Makino) trees growing in an urban street (Omasa et al. 1990). Environmental conditions: air temperature, 26.5°C; light intensity, about 500 μmol photon m^{-2} s^{-1}

Figure 6 shows the relationship between stomatal conductance and photosynthetic photon flux density (PPFD) for some species of healthy trees. These data were measured from 1300 to 1800h. The lowered PPFD tended to decrease stomatal conductance irrespective of species, although the data vary with each leaf site. In particular, stomatal conductance decreased rapidly below 200 μmol photon m^{-2} s^{-1}. However, the rate of decrease was low above 300 μmol photon m^{-2} s^{-1} and reached different steady-state conditions in each species. Stomatal conductance is a reciprocal of stomatal resistance and an indicator of stomatal opening. Therefore, the decrease of stomatal conductance shown in Fig.6 indicates stomatal closure and a decrease in the rates of transpiration and photosynthesis. The stomata of healthy leaves opened rapidly after sunrise and closed slowly in the afternoon. The stomata of injured and water-stressed leaves did not open properly in the daytime. When it was cloudy, the PPFD was below 500 μmol photon m^{-2} s^{-1}. Therefore, the thermal image used for diagnosis should be measured under a cloudy sky with PPFD of 300 to 500 μmol photon m^{-2} s^{-1}. The effects of stomatal closure caused by water stress in the daytime are avoided by measurements under such conditions.

4.3 Diagnosis of Trees from a Helicopter

Woods and forests in urban areas and the neighboring mountains have been injured by the various environmental changes just described. However, it is difficult to diagnose damage to individual trees throughout the woods and forests by means of measurements made on the ground. Recently, thermal remote sensing from satellites and airplanes has been shown to estimate the function of woods and forests (Hobbs and Moony 1990; Omasa et al. 1993). Because helicopters can

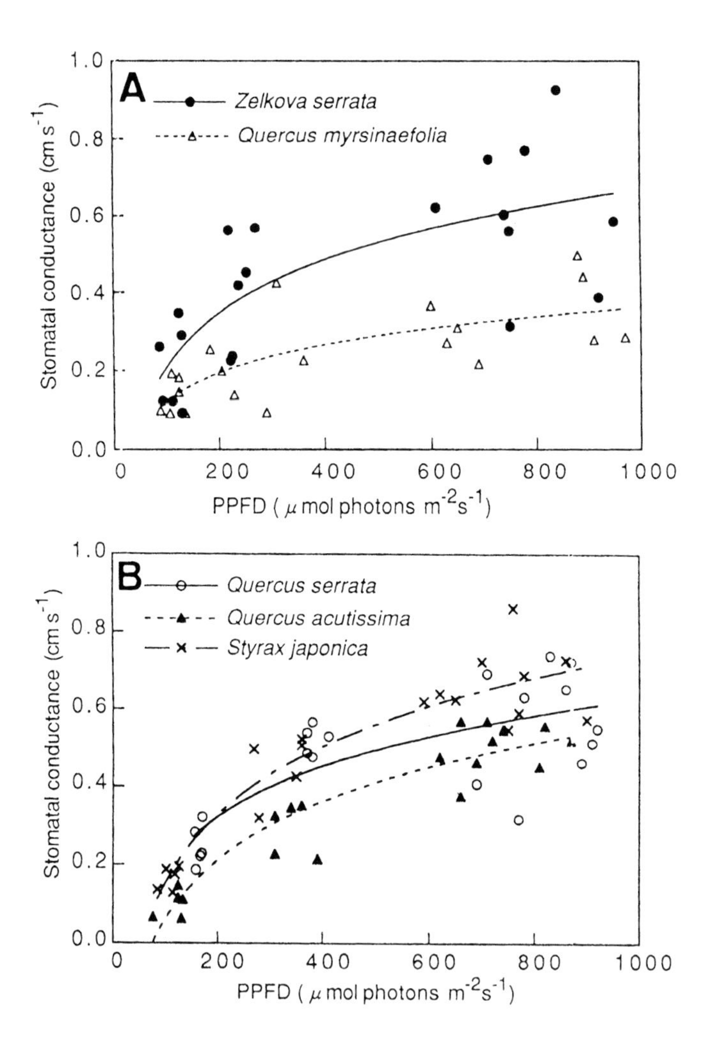

Fig. 6A, B. Relationship between stomatal conductance and photosynthetic photon flux density (PPFD) for some species of healthy tree (Omasa et al. 1993). Environmental conditions: air temperature, 30°-33°C; relative humidity, 50%-60%

approach a height of only several tens of meters, remote sensing from a helicopter makes it possible to diagnose individual trees (Omasa et al. 1993).

Figure 7 shows an aerial photograph and a thermal image of temple woods and the adjacent area in the suburbs of Tokyo. These were measured by the portable thermal camera from a helicopter at a height of about 300 m under cloudy and breezy conditions. In the temple woods, Japanese red pines (*Pinus densiflora* Sieb. et Zucc.) (*a* in Figs. 7A and 7B) were standing almost dead and many

Japanese red cedars had died back at their tops. These injuries were also observed in the aerial photograph. The leaf temperature of these trees was higher than that of other trees because of the decrease in stomatal conductance and transpiration. Although injuries such as abnormal leaf shape were found in a zelkova tree (*b*) by survey on the ground, it was not observed in the aerial photograph. The leaf temperature of the zelkova tree, however, was higher than that of Japanese white oak (*Quercus myrsinaefolia* Bulume) (*c*) shown as a species with lower conductance in Fig. 6. This result means that it is possible to reliably diagnose slight damage to trees not observed in aerial photographs from the thermal image measured under a cloudy sky above about 300 μmol photon m^{-2} s^{-1}. The surface temperatures of houses (*d*), roads, and parking lots were above 35°C.

The measured temperature is influenced by absorption and radiation by the atmosphere, although the influence is small in the wavelength range 8 to 13 μm. Figure 8 shows a thermal image, measured from a height of about 700 m, of the same area as that shown in Fig. 7 at about the same time. Points a and d in Fig. 8 correspond to those in Fig. 7B. The leaf temperature measured from ca. 700 m showed a reduction of 0.8°C in comparison with that from ca. 300 m. It must be noted that the extent of this reduction is influenced by atmospheric conditions. A rise in height also causes an error according to the spatial averaging shown in Fig. 3. When the temperature of a tree of 3- to 5- m diameter was measured within 5% error, the height of the helicopter is about 300 to 500 m. The error in the tree's temperature measured by the portable thermal camera increases markedly according to the rise in height.

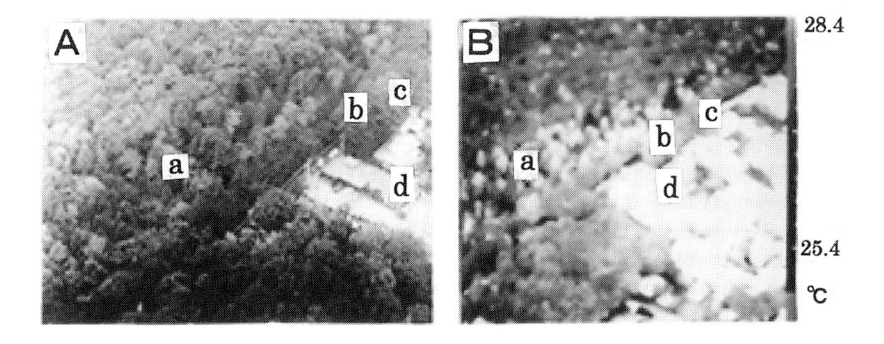

Fig. 7A, B. Aerial photograph (**A**) and thermal image (**B**) of temple woods and the adjacent area in the suburbs of Tokyo from a helicopter at a height of about 300 m (Omasa et al. 1993). Sites *a* to *d* in **A** correspond to those in **B**. Environmental conditions were not measured at the temple, but at a position about 10 km from it; air temperature and PPFD measured after 30 min were about 29°C and 400 μmol photon m^{-2} s^{-1}, respectively

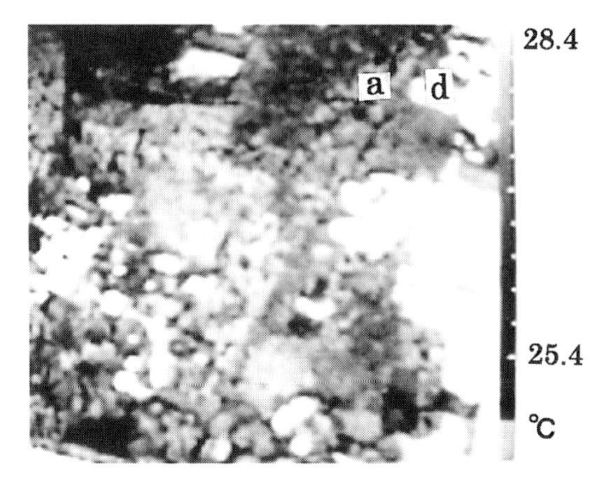

Fig. 8. Thermal image of temple woods and the adjacent area measured from a helicopter at a height of about 700 m (Omasa et al. 1993). Sites *a* and *d* in Fig. 8 correspond to those in Fig. 7. This image was taken at about 100 s before the measurement for Fig. 7

5. Conclusion

Changes in leaf temperature depend on those in transpiration rate from the leaf via stomata under constant thermal conditions; consequently, the leaf temperature becomes an indicator of stomatal response and absorption of air pollutants and CO_2. Therefore, the measurement of leaf temperature by the portable thermal camera can remotely provide spatial information for early detection of plant stresses, because stomatal closure occurs before the appearance of visible injury, and for screening of plants with high growth and high levels of air pollutants. For example, the evidence of spatially different responses of stomata in attached leaves of Chinese laurestine and Japanese red cedar to O_3 exposure was shown in this chapter. Although it was difficult to measure spatial differences in the response of stomata of trees with attached needle leaves using ordinary porometers, the use of the thermal camera made it possible to obtain such spatial information easily.

The thermal camera was also applied to the diagnosis of zelkova trees growing in an urban street and of some species in urban temple woods from the ground and from a helicopter. It was possible to diagnose reliably slight damage to trees that was not observed in photographs from the thermal image measured under a cloudy sky above approximately 300 μmol photon m^{-2} s^{-1}. It was necessary to measure at a height less than 300 to 500 m for remote sensing from helicopter to obtain the exact temperature of individual trees.

Recent advance in development of thermal array sensors is remarkable. Consequently, more small-sized, convenient thermal camera with an array sensor such as uncooled FPA are appearing in the market. In the near future, we may easily come to use it like an ordinary video camera.

References

Colwell RN (ed) (1983) Manual of remote sensing, 2nd edn, vols I, II. American Society of Photogrammetry, Falls Church, VA

Cowling EB (1989) Recent changes in chemical climate and related effects on forests in North America and Europe. Ambio 18:167-171

Daley PF, Raschke K, Ball JT, et al (1989) Topography of photosynthetic activity in leaves obtained from video images of chlorophyll fluorescence. Plant Physiol 90:1233-1238

Environment Agency of Japan (1991-1993) Global environment research in 1990-1992. C1-C3

Fuchs M, Tanner CB (1966) Infrared thermometry of vegetation. Agron J 58:597-601

Gates DM, Keegan HJ, Schleter JC, et al (1965) Spectral properties of plants. Appl Opt 4:11-20

Guderian R (ed) (1985) Air pollution by photochemical oxidants: formation, transport, control, and effects on plants. Ecological studies, vol 52. Springer, Berlin

Hashimoto Y, Ino T, Kramer PJ, et al (1984) Dynamic analysis of water stress of sunflower leaves by means of a thermal image processing system. Plant Physiol 76:266-269

Hashimoto Y, Kramer PJ, Nonami H, et al (eds) (1990): Measurement techniques in plant science. Academic Press, San Diego, pp 343-431

Hobbs RJ, Mooney HA (eds) (1990) Remote sensing of biosphere functioning. Ecological studies, vol 79. Springer, New York

Horler DNH, Barber J, Barringer AR (1980) Effects of cadmium and copper treatments and water stress on the thermal emission from peas (*Pisium sativum* L.): controlled environment experiments. Remote Sens Environ 10:191-199

Inoue Y, Kimball BA, Jackson RD, et al (1990) Remote estimation of leaf transpiration rate and stomatal resistance based on infrared thermometry. Agric For Meteorol 51:21-33

Jones HG (1992) Plants and microclimate, 2nd edn. Cambridge University Press, Cambridge

Jones HG (1999) Use of thermography for quantitative studies of spatial and temporal variation of stomatal conductance over leaf surface. Plant Cell Environ 22:1043-1055

Monteith JL (1973) Principles of environmental physics. Arnold, London

Nemani RR, Running SW (1989) Estimation of regional surface resistance to evapotranspiration from NDVI and thermal-IR AVHRR data. J Appl Meterol 28:276-284

Omasa K (1979) Sorption of air pollutants by plant communities. Analysis and modelling of phenomena (in Japanese). Res Rep Natl Inst Environ Stud Jpn 10:367-385

Omasa K (1990a) Image instrumentation methods of plant analysis. In: Linskens HF, Jackson JF (eds) Modern methods of plant analysis. New series 11. Springer, Berlin, pp 203-243

Omasa K (1990b) Image analysis of chlorophyll fluorescence in leaves. In: Hashimoto Y, Kramer PJ, Nonami H, et al (eds) Measurement techniques in plant science. Academic

Press, San Diego. pp 387-401
Omasa K (1994) Diagnosis of trees by portable thermographic system. In: Kuttler W, Jochimsen M (eds) Immissionsökologische Forschung im Wandel der Zeit. Essener Ökologische Schriften, pp 141-152
Omasa K (1996) Image instrumentation of living plants (in Japanese). Biosci Ind. 54:545-546, 569-571.
Omasa K (2000) Phytobiological IT in agricultural engineering. Proc the XIV Memorial CIGR World Congress 2000. pp 125-132
Omasa K, Aiga I (1987) Environmental measurement: Image instrumentation for evaluating pollution effects on plants. In: Singh MG (ed) Systems and control encyclopedia. Pergamon Press, Oxford, pp 1516-1522
Omasa K, Croxdale JG (1992) Image analysis of stomatal movements and gas exchange. In: Häder DP (ed) Image analysis in biology, CRC Press, Boca Raton, pp 171-197
Omasa K, Abo F, Hashimoto Y, et al (1979) Measurement of the thermal pattern of plant leaves under fumigation with air pollutant. Res Rep Natl Inst Environ Stud Jpn 10:259-267 (in Japanese and English summary); 11:239-247(1980) (in English translation)
Omasa K, Abo F, Aiga I, et al (1981a) Image instrumentation of plants exposed to air pollutants: quantification of physiological information included in thermal infrared images. Trans Soc Instrum Control Eng. 17:657-663 (in Japanese and English summary); Res Rep Natl Inst Environ Stud Jpn. 66:69-79 (1984) (in English translation)
Omasa K, Hashimoto Y, Aiga I (1981b) A quantitative analysis of the relationships between SO_2 or NO_2 sorption and their acute effects on plant leaves using image instrumentation. Environ Control Biol 19:59-67
Omasa K, Hashimoto Y, Aiga I (1981c) A quantitative analysis of the relationships between O_3 sorption and its acute effects on plant leaves using image instrumentation. Environ Control Biol 19:85-92
Omasa K, Shimazaki K, Aiga I, et al (1987) Image analysis of chlorophyll fluorescence transients for diagnosing the photosynthetic system of attached leaves. Plant Physiol 84:748-752
Omasa K, Tajima A, Miyasaka K (1990) Diagnosis of street trees by thermography. Zelkova trees in Sendai City (in Japanese and English summary). J Agric Meteorol 45:271-275
Omasa K, Shimizu H, Ogawa K, et al (1993) Diagnosis of trees from helicopter by thermographic system (in Japanese and English summary). Environ Control Biol 31:161-168
Omasa K, Kai H, Taoda Z, et al (eds) (1996) Climate change and plants in East Asia. Springer, Tokyo
Osmond CB, Dayley PF, Badger MR, et al (1998) Chlorophyll fluorescence quenching during photosynthetic induction in leaves of *Abutilon striatum* disks. Infects with abutilon mosaic virus, observed with a field-portable imaging system. Bot Acta 111:390-397
Paoletti E (ed) (1998) Stress factors and air pollution. Chemosphere 36:625-1166
Rencz AN (ed) (1999) Manual of remote sensing, 3rd edn, Vol 3. Wiley , New York
Sakai A, Larcher W (1987) Frost survival of plants. Springer, Berlin
Sandermann H, Wellburn AR, Heath RL (eds) (1997) Forest decline and ozone. Springer, Berlin
Schulze E-D, Lange OL, Oren R (eds) (1989) Forest decline and air pollution. Ecological studies, vol 77. Springer, Berlin

Schurer K (1975) Thermography in agricultural engineering. Proc. 1st Eur. Congr. Thermography 1994, Bibl Radiol No. 6, pp 249-254

Shimizu H, Fujinuma Y, Kubota K, et al (1993) Effects of low concentrations of ozone (O_3) on the growth of several woody plants. J Agric Meteorol 48:723-726

Taconet O, Olioso A, Ben Mehrez M, et al (1995) Seasonal estimation of evaporation and stomatal conductance over a soybean field using surface IR temperatures. Agric For Meteorol 73:321-337

Waring RH, Running SW (1998) Forest ecosystems: analysis at multiple scales. Academic Press, San Diego

IV. Generation of Transgenic Plants

19
Manipulation of Genes Involved in Sulfur and Glutathione Metabolism

Shohab Youssefian

Biotechnology Institute, Faculty of Bioresource Sciences, Akita Prefectural University, Minami 2-2, Ohgata-mura, Akita 010-0444, Japan

1. Introduction

Sulfur is an essential nutrient for plant growth and development and, under conditions of sufficient sulfur availability, is taken up from the soil by roots in the form of sulfate and transferred to the shoots. The sulfate is subsequently activated and then reduced to sulfite and then to sulfide, in what is generally referred to as the sulfur reduction assimilatory pathway; the main function and ultimate product of which is the biosynthesis of cysteine (Cys). The Cys thus formed serves as a precursor for several other reduced sulfur-containing metabolites, notably methionine and glutathione (GSH), the latter constituting the principal storage and transport form of reduced sulfur in plants, as well as a key factor controlling plant responses to a variety of biotic and abiotic stresses (Noctor et al. 1998a; May et al. 1998a).

When plants are grown under conditions of sulfur deficiency, however, they can absorb a greater part of their sulfur requirements directly from atmospheric pollutants, notably SO_2, but also H_2S and sulfates, which can serve as major sources of sulfur fertilization in industrialized areas (Schnug 1998). The absorption of SO_2 and H_2S by plants is affected by many physiological factors, such as stomatal opening, as well as by the solubility, reactivity with cellular components, and metabolism of the pollutants (De Kok 1990; Kondo, this volume). Although the exact processes involved are still unclear, the absorbed

Air Pollution and Plant Biotechnology
–Prospects for Phytomonitoring and Phytoremediation–
Edited by K. Omasa, H. Saji, S. Youssefian, and N. Kondo
© Springer -Verlag Tokyo 2002

SO_2 and H_2S result in an accumulation of sulfate, which is predominantly stored in vacuoles, and in the formation of the nonprotein thiols Cys and GSH (Rennenberg and Herschbach 1996).

Conversely, when plants are grown in heavily polluted environments, which for SO_2 and H_2S may exceed clean air rural area levels by up to 2000- and 5000 fold, respectively, the plants unavoidably receive surplus sulfur. Although such plants can form sinks for these pollutants as just noted, excess exposure severely interferes with processes that modulate the plant's sulfur status and results in a retardation of growth and a reduction in grain yields as well as in acute damage and even death (De Kok et al., this volume). The physiology of such pollutant-induced plant damage remains unresolved, but the phytotoxicity of both SO_2 and H_2S is no doubt related to their reactions, or those of their secondary products, with cellular components such as enzymes and membranes rather than from a disturbance in the sulfur status of the plants. In the case of SO_2, the production of free radicals or reactive oxygen species (ROS; including $O_2^{-\cdot}$, $OH^{\cdot}$, $SO_3^{-\cdot}$, and H_2O_2), resulting from the oxidation of SO_2-derived sulfite to sulfate in leaves, possibly together with cellular acidification, appear to be the main factors responsible for its phytotoxicity (Rennenberg and Herschbach 1996; Kondo, this volume).

Despite the fall in SO_2 emissions in most developed countries as the result of stringent air pollution control measures, the global levels of man-made emissions are still increasing by about 4% annually because of the rapid industrialization of developing nations (Yunus et al. 1996). Furthermore, because of its role in acid rain, which due to its low pH has drastic effects over wide geographic areas, additional attention is being focused on this pollutant (Wellburn 1988). To restrict the continued buildup of such pollutants in the atmosphere and to limit the extent of pollution-related damage, various methods of control, including biological means, are now being explored. Variation between plant species in their responses to, and accumulation of, such pollutants suggests that available genetic resources do exist and can be manipulated so as to develop new model plants with such beneficial characteristics that cannot be easily obtained through mutation or conventional breeding strategies.

This review attempts to briefly summarize our current understanding of the plant sulfur and GSH biosynthetic pathways, to examine the characteristics of genes that encode several enzymes of these pathways, and to present examples of how overexpression of such genes has been used to modify the metabolism of sulfur and GSH in transgenic plants, with special emphasis on their responses to sulfurous pollutants or ozone (O_3). However, it is not the aim of this chapter to discuss the detailed mechanisms involved in these pathways, which have been recently addressed by several excellent reviews (Hell 1997; Noctor et al. 1998a; May et al. 1998a; Leustek and Saito 1999), but rather to demonstrate the potential use of such molecular strategies, both to further our understanding of the regulation of plant responses to these pollutants and to develop novel plants with enhanced pollutant tolerance or for use in phytomonitoring and phytoremediation.

2. Molecular Regulation of Sulfur Assimilation and Glutathione Biosynthesis

A simplified metabolic pathway for the biosynthesis of Cys and GSH in plants, together with some of the genes isolated to date as well as some of the factors responsible for their expression, is presented in Fig. 1.

2.1 Sulfur Assimilation and Cysteine Biosynthesis

The process of sulfate uptake from the rhizosphere into roots, or from the apoplastic space into cells, together with loading into the vascular system for long-distance translocation and subsequent unloading in leaves, appears to be mediated by a family of plasma membrane sulfate transporters. Genes isolated from *Stylosanthes hamata*, *Arabidopsis thaliana*, *Hordeum vulgare*, and *Zea mays*, as well as from several other plants (see Leustek and Saito 1999), suggest that the various transporters have different affinities for sulfate, and also specialized functions (Smith et al. 1995, 1997; Takahashi et al. 1997). High-affinity transporters appear to be root specific with expression of the genes being rapidly induced by sulfur deficiency or *O*-acetylserine (OAS), a candidate positive regulator of sulfur metabolism, but suppressed by the negative regulators, GSH and Cys (Smith et al. 1997). Such transporters may, therefore, by responding to the internal sulfur status, mediate the uptake of sulfate from the soil into the plant. In contrast, low-affinity transporters are present primarily in leaves but also in roots, with gene transcripts being only slightly inducible in roots but repressed in leaves upon sulfate deprivation, suggesting that they could well be involved in inter- and intracellular transport of sulfate. As these transporters are essential components of long-distance translocation of sulfate and reduced thiols, and possibly of their influx and efflux from chloroplasts and vacuoles, they may be expected to control the development of sink tissues, especially on exposure to high sulfur environments. Although there are no reports describing the manipulation of these transporter genes in transgenic plants, it is clear that the normal mechanisms regulating sulfate uptake, assimilation, partitioning, and storage could be selectively dissociated in such plants, which would therefore serve as ideal candidates for phytoremedial use.

The inorganic sulfate taken up by these processes is chemically stable and so has to be activated before reductive assimilation to sulfite; this is achieved by the enzyme ATP sulfurylase, which generates adenosine 5'-phosphosulfate (APS). Most plants possess two ATP sulfurylase isoforms, a major plastidic form and a minor cytosolic form (Lunn et al. 1990). Genes encoding both isoforms have been isolated from potato (Klonus et al. 1994), whereas only those encoding plastidic forms have been isolated from Arabidopsis (Leustek et al. 1994; Klonus et al. 1995). While the functions of the cytosolic form are still unresolved, the plastidic isoform, which is found in leaves and roots, is thought to play a role in regulating

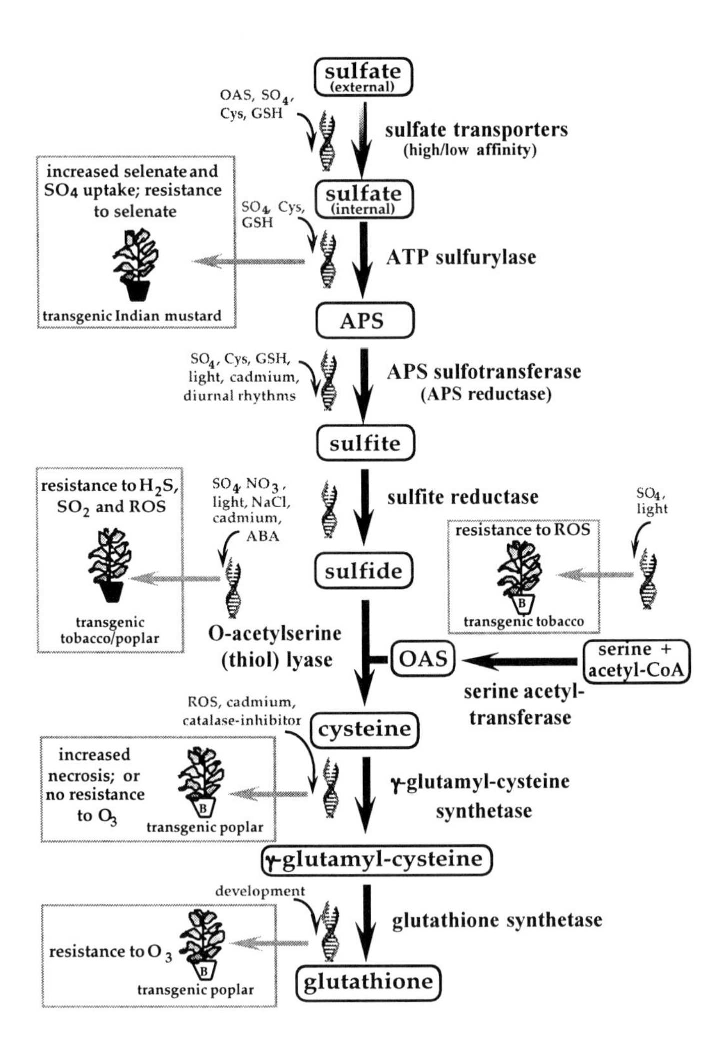

Fig. 1. Schematized sulfur assimilatory reduction pathway in plants, showing the genes isolated and the transgenic plants generated to date. Isolated genes, encoding different enzymes of the pathway, from various plant species, are represented by doublehelices. Factors that positively or negatively affect gene expression are indicated by *arrows* pointing to each helix: *OAS*, *O*-acetylserine; *Cys*, cysteine; *GSH*, glutathione; *SO_4*, sulfate; *NO_3*, nitrate; *ROS*, reactive oxygen species; *ABA*, abscisic acid. Transgenic plants generated to overexpress these genes are *boxed*, together with their new resistant characteristics. *Black plant pots* indicate the use of plant transgenes; *clear pots* with the letter *B* denote the use of bacterial transgenes

reductive sulfur assimilation. Indeed, transcripts and activity of ATP sulfurylases are somewhat enhanced by sulfur deprivation, but reduced when treated with the reduced thiols Cys or GSH (Lappartient et al. 1999; Bolchi et al. 1999). The effects of overexpressing ATP sulfurylases on sulfate assimilation in transgenic plants are described in Section 3.1.1.

The exact reductive pathway leading from APS to sulfide has been the subject of some controversy, with proposed routes including a 'bound intermediate' pathway and a 'free intermediate' pathway, as outlined by Hell (1997). However, a more critical appraisal of the enzymes and genes that have been conclusively demonstrated (Leustek and Saito 1999) concludes that the APS is converted by an APS sulfotransferase (APS reductase) to a thio-sulfonate, such as *S*-sulfoglutathione, that is then readily reduced in the presence of excess GSH to sulfite, the latter subsequently being reduced to sulfide by a ferredoxin-dependent sulfite reductase. Both APS sulfotransferase and sulfite reductase are plastid localized but, although the former is encoded by a gene family in Arabidopsis (Bick and Leustek 1998), the sulfite reductase appears to be encoded by a single gene (Bork et al. 1998). A substantial amount of evidence suggests that the APS sulfotransferase is a major point of regulation of the sulfur assimilatory pathway (Brunold and Rennenberg 1997). Not only is its enzymatic activity directly affected by sulfur starvation and exposure to reduced thiols, but also by stress situations, such as heavy metals, that require increased cellular sulfate. Indeed, transcripts of APS sulfotransferase genes appear to be induced by both sulfur starvation (Gutierrez-Marcos et al. 1996; Takahashi et al. 1997) and heavy metals (Lee and Leustek 1999), as well as by light and diurnal rhythms (Kopriva et al. 1999). In contrast, sulfite reductase appears not to be significantly regulated at the transcript level (Bork et al. 1998). Although the generation of transgenic plants expressing either APS sulfotransferase or sulfite reductase has not as yet been reported, it could well be expected that elevated activities of these key enzymes in transgenic plants would, by enhancing the metabolism of sulfurous air pollutants, especially SO_2, constitute important components of the increased resistance of these plants to such pollutants.

The final and a key step in sulfur assimilation in plants is the incorporation of sulfide into Cys, catalyzed by the enzyme *O*-acetylserine(thiol) lyase (OASTL), and using *O*-acetylserine (OAS) produced from the reaction catalyzed by serine acetyl transferase (SAT). Both OASTL and SAT have been identified in the cytosol, plastids, and mitochondria from various plants (Lunn et al. 1990; Rolland et al. 1992; Ruffet et al. 1994), suggesting that Cys biosynthesis takes place in all cellular compartments in which protein synthesis occurs, perhaps because of the inability of these organelles to transport Cys across their membranes (Lunn et al. 1990). Although the free forms of both OASTL and SAT have catalytic activity, the two enzymes can exist in a bi-enzyme complex, known as Cys synthase, whose stability is affected by its substrates, OAS and sulfide, and in which the activity of OASTL is drastically reduced (Droux et al. 1998). It has been proposed, at least for chloroplasts, that the large excess, almost 400 fold (Ruffet et al. 1994), of free unbound auxiliary OASTL present is essential to convert the buildup of OAS and to achieve full Cys synthesis capacity (Droux et al. 1998). Accumulating evidence suggests that OAS availability acts as a positive regulator, and Cys and GSH act as negative regulators, of the sulfur assimilatory pathway.

Genes encoding the different isoforms of these two enzymes from various plant species have been cloned and characterized (Hell 1997 and references therein). Transcripts of many of the SAT and OASTL genes isolated to date are regulated by different combinations of sulfate and nitrate availability as well as by light, possibly an indication of the coregulatory nature of the nitrate and sulfate assimilatory pathways. For example, SAT transcripts in Arabidopsis accumulate twofold in response to sulfate and light above an already high basal expression level (Bogdanova et al. 1995). For OASTL, transcripts of cytosolic and chloroplastic isoforms from spinach increase marginally in response to sulfate and nitrate deprivation whereas those of the mitochondrial form increase fivefold (Takahashi and Saito 1996). Similarly, transcript levels of several OASTL genes from rice are either increased by sulfur deprivation, or reduced by dark or by dark and sulfur and nitrogen deprivation (Nakamura et al. 1999). In contrast, transcripts of a wheat cytosolic OASTL accumulate tenfold, but only in the presence of nitrates in the dark (Youssefian and Nakamura, unpublished).

Despite the detailed characterization of numerous genes encoding OASTL and SAT, the exact roles of the various isoforms are still unclear. While they are most certainly involved in regulating Cys biosynthesis in the different compartments, several have also been implicated in various detoxification mechanisms. For example, transcripts of a cytosolic OASTL in Arabidopsis have most recently been shown to be induced by salt and cadmium and mediated by abscisic acid (Barroso et al. 1999), while the identity of an OASTL isoform with β-cyanoalanine synthase, which plays a pivotal role in cyanide fixation, has been recently proposed (Maruyama et al. 1998). In addition, differential responses of transgenic tobacco plants, overexpressing a wheat OASTL or a bacterial SAT, to H_2S and SO_2, and particularly to oxidative stress, have recently been demonstrated (see Section 3.1, following).

2.2 Glutathione Biosynthesis

In addition to serving as a regulator of plant sulfur nutrition and as the main storage and transport form of reduced sulfur in plants, observations that the enhanced synthesis of GSH constitutes an intrinsic response of plants to stress, and that elevated levels correlate with environmental stress tolerance, confirm the critical role that this thiol plays in the redox regulation of cellular processes. Such regulation is thought to be largely dependant both on the levels of glutathione and on the ratio between GSH and its oxidized form, GSSG. While substrate availability and the activity of the GSH biosynthetic enzymes are thought to be the primary determinants of cellular GSH homeostasis, the complex interplay with other metabolic factors, including GSH distribution, oxidation, and degradation, also appears to affect the cellular responses. Although the focus here is placed on the GSH biosynthetic enzymes and the responses of transgenic plants to O_3, a more detailed overview of GSH synthesis, regulation, and functions can be obtained from several excellent recent reviews (Rennenberg 1997; Noctor et al. 1998a; May et al. 1998a).

GSH is a tripeptide (γ-glutamylcysteinyl glycine) synthesized from its constituent amino acids in an ATP-dependent two-step reaction. In the first step, γ-glutamylcysteine (γ-EC) is synthesized from Cys and Glu by the action of γ-glutamylcysteine synthetase (γ-ECS), and in the second reaction GSH is formed by coupling Gly to the γ-EC using glutathione synthetase (GS). These two enzymes have been identified in both the cytosol and plastids of leaves and roots where GSH biosynthesis takes place (Bergmann and Rennenberg 1993; Hell and Bergman 1990). The emerging consensus for the observed increases in GSH pool sizes in response to various stress conditions, including air pollutants (Sen Gupta et al. 1991; Madamanchi and Alscher 1991), is that these are mediated through increased γ-ECS activities, as shown for responses to cadmium, herbicide safeners, and chilling (see references in May et al. 1998a). Furthermore, accumulating evidence suggests that γ-ECS is the rate-limiting step in GSH biosynthesis, although the availability of Cys, and feedback inhibition of γ-ECS by GSH (Hell and Bergman 1990), also appear to provide additional levels of control.

Genes encoding putative γ-ECS and GS enzymes have been isolated from several different plant species, including *A. thaliana* (May and Leaver 1994; Rawlins et al. 1995), *Brassica juncea* (Schäfer et al. 1998), and *Medicago truncatula* (Frendo et al. 1999). Accumulating molecular evidence suggests that both transcriptional and posttranscriptional mechanisms contribute to the regulation of γ-ECS in plants. For example, rapid post-/transcriptional activation of γ-ECS genes is observed following treatment of plants with cadmium (Schäfer et al. 1997), or with a ROS generator, a catalase inhibitor, or a safener (May et al. 1998b). Such studies, together with those that have used transgenic plants overexpressing bacterial genes for γ-ECS or GS (see Section 3.2), have not only increased our understanding of the molecular mechanisms that regulate the biosynthesis and homeostasis of GSH in plants, but have provided a means by which plants with ameliorated responses to stress can be generated.

3. Manipulation of Genes and Production of Transgenic Plants

Of the relatively large number of characterized plant genes that have been shown to be involved in sulfur and GSH metabolism, only a few have so far been used to generate transgenic plants and, of these, only a limited number have been used to examine their effects on plant responses to either air pollutants or to other stress factors. The characteristics of these transgenic plants are described in the following sections, and are briefly summarized in Fig. 1.

3.1 Manipulation of Genes Involved in Sulfur Metabolism

Several sets of transgenic tobacco, poplar, and brassica plants, overexpressing either plant- or bacterial-derived genes encoding enzymes of the sulfur assimilatory pathway, have recently been generated. Analysis of these plants has shed light not only on the regulation of this pathway but also on the importance of Cys biosynthesis in stress amelioration.

3.1.1 ATP Sulfurylase

ATP sulfurylase is considered to constitute one of the rate-limiting and regulatory enzymes of the sulfur assimilation pathway (Leustek 1996), and so transgenic plants overexpressing this gene could well be expected to have a modified metabolism and accumulation of sulfur compounds. Transformation of BY2 tobacco cells with an Arabidopsis cDNA encoding a putative plastidic ATP sulfurylase isoform, placed under control of the constitutive 35S CaMV promoter, was found to result in an eightfold increase in ATP sulfurylase enzymatic activities over control cells. However, after subjecting the cells to several different sulfur sources, it was concluded that ATP sulfurylase abundance did not regulate either sulfur uptake or content, and was not limiting for cell metabolism under the conditions used (Hatzfeld et al. 1998). In contrast, *Brassica juncea* transgenic plants, which overexpressed an Arabidopsis plastidic ATP sulfurylase, showed twofold higher shoot and root activities than controls, as well as increased uptake and reduction of, and tolerance to, the toxic sulfate analogue, selenate. In addition, the plants could uptake more sulfate and accumulate significantly higher levels of thiols and GSH than their control counterparts (Pilon-Smit et al. 1999) suggesting, in contrast to the report on BY2 cells, that ATP sulfurylase is limiting for GSH biosynthesis. There are currently no reports of the responses of these plants to air pollutants; however, it is conceivable that such transgenic plants could, by mechanisms similar to the foregoing, accumulate and possibly metabolize higher levels of sulfate that result, for example, from acid deposition.

3.1.2 O-Acetylserine(thiol) Lyase

The first reported case of transgenic plants containing a gene of either Cys or GSH biosynthesis was that of a set of tobacco plants overexpressing the *cys1* gene, encoding a wheat OASTL (Youssefian et al. 1993). The *cys1* gene, now considered to encode a cytosolic OASTL isoform, was placed under control of the constitutive CaMV 35S promoter in either sense or antisense orientations. Sense transformants showed three- to fivefold higher OASTL activities than both control and antisense plants, but no differences in either growth or development. Measurements of sulfhydryl levels in sense transgenic tobacco plantlets demonstrated, however, that the Cys, total thiol (SH), and GSH contents were elevated by over 120%, 20%, and 10%, respectively, over those of control plant levels (Youssefian et al. 2001). Similar increases in Cys and GSH were also observed in transgenic tobacco overexpressing spinach OASTL genes, although these effects were apparently not significant (Saito et al. 1994), and in a set of transgenic *cys1* hybrid poplar (*Populus sieboldi* x *P. grandidentata*) plants, with threefold higher OASTL activities (Nakamura and Youssefian, unpublished).

To evaluate changes in the thiol contents of these transgenic plants in response to sulfurous inputs, and yet overcome any feedback mechanisms that could affect root sulfate uptake, the transgenic poplar plants were fumigated with H_2S gas, a substrate of OASTL that is thought to be rapidly metabolized to Cys because of the high affinity of OASTL for H_2S (see De Kok et al., this volume). Exposure of the poplar plants to 0.2 μl l^{-1} H_2S for 3 or 6 days demonstrated that whereas Cys levels in control plants increased by 60% and 43%, respectively, over their unexposed control counterparts, Cys levels in transgenic poplar plants increased by 71% and 290%, respectively, over the unexposed transgenic plant levels. Furthermore, while GSH levels increased only marginally in control plants after 3 and 6 days of exposure, levels in transgenic plants increased by 24% and 74%, respectively, over unexposed plant levels (Fig. 2).

Similarly, in experiments in which we measured the thiol contents of the sixth distal leaf of 5-month-old transgenic tobacco plants exposed to 1.0 μl l^{-1} SO_2 for 48 h, at the end of which neither control nor transgenics showed any visible exposure-related damage, the transgenics demonstrated clear differential responses, with

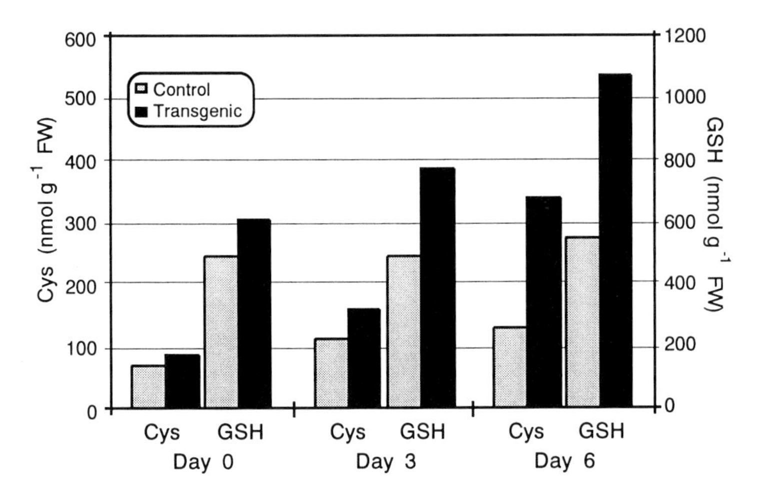

Fig. 2. Increased thiol contents in transgenic poplar plants exposed to H_2S gas. The fifth most distal (youngest mature) leaf from control (*shaded*) or *cys1* transgenic (*black*) poplar plants was sampled before (*Day 0*) or after 3 (*Day 3*) or 6 (*Day 6*) days of fumigation with 0.2 μl l^{-1} H_2S, and total nonprotein cysteine (*Cys*) and glutathione (*GSH*) levels were determined. No foliar damage was observed at the end of the treatment periods (from Nakamura and Youssefian, unpublished)

resultant thiol increases of 25% in controls and 61% in transgenics, consisting of Cys increases of 22% in controls and 65% in transgenics, and GSH increases of 21% in controls and 51% in transgenics (Youssefian et al. 2001).

Taken together, the results from our poplar and tobacco plants suggest that while the transgenic plants may have shown small but consistently higher levels of thiols than control plants under normal growth conditions, both sets of transgenic plants demonstrated greater capacities than controls to synthesize and accumulate Cys and GSH after exposure to the sulfurous gases. Saito et al. (1994) also showed that treatment of tobacco leaf disks with sulfite resulted in over twofold increases in the Cys levels of OASTL overexpressing transgenic plants but not of control plants.

To evaluate whether the increased potential of the transgenic plants to generate thiols was at all associated with an enhanced resistance to H_2S or SO_2, the extent of foliar and cellular damage was evaluated after exposure of the plants to acute doses of these pollutants. Exposure of the transgenic tobacco plants to an acute dose (at a peak level of 100 $\mu l\ l^{-1}$) of H_2S for either 2 h, 24 h, or 6 days, followed by a period of recovery, had demonstrated that after 2-h exposure, the sense plants displayed no obvious symptoms whereas the control plants developed small necrotic regions during the recovery period. After 24 h of exposure, all emerged leaves of the control plants showed severe necrosis resulting in final plant death, whereas only the older leaves of the sense plants became necrotic, with the younger leaves recovering and continuing to expand and develop. At the end of the 6-day exposure period, all plants died (Youssefian et al. 1993). Although these results clearly demonstrated the enhanced resistance of the transgenic plants to H_2S, the unrealistically high levels used and the unknown mechanisms of H_2S toxicity led us to evaluate the plant responses to SO_2, for which more phytotoxic information is available.

After exposure of young tobacco plantlets to 2 $\mu l\ l^{-1}$ SO_2 for 8 h, control plants could be clearly distinguished by the extensive and visible foliar damage, whereas transgenic plants were almost nonsymptomatic. These differences, which were not related to variations in stomatal resistance, were confirmed by measuring the effects of SO_2 on the photosynthetic performance of the plants on the basis of chlorophyll fluorescence, a measure of the effective quantum yield of photosystem II (Genty et al. 1989). In the control and transgenic plants, neither of which showed any visible foliar damage after fumigation with 1 $\mu l\ l^{-1}$ SO_2 for 6 h, the quantum yields were reduced by 21% and 11%, respectively. Exposure to 2 $\mu l\ l^{-1}$ SO_2 which, after the recovery period, caused visible necrotic patches in the control plants but not in the transgenic plants, resulted in more drastic reductions in quantum yields of 77% in the controls and only 37% in the transgenic plants in comparison with unfumigated plants (Youssefian et al. 2001). These results clearly demonstrated that the increased OASTL activities in the transgenic plants not only led to increased thiol levels, but also to enhanced tolerance to SO_2, through an increase in SO_2 metabolism and/or the detoxification of factors responsible for the SO_2-induced damage.

Indeed, SO_2 exposure is known not only to elevate sulfate and thiol contents, but also to increase cellular acidification, which could inactivate Calvin Cycle enzymes, and to enhance the production of ROS, which are thought to be the main proponents of SO_2-induced damage to the photosynthetic apparatus (Shimazaki et al. 1980). The possibility that the transgenic plants could tolerate ROS more effectively was examined through the use of methyl viologen (MV, paraquat) which, by functioning as a redox cycler, generates ROS independently of the sulfur pathway. Experiments with leaf disks infiltrated with various MV concentrations (0.1 to 5.0 μM), and incubated under light conditions for 18 to 24 h, demonstrated that leaf disks from control plants showed signs of chlorosis at about 0.2 mM MV, and almost complete chlorotic damage at 0.5 mM MV, whereas leaf disks from transgenic plants showed almost no chlorosis at 0.2 mM MV, limited signs of chlorosis at 0.5 mM MV, and complete chlorotic damage only at 2.0 mM MV (Youssefian et al. 2001). The extent of light-dependent, MV-induced membrane damage, used as an indicator of the extent of cellular injury, was determined by measuring the leakage of ionic solutes out of the cells. Whereas control and transgenic plants showed only minimal membrane damage in the dark, or in the light with water alone, both sets of plants were damaged by MV treatment in the light. However, in repeated experiments, control plants consistently showed greater rates of electrolyte leakage than the transgenic plants, indicative of the increased resistance of the transgenic plants to oxidative stress-induced membrane damage.

As oxidative stress is known to induce changes in the antioxidant defense system (for example, see Bowler et al. 1992), the possibility was examined that transcript levels or activities of the scavenging enzymes in the transgenic plants were also modified so as to further enhance their oxidative stress tolerance. Transcript levels of endogenous tobacco genes encoding various scavenging enzymes, including several isoforms of superoxide dismutase (SOD), catalase, ascorbate peroxidase, and glutathione reductase (GR), were thus analysed. Of these, transcripts of only a cytosolic Cu/ZnSOD were found to be consistently two- to threefold higher in the transgenic than in the control plants, under both normal conditions and those known to increase internal sulfate contents. However, under conditions that promoted Cys biosynthesis, transcript levels of the control plants attained those of the transgenic plants. Interestingly, the promoter of this same cytosolic Cu/ZnSOD was previously shown to be inducible only by reduced thiols, including Cys and GSH (Hérouart et al. 1993). Furthermore, an analysis of the transgenic plants demonstrated that the relative SOD enzyme activities of the transgenic plants were up to twofold higher than those of the controls under the conditions tested (Youssefian et al. 2001).

Overall, the results from these transgenic *cys1* plants suggest that these plants have greater potential to generate Cys than control plants, especially in response to SO_2, H_2S and possibly other oxidative stress conditions. As Cys availability appears to be limiting for GSH biosynthesis (see following section on GSH), the higher Cys levels could allow for increased GSH biosynthesis, especially under particular stress conditions that require elevated GSH levels. These increased thiol levels could then increase the reducing capacity of the cells or, by increasing SOD activities, directly detoxify the ROS generated by either SO_2 or MV. Whether overexpression of OASTL affects other cellular processes that could alter the responses of these plants to such stresses has yet to be determined. It is conceivable, for example, that in the transgenic plants the preferable reduction of SO_2-derived sulfite to sulfide rather than oxidation to sulfate in chloroplasts not only reduces the amount of ROS generated but also limits cellular acidification. Interestingly, an independent set of transgenic tobacco plants expressing the wheat *cys1* gene, were found to be considerably more resistant to sulfuric acid (pH 3-4) than control plants (Akihama et al., unpublished).

3.1.3 Serine Acetyltransferase

Cys biosynthesis is thought to be primarily regulated by the availability of OAS (Rennenberg 1984), which acts as a positive regulator of the pathway, through the regulated activity of serine acetyltransferase (SAT). Therefore, overexpression of SAT in transgenic plants would be expected to dramatically enhance Cys and GSH levels and to affect various cellular responses.

Recently, overexpression of a bacterial SAT, encoded by the cysE gene, was achieved in transgenic tobacco (Blaszczyk et al. 1999). The gene products, targeted to either the chloroplast or cytosol, resulted in an average two- to threefold increase in SAT activity as well as up to fourfold elevated Cys levels and, in several samples, twofold higher GSH levels under nonstressed conditions. Although there was no direct positive correlation between SAT activities and Cys levels, a correlation between Cys and GSH was observed. Most interestingly, these levels of GSH could be correlated with increased resistance to oxidative stress, as induced by 1 mM H_2O_2, and measured by the amount of chlorophyll remaining after the stress treatment. Indeed, T1 progenies, especially with SAT targeted to the cytosol, showed a higher level of resistance to 0.5 mM H_2O_2 than control plants. Such findings confirm the importance of OAS for Cys and GSH biosynthesis in plants, and hence for the subsequent resistance of the transgenic plants to various abiotic stress conditions. Whether such plants also show enhanced resistance to sulfurous pollutants or other atmospheric oxidants has yet to be evaluated.

3.2 Manipulation of Genes Involved in Glutathione Metabolism

Different series of transgenic tobacco and poplar plants have been generated to overexpress bacterial genes encoding γ-glutamylcysteine synthetase (γ-ECS) or glutathione synthetase (GS) in either their cytosol or chloroplasts, and several hybrids containing both these transgenes have been developed and used to analyze changes in the regulation of GSH homeostasis and in the sensitivity of the transgenic plants to oxidative stress exposure, including O_3.

3.2.1 γ-Glutamylcysteine Synthetase

The coding sequence of the *E. coli* gsh1 gene, encoding γ-ECS, was placed under control of the CaMV 35S promoter and used to transform poplar plants (Noctor et al. 1996; Arisi et al. 1997). The γ-ECS targeted to the cytosol resulted in an estimated 30 fold increase in γ-ECS activities (Noctor et al. 1997), and an average tenfold elevation in γ-glutamylcysteine (γ-EC), a threefold increase in GSH, and an almost twofold increase in Cys levels over control plants, without affecting the reduction state of the glutathione pool. Furthermore, feeding with Cys resulted in increased GSH levels in both transgenic and controls, and further increases in transgenic γ-EC levels. Overall, these findings suggested that γ-ECS overexpression could overcome the homeostatic restrictions on GSH biosynthesis, and that Cys availability, regardless of γ-ECS levels, was limiting for γ-EC and GSH synthesis, although of course the amount of γ-ECS then became a limiting factor. Exposure of such transgenic plants, with twofold the GSH levels of controls, to 94 nl l^{-1} O_3 for 4 weeks, however, demonstrated that the transgenic plants were no more resistant than controls in terms of the extent of visible damage (Will et al. 1997). In addition, leaf disks from control and transgenic poplar, expressing γ-ECS in the cytosol, showed only a marginal reduction in sensitivity to MV (Noctor et al. 1998a).

Similarly, overexpression of γ-ECS in the chloroplast of transgenic poplar, using the pea rbcS transit peptide sequence under control of the 35S CaMV promoter, also led to a 150 fold increase in the γ-EC level and to a two- to fourfold higher GSH pool in the transgenic plants without affecting the GSH reduction state or imposing any detrimental effects on plant growth (Noctor et al. 1998b). In contrast, transgenic tobacco with γ-ECS similarly targeted to the chloroplast had been found to result in necrotic lesions and impaired growth (Creissen et al. 1996). Further studies suggested that these transgenic tobacco plants, despite their fivefold higher GSH and 25 fold elevated γ-EC levels than control plants, were subject to increased oxidative stress, attributable to the low redox state of their GSH and γ-EC pools, possibly by a disturbance of their chloroplastic redox sensing balance (Creissen et al. 1999). Indeed, these transgenic plants were found to be highly sensitive to exposure to both MV and atmospheric O_3, based on changes in chlorophyll contents (Wellburn et al. 1998), and on ethylene emissions, considered a consistent discriminator of O_3 tolerance and sensitivity (Wellburn and Wellburn 1996). Hence, on the basis of these and similar findings, it is generally accepted that the most important determinant of oxidative stress resistance is not solely an increase in the total GSH pool size, but rather the capacity to regenerate reduced GSH pools, such as in transgenic plants overexpressing GR (Noctor et al. 1998a; see also Aono, this volume). In this regard, the poplar plants expressing γ-ECS in their chloroplasts appear to be able to maintain their high-content GSH pools under oxidizing conditions, an attribute considered important for future attempts to engineer multicomponent, stress-tolerant plants (Noctor et al. 1998a).

3.2.2 Glutathione Synthetase

Transgenic poplar plants expressing the *E. coli* gshII gene, encoding glutathione synthetase (GS), placed under the control of the CaMV 35S promoter, have been generated. Targeting of the GS to the cytosol resulted in 15- to 60 fold higher (Strohm et al. 1995) or up to 300 fold higher (Foyer et al. 1995) GS activities than control wild-type plants, yet their total foliar glutathione contents remained unaltered. As in the cytosolic-targeted γ-ECS transgenic poplar plants (Noctor et al. 1996; Arisi 1997), feeding leaf disks from GS and control plants with Cys resulted in twofold increases in GSH contents in both plants, but the transgenic leaf disks could sustain a twofold-higher rate of total glutathione synthesis over controls when supplied with exogenous γ-EC (Strohm et al. 1995). Taken together with the results of subsequent experiments, these findings reconfirmed that GSH biosynthesis is primarily limited by the availability of Cys and, when this limitation is overcome, control is exerted first by γ-ECS and then by GS activity (Noctor et al. 1998a). To evaluate the significance of such 200- to 300 fold higher GS levels in the cytosol of transgenic poplar plants on their tolerance to acute O_3, the transgenic plants were exposed to 100, 200, or 300 nl l^{-1} O_3 for 3 days at 7 h/d. Exposure to 100 nl l^{-1} resulted in negligible leaf injury, but 200 or 300 nl l^{-1} exposure resulted in up to 60% visible injury on mature leaves of both control and transgenic lines. From these studies, it was concluded that improved tolerance of hybrid poplar to acute O_3 stress could not be achieved simply by elevating cytosolic GS activities, but that such resistance was rather associated with leaf development (Strohm et al. 1999).

Similarly, transgenic poplar or tobacco plants overexpressing GS in the chloroplast, using the *E. coli* gshII gene together with the pea rubisco (rbcS) transit sequence placed under control of the CaMV 35S promoter, did not affect the total GSH contents compared to control plants despite their high chloroplastic GS activities (Noctor et al. 1998b; Wellburn et al. 1998). Although there are currently no reports of the responses of these poplar transformants to stress conditions, the tolerance of transgenic tobacco plants overexpressing chloroplastic GS to both MV and O_3 were examined in the same experiment in which tobacco plants overexpressing chloroplastic γ-ECS were evaluated (Wellburn et al. 1998). Here, the T2 progeny of a cross between plants overexpressing either γ-ECS or GS in their chloroplasts were exposed to a total of 480 nl l^{-1} atmospheric O_3 for 2 days, at a peak concentration of 120 nl l^{-1}. Almost 50% of the progeny carrying only the GS transgene showed tolerance to O_3, based on reduced ethylene emissions and increased or unchanged total chlorophyll contents. These plants were, however, not tolerant to MV. In contrast, only one plant carrying both γ-ECS and GS showed limited O_3 tolerance and, together with another hybrid, some level of MV tolerance. Of the progeny of another two independent GS transformed lines that included both transgenic and wild-type plants which were also challenged by O_3 exposure, again 20% to 40% of the plants overexpressing GS were found to be O_3 tolerant. The discrepancies between these latter results, showing tolerance of some transgenic tobacco to O_3, and those of Strohm et al. (1999), in which poplar plants overexpressing GS in cytosol were found to be no more O_3 tolerant than control plants, could be related to the species used, the localization of the transgene product, the fumigation conditions employed, or the methods of assessing damage and tolerance.

Despite the various discrepancies in results, the fact that emerges most clearly from all these transgenic studies is that the GSH pools can be effectively modified. Possibly through the concomitant alteration of other ROS-scavenging enzymes, especially GR, SOD, and ascorbate peroxidase, an even greater level of tolerance to O_3 may be achieved in such transgenic plants.

4. Concluding Remarks

The recent and rapid advances in our understanding of the molecular mechanisms regulating sulfur and GSH metabolism have brought within our reach the possibility of manipulating specific aspects of these pathways to generate plants with altered responses, not only to sulfurous pollutants and photochemical oxidants such as SO_2 and O_3 but also to other abiotic and biotic stress responses, including oxidative stress, heavy metals, and even pathogens, with which these pathways are intricately associated and in which Cys and GSH apparently play protective roles.

Yet, despite the continuing progress in the isolation and characterization of a growing number of genes involved in these pathways, numerous fundamental questions still remain to be addressed, especially with regards to the exact functions of the various isoforms and, most importantly, to the precise molecular mechanisms by which the various enzymes are coordinately regulated and the overall pathways finely tuned to respond to environmental changes. Progress in these areas, as well as in those that aim to identify the various signaling factors involved, such as protein kinases and secondary messengers, will allow the overall coordinated responses of such plants to be modified. Furthermore, although the results from current analyses of transgenic plants generally demonstrate enhanced levels of resistance to various pollutants, the potential use of transgenic plants for bioremediation will require, in addition to increased pollutant metabolism, an improvement in the transport and storage of the pollutant secondary products. Therefore, a multicomponent strategy, with emphasis on these characteristics, will be an essential aspect of generating plants useful for phytoremediation. Taken together, these very preliminary studies of transgenic plants clearly serve as simple indicators of the tremendous potential of generating transgenic plants that will be essential components of future strategies for air pollution monitoring and control.

Acknowledgments. My grateful thanks to all those colleagues who sent reprints and manuscripts before publication.

References

Arisi ACM, Noctor G, Foyer C, Jouanin L (1997) Modifications of thiol contents in poplars (*poplar tremula* x *P. alba*) overexpressing enzymes involved in glutathione synthesis. Planta 203:362-372

Barroso C, Romero LC, Cejudo FJ, Vega JM, Gotor C (1999) Salt-specific regulation of the cytosolic *O*-acetylserine(thiol)lyase gene from Arabidopsis is dependent on abscisic acid. Plant Mol Biol 40:729-736

Bergmann L, Rennenberg H (1993) Glutathione metabolism in plants. In: De Kok LJ, Stulen I, Rennenberg H et al. (eds) Sulphur nutrition and assimilation in higher plants. SPB Academic, The Hague, pp 107-123

Bick JA, Leustek T (1998) Plant sulfur metabolism: the reduction of sulfate to sulfite. Curr Opin Plant Biol 1:240-244

Blaszczyk A, Brodzik R, Sirko A (1999) Increased resistance to oxidative stress in transgenic tobacco plants overexpressing bacterial serine acetyltransferase. Plant J 20:237-243

Bogdanova N, Bork C, Hell R (1995) Cysteine biosynthesis in plants: isolation and functional characterization of a cDNA encoding serine acetyltransferase from *Arabidopsis thaliana*. FEBS Lett 358:43-47

Bolchi A, Petrucco S, Tenca PL, Foroni C, Ottonello S (1999) Coordinate modulation of maize sulfate permease and ATP sulfurylase mRNAs in response to variations in sulfur nutritional status: stereospecific down-regulation by L-cysteine. Plant Mol Biol 39:527-537

Bork C, Schwenn JD, Hell R (1998) Isolation and characterization of a gene for assimilatory sulfite reductase from *Arabidopsis thaliana*. Gene 212:147-153

Bowler C, Van Montagu M, Inze D (1992) Superoxide dismutase and stress tolerance. Annu Rev Plant Physiol Plant Mol Biol 43:83-116

Brunold C, Rennenberg H (1997) Regulation of sulphur metabolism in plants: first molecular approaches. Prog Bot 58:164-186

Creissen G, Broadbent P, Stevens R, Wellburn AR, Mullineaux P (1996) Manipulation of glutathione metabolism in transgenic plants. Biochem Soc Trans 24:465-469

Creissen G, Firmin J, Fryer M, Kular B, Leyland N, Reynolds H, Pastori G, Wellburn F, Baker N, Wellburn A, Mullineaux P (1999) Elevated glutathione biosynthetic capacity in the chloroplasts of transgenic tobacco plants paradoxically causes increased oxidative stress. Plant Cell 11:1277-1291

De Kok LJ (1990) Sulfur metabolism in plants exposed to atmospheric sulphur. In: Rennenberg H, Brunold C, De Kok LJ, Stulen I (eds) Sulfur nutrition and sulfur assimilation in higher plants. SPB Academic, The Hague, pp 111-130

Droux M, Ruffet ML, Douce R, Job D (1998) Interactions between serine acetyltransferase and *O*-acetylserine (thiol) lyase in higher plants: structural and kinetic properties of the free and bound enzymes. Eur J Biochem 255:235-245

Foyer CH, Souriau N, Perret S, Lelandais M, Kunert K-J, Pruvost C, Jouanin L (1995) Overexpression of glutathione reductase but not glutathione synthetase leads to increases in antioxidant capacity and resistance to photoinhibition in poplar trees. Plant Physiol 109:1047-1057

Frendo P, Gallessi D, Turnbull R, Van de Sype G, Hérouart D, Puppo A (1999) Localization of glutathione and homoglutathione in *Medicago trunculata* is correlated to a differential expression of genes involved in their synthesis. Plant J 17:215-219

Genty BE, Briantais JM, Baker NR (1989) The relationship between the quantum yield of photosynthetic electron transport and quenching of chlorophyll fluorescence. Biochim

Biophys Acta 990:87-92

Gutierrez-Marcos JF, Roberts MA, Campbell EI, Wray JL (1996) Three members of a novel small gene family from *Arabidopsis thaliana* able to complement functionally an *Escherichia coli* mutant defective in PAPS reductase activity encode proteins with a thioredoxin-like domain and "APS reductase" activity. Proc Natl Acad Sci USA 93:13377-13382

Hatzfeld Y, Cathala N, Grignon C, Davidian J-C (1998) Effect of ATP sulfurylase overexpression in Bright Yellow 2 tobacco cells. Regulation of ATP sulfurylase and SO_4^{2-} transport activities. Plant Physiol 116:1307-1313

Hell R (1997) Molecular physiology of plant sulphur metabolism. Planta 202:138-148

Hell R, Bergman L (1990) γ-Glutamylcysteine synthase in higher plants: catalytic properties and subcellular localization. Planta 180:603-612

Hérouart D, Van Montagu M, Inze D (1993) Redox-activated expression of the cytosolic copper/zinc superoxide dismutase gene in *Nicotiana*. Proc Natl Acad Sci USA 90:3108-3112

Klonus D, Hofgen R, Willmitzer L, Riesmeier JW (1994) Isolation and characterization of two cDNA clones encoding ATP-sulfurylases from potato by complementation of a yeast mutant. Plant J 6:105-112

Klonus D, Riesmeier JW, Willmitzer L (1995) A cDNA clone for an ATP-sulfurylase from *Arabidopsis thaliana*. Plant Physiol 107:653-654

Kopriva S, Muheim R, Koprivova A, Trachsel N, Catalano C, Suter M, Brunold C (1999) Light regulation of assimilatory sulphate reduction in *Arabidopsis thaliana*. Plant J 20:37-44

Lappartient AG, Vidmar JJ, Leustek T, Glass ADM, Touraine B (1999) Inter-organ signalling in plants: regulation of ATP-sulfurylase and sulfate transporter genes expression in roots mediated by phloem-translocated compound. Plant J 18:89-95

Lee S, Leustek T (1999) The effect of cadmium on sulfate assimilation enzymes in *Brassica juncea*. Plant Sci 141:201-207

Leustek T (1996) Molecular genetics of sulfate assimilation in plants. Physiol Plant 97:411-419

Leustek T, Saito K (1999) Sulphate transport and assimilation in plants. Plant Physiol 120:637-643

Leustek T, Murillo M, Cervantes M (1994) Cloning of a cDNA encoding ATP sulfurylase from *Arabidopsis thaliana* by functional complementation in *Saccharomyces cerevisae*. Plant Physiol 105:897-902

Lunn JE, Droux M, Martin J, Douce R (1990) Localization of ATP sulfurylase and *O*-acetyl-L-serine(thiol)lyase in spinach leaves. Plant Physiol 94:1345-1352

Madamanchi NR, Alscher RG (1991) Metabolic bases for differences in sensitivity of two pea cultivars to sulfur dioxide. Plant Physiol 97:88-93

Maruyama A, Ishizawa K, Takagi T, Esashi Y (1998) Cytosolic β-cyanoalanine synthase activity attributed to cysteine synthases in cocklebur seeds. Purification and characterization of cytosolic cysteine synthases. Plant Cell Physiol 39:671-680

May MJ, Leaver CJ (1994) *Arabidopsis thaliana* γ-glutamylcysteine synthase is structurally unrelated to mammalian, yeast and *E. coli* homologs. Proc Natl Acad Sci USA 91:10059-10063

May MJ, Vernoux T, Leaver C, Van Montagu M, Inze D (1998a) Glutathione homeostasis in plants: implications for environmental sensing and plant development. J Exp Bot 49:649-667

May MJ, Vernoux T, Sanchez-Fernandez R, Van Montagu M, Inze D (1998b) Evidence for posttranscriptional activation of γ-glutamylcysteine synthase during plant stress

responses. Proc Natl Acad Sci USA 95:12049-12054
Nakamura T, Yamaguchi Y, Sano H (1999) Four rice genes encoding cysteine synthase: isolation and differential responses to sulfur, nitrogen and light. Gene 229:155-161
Noctor G, Strohm, M, Jouanin L, Kunert KJ, Foyer CH, Rennenberg H (1996) Synthesis of glutathione in leaves of transgenic poplar (*Populus tremula* x *P. alba*) overexpressing γ-glutamylcysteine synthetase. Plant Physiol 112:1071-1078
Noctor G, Jouanin L, Foyer CH (1997) The biosynthesis of glutathione explored in transgenic plants. In: Hatzios K (ed) Regulation of enzymatic systems detoxifying xenobiotics in plants. Nato ASI series. Kluwer, Dordrecht, pp 109-124
Noctor G, Arisi ACM, Jouanin L, Kunert KJ, Rennenberg H, Foyer CH (1998a) Glutathione: biosynthesis, metabolism and relationship to stress tolerance explored in transformed plants. J Exp Bot 49:623-647
Noctor G, Arisi ACM, Jouanin L, Foyer CH (1998b) Manipulation of glutathione and amino acid biosynthesis in the chloroplast. Plant Physiol 118:471-482
Pilon-Smits EHA, Hwang S, Lytle CM, Zhu Y, Tai JC, Bravo RC, Chen Y, Leustek T, Terry N (1999) Overexpression of ATP sulfurylase in Indian mustard leads to increased selenate uptake, reduction and tolerance. Plant Physiol 119:123-132
Rawlins MR, Leaver CJ, May MJ (1995) Characterization of a cDNA encoding *Arabidopsis* glutathione synthetase. FEBS Lett 376:81-86
Rennenberg H (1984) The fate of excess sulphur in higher plants. Annu Rev Plant Physiol 35:121-153
Rennenberg H (1997) Molecular approaches to glutathione biosynthesis. In: Cram WJ, De Kok LJ, Stulen I, et al. (eds) Sulphur metabolism in higher plants, molecular, ecophysiological and nutritional aspects. Backhuys, Leiden, pp 59-70
Rennenberg H, Herschbach C (1996) Responses of plants to atmospheric sulphur. In: Yunus M, Iqbal M (eds) Plant response to air pollution. J Wiley, London, pp 285-293
Rolland N, Droux M, Douce R (1992) Subcellular distribution of *O*-acetylserine(thiol)lyase in cauliflower (*Brassica oleracea* L.) inflorescence. Plant Physiol 98:927-935
Ruffet M-L, Droux M, Douce R (1994) Purification and kinetic properties of serine acetyltransferase free of *O*-acetylserine(thiol)lyase from spinach chloroplasts. Plant Physiol 104:597-604
Saito K, Kurosawa M, Tatsuguchi K, Takagi Y, Murakoshi I (1994) Modulation of cysteine biosynthesis in chloroplasts of transgenic tobacco overexpressing cysteine synthase [*O*-acetylserine(thiol)-lyase]. Plant Physiol 106:887-895
Schäfer HJ, Greiner S, Rausch T, Haag-Kerwer A (1997) In seedlings of the heavy metal accumulator *Brassica juncea* Cu^{2+} differentially affects transcript amounts for γ-glutamylcysteine synthase (γ-ECS) and metallothionein (MT2). FEBS Lett 404:216-220
Schäfer HJ, Haag-Kerwer A, Rausch T (1998) cDNA cloning and expression analysis of genes encoding GSH synthesis in roots of the heavy-metal accumulator *Brassica juncea* L. evidence for Cd-induction of a putative mitochondrial γ-glutamylcysteine synthase isoform. Plant Mol Biol 37:87-97
Schnug E (1998) Response of plant metabolism to air pollution and global change - impact on agriculture. In: De Kok LJ, Stulen I (eds) Responses of plant metabolism to air pollution and global change. Backhuys, Leiden, pp 15-22
Sen Gupta A, Alscher RG, McCune D (1991) Response of photosynthesis and cellular antioxidants to ozone in Populus leaves. Plant Physiol 96:650-655
Shimazaki K-I, Sakaki T, Kondo N, Sugahara K (1980) Active oxygen participation in chlorophyll destruction and lipid peroxidation in SO_2-fumigated leaves of spinach. Plant Cell Physiol 21:1193-1204
Smith FW, Ealing PM, Hawkesford MJ, Clarkson DT (1995) Plant members of a family of

sulphate transporters reveal functional subtypes. Proc Natl Acad Sci USA 92:9373-9377

Smith FW, Hawkesford MJ, Ealing PM, Clarkson DT, Vanden Berg PJ, Belcher AR, Warrilow AGS (1997) Regulation of expression of a cDNA encoding a high affinity sulphate transporter. Plant J 12:875-884

Strohm M, Jouanin L, Kunert, KJ, Pruvost C, Polle A, Foyer CH, Rennenberg H (1995) Regulation of glutathione synthesis in leaves of transgenic poplar (*Populus tremula* x *P. alba*) overexpressing glutathione synthetase. Plant J 7:141-145

Strohm M, Eiblmeier M, Langebartels C, Jouanin L, Polle A, Sandermann H, Rennenberg H (1999) Responses of transgenic poplar (*Populus tremula* x *P. alba*) overexpressing glutathione synthetase or glutathione reductase to acute ozone stress: visible injury and leaf gas exchange. J Exp Bot 50:363-372

Takahashi H, Saito K (1996) Subcellular localization of spinach cysteine synthase isoforms and regulation of their gene expression by nitrogen and sulfur. Plant Physiol 112:273-280

Takahashi H, Yamazaki M, Sasakura N, Watanabe A, Leustek T, de Almeida Engler J, Engler G, Van Montagu M, Saito K (1997) Regulation of sulphur assimilation in higher plants: a sulphur transporter induced in sulfate-starved roots plays a central role in *Arabidopsis thaliana.* Proc Natl Acad Sci USA 94:11102-11107

Wellburn AR (1988) Air pollution and acid rain: the biological impact. Longman, England, pp 1-274

Wellburn FAM, Wellburn AR (1996) Variable patterns of antioxidant protection but similar ethylene differences in several ozone-sensitive and -tolerant selections. Plant Cell Environ 19:754-760

Wellburn FAM, Creissen GP, Lake JA, Mullineaux PM, Wellburn AR (1998) Tolerance to atmospheric ozone in transgenic tobacco overexpressing glutathione synthetase in plastids. Physiol Plant 104:623-629

Will B, Eiblmeier M, Langebartels C, Rennenberg H (1997) Consequences of chronic ozone exposure in transgenic poplars overexpressing enzymes of the glutathione metabolism. In: Cram WJ, De Kok LJ, Stulen I, et al. (eds) Sulphur metabolism in higher plants, molecular, ecophysiological and nutritional aspects. Backhuys, Leiden, Netherlands, pp 257-259

Youssefian S, Nakamura M, Sano H (1993) Tobacco plants transformed with the *O*-acetylserine (thiol) lyase gene of wheat are resistant to toxic levels of hydrogen sulphide gas. Plant J 4:759-769

Youssefian S, Nakamura M, Orudgev E, Kondo N (2001) Increased cysteine biosynthesis capacity of transgenic tobacco overexpressing an O-acetylserine(thiol) lyase modifies plant responses to oxidative stress. Plant Physiol 126 (in press)

Yunus M, Singh N, Iqbal M (1996) Global status of air pollution. In: Yunus M, Iqbal M (eds) Plant response to air pollution. J Wiley, London, pp 1-34

20
Manipulation of Genes for Nitrogen Metabolism in Plants

Hiromichi Morikawa, Misa Takahashi, and Gen-Ichiro Arimura

Department of Mathematical and Life Sciences, Graduate School of Science, Hiroshima University, Kagamiyama 1-3-1, Higashi-Hiroshima, Hiroshima 739-8526, Japan

1. Introduction

A 1980 estimate puts the total natural and anthropogenic emissions of nitrogen oxides, which include nitric oxide (NO) and nitrogen dioxide (NO_2) as the major components, at 150 million tons per year. Road transport, the major anthropogenic source of nitrogen oxides in many developed countries, accounted for up to 75% of the nitrogen oxides in some metropolitan cities in 1984. This value is still rising because of the increase in volume of road traffic. In many developing countries, petrofueled motor vehicles are also reported to be the principal source of nitrogen oxides (Yunus et al. 1996).

In mammalian cells, NO is known to be a short-lived vital intercellular messenger and has been implicated in the regulation of blood pressure, platelet adhesion, neutrophil aggregation, and synaptic plasticity in the brain (for example, see Lancaster and Stuehr 1996). The abundant presence of this chemical in the environment (atmospheric concentration of NO is usually similar to that of NO_2) must necessarily affect human and animal health, although no reports on the quantitative analysis of the effects of environmental NO on human health have been published to date. NO_2 is known to induce oxidative modification of nucleic acid bases (Suzuki et al. 1996). The physiological effects of nitrogen oxides on plant cells, however, are unclear, apart from the pioneering studies of their inhibitory effects on plant growth (Srivastava et al. 1994; Wellburn 1990; Lea et

Air Pollution and Plant Biotechnology
–Prospects for Phytomonitoring and Phytoremediation–
Edited by K. Omasa, H. Saji, S. Youssefian, and N. Kondo
© Springer -Verlag Tokyo 2002

al. 1996). We recently have found that the uptake of CO_2 in plant leaves is inhibited by nitrogen oxides (Goshima et al., unpublished results).

Plants are reported to assimilate the nitrogen of NO_2 into organic compounds (Durmishidze and Nutsubidze 1976; Rogers et al. 1979; Wellburn 1990, 1994; Yoneyama and Sasakawa 1979), including amino acids. Therefore, natural and artificial (agricultural) vegetation can act as a major "sink" for atmospheric pollutants in terrestrial ecosystems, although quantitative data on the contribution of worldwide vegetation as a global sink for nitrogen oxides are not available.

The fact that the atmospheric level of nitrogen oxides is rising all over the world suggests that the capacity of "naturally occurring sinks," such as plants, may be already becoming saturated. Therefore, innovative methods to decrease the atmospheric level of nitrogen oxides, or improvement of existing sinks for these gases, is an issue of considerable urgency.

Trees (such as phenol releasers) that provide a pertinent rhizospheric environment for microorganisms which degrade petrochemical contaminants in soils have been explored (J. Fletcher, personal communication; Donnelly and Fletcher 1995; Schwab and Banks 2000). Also, hyperaccumulators, such as *Brassica juncea*, that have a high ability to accumulate heavy metals from tainted soils have been investigated (for example, Baker et al. 1998; Salt et al. 1995). In contrast to such remediation of polluted soils, the remediation of air is far more difficult to evaluate, at least on a field-test scale. However, screening for "hyperassimilators" that have a high capability to clean up air pollutants is a very intriguing and important concept.

With the eventual aim of improving the ability of plants to act as sinks for atmospheric nitrogen oxides, we have been studying the production of a novel " NO_2-philic plant" that can grow with atmospheric NO_2 as its sole nitrogen source (Morikawa et al. 1992). We have discovered that among 217 plant taxa tested, the ability to assimilate nitrogen dioxide varies more than 600 fold, and that nine plant species, including woody street trees, can be considered as NO_2-philic because of their high ability to utilize nitrogen dioxide nitrogen as a nitrogen source (Morikawa et al. 1998a,b).

Because most, if not all, the nitrogen of nitrogen oxides taken up into plant cells is metabolized through the primary nitrogen pathway (Yoneyama and Sasakawa 1979), it may be most pertinent to study and manipulate those genes involved in nitrate assimilation to increase the ability of the plants to assimilate nitrogen dioxide, and to thus act as efficient sinks of nitrogen dioxide. Such engineered plants will be vital for phytoremediation of the global atmospheric environment.

In this review we focus on studies in which genes related to nitrate metabolism in plants have been manipulated. We then briefly discuss some of the experimental results recently obtained in our laboratory, in relation to the manipulation of nitrite reductase and glutamine synthetase genes for improving the ability of plants to assimilate nitrogen oxides.

2. Genetic Manipulation of NR Genes

Nitrate reductase (NR) in plants is a homodimeric enzyme, with each subunit consisting of a 110- to 120- kDa polypeptide. The enzyme is a metalloprotein and each subunit is associated with three redox prosthetic groups: flavin adenine dinucleotide (FAD), heme, and a molybdenum cofactor (MoCo) (stoichiometry, 1:1:1). NR catalytically transfers two electrons from NAD(P)H to nitrate via these three redox centers, in this order. NR has two active sites, one where NADH donates electrons to FAD to begin the transport of electrons via the heme-Fe to the Mo/Mo-pterin, in the second active site, where nitrate is reduced to nitrite. The number of NR genes in plants is believed to vary from one to four depending on the plant species (Caboche and Rouze 1990; Campbell 1996).

Nitrate reduction is an energy-consuming process, and as much as 25% of the energy generated by photosynthesis is consumed in driving nitrate assimilation (Solomonson and Barber 1990). Furthermore, incorporation of the ammonium ion, an intermediate of the nitrate reduction process, into amino acids requires carbon skeletons. In addition, the nitrite and ammonium ions are toxic to the cell when accumulated. Therefore, it is assumed that to prevent excessive use of the plant carbohydrate reserves and to avoid the accumulation of toxic intermediates, NOR gene expression is tightly regulated at the transcriptional and posttranscriptional levels. When shifted to the dark, the level of NOR activity in leaves of tobacco and other plants is decreased posttranscriptionally to about 50% of that in the light. Even the expression of transgenes, carrying NR sequences driven by a constitutively expressed promoter, is known to be suppressed, although not always, by the induction of host NR gene expression (Vaucheret et al. 1992b; Vaucheret and Caboche 1992). However, it is intriguing that mutations that reduce NR activity 10 fold have little observable effect on the growth of plants fed only nitrate (NO_3^-) (Crawford 1995).

Transgenic overexpressors of NR, in which the NR enzyme activity was elevated about 4 fold (Dorlhac de Borne et al. 1994) to 5 fold (Foyer et al. 1994) or 1.25- to 2.5 fold (Quillere et al. 1994) that of the wild type, have been reported. In these studies, the tobacco NR (*nia*-2) cDNA, driven by a cauliflower mosaic virus (CaMV) 35S promoter, 35S-NR, was introduced into wild-type plants of *Nicotiana tabacum* (Dorlhac de Borne et al. 1994) or into an NR-deficient line (E23) of *Nicotiana. plumbaginifolia* (Foyer et al. 1994; Quillere et al. 1994). Light/dark responses of the enzyme activities were similar to that of the wild-type control plant, suggesting that while NR transgene expression was constitutive in these transgenic plants, the translational regulation and posttranslational modification of the product proceeded normally (Foyer et al. 1994).

A fivefold increase in NR activity is substantial, considerably larger than the observed twofold increase in this enzyme activity in response to light in tobacco and other plants. However, the increased enzyme activity in transgenic plants did not lead to a direct increase in their total protein content, photosynthetic activity, or biomass yield (Foyer et al. 1994; Quillere et al. 1994).

Interestingly, the foliar nitrate content in the overexpressors was decreased to 32% to 47% that of controls (Quillere et al. 1994), although it is unclear whether this decrease occurred in the metabolic or nonmetabolic (storage) pools (Aslam et al. 1976). Furthermore, the ammonia content increased slightly (28%) and the level of glutamine also increased by about 2.3 fold in the overexpressors (Foyer et al. 1994; Quillere et al. 1994). This result clearly indicates that overexpression of NR influences the intracellular steady -state levels of intermediates of nitrate metabolism, such as ammonia and glutamine. Because glutamine is considered to be a signal molecule in plants, an increase in root glutamine content may be expected to cause a decrease in NO_3^- uptake (Ferrario-Mery et al. 1997a,b).

Gojon et al. (1998) have investigated nine transformant NR underexpressors and overexpressors of *Nicotiana plumbaginifolia* and *N. tabacum.* In the overexpressors, total $^{15}NO_3^-$ assimilation was not significantly increased when compared with wild-type plants, because the higher $^{15}NO_3^-$ reduction efficiency was offset by a lower $^{15}NO_3^-$ uptake by the roots. This inhibition of NO_3^- uptake appeared to be the result of negative feedback regulation of NO_3^- influx and was interpreted as an adjustment of NO_3^- uptake to prevent excessive amino acid synthesis. Gojon et al. also concluded that the products of NO_3^- assimilation are not the only factors responsible for downregulation of the NO_3^- uptake system. Net uptake of the NO_3^- ion is balanced by efflux and influx, both of which require metabolic energy and appear to be separate systems that are regulated independently (Travis et al. 1998).

Culture of plants under elevated CO_2 results in an increase in the C:N ratio of plants (Purvis et al. 1974; Cure et al. 1988; Hocking and Meyer 1991; Rogers et al. 1993; McKee and Woodward 1994). Therefore, enrichment of the plant culture atmosphere with CO_2 would be expected to lead to formation of a novel nitrogen sink. However, the biomass yield and the C:N ratio in ^{35}S-NR tobacco plants were not altered by culture in elevated CO_2 (Ferrario-Mery et al. 1997b). Stimulation of the translocation of glutamine or creation of new nitrogen sinks may well, therefore, improve the ability of NR overexpressors to metabolize nitrate. In this context, screening "sink proteins," if any, that preferentially accumulate NO_2-derived nitrogen in plant leaves is a potential area of future study. No research has been published on the ability to assimilate nitrogen ^{15}N in these NR overexpressors. Therefore, the question of whether these overexpressors have a high ability to assimilate nitrate and nitrogen still remains open.

3. Genetic Manipulation of NiR Genes

Nitrite reductase (NiR) is the second enzyme involved in the primary metabolism of nitrate, and catalyzes the six-electron reduction of nitrite to ammonia. NiR is a monomeric metalloprotein of 63 kDa containing a siroheme, to which nitrite binds, and a 4 Fe/4 S center, which is probably the initial electron acceptor, as prosthetic groups (Ip et al. 1990; Siegel and Wilkerson 1989). The number of NiR genes in plants varies from one to four depending on the plant species (Kronenberger et al. 1993; Tanaka et al. 1994).

In green leaves, NiR is located in the chloroplasts (Dalling et al. 1972), whereas in roots it is located in the plastids. However, the presence of an extrachloroplastic form has been proposed in the cotyledon of mustard (Schuster and Mohr 1990). The direct electron donor in the chloroplast is reduced ferredoxin while in roots a ferredoxin-like protein has been implicated (Suzuki et al. 1985), which probably obtains reducing power from NADH generated in the oxidative pentose phosphate (OPP) pathway in plastids (Bowsher et al. 1989; Oji et al. 1985). Recently, Jin et al. (1998) have provided direct evidence for electron transport from glucose-6-phosphate dehyrogenase (G6PDH), the key regulatory enzyme of the OPP pathway, to support NO_2^- reduction using in vitro reconstitution system composed of G6PDH, ferredoxin (Fd) NADP+ oxidoreductase (FNR), Fd, and NiR from *Chlamydomonas reinhardtii*.

Regulated expression of the NiR gene is believed to share similarities with that of NR, and thus NiR expression is tightly regulated because the reaction catalyzed by NiR is also an energy- and carbon skeleton consuming, toxic compound-forming, process. Coordinated expression of the gene encoding the NiR apoenzyme, as well as of the prosthetic groups, is essential for the formation of the NiR holoenzyme. Therefore, it is generally considered that a simple overexpression of the NiR gene itself will not lead to increased in vivo NiR activity, even though it may increase the in vivo level of the NiR apoenzyme. Nevertheless, the in vitro enzyme activity was found to increase in both *Nicotiana plumbaginifolia* and *Arabidopsis thaliana* by introduction of the NiR cDNA driven by the CaMV 35*S* promoter (Crete et al. 1997). We also have obtained similar results with *Arabidopsis thaliana* (Takahashi et al. 2001). These findings suggest that synthesis of the NiR prosthetic groups is somehow induced upon synthesis of the NiR apoenzyme or is redundantly regulated. In both NiR overexpressors, including ours, the in vitro NiR activity of the NiR overexpressors was no more than threefold that of the wild type. It is possible that high levels of NiR activity are lethal for plants, and that during the regeneration process we inadvertently counterselected the highest overexpressing transformants (Crete et al. 1997). Whether these NiR overexpressors retain high in vivo NiR activity is uncertain, because overexpression of NiR gene per se is insufficient to increase in vivo NiR activity and a supply of relevant reduced ferredoxin is also essential. No reports, however, are available on the ability of NiR overexpressors to reduce nitrate or nitrite except for an analysis of their ability to reduce nitrogen dioxide.

It has been difficult isolating NiR-defective mutants in plants because inhibition of nitrite reduction would lead to the accumulation of toxic nitrite (Duncanson et al. 1993). A reduction of NiR levels has been achieved in tobacco by expressing NiR antisense RNA (Vaucheret et al. 1992a). These antisense transformants, with no detectable NiR activity, displayed impaired development and chlorotic leaves when fed nitrate and had to be maintained on ammonia as the sole nitrogen source. Interestingly, these transformants had higher and still inducible levels of NR transcripts and activity, probably because of the reduced levels of a nitrogen metabolite (such as glutamine) or because of the increased level of nitrate in their metabolic pools.

4. Genetic Manipulation of GS Genes

Ammonia is incorporated into glutamine and glutamate through the collaborative action of two enzymes, glutamine synthetase (GS) and a ferredoxin-dependent glutamate synthase (Fd-GOGAT) or an NADH-dependent glutamate synthase (NADH-GOGAT) (Lea et al. 1990). GS catalyzes the incorporation of ammonia into the amide position of glutamate, producing glutamine. GOGAT then catalyzes the reductive transfer of the amide group of glutamine to the α-keto position of 2-oxoglutarate, forming two molecules of glutamate. Glutamate dehydrogenase (GDH), which serves in the primary route of nitrogen metabolism in microorganisms, functions largely in glutamate catabolism in higher plants (Lea et al. 1990; Robinson et al. 1991; see also following).

GS in plants occurs as a number of isoenzymes, and is encoded by a small multigene family. GS is an octameric enzyme composed of 40-kDa subunits in the GS1 and 45-kDa subunits in the GS2 isoform. The GS1 and GS2 polypeptides, which can be distinguished by ion-exchange chromatography, can be broadly categorized into cytosolic (GS1) and plastidic (GS2) forms on the basis of their subcellular localization. In all species published to date, a single nuclear gene encodes the chloroplastic GS2. All plants have multiple genes encoding the cytosolic GS1 (Temple and Sengupta-Gopalan 1997).

GS2 appears to function to assimilate primary ammonia reduced from nitrate, and also to reassimilate ammonia released during photorespiration in mesophyll cells (Forde et al. 1989; Edwards et al. 1990; Carvalho et al. 1992; Hirel et al. 1992). Ammonia produced via photorespiration exceeds primary nitrogen assimilation by 10 fold (Keys et al. 1978), and thus the ability of GS2 to reassimilate this ammonia is essential for plant growth.

Histochemical studies have shown that cytosolic GS1 genes are not expressed in mesophyll cells but rather in phloem cells of vascular tissues (Edwards et al. 1990; Carvalho et al. 1992; Kamachi et al. 1992). Therefore, GS1 appears to be involved in translocation of glutamine via phloem tissues. Glutamine and glutamate serve as nitrogen-transport compounds and nitrogen donors in the biosynthesis of an enormous number of compounds, including essentially all amino acids. Nitrogen from glutamine and glutamate is converted to aspartate by aspartate aminotransferase (AspAT), or to asparagine by asparagine synthetase (AS). Glutamine, glutamate, asparagine, and aspartate are the predominant amino acids found in most higher plants (Lam et al. 1995), although their relative concentrations vary considerably; the xylem contains low concentrations (for example, 3 to 20 mM in *Urtica*), and the phloem contains higher concentrations (for example, 60 to 140 mM in sugar beet) (Fisher et al. 1998).

The first report of transgenic GS1-overexpressing plants was made by Eckes et al. (1989), who introduced an alfalfa-derived genomic fragment of a promoter-less GS1 gene, under control of the CaMV 35S promoter, into tobacco. The resulting transgenic tobacco plants showed a ten-fold increase in GS1 mRNA levels and a five-fold increase in specific GS activity. The GS1 polypeptide accounted for up to 5% of the total soluble protein in these transgenic plants. The amino acid composition of the overexpressors was not altered significantly, although there was a seven-fold reduction in free ammonia.

However, in a similar experiment using a full-length alfalfa-derived GS1 cDNA, under control of the CaMV 35S promoter the resulting transgenic tobacco plants showed only a 10%-25% increase in total GS activity in spite of considerable accumulation of the transgene transcripts (Temple et al. 1993).

Hemon et al. (1990) fused an *N. plumbaginifolia* mitochondrial targeting sequence to a *Phaseolus vulgaris* GS1 cDNA, driven by the CaMV 35S promoter, and introduced the costruct into *N. plumbaginifolia*. Under tissue culture conditions, the GS1 isozyme in the transformants made up to 25% of the total GS activity, and this activity was associated with the mitochondria. In greenhouse-grown plants, however, expression of the introduced GS gene at the protein and enzymatic level was greatly reduced, although high transgene mRNA levels were maintained.

Hirel et al. (1992) introduced a soybean GS1 genomic fragment containing the coding region and 3'-untranslated region, fused to the CaMV 35S promoter, into tobacco. The resulting soil-grown mature plants expressed functional soybean GS1 enzyme in the cytoplasm of leaves where this enzyme is not normally present. Anion-exchange chromatography indicated that GS1 represented 25%-30% of the total GS activity. Expression of the soybean GS1 gene in tobacco leaves forced concomitant expression (transcription) of an endogenous tobacco gene encoding cytosolic GS1 in leaves, where this gene is not normally expressed.

The cDNA clones, pGS100 and pGS13, representing two major classes of GS1 genes in alfalfa, have been characterized, and constructs of these cDNAs, driven by various promoters such as the CaMV 35S or vasculature-specific (acidic chitinase from *Arabidopsis*) promoter, have been independently introduced into alfalfa (Temple and Sengupta-Gopalan 1997). Alfalfa transformants containing the CaMV 35S promoter driving the pGS100 cDNA showed no change in GS1 polypeptide levels. Two-dimensional SDS-PAGE western analysis indicated that the specific distribution of all GS polypeptides remained essentially unchanged. This same gene construct, when introduced into tobacco, however, is reported to have resulted in a significant accumulation of the corresponding transcript and peptide. Greenhouse-grown, nodulated alfalfa transformants, containing pGS100 cDNA driven by the vasculature-specific promoter, showed a significant decrease in the level of pGS100-related transcripts in the nodules and a significant increase in the stem (Temple and Sengupta-Gopalan 1997). Alfalfa plants, transformed with a construct in which the vasculature-specific promoter was used to drive the pGS100 cDNA in antisense orientation, were found to have an extremely high level of mortality during the later stages of tissue culture, and also bleached leaves. This suggested that vascular-localized GS isozymes play a very crucial role in nitrogen assimilation (Temple and Sengupta-Gopalan 1997). Involvement of chaperones related to Gro EL or Gro EL-like proteins in the folding/assembly of higher plant GS enzymes have been suggested. If the assembly or activation of GS is mediated by a chaperone-like assembly factor(s), it may be advantageous to coexpress a chaperone gene with a suitable GS gene to obtain high levels of overexpression (Temple and Sengupta-Gopalan 1997).

Kozaki and Takeba (1996) have reported that transgenic plants with twice the normal amount of GS2 had an improved capacity for photorespiration and an increased tolerance to high-intensity light, whereas those with a reduced amount of GS2 had a diminished capacity for photorespiration and were severely photoinhibited by high-intensity light compared with control plants. They concluded that photorespiration protects C3 plants from photoinhibition.

Despite the reports of various GS overexpressors, there have been no descriptions to date of their ability to assimilate ammonia or nitrate. In general, introduction of a single gene, such as GS, may not be expected to result in a direct effect on plant growth, but may result in an increased ability to assimilate ammonia, although the studies of Lam et al. (1995) demonstrated that transgenic tobacco plants, expressing pea chloroplastic and cytosolic GS genes under control of the CaMV 35S promoter, showed a considerable growth advantage as compared with wild-type plants. Therefore, it has been intriguing for us to study whether such GS overexpressors contain any "hyperassimilators of nitrogen dioxide" using ^{15}N.

Differences in plant background, perhaps relative to that of the transgene, plant age, and growth conditions, all appear to strongly affect the level of overexpression. The lack of correlation between the transgene transcript level, the associated polypeptide, and the activity attributable to the transgene, support the idea of translational, posttranslational, or assembly-mediated control (Temple and Sengupta-Gopalan 1997). GS enzyme activities in transgenic plants reported to date all represent the in vitro activity, and compartmentation of the enzyme activity in cells is unknown. If the efficiency of loading glutamine into phloem cells was increased by 50%, for example, due to an increase in GS1 activity in transgenic plants, the nitrogen translocation and hence nitrogen metabolism in these plants would be expected be greatly influenced. Therefore, more direct measurements of the assimilation of ammonia using ^{15}N in these transgenic plants are required.

Modulation of other nitrogen assimilatory enzymes, such as GOGAT, AS, and GDH, have been recently reviewed (Temple et al. 1998; Temple and Sengupta-Gopalan 1997). In addition, a bacterial GDH driven by the CaMV 35S promoter was most recently introduced into tobacco, and the resulting transformants were shown to have a high resistance to phosphinothricin, a glutamate analogue (Ameziane et al. 1998). The authors concluded that GDH/GOGAT activity can partially replace GS/GOGAT activity.

5. Nitrogen Dioxide Assimilation in Transgenic Plants Containing Chimeric NiR cDNA, and GS1 and GS2 cDNA

A chimeric gene containing spinach NiR cDNA (Back et al. 1988) and *hpt* gene (Gritz and Davies 1983), under control of the 35S promoter, was constructed and introduced by particle bombardment into root sections of *Arabidopsis thaliana* ecotype C24, and transgenic plant lines were selected for hygromycin resistance (Takahashi and Morikawa 1996; Takahashi et al. 2001). T_1 to T_3 plants were analyzed further. The presence of the introduced spinach NiR cDNA in transgenic plants was confirmed by PCR analysis using specific primers, which define a 352-bp fragment in the 5'-region of spinach NiR cDNA. Northern blot analysis, using specific *Arabidopsis thaliana* and spinach gene probes, clearly showed hybridization bands of about 2 kb for spinach NiR and 1.8 kb for the endogenous NiR gene in transgenic plants, indicating that these plants successfully expressed the introduced spinach NiR cDNA.

Two-dimensional gel electrophoresis followed by immunoassay showed that the transgenic plants contained several additional protein spots, with approximate molecular weights of 60 kDa and pIs of 6, corresponding to the translates of the introduced spinach NiR cDNA, evidence that spinach NiR polypeptides were formed in the transgenic plants and that cosuppression was unlikely to have occurred, at least in the tested transformant lines.

Twelve transgenic lines of T_1 to T_3 plants, which had been confirmed to contain the 352-bp PCR bands specific to spinach NiR cDNA, were further analyzed for NiR enzyme activity and NO_2 assimilation. The NiR enzyme activity was determined according to the method of Wray and Fido (1990) using methyl viologen as an electron donor. Typical NiR activity of the control plants was 150.6 ± 30.7 (nmol NO_2^-/min/mg protein; average of five samples $\pm$ SD), whereas activities of transgenic plants varied from about 0.8- to 1.8-fold the control values, depending on the individual transformant lines.

Transgenic tobacco and *Arabidopsis thaliana* plants carrying a chimeric NiR tobacco gene were recently reported (Crété et al. 1997). The NiR activities in these transgenic plants were never more than three-fold that of the wild type, as also suggested by our results. It therefore appears that some regulatory mechanism may control the maximum NiR enzyme activity in plants in which NiR mRNA is constitutively expressed. Crété et al. (1997) suggested the presence of a posttranscriptional mechanism for regulation of NiR enzyme levels in plants. Targeting of the NiR cDNA into chloroplasts will be an interesting challenge in the future as regulation of NiR gene expression may well differ between the nucleus and plastid.

NO_2 assimilation in our NiR transgenic *Arabidopsis thaliana* plants was analyzed using ^{15}N-labeled NO_2 and mass spectrometry as reported elsewhere (Morikawa et al. 1998a). Briefly, plants were fumigated with 4 ± 0.1 ppm $^{15}NO_2$ (51.6% $^{15}NO_2$) for 8 h at 22℃ under fluorescent lights (70 μmol/m^2/s), with a

70% humidity, and a CO_2 concentration at atmospheric levels (0.03%-0.04%). Leaves were harvested from fumigated plants, washed with distilled water, and dried at 80oC for 3 days. Total reduced nitrogen contents in fine powder of the plants were determined by the Kjeldahl method (Morikawa et al. 1998a). The ^{15}N contents in samples were determined by mass spectrometry. NO_2-derived reduced nitrogen content was defined as the " NO_2-N content." Typical values for the NO_2-N contents were 1.18 $\pm$ 0.08 (mg N/g dw; average of five samples $\pm$ SD) for the control plants, and between 0.8- and 1.4-fold these values for the transgenic plants, depending on the lines. A 40% increase in the ability of these transgenic plants to assimilate NO_2 is statistically significant and may be providing a basis for further improvement of the NiR gene by bioengineering, prone to be very important as a future phytoremediation strategy.

Figure 1 shows the NO_2-N contents of 12 transgenic lines plotted against their NiR activities with the control sample values being taken as 100. The NO_2-N content increased almost linearly with the increase in NiR activity, and showed a positive correlation (r=0.86). These results clearly indicate that increases in NiR enzyme activities can enhance the NO_2-assimilation ability of plants, and provide further support and encouragement to our eventual aim of producing a novel " NO_2-philic plant" that can grow with atmospheric NO_2 as its sole nitrogen source (Morikawa et al. 1992; Kamada et al. 1992).

When the regression of Fig. 1 is extrapolated to zero NiR activity, the NO_2-N content does not reach a zero level, and suggests that, in theory, a plant with zero NiR activity can still produce reduced nitrogen. Vaucheret et al. (1992) previously reported the production of transgenic tobacco plants that lack or possess much reduced NiR activity because of expression of the NiR cDNA in an antisense orientation. Using these transgenic tobacco (clone 271) plants, we performed fumigations with ^{15}N-labeled NO_2 under the conditions described earlier and analyzed their NO_2-N content. We have discovered that these transgenic plants have a surprisingly high ability to reduce both NO_2 and NO_3^- to NH_3 and organic nitrogen (Takahashi et al. 1998), and that their postfumigation NO_2-N content was 40%-50% that of control untransformed tobacco. Typical values of NiR activity and NO_2-N content in the control untransformed tobacco plants (11 weeks old) were 129.1 $\pm$ 25.5 (nmol NO_2^-/min/mg protein; average of three samples $\pm$ SD), and 1.31 $\pm$ 0.33 (mg N/g dw; average of three samples $\pm$ SD), respectively. In addition, the ability of clone 271 plants to reduce nitrate was almost 100% that of the control, and formation of ammonia in these plants was directly confirmed by mass spectrometry (Takahashi et al. 1998). The clone 271 plants we used showed much reduced NiR activity (>5% that of the control) when methyl viologen was used as an electron donor. Further studies, including determination of NiR activity using reduced ferredoxin as an electron donor, are currently underway.

In *Escherichia coli*, in addition to NADH-NiR, NiRs that use formate, pyruvate, or ethanol as an electron donor have been reported (Page et al. 1990). Formate dehydrogenase has also been found in plant mitochondria (Colas des Francs-Small et al. 1994). Therefore, our present findings that plants lacking NiR activity can reduce nitrate and NO_2 to ammonia strongly suggest the existence of an alternative pathway, independent of NiR, for nitrate reduction in plants. While the involvement of sulfite reductase (SiR) in the observed reduction of nitrate and NO_2 in clone 271 plants cannot be currently excluded, the SiRs are reported to have more than a 250-fold-lower affinity for nitrite than for sulfite in spinach (Krueger

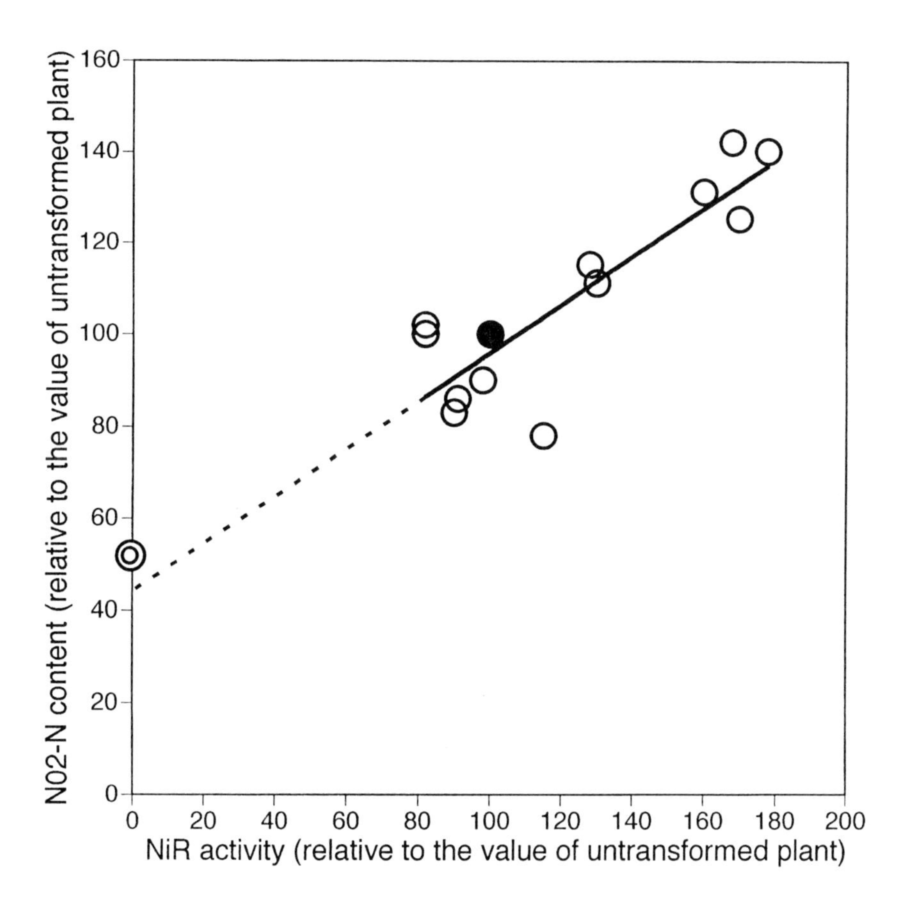

Fig. 1. NiR activity and NO_2 assimilation ability of transgenic plants of *Arabidopsis thaliana* containing spinach NiR cDNA. The NO_2-N content (NO_2-derived reduced nitrogen content) was plotted against the NiR enzyme activity of 12 transgenic plant lines (*open circles*) and an untransformed control plant (*closed circle*) of *Arabidopsis thaliana*. Leaves of plants, fumigated with 4 ppm ^{15}N-labeled NO_2 for 8 h, were harvested and digested by the Kjeldahl method; the ^{15}N content in the reduced nitrogen fraction then analyzed by mass spectrometry (values relative to that of the control). The relative value of clone 271 transgenic tobacco (*double circle*), that has much reduced NiR activity, is also shown

and Siegel 1982). We are currently studying the detailed mechanisms by which nitrate is reduced in these NiR-lacking transgenic plants.

More recently, we have cloned cDNAs encoding the GS1 and GS2 of *Arabidopsis thaliana* ecotype C24. Their coding regions show approximately 70% identity, while the identity of their untranslated regions (UTR) was less than 40%. Using these cDNAs, four constructs were developed, consisting of full-length or fragments of the GS1 or GS2 cDNAs, in either sense or antisense orientation, under control of the CaMV 35S promoter and with the NOS polyadenylation signal (see legend of Fig. 2). Each construct was separately introduced, together with plasmid pCH having the *hpt* gene under control of the

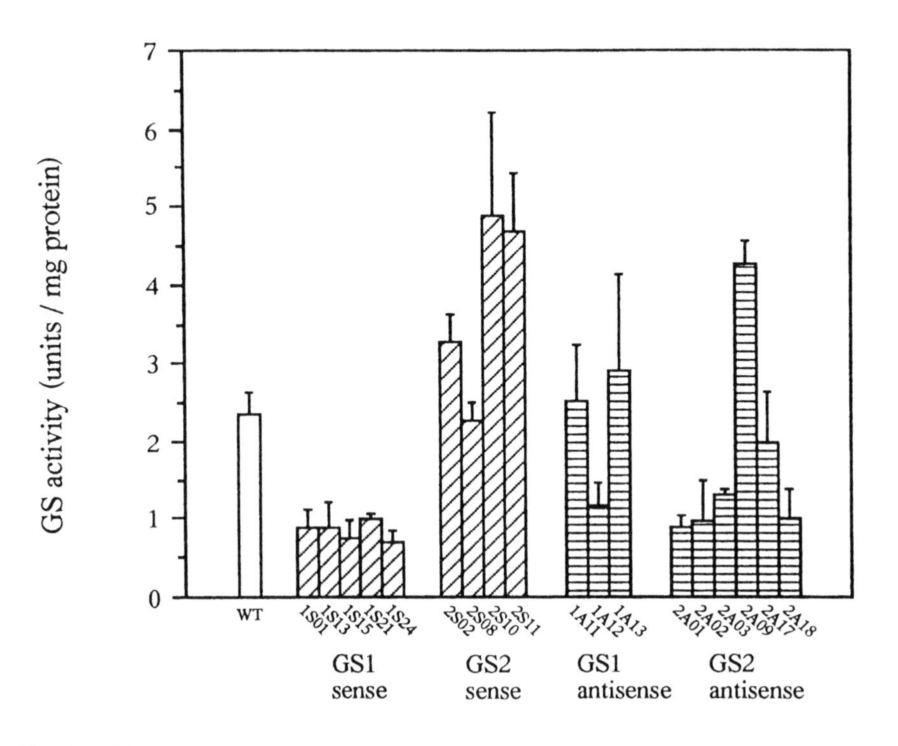

Fig. 2. GS enzyme activity in various transgenic plant lines of *Arabidopsis thaliana*. Constructs introduced are pAGS1, containing the full-length 1335-bp GS1 cDNA in sense orientation; pAGS2, containing the full-length 1545-bp GS2 cDNA in sense orientation; pASG1, containing the 153-bp UTR of GS1 cDNA in antisense orientation; and pASG2, containing the 164-bp UTR of GS2 cDNA in antisense orientation. Each cDNA fragment was placed under control of the CaMV 35S promoter and fused to the nopaline synthase (NOS) polyadenylation signal. Columns represent means of results for two to four plants; bars represent the SD

CaMV 35S promoter and NOS polyadenylation signal, into root sections of *Arabidopsis thaliana* ecotype C24 by particle bombardment as described previously (Takahashi and Morikawa 1996). Transgenic plants, selected for hygromycin resistance, were analyzed by PCR using primers specific to the introduced cDNA sequences to confirm the presence of the GS cDNA. T_2 plants were analyzed for total GS (transferase) activity according to Rhodes et al. (1975), with one unit of GS activity being defined as 1 μ mol γ -glutamylhydroxamate formed per minute (Fig. 2).

The mean GS activity of the untransformed control plants was 2.35 ± 0.31 (units/mg protein), whereas activities of the transformant lines containing GS2 cDNA in sense or antisense orientation varied greatly from 2.28 to 4.88 and 0.91 to 4.28 (units/mg protein), respectively. Therefore, the introduction of constitutively expressing GS2 cDNA, in sense or antisense orientations, seems to be useful in the production of transformants with altered in vitro GS transferase activities. On the other hand, lines containing GS1 cDNA in sense or antisense orientation (except line 2A09), showed much lower GS activities, with average values of 0.70 to 1.01 and 1.18 to 2.92 (units/mg protein), respectively. Cosuppression may be a possible cause of the suppressed GS activities in transformants with GS1 in sense orientation. Further analyses, including northern blots, are required to elucidate the differences in action between GS1 and GS2 cDNA.

Lines 2A09, 2S10, and 2S11, which had more than two-fold higher GS activities (4.28 ± 0.29, 4.88 ± 1.37 and 4.70 ± 0.75 units/mg protein, respectively) than the controls are targets for detailed analysis of their ability to assimilate nitrogen dioxides and nitrate using ^{15}N (results will be published elsewhere). Furthermore, crosses between the NiR overexpressors and GS overexpressors may possibly allow the production of NO_2 hyperassimilators.

The metabolism of nitrogen in plants has long been an area of intensive study; nevertheless, our current knowledge is too limited for us to fully understand the detailed mechanisms underlying the regulation and metabolisms of this element in plant cells. When we fumigate Arabidopsis thaliana plants with $^{15}NO_2$ and analyze the excess ^{15}N levels in the total, reduced, ammonium, and nitrate/nitrite nitrogen (N) fractions of the leaves, we have discovered a lack of stoichiometry in NO_2-derived N between the total N and the sum of reduced (Kjeldahl) N and nitrate/nitrite N, with 20% to 30% of the total N being in the form of unidentified nitrogen (UN) (Arimura et al. 1998). In contrast, in the leaves of plants fed $K^{15}NO_3$, between 40% and 60% of the nitrate-derived total N is found in the nitrate/nitrite N fraction, more than 10-fold the value for NO_2-derived nitrogen. These results indicate that the metabolic fate of nitrogen dioxide nitrogen distinctly differs from that of nitrate. In addition, more recently, we have identified a nitrosoamine compound as a UN-bearing compound (Arimura et al. unpublished results); and this may be harmful to both plant and animal cells. Details of the metabolic fate of the nitrogen of nitrogenous oxides are to be the subject of future study.

Although the ability to assimilate nitrogen dioxide was found to vary by more than 600- fold between 217 plant taxa tested (Morikawa et al. 1998a,b), very little is known about the genes involved in these observed differences in assimilation. Whether such "hyperassimilators," which have a high ability to assimilate nitrogen dioxide, have any specific transporters to uptake NO_2 or "sink proteins" for NO_2 nitrogen in their cells is another area of future study. Indeed, identification of genes that are specifically expressed in response to air pollutants is an essential area of study, and we have recently found that germin-like proteins are induced in the leaves of *Rhododendron mucronatum* in response to fumigation with nitrogen dioxide gas (Umehara et al. unpublished results).

Manipulation of genes involved in the uptake, metabolism, and tolerance of plants to various toxic chemicals is a vital approach to the production of "wonder plants" that can clean up, and serve as powerful sinks of, air pollutants. Not only will such engineered plants be able to maintain a sustainable environment for future generations, but they also will provide a novel and essential experimental tool for basic research into the physiology of plants.

We would like to dedicate this chapter to Professor Alan Wellburn who deceased in May, 1999.

Acknowledgments. This research was supported in part by the Research for the Future, Japanese Society for the Promotion of Science (JSPS-RFTF96L00604) program. We thank Ms. Makiko Fujii and Dr. Naoki Goshima of Hiroshima University for their technical help in clonig Arabidopsis thaliana GS genes.

References

Ameziane R, Bernhard K, Bates R, Lightfoot DA (1998) Metabolic engineering of C and N metabolism with NADPH-glutamate dehydrogenase. Abstracts, Annual Meeting of the American Society of Plant Physiologists, June 27-July 1, 1998, Madison, WI, p 15

Arimura G, Takahashi M, Goshima N, Morikawa H (1998) Metabolic fate of nitrogen dioxide nitrogen differs from that of nitrate nitrogen in plant leaves. Abstracts, Annual Meeting of the American Society of Plant Physiologists, June 27-July 1, 1998, Madison, WI, p 59

Aslam M, Oaks A, Huffaker RC (1976) Effect of light and glucose on the induction of nitrate reductase and on the distribution of nitrate in totalled barley leaves. Plant Physiol 58:588-591

Back E, Burkhart W, Moyer M, Privalle L, Rothstein S (1988) Isolation of cDNA clones coding for spinach nitrite reductase: complete sequence and nitrate induction. Mol Gen Genet 212:20-26

Baker AJM, McGrath SP, Reeves RD, Smith JAC (2000) Metal hyperaccumulator plants: a review of the ecology and physiology of a biological resource for phytoremediation of metal-polluted soils. In: Terry N, Banuelos GS (eds) Phytoremediation of contaminated soils and water . CRC press, FL

Bowsher CG, Hucklesby DP, Emes MJ (1989) Nitrite reduction and carbohydrate metabolism in plastids purified from roots of *Pisum sativum* L. Planta 177:359-366

Caboche M, Rouze P (1990) Nitrate reductase: a target for molecular and cellular studies in higher plants. Trends Genet 6:187-192

Campbell WH (1996) Nitrate reductase biochemistry comes of age. Plant Physiol 111:355-361

Carvalho E, Pereira S, Sunkel C, Salema R (1992) Detection of cytosolic glutamine synthetase in leaves of *Nicotiana tabacum* L. by immunocytochemical methods. Plant Physiol 100:1591-1594

Colas des Francs-Small C, Ambard-Bretteville F, Small ID, Rémy R (1994) Identification of a major soluble protein in mitochondria from nonphotosynthetic tissues as NAD-dependent formate dehydrogenase. Plant Physiol 102:1171-1177

Crawford NM (1995) Nitrate: nutrient and signal for plant growth. Plant Cell 7:859-868

Crawford NM, Wilkinson JQ, LaBrie ST (1992) Metabolic control of nitrate reduction in *Arabidopsis thaliana*. Aust J Plant Physiol 19:377-385

Crété P, Caboche M, Meyer C (1997) Nitrite reductase expression is regulated at the post-transcriptional level by the nitrogen source in *Nicotiana plumbaginifolia* and *Arabidopsis thaliana*. Plant J 11:625-634

Cure JD, Israel DW, Rufty TW Jr. (1988) Nitrogen stress effects on growth and seed yield of non-nodulated soybean exposed to elevated carbon dioxide. Crop Sci 28:671-677

Dalling MJ, Tolbert NE, Hageman RH (1972) Intracellular location of nitrate reductase and nitrite reductase. Biochim Biophys Acta 283:505-512

Donnelly PK, Fletcher JS (1995) PCB metabolism by ectomycorrhizal fungi. Bull Environ Contam Toxicol 54:507-513

Dorbe M-F, Caboche M, Daniel-Vedele F (1992) The tomato *nia* gene complements a *Nicotiana plumbaginifolia* nitrate reductase-deficient mutant and is properly regulated. Plant Mol Biol 18:363-375

Dorlhac de Borne F, Vincentz M, Chupeau Y, Vaucheret H (1994) Co-suppression of nitrate reductase host genes and transgenes in transgenic tobacco plants. Mol Gen Genet 243:613-621

Duncanson E, Gilkes AF, Kirk DW, Sherman A, Wray JL (1993) *nir 1*, a conditional-lethal mutation in barley causing a defect in nitrite reduction. Mol Gen Genet 236: 275-282

Durmishidze SV, Nutsubidze NN (1976) Absorption and conversion of nitrogen dioxide by higher plants. Dokl Biochem 227:104-107

Eckes P, Schmitt P, Daub W, Wengenmayer F (1989) Overproduction of alfalfa glutamine synthetase in transgenic tobacco plants. Mol Gen Genet 217:263-268

Edwards JW, Walker EL, Coruzzi GH (1990) Cell-specific expression in transgenic plants reveals nonoverlapping roles for chloroplast and cytosolic glutamine synthetase. Proc Natl Acad Sci U S A 87:3459-3463

Ferrario S, Valandier M-H, Morot-Gaundry J-F, Foyer CH (1995) Effects of constitutive expression of nitrate reductase in transgenic *Nicotiana plumbaginifolia* L. in response to varying nitrogen supply. Planta 196:288-294

Ferrario-Méry S, Murchie E, Galtier N, Quick WP, Foyer CH (1997a) Manipulation of the pathways of sucrose synthesis and nitrogen assimilation in transformed plants to improve photosynthesis and productivity. In: Foyer CH, Quick WP (eds) A molecular approach to primary metabolism in higher plants. Taylor & Francis, London, pp 125-153

Ferrario-Méry S, Thebaud MD, Betsche T, Valadier M-H, Foyer CH (1997b) Modulation of carbon and nitrogen metabolism, and of nitrate reductase, in untransformed and transformed *Nicotiana plumbaginifolia* during CO_2 enrichment of plants grown in pots and in hydroponic culture. Planta 202:510-521

Fisher W-F, Andre B, Rentsch D, Krolkiewicz S, Tegeder M, Breitkreuz K, Frommer WB (1998) Amino acid transport in plants. Trend Plant Sci 3:188-195

Forde BG, Day HM, Turton JF, Shen WJ, Cullimore JV, Oliver JE (1989) Two glutamine synthetase genes from *Phaseolus vulgaris* L. display contrasting developmental and spatial patterns of expression in transgenic *Lotus corniculatus* plants. Plant Cell 1:391-401

Foyer CH, Lescure JC, Lefebvre C, Vincentz M, Vaucheret H (1994) Adaptations of photosynthetic electron transport, carbon assimilation and carbon partitioning in transgenic *Nicotiana plumbaginifolia* plants to changes in nitrate reductase activity. Plant Physiol 104:171-178

Gojon A, Dapoigny L, Lejay L, Tillard P, Rufty TW (1998) Effects of genetic modification of nitrate reductase expression on $^{15}NO_3^-$ uptake and reduction in Nicotiana plants. Plant Cell Environ 21:43-53

Gritz L, Davies J (1983) Plasmid-encoded hygromycin B resistance: the sequence of hygromycin B phosphotransferase gene and its expression in *Escherichia coli* and *Saccharomyces cerevisiae*. Gene 25:179-188

Hemon P, Robbins MP, Cullimore JV (1990) Targeting of glutamine synthetase to the mitochondria of transgenic tobacco. Plant Mol Biol 15:895-904

Hirel B, Marsolier MC, Hoarau A, Hoarau J, Brangeon J, Schafer R, Verma DPS (1992) Forcing expression of a soybean root glutamine synthetase gene in tobacco leaves induces a native gene encoding cytosolic enzyme. Plant Mol Biol 20:207-218

Hocking PJ, Meyer CP (1991) Effects of CO_2 enrichment and nitrogen stress on growth and partitioning of dry matter and nitrogen in wheat and maize. Aust J Plant Physiol 18:339-356

Ida S (1987) Immunological comparisons of ferredoxin-nitrite reductases from higher plants. Plant Sci 49:111-116

Ida S, Mikami B (1986) Spinach ferredoxin-nitrite reductase: a purification procedure and characterization of chemical properties. Biochim Biophys Acta 871:167-176

Ip SM, Kerr J, Ingledew WJ, Wray JL (1990) Purification and characterization of barley leaf nitrite reductase. Plant Sci 66:155-165

Jin T, Huppe HC, Turpin DH (1998) In vitro reconstitution of electron transport from glucose-6-phosphate and NADPH to nitrite. Plant Physiol 117:303-309

Kamachi K, Yamaya T, Hayakawa T, Mae T, Ojima K (1992) Vascular bundle-specific localization of cytosolic glutamine synthetase in rice leaves. Plant Physiol 99:1481-1486

Keys AJ, Bird IF, Cornelius MJ, Lea PJ, Wallsgrove RM, Miflin BJ (1978) Photorespiratory nitrogen cycle. Nature 275:741-743

Kozaki A, Takeba G (1996) Photorespiration protects C3 plants from photooxidation. Nature 384:557-560

Kronenberger J, Lepingle A, Caboche M, Vaucheret H (1993) Cloning and expression of distinct nitrite reductases in tobacco leaves and roots. Mol Gen Genet 236:2030-2038

Krueger RJ, Siegel LM (1982) Spinach siroheme enzymes: isolation and characterization of ferredoxin-sulfite reductase and comparison of properties with ferredoxin-nitrite reductase. Biochemistry 21:2892-2904

Lam HM, Coschigano K, Schultz C, Melo-Oliveira R, Tjaden G, Oliveira I, Ngai N, Hsieh

MH, Coruzzi G (1995) Use of *Arabidopsis* mutants and genes to study amide amino acid biosynthesis. Plant Cell 7:887-898

Lancaster J Jr, Stuehr DJ (1996) The intracellular reactions of nitric oxide in the immune system and its enzymatic synthesis. In: Lancaster J Jr (ed) Nitric oxide. Principles and actions. Academic Press, San Diego, pp 139-175

Lea PJ, Robinson SA, Stewart GR (1990) The enzymology and metabolism of glutamine, glutamate and asparagine. In: Miflm BJ, Lea PJ (eds) The biochemistry of plants. Intermediary nitrogen metabolism, vol 16. Academic Press, San Diego, pp 121-159

Lea PJ, Rowland-Bamford AJ, Wolfenden J (1996) The effect of air pollutants and elevated carbon dioxide on nitrogen metabolism. In: Yunus M, Iqbal M (eds) Plant response to air pollution. Wiley, New York, pp 319-352

McKee IF, Woodward FI (1994) CO_2 enrichment responses of wheat: interactions with temperature, nitrate and phosphate. New Phytol 127:447-453

Morikawa H, Higaki A, Nohno M, Kamada M, Nakata M, Toyohara G, Fujita K, Irifune K (1992) "Air-pollutant-philic plants" from nature. In: Murata N (ed) Research in photosynthesis, vol IV, Kluwer, Dordrecht, pp 79-82

Morikawa H, Higaki A, Nohno M, Takahashi M, Kamada M, Nakata M, Toyohara G, Okamura Y, Matsui K, Kitani S, Fujita K, Irifune K, Goshima N (1998a) More than a 600-fold variation in nitrogen dioxide assimilation among 217 plant taxa. Plant Cell Environ 21:180-190

Morikawa H, Takahashi M, Irifune K (1998b) Molecular mechanism of the metabolism of nitrogen dioxide as an alternative fertilizer in plants. In: Satoh K, Murata N (eds) Stress responses of photosynthetic organisms. Elsevier, Amsterdam, pp 227-237

Oji Y, Watanabe M, Wakiuchi N, Okamoto S (1985) Nitrite reduction in barley root plastids: dependence on NADPH coupled with glucose-6-phosphate and 6-phosphogluconate dehydrogenases and possible involvement of an electron carrier and a diaphorase. Planta 165:85-90

Page L, Griffiths L, Cole JA (1990) Different physiological roles of two independent pathways for nitrite reduction to ammonia by enteric bacteria. Arch Microbiol 154:349-354

Peterman TK, Goodman HM (1991) The glutamine synthetase gene family of *Arabidopsis thaliana*: light-regulation and differential expression in leaves, roots and seeds. Mol Gen Genet 230:145-154

Purvis AC, Peters DB, Hageman RH (1974) Effect of nitrogen supply on the accumulation of photosynthesis to elevated CO_2. Photosynth Res 39:389-400

Quilleré I, Dufossé C. Roux Y, Foyer CH, Caboche M, Morot-Gaudry JF (1994) The effect of the deregulation of NR gene expression on growth and nitrogen metabolism of winter-grown *Nicotiana pumbaginifolia*. J Exp Bot 45:1205-1212

Rhodes D, Rendon GA, Stewart GR (1975) The control of glutamine synthetase level in *Lemna minor* L. Planta 125:201-211

Robinson SA, Slade AP, Fox CG, Phillips R, Ratciffe RG, Stewart GR (1991) The role of glutamate dehydrogenase in plant nitrogen metabolism. Plant Physiol 95:509-516

Rogers HH, Campbell JC, Volk RJ (1979) Nitrogen-15 dioxide uptake and incorporation by *Phaseolus vulgaris* (L.). Science 206:333-335

Rogers GS, Payne L, Milham P, Conroy J (1993) Nitrogen and phosphorus requirements of cotton and wheat under changing atmospheric CO_2 concentrations. Plant Soil 155/156:231-234

Salt DE, Blaylock M, Kumar PBAN, Dushenkov V, Ensley BD, Chet I, Raskin I (1995)

Phytoremediation: a novel strategy for the removal of toxic metals from the environment using plants. Biotechnology 13:468-474

Schuster C, Mohr H (1990) Appearance of nitrite reductase mRNA in mustard seedling cotyledons is regulated by phytochrome. Planta 181:327-334

Schwab AP, Banks MK (2000) Phytoremediation of petroleum contaminated soils. In: Fiorenza, S., Oubre, C. L. and Ward, C. H. (Eds.) Phytoremediation of Hydrocarbon-Contaminated Soil. Lewis Publishers, Boca Raton, NY & Washington D.C., USA.

Siegel LM, Wilkerson JQ (1989) Structure and function of spinach ferredoxin-nitrite reductase. In: Wray JL, Kinghorn JR (eds) Molecular and genetic aspects of nitrate assimilation. New York, Oxford University Press, pp 263-283

Solomonson LP, Barber MJ (1990) Assimilatory nitrate reductase: functional properties and regulation. Annu Rev Plant Physiol Plant Mol Biol 41:225-253

Srivastava H, Wolfenden J, Lea PJ, Wellburn A (1994) Differential responses of growth and nitrate reductase activity in wild type and NO_2-tolerant barley mutants to atmospheric NO_2 and nutrient nitrate. J Plant Physiol 143:738-743

Suzuki A, Oaks A, Jacquot JP, Vidal J, Gadal P (1985) An electron transport system in maize roots for reactions of glutamate and nitrite synthase. Plant Physiol 78:374-378

Suzuki T, Yamaoka R, Nishi M, Ide H, Makino K (1996) Isolation and characterization of a novel product, 2'-deoxyoxanosine, from 2'-deoxyguanosine, oligodeoxynucleotide, and calf thymus DNA treated by nitrous acid and nitric oxide. J Am Chem Soc 118:2515-2516

Takahashi M, Hara K, Caboche M, Morikawa (1998) Reduction of nitrate and nitrogen dioxide in plants that lack nitrite reductase activity. Abstracts, Annual meeting of the American Society of Plant Physiologists, June 27-July 1, 1998, Madison, WI, p 59

Takahashi M, Morikawa H (1996) High frequency stable transformation of *Arabidopsis thaliana* by particle bombardment. J Plant Res 109:331-334

Takahashi M, Sasaki Y, Ida S and Morikawa H. Nitrite reductase gene enrichment improves assimilation of nitrogen dioxide in Arabidopsis. Plant Physiol 126:731-741 (2001)

Tanaka T, Ida S, Irifune K, Oeda K, Morikawa H (1994) Nucleotide sequence of a gene for nitrite reductase from *Arabidopsis thaliana*. DNA Seq 5:57-61

Temple SJ, Sengupta-Gopalan C (1997) Manipulating amino acid biosynthesis. In: Foyer CH, Quick WP (eds) A molecular approach to primary metabolism in higher plants. Taylor & Francis, London, pp 155-177

Temple SJ, Knight TJ, Unkefer PJ, Sengupta-Gopalan C (1993) Modulation of glutamine synthetase gene expression in tobacco by the introduction of an alfalfa glutamine synthetase gene in sense and antisense orientation: molecular and biochemical analysis. Mol Gen Genet 236:315-325

Temple SJ, Vance CP, Gantt JS (1998) Glutamate synthase and nitrogen assimilation. Trends Plant Sci 3:51-56

Travis RL, Aslam M, Fritschi F, Rains DW (1998) Metabolic regulation of nitrate efflux and net uptake in *Acala* and *Pima* cotton. Abstracts, Annual meeting of the American Society of Plant Physiologists, June 27-July 1, 1998, Madison, WI, pp 134

Vaucheret H, Kronenberger J , Lepingle, Vilaine F, Boutin J-P, Caboche M (1992a) Inhibition of tobacco nitrite reductase activity by expression of antisense RNA. Plant J 2:559-569

Vaucheret H, Marion-Poll A, Meyer C, Faure J-M, Marin E, Caboche M (1992b) Interest in and limits to the utilization of reporter genes for the analysis of transcriptional regulation of nitrate reductase. Mol Gen Genet 235:259-268

Vaucheret H, Caboche M (1992) Induction of nitrate reductase host gene expression has a negative effect on the expression of transgenes driven by the nitrate reductase promoter. Plant Sci 107:95-104

Vincentz M, Moureaux T, Leydecker MT, Vaucheret H, Caboche M (1993) Regulation of nitrate and nitrite reductase expression in *Nicotiana plumbaginifolia* leaves by nitrogen and carbon metabolites. Plant J 3:315-324

Warner RL, Kleinhofs A (1981) Nitrate reductase-deficient mutants in barley. Nature 269:406-407

Wellburn AR (1990) Why are atmospheric oxides of nitrogen usually phytotoxic and not alternative fertilizers? New Phytol 115:395-429

Wellburn AR (1994) Air pollution and climate change: the biological impact. Longman, Essex

Wilkinson JQ, Crawford NM (1991) Identification of the *Arabidopsis CHL3* gene as the nitrate reductase structural gene *NIA2*. Plant Cell 3:461-471

Wray JL, Fido RJ (1990) Nitrate reductase and nitrite reductase. In: Dey PM, Harborne JB (eds) Methods in plant biochemistry, vol 3. Academic Press, London, pp 241-256

Yoneyama T, Sasakawa H (1979) Transformation of atmospheric NO_2 absorbed in spinach leaves. Plant Cell Physiol 20:263-266

Yunus M, Singh N, Iqbal M (1996) Global status of air pollution: an overview. In: Yunus M, Iqbal M (eds) Plant response to air pollution, Wiley, New York. pp 1-34

21 Manipulation of Genes for Antioxidative Enzymes

Mitsuko Aono

Environmental Biology Division, National Institute for Environmental Studies, Onogawa 16-2, Tsukuba, Ibaraki 305-8506, Japan

1. Introduction

As the earth's population and industrial productivity increase, new means of improving plant resistance against air pollutants must be developed to protect the plants from a contaminated atmospheric environment as well as to increase agricultural productivity. One strategy to achieve this is to develop plants that are more tolerant to air pollutants. Plants are stressed and resultantly damaged by air pollutants (Shimazaki et al. 1980), as well as by various other environmental factors such as some herbicides (Dodge 1975), drought (Smirnoff 1993), and low temperatures (Schöner and Krause 1990) under existing light and oxygen. Such stress and resulting damage under photooxidative conditions are called photooxidative stress and photooxidative damage, respectively. The generation of active oxygen species (AOS), such as $^1O_2$, $O_2^{\cdot -}$, H_2O_2, and $HO^{\cdot}$, are thought to be promoted in plants during photooxidative stress induced by such environmental factors as just described (Shimazaki et al. 1980; Dodge 1975; Smirnoff 1993; Schöner and Krause 1990).

AOS are also generated as by-products at many biological reactions and, unless removed rapidly by the scavenging system of cells, may destroy various cellular components or inactivate metabolism. Plants have enzymes, including glutathione reductase (GR), ascorbate peroxidase (APX), superoxide dismutase (SOD), and catalase, that are responsible for scavenging AOS (Fig. 1), and these enzymes have been implicated in the tolerance of plants to photooxidative stress (Foyer and

Air Pollution and Plant Biotechnology
–Prospects for Phytomonitoring and Phytoremediation–
Edited by K. Omasa, H. Saji, S. Youssefian, and N. Kondo
© Springer -Verlag Tokyo 2002

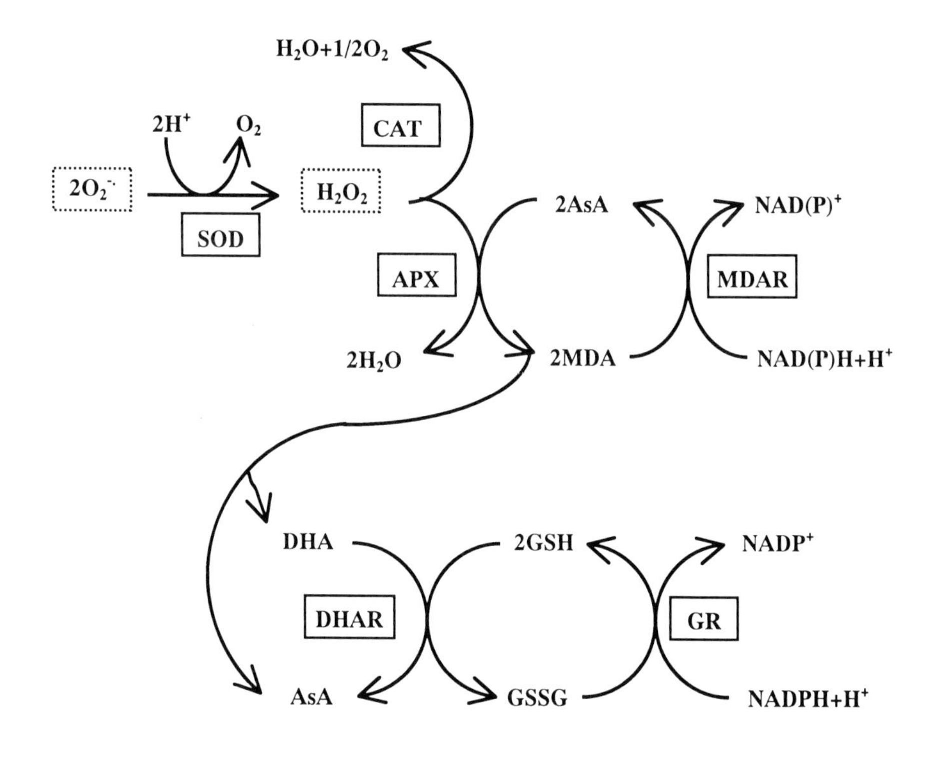

Fig. 1. A system for scavenging active oxygen species in plants. APX, ascorbate peroxidase; AsA, ascorbate; CAT, catalase; DHA, dehydroascorbate; DHAR, dehydroascorbate reductase; GR, glutathione reductase; GSH, reduced glutathione; GSSG, oxidized glutathione; MDA, monodehydroascorbate; MDAR, monodehydroascorbate reductase; SOD, superoxide dismutase. ⬚, active oxygen species; ▯, enzymes

Halliwell 1976; Smith et al. 1990; Bowler et al. 1994). Hence, manipulation of genes for these antioxidative enzymes can be used as a specific means of achieving improved photooxidative stress tolerance in plants.

As shown in Fig. 1, SOD (EC 1.15.1.1) catalyzes the dismutation of two superoxide radicals ($O_2^{-\cdot}$) into oxygen and hydrogen peroxide (H_2O_2). SOD and peroxidase/catalase function cooperatively to remove AOS, with SOD removing superoxide radicals and the peroxidase/catalase removing the produced peroxide. APX (EC 1.11.1.11) is the primary enzyme for H_2O_2 scavenging in the chroloplasts and cytosol of plant cells and has a high substrate specificity for ascorbate (Asada 1992), while catalase decomposes H_2O_2 in peroxisome without consumption of reducing power and has a relatively low substrate specificity. GR (EC 1.6.4.2) catalyzes the reduction of the oxidized form of glutathione (GSSG) with an accompanying oxidation of the electron donor, NADPH. By supplying the

Table 1. Introduction of genes, encoding antioxidative enzymes, into transgenic plants and evaluation of their air pollutant tolerance

Gene	Host (localization)	Tolerance to air pollutant	References
APX	Tobacco (cytosol)	O_3 (no change)	Saji et al. 1996
GR	Tobacco (cytosol)	O_3 (no change)	Aono et al. 1991
GR	Tobacco (chloroplast)	O_3 (no change)	Aono et al. 1993
		SO_2 (tolerance)	Aono et al. 1993
GR	Tobacco (chloroplast)	O_3 (no change)	Broadbent et al. 1995
GR	Tobacco (cytosol)	O_3 (no change)	Broadbent et al. 1995
GR	Tobacco (chloroplast/ mitochondoria)	O_3 (tolerance)	Broadbent et al. 1995
Cu/Zn-SOD	Tobacco(chloroplast)	O_3 (no change)	Pitcher et al. 1991
Cu/Zn-SOD	Tobacco (cytosol)	O_3 (no change)	Aono et al. 1998
Mn-SOD	Tobacco (chloroplast)	O_3 (tolerance)	Van Camp et al. 1994
Mn-SOD	Tobacco (mitochondoria)	O_3 (tolerance)	Van Camp et al. 1994
APX and GR	Tobacco (cytosol)	O_3 (tolerance)	Aono et al. 1998
GR and SOD	Tobacco (cytosol)	O_3 (no change)	Aono et al. 1998
APX and SOD	Tobacco (cytosol)	O_3 (no change)	Aono et al. 1998

APX, ascorbate peroxidase; GR, glutathione reductase; Cu/Zn-SOD, copper/zinc superoxide dismutase; Mn-SOD, manganese superoxide dismutase.

reduced form of glutathione (GSH), this enzyme has been postulated to function in the regeneration of ascorbate from dehydroascorbate, a critical reaction in the scavenging system of AOS in plant cells.

Recently, various genes that apparently confer photooxidative stress tolerance have been introduced into higher plants, and some of the resulting transgenic plants have been tested for air pollutant tolerance (Table 1). This section focuses on recent attempts, using molecular techniques, to protect plants from photooxidative stress caused by air pollutants.

2. Manipulation of Genes for Superoxide Dismutases

A number of experiments have shown that increased SOD activity enhances resistance to (photo)oxidative stress (Rennenberg and Polle 1994; Allen 1995; Holmberg and Bülow 1998). The SODs are essential components of the antioxidative defense system in almost all plants. The SOD isozymes are categorized into three different classes according to their metal cofactor copper/zinc (Cu/Zn), manganese (Mn), and iron (Fe). Fe-SOD is localized in chloroplasts, Mn-SOD in mitochondria, and Cu/Zn-SOD in both chloroplasts and cytosol (Bowler et al. 1994).

Although attempts to manipulate the levels of Cu/Zn SOD in plant cells by gene transfer have been successful, there are apparent inconsistencies in the responses of the resulting transgenic plants to oxidative stress. Transgenic tobacco plants expressing a petunia cDNA for chloroplastic Cu/Zn SOD showed no differences from control plants in the levels of inhibition of CO_2 fixation by the redox-cycling herbicide, paraquat, or in the extent of leaf injury after 0.3 ppm ozone fumigation for 6 h (Pitcher et al. 1991). In contrast, transgenic tobacco plants expressing a rice cDNA for cytosolic Cu/Zn SOD were more resistant than control plants to light-mediated paraquat-induced damage (Aono et al. 1995; Sakamoto and Tanaka 1993), but were not different from control plants in the quantum yields of PSII during exposure to 0.2 ppm ozone for 2 days (Aono et al. 1998). Increased paraquat tolerance in transgenic tobacco expressing a pea chloroplastic Cu/Zn SOD (Sen Gupta et al. 1993) and in transgenic potato expressing cytosolic or chloroplastic Cu/Zn SOD derived from tomato (Perl et al. 1993) has also been reported. In these studies, the increased activity of Cu/Zn SOD alone was thus insufficient to provide a detectable change in photooxidative stress tolerance, with the protective effects being found only over a narrow range of paraquat concentrations. These results suggest that dismutation of $O_2^{\cdot-}$ is not the rate-limiting step in the AOS scavenging system in chloroplasts or in cytosol, especially during ozone exposure in high concentration. Alternatively, Cu/Zn SOD may, because of its sensitivity to H_2O_2, be deactivated under stress conditions that enhance H_2O_2 accumulation. Whichever the case, overexpression of other components of the antioxidative systems in addition to SOD appears to be necessary for increased protection against ozone-induced photooxidative stress.

Overexpression of Mn-SOD in chloroplasts or mitochondria of transgenic tobacco has been achieved and ozone sensitivity of the transgenic plants was estimated (Van Camp et al. 1994). Transgenic tobacco plants in which Mn-SOD was targeted to the chloroplasts showed a threefold and twofold reduction in visible leaf injury as compared with wild-type and mitochondrial-targeted transgenic tobacco, respectively, after exposure to near-ambient ozone fluctuations of 0.04 - 0.12 ppm with an average dose of 0.06 ppm over 7 days. This result implies the efficiency of removing $O_2^{\cdot-}$ from chloroplasts to avoid photooxidative damage caused by the near-ambient level of ozone. Paraquat-mediated photooxidative damage to membranes and chlorophyll were also measured in these chloroplast-targeted transgenic plants by the accumulation of pheophytin, and levels of pheophytin were found to be significantly decreased in these transgenic plants as compared to control plants (Bowler et al. 1991). Elevated activity of H_2O_2-resistant Mn-SOD in the chloroplast stroma thus apparently provides enhanced protection from photooxidative damage induced by ozone and paraquat. However, the reason why increased SOD activity in the mitochondria enhanced ozone tolerance in this study remains obscure.

Another SOD isozyme, Fe-SOD, from *Arabidopsis thaliana* has also been expressed in transgenic tobacco (Van Camp et al. 1996) and in transgenic maize (Van Breusegem et al. 1999). This enzyme protected, at least slightly, both the cell membrane and PSII against paraquat-mediated photooxidative damage when it was targeted to chloroplasts. However, the tolerance of these plants to air pollutants has not yet been tested.

Although overexpression of SOD and increased SOD activities in chloroplasts clearly provide a certain increase in the level of protection from photooxidative stress, the protection afforded may not be sufficient to fully protect the plants from damage. The type of SOD that is increased may also be a critical factor; differences in the protective functions provided by Cu/Zn SOD and Mn SOD in chloroplasts may be due to the different biochemical characteristics and original subcellular localization of these enzymes.

3. Manipulation of Genes for Glutathione Reductase

In leaf cells, GR activity is both intra- and extrachloroplastic (Gillham and Dodge 1986) and, in pea leaves, GR has been localized to the chloroplastic, mitochondrial, and cytosolic cellular compartments (Edwards et al. 1990). The GR *Escherichia coli* gene (*gor*), placed under control of the CaMV 35S promoter, has been used to increase the levels of GR activity in transgenic plants (Aono et al. 1991; Foyer et al. 1991). Total GR specific activity in leaves of transgenic tobacco plants expressing the chimeric *gor* gene was almost 3.5 fold that of control nontransgenic plants and, because the bacterial protein lacked a transit peptide, this activity was presumably confined to the cytosol. Although transgenic plants expressing high levels of GR from *E. coli* showed no increased resistance to an ozone exposure of 0.5 ppm for 4 h, either in terms of leaf injury or with respect to photosynthetic activity, the plants were to some extent more tolerant to paraquat (Aono et al. 1991) and were able to maintain the reduced state of their ascorbate pools more effectively then control plants (Foyer et al. 1991). By targeting *E. coli* GR to the chloroplasts of transgenic tobacco plants, the GR activity was increased to almost 3 fold that of control plants. These plants thus showed an increased tolerance not only to paraquat but also to sulfur dioxide (1.0 ppm for 2 days) in terms of leaf injury, but exhibited no change in their ozone tolerance (0.5 ppm for 18 h) compared to control plants (Aono et al. 1993).

An attempt to generate transgenic poplar trees that overproduce the *gor* gene in their chloroplasts or cytosol has been recently reported (Foyer et al. 1995). When the *gor* gene was targeted to chloroplasts, leaf GR activities were up to 1000 fold those of the control, whereas *gor* targeted to the cytosol resulted in only 2- to 10 fold higher GR activities. The high chloroplastic GR expressors showed increased resistance to photoinhibition; although paraquat inhibited CO_2 assimilation in both transgenic and control lines, the increased foliar levels of glutathione and ascorbate in the transgenic plants persisted despite this treatment. These results suggest that overexpression of GR in the chloroplast increases the antioxidant capacity of the leaves of transgenic poplar trees and that this improves the capacity to withstand photooxidative stress. Moreover, as trees probably suffer the most from polluted environments, the need for air pollutant-resistant transgenic trees is now becoming of critical importance (see chapter by Endo et al., this volume).

Transgenic tobacco plants expressing pea GR in both chloroplasts and mitochondria have been reported to possess increased tolerance to an ozone exposure of 0.2 ppm for 8 h per day for up to 2 days in terms of photosynthetic activity but not to paraquat (Broadbent et al. 1995). In contrast, transgenic tobacco plants expressing the pea GR either in cytosol or chloroplasts showed tolerance to paraquat but not to ozone (Broadbent et al. 1995). Despite the conflicting data that have emerged so far from such experiments, it is clear that there is a potential to manipulate GR levels in transgenic plants and thus alter plant responses to oxidative stress.

4. Manipulation of Genes for Other Antioxidative Enzymes

APXs and catalases are enzymes that are also involved in the scavenging of AOS, and so attempts have been made to generate transgenic plants with elevated levels of these enzymes.

Transgenic tobacco plants overexpressing APX have been developed (Pitcher et al. 1994; Saji et al. 1996). Increased tolerance to paraquat damage has been reported in Bel W3 tobacco plants overexpressing cytosolic APX but not in those expressing a chloroplast-targeted isoform (Pitcher et al. 1994). However, in our more recent study, transgenic tobacco plants that overexpressed an *Arabidopsis* cytosolic APX (Kubo et al. 1992), when placed under control of the ribulose-1,5-bisphosphate carboxylase small subunit promoter, showed increased cytosolic APX activity but demonstrated no increase in their paraquat tolerance (Saji et al. 1996). Moreover, the transgenic plants showed no more resistance to ozone exposure, at 0.2 ppm ozone under light (400 $\mu E/m^2/s$), than the nontransgenic control plants in the extent of visible foliar damage or photosynthetic activity (Aono et al. 1998). These results may indicate that, although APXs are clearly essential for scavenging H_2O_2 in plant cells, manipulation of the levels of APX activity alone may be an ineffective means of improving the resistance of plants to air pollutants.

Although catalases are also capable of scavenging large quantities of H_2O_2, their ability to maintain H_2O_2 concentrations low enough to prevent chloroplastic damage is limited because they are located in peroxisomes and have a high K_m. Nevertheless, transgenic tobacco plants overproducing an *E. coli* catalase in their chloroplasts were found to have increased resistance to paraquat and drought stress at a high light intensity (1600 $\mu E/m^2/s$) (Shikanai et al. 1998). Although these transgenic plants have not been tested for their air pollutant resistance, this enzyme may possibly be used for improving air pollutant tolerance of plants. In addition, this result shows that by genetic engineering an exogenous catalase can function at chloroplasts, a subcellular location different from its original location, i.e., peroxisomes, for scavenging H_2O_2 during photooxidative stress.

5. Manipulation of more than One Gene Encoding Antioxidative Enzymes

As presented so far, the enhanced activity of only one antioxidative enzyme in plant cells does not always confer increased stress resistance on transgenic plants. One approach to further increase tolerance to air pollutants is to express two different antioxidative enzymes in the transgenic plants.

To evaluate the potential of this strategy, we developed transgenic tobacco with simultaneously enhanced activities of GR and SOD (GR-SOD hybrids) by cross-fertilization of parents containing an *E. coli* GR or a rice SOD gene. The GR-SOD hybrids showed extremely high tolerance to paraquat, even higher than that of their parents, in the level of electrolyte leakage from paraquat-treated leaf disks (Aono et al. 1995), whereas their level of tolerance to ozone was comparable to that of the control nontransgenic plants (Aono et al. 1998). These results suggest that the mechanism of ozone-induced injury may differ from that induced by paraquat, and that the cooperative function of antioxidant enzymes is an important means by which plants tolerate photooxidative stress.

In addition to these transgenic plants, two more types of hybrid transgenic tobacco plants with simultaneously increased activities of APX and GR, and with APX and SOD, were generated by cross-fertilization of parents expressing only one of the genes. Leaves of the hybrids APX-GR, APX-SOD, and GR-SOD exhibited 2.5- to 5 fold higher activities of the appropriate enzymes than those of the control nontransgenic SR1 tobacco plants. During exposure to 0.2 ppm ozone, APX-GR hybrids showed the highest photosynthetic activity (quantum yield of PS II electron transport) among the hybrids, their parents, and the control plants (Aono et al. 1998) (Fig. 2). These findings suggest that the cooperative function of APX and GR is an essential component of plant tolerance to ozone toxicity.

We consider that these elevated APX and GR levels in the transgenic plants both enhance the scavenging of H_2O_2 and concomitantly regenerate ascorbate so that the capacity to scavenge AOS becomes high enough to maintain the photosynthetic activity during ozone-induced photooxidative damage. The observations that paraquat-resistant transgenic tobacco, overexpressing Cu/Zn SOD and/or GR, did not exhibit resistance to ozone again imply differences between the mechanisms of ozone- and paraquat-induced injury. When the regeneration of ascorbate is sufficient, the scavenging of H_2O_2 in the cytosol appears to be more efficient than the removal of $O_2^{\cdot -}$ for protecting cells from to be most effective against paraquat toxicity.

Despite the fact that photosynthetic activity was maintained in the transgenic APX-GR hybrid during ozone exposure, there was no difference in the extent of visible foliar damage in these or other transgenic plants and the control plants during and after ozone exposure. Hence, AOS, even in the APX-GR transgenic cells, may not have been entirely removed and thus finally resulted in cell death.

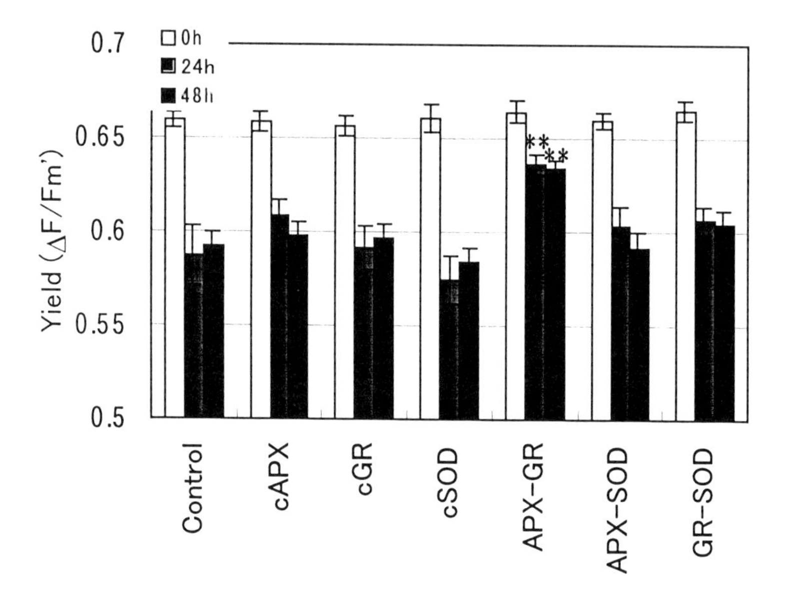

Fig. 2. Photosynthetic activity in leaves of transgenic tobacco plants during ozone exposure. The 4- to 5-week old tobacco plants were exposed to 0.2 ppm ozone at 25ºC under cycles of 14 h light (400 $\mu E/m^2/s$) and 10 h darkness in a growth cabinet. Quantum yields ($\Delta F/Fm'$) of PSII electron transport in the young expanded leaves were measured in the light using a chlorophyll fluorometer (PAM-2000, H. Waltz, Germany) in the middle of the photoperiod. Control, nontransgenic SR1; cAPX, transgenic tobacco with enhanced cytosolic ascorbate peroxidase (APX) activity (Saji et al. 1996); cGR, transgenic tobacco with enhanced cytosolic glutathione reductase (GR) activity (Aono et al. 1995); cSOD, transgenic tobacco with enhanced superoxide dismutase (SOD) activity (Sakamoto and Tanaka 1993); APX-GR, a hybrid between cAPX and cGR (Aono et al. 1998); APX-SOD, a hybrid between cAPX and cSOD (Aono et al. 1998); GR-SOD, a hybrid between cGR and cSOD (Aono et al. 1995). The mean values ± SE obtained from five experiments are shown (Control, cAPX, cGR, cSOD, and APX-SOD, n=24; APX-GR, n=18; GR-SOD, n=23). **, $P<0.01$ (determined by t-test) (from Aono et al. 1998, with permission)

6. Conclusions and Perspectives

Genetic engineering is considered to have tremendous potential for developing air pollutant-tolerant plants. However, more detailed studies are required before such transgenic plants can be used for practical applications such as preservation and improvement of the environment. For such purposes, it is essential that the effects of antioxidative enzyme combinations are extensively investigated and that the expression of the introduced gene and the subcellular localization of its product are appropriately regulated.

In addtion to the antioxidative enzymes, antioxidants such as ascorbate and glutathione play important roles in the scavenging system of AOS (see Fig. 1). It has been reported that the content of ascorbate decreased by ozone in spinach leaves with a corresponding increase in dehydroascorbate, an oxidized product of ascorbate, and the total amount of ascorbate and dehydroascorbate decreased during ozone exposure (Sakaki et al. 1983). This result suggests that not only regeneration of ascorbate but also its synthesis is a critical process in mechanisms of photooxidative stress tolerance. Hence, genetic manipulation of enzymes that are involved in the ascorbate synthesis pathway, such as GDP-mannnose pyrophosphorylase, may possibly be an effective means to increase the ascorbate level in plants and thus to confer air pollutant tolerance to plants.

Reference

Allen RD (1995) Dissection of oxidative stress tolerance using transgenic plants. Plant Physiol 107:1049-1054

Aono M, Kubo A, Saji H, Natori T, Tanaka K, Kondo N (1991) Resistance to AOS toxicity of transgenic *Nicotiana tabacum* that expresses the gene for glutathione reductase from *Escherichia coli*. Plant Cell Physiol 32:691-697

Aono M, Kubo A, Saji H, Tanaka K, Kondo N (1993) Enhanced tolerance to photooxidative stress of transgenic *Nicotiana tubacum* with high chloroplastic glutathione reductase activity. Plant Cell Physiol 34:129-135

Aono M, Saji H, Sakamoto A, Tanaka K, Kondo N, Tanaka K (1995) Paraquat tolerance of transgenic *Nicotiana tabacum* with enhanced activities of glutathione reductase and superoxide dismutase. Plant Cell Physiol 36:1687-1691

Aono M, Ando M, Nakajima N, Kubo A, Kondo N, Tanaka K, Saji H (1998) Response to photooxidative stress of transgenic tobacco plants with altered activities of antioxidant enzymes. In: De Kok LJ, Stulen I (eEds) Responses of plant metabolism to air pollution and global change. Backhuys Publishers, Leiden, pp 269-272

Asada K (1992) Ascorbate peroxidase - a hydrogen peroxide scavenging enzyme in plants. Plant Physiol 85:235-241

Bowler C, Slooten L, Vandenbranden S, De Rycke R, Botterman J, Sybesma C, Van Montagu M, Inzé D (1991) Manganese superoxide dismutase can reduce cellular damage mediated by oxygen radicals in transgenic plants. EMBO J 10:1723-1732.

Bowler C, Van Camp W, Van Montague M, Inzé D (1994) Superoxide dismutase in plants.

Crit Rev Plant Sci 13:199-218
Broadbent P, Creissen GP, Kular B, Wellburn AR, Mullineaux P (1995) Oxidative stress responses in transgenic tobacco containing altered levels of glutathione reductase activity. Plant J 8:247-255
Dodge AD (1975) Some mechanisms of herbicide action. Sci Prog 62:447-466
Edwards EA, Rawsthorne S, Mullineaux PM (1990) Subcellular distribution of mutiple forms of glutathione reductase in leaves of pea (*Pisum sativum* L.). Planta 180:278-284
Foyer CH, Halliwell B (1976) The presence of glutathione and glutathione reductase in chloroplasts: a proposed role in ascorbic acid metabolism. Planta 133:21-25
Foyer CH, Lelandais M, Galap C, Kunert KJ (1991) Effects of elevated cytosolic glutathione reductase activity on the cellular glutathione pool and photosynthesis in leaves under normal and stress conditions. Plant Physiol 97:863-872
Foyer CH, Souriau N, Perret S, Lelandais M, Kunert KJ, vost C, Jouanin L (1995) Overexpression of glutathione reductase but not glutathione synthetase leads to increases in antioxidant capacity and resistance to photoinhibition in poplar trees. Plant Physiol 109:1047-1057
Gillham DJ, Dodge AD (1986) Hydrogen-peroxide-scavenging systems within pea chloroplasts. A quantitative study. Planta 167:246-251
Holmberg N, Bülow L (1998) Improving stress tolerance in plants by gene transfer. Trends Plant Sci 3:61-66
Kubo A, Saji H, Tanaka K, Tanaka K, Kondo N (1992) Cloning and sequencing of a cDNA encoding ascorbate peroxidase from *Arabidopsis thaliana*. Plant Mol Biol 18:691-701
Perl A, Perl-Treves R, Galili S, Aviv D, Shalgi E, Malkin S, Galun E (1993) Enhanced oxidative-stress defence in transgenic potato expressing tomato Cu, Zn superoxide dismutases. Theor Appl Genet 85:568-576
Pitcher LH, Brennan E, Hurley A, Dumsmuir P, Tepperman JM, Zilinskas BA (1991) Overproduction of petunia copper/zinc superoxide dismutase does not confer ozone tolerance in transgenic tobacco. Plant Physiol 97:452-455
Pitcher LH, Repetti P, Zilinskas BA (1994) Overproduction of ascorbate peroxidase protects transgenic tobacco plants against oxidative stress (abstract 623). Plant Physiol 105:S-169
Rennenberg H, Polle A (1994) Protection from oxidative stress in transgenic plants. Biochem Soc Trans 22:936-940
Saji H, Aono M, Kubo A, Tanaka K, Kondo N (1996) Paraquat sensitivity of transgenic *Nicotiana tabacum* plants that overproduce a cytosolic ascorbate peroxidase. Environ Sci 9:241-248
Sakaki T, Kondo N, Sugahara K (1983) Breakdown of photosynthetic pigments and lipids in spinach leaves with ozone fumigation: role of active oxygens. Physiol Plant 59:28-34
Sakamoto A, Tanaka K (1993) Expression of superoxide dismutase genes and stress tolerance of transgenic tobacco. In: Phenotypic expression and mechanisms of environmental adaptation in plants. IGE series 17. Institute of Genetic Ecology, Tohoku University, Sendai, pp 111-121
Schöner S, Krause GH (1990) Protective systems against active oxygen species in spinach: response to cold acclimation in excess light. Planta 180:383-389
Sen Gupta A, Heinen JL, Holaday AS, Burke JJ, Allen RD (1993) Increased resistance to oxidative stress in transgenic plants that over-express chloroplastic Cu/Zn superoxide dismutase. Proc Natl Acad Sci USA 90:1629-1633

Shikanai T, Takeda T, Yamauchi H, Sano S, Tomizawa K, Yokota A, Shigeoka S (1998) Inhibition of ascorbate peroxidase under oxidative stress in tobacco having bacterial catalase in chloroplasts. FEBS Lett 428:47-51

Shimazaki K, Sakaki T, Kondo N, Sugahara K (1980) Active oxygen participation in chlorophyll destruction and lipid peroxidation in SO_2-fumigated leaves of spinach. Plant Cell Physiol 21:1193-1204

Smirnoff N (1993) The role of active oxygen in the response of plants to water deficit and desiccation. New Phytol 125:27-58

Smith I, Polle A, Rennenberg H (1990) In: Alscher RG, Cumming JR (eds) Stress response in plants: adaptation and acclimation mechanisms. Wiley-Liss, New York, pp 201-215

Van Breusegem F, Slooten L, Stassart J-M, Moens T, Botterman J, Van Montagu M, Inzé D (1999) Overproduction of *Arabidopsis thaliana* FeSOD confers oxidative stress tolerance to transgenic maize. Plant Cell Physiol 40:515-523

Van Camp W, Willekens H, Bowler C, Van Montague M, Inzé D (1994) Elevated levels of superoxide dismutase protect transgenic plants against ozone damage. Bio/Technology 12:165-168

Van Camp W, Capiau K, Van Montagu M, Inzé D, Slooten L (1996) Enhancement of oxidative stress tolerance in transgenic tobacco plants overproducing Fe-superoxide dismutase in chloroplasts. Plant Physiol 112: 1703-1714

22
Application of Genetic Engineering for Forest Tree Species

Saori Endo, Etsuko Matsunaga, Keiko Yamada-Watanabe, and Hiroyasu Ebinuma

Pulp and Paper Research Laboratory, Nippon Paper Industries Co., LTD, Oji 5-21-1, Kita-ku, Tokyo 114-0002, Japan

1. Introduction

Forest tree species are the most important component of the earth's biomass and have important ecological and economical roles. In addition to those mechanisms, they are able to neutralize toxic substances, including CO_2 and other air pollutants, that pose a serious threat to human life and the ecosystem. They also have important role in environmental protection. The evolutionary survival of forest areas is an important aspect of environmental protection concerns. However, in near future, natural forest areas will in turn decline because the rapidly increasing global population will require more land devoted to agriculture to support the demand for increased food production. Moreover, forest areas for plantations will be limited to those highly unfavorable environments that pose various forms of stress, including drought, mineral deficiency, acidity, and increased air pollutants concentration on the trees. For the survival of forest trees, genotypes that have the ability to grow under unfavorable environments will be essential.

Recently, in plant breeding, transformation of plants has come to be thought of as a useful tool for improving numerous traits. This technology can generate genetic variability in a way that cannot be achieved by classical breeding methods. Various transgenic crops with new properties have been generated in the past 10 years and over 40 major genetically modified crop species are now in the process of, or have completed, field trials (Teuber 1996). The basic techniques of tissue

Air Pollution and Plant Biotechnology
–Prospects for Phytomonitoring and Phytoremediation–
Edited by K. Omasa, H. Saji, S. Youssefian, and N. Kondo
© Springer -Verlag Tokyo 2002

culture, genetic engineering and molecular biology, developed in major crops, are now being applied to forest tree species with varying degrees of success. Further development and adaptation of these biotechnological methods to forest tree species, which have long life cycles and high levels of heterogyzosity, are an essential prerequisite to their genetic improvement. Several such advances have shown tremendous promise; these include rapid clonal propagation, gene transfer, and the isolation, cloning, and expression of genes.

In this review, we introduce the application of genetic engineering to forest tree species, and propose a strategy for the use of transformation in the improvement of forest tree species. In addition, we discuss the generation of transgenic hybrid aspen, into which a stress-tolerance related gene has been introduced as an approach for the generation of stress tolerant forest species.

2. Genetic Improvement of Forest Tree Species

Forest trees have been gaining in economic importance because they are important for the production of woody products, such as pulp and paper products. Breeding programs for the genetic improvement of economically valuable traits of them have been successful. However, the possibilities of using genetic analysis for the improvement of forest trees are limited compared to major crops.

Forest tree species are characterized by large size, long life cycle, long germination time, high heterozygosity, and high inbreeding depression, and so it is difficult to(1) hybridize under open pollination; (2) generate genetic linkage maps; (3) select individuals with specific traits within the large number of progeny; and (4) effectively select parents for breeding programs. Such limitations complicate the genetic analysis of forest tree species compared to major crops that have been under artificial selection for hundreds or thousands of years. Methods to accelerate breeding programs of forest tree species are therefore needed.

Recently, genetic engineering has become an extremely useful tool in plant breeding and has been applied to various plant species, including most major economic crops, vegetables, ornamentals, fruit plants, and trees. Much of the support for the application of genetic engineering for the improvement of plant species has been provided because of some exceptional advantages. These advantages are of this approach, which may (1) improve a single gene involving a desirable trait without altering other traits, (2) identify the commercial value of an improved plant line more efficiently than classical breeding programs, and (3) reduce the length of the improvement program and the cost per trait. Because of these advantages, transgenic crops expressing foreign genes are being rapidly developed. These expectations will be even more effective with forest tree species compared to the major crop species that have been under artificial selection for hundreds or thousands of years. The application of genetic engineering may significantly accelerate the genetic improvement of forest tree species.

3. Transformation Studies in Forest Tree Species

3.1 Strategy to Achieve Transformation

Transformation of forest tree species is at a threshold. Woody plants, including forest tree species, are commonly considered to be recalcitrant to transformation. The process for producing transgenic plants includes 5 main components (Fig. 1): (1)Vector constructs: the desirable genes are identified for transformation: (2)Gene transfer: the desirable gene integrate into plant genome: (3)Selection of transformed cells: transformed cells are selected from nontransformed cells: (4)Regeneration of transformed cells: transformed cells are regenerated with exogenous hormone: (5)Confirmation: the presence or function of transgenes in the genome of transgenic plants is confirmed. In various forest tree species,

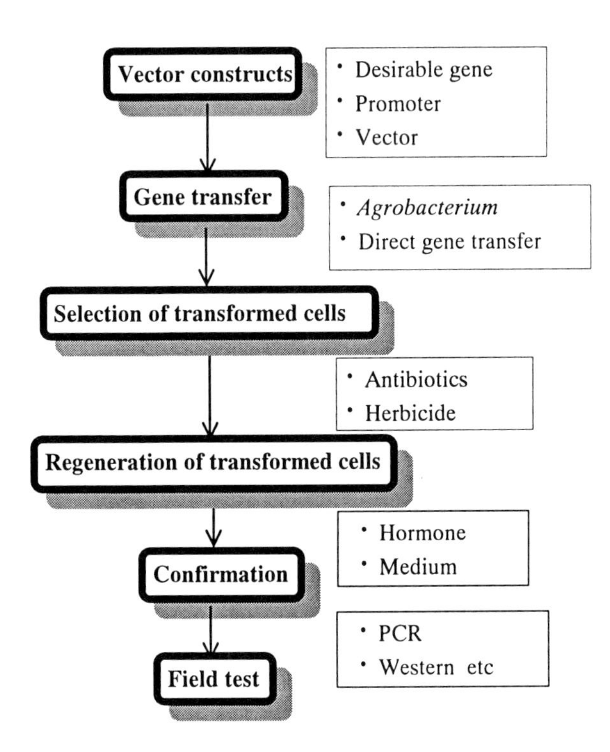

Fig. 1. The strategy and approach of improvement program through transformation methods

success in plant regeneration has been reported, however, the development of stable transformants has occurred in a limited number of species. The major limitations to the development of efficient transformation methods are the very low frequency of regenerants from mature tissues in vitro, the lack of a stable gene transfer system, and the lack of a suitable system of selecting transformed cells from the mass of nontransformed cells. We now focus on the various gene transfer methods available, the regeneration system for forest tree species.

3.2 Gene Transfer Methods Applicable to Forest Tree Species

Over the past decade, a wide range of methods and different approaches for transfering genes into plant cells have been explored. Among these current gene transfer methods, *Agrobacterium*-mediated gene transfer, protoplast-based direct gene transfer, and biolistic DNA transfer (particle bombardment) constitute the major techniques. Comprehensive reviews of the transformation methods have been compiled by Potrykus (1991), Christou (1996), and Siemens and Schieder (1996). Numerous studies have reported transformation of forest trees by these methods, especcially *Agrobacterium*-mediated transformation and biolistic transformation, through which the highest levels of stable transformation have been achieved to date.

3.2.1 *Agrobacterium*-Mediated Gene Transfer

The single most important advantage of *Agrobacterium*-mediated gene transfer over other methods is that transgenes are targeted to the nucleus and are integrated into the host DNA. *Agrobacterium* vectors are the most frequently used for high levels of stable forest tree transformation.

Agrobacterium is a soil-borne bacterium responsible for the development of crown gall disease or hairy roots on a range of dicotyledonous plants (Christey 1997). *A. tumefaciens* possesses a tumor-inducing (Ti) plasmid that is responsible for DNA transfer and results in tumor formation (recently reviewed by Zupan and Zambryski 1995). The transferred DNA (T-DNA) region, which is one part of the Ti plasmid, is integrated in the nuclear genome of the host and stably maintained in the transformed cells (Chilton et al. 1977). The T-DNA contains oncogenes that encode, or regulate, enzymes involved in the synthesis of the phytohormones, auxin(s), and cytokinin(s). The removal of oncogenes from the Ti plasmid results in disarmed strains taht are used in *A. tumefaciens*-mediated transformation (Klee et al. 1987). For transformation of forest tree species, LBA 4404, EHA 101, A281, and C58 are the disarmed strains most widely used. *A. tumefaciens*-mediated transformation has been used to transform several forest species, including populus, fir, pine, and eucalyptus (Ahuja 1986; Parson et al. 1986; Ho et al. 1997). However, because *Agrobacterium*-mediated transformation can only be applied to cultivars for which an adventitous regeneration system has been developed, the method has several limitations for use with most forest tree species for which a high-frequency adventitious regeneration system has not been developed.

Recently, a shoot regeneration system from *A. rhizogenes*-induced hairy roots has been developed for a wide range of plants, and several transgenic plants have been obtained from *A. rhizogenes*-mediated transformation (Christey 1997). *A. rhizogenes* strain possess a root inducing the (Ri) plasmid, which is responsible for DNA transfer and the resulting formation of hairy roots (Tepfer 1984). In several tree species, including Casuasinacae trees and hybrid poplar, transgenic plants have been obtained by *A. rhizogenes*-mediated transformation were obtained (Duhoux et al. 1996). The advantage of *A. rhizogenes*-mediated transformation is that the transformants can be selected on the basis of their Ri-induced characteristics, without the need of chemicals, such as antibiotics for selection. In general, woody plants are sensitive to antibiotics, and so regeneration is inhibited by their presence. The *A. rhizogenes*-mediated transformation methods are thought to have considerable potential for transformation of forest tree species and plants with a high sensitivity to antibiotics.

3.2.2 Direct Gene Transfer; Particle Bombardment

Particle bombardment is a proven method for delivery of DNA into plant cells for both transient gene expression studies and stable transformation (Klein et al. 1988). The main advantage of this method is that intact plant tissue can serve as the target. Cell suspensions, calli, meristems, embryos, somatic tissues of plants grown in vitro and even in vivo may also be targeted by particle bombardment. For forest tree species, embryogenic cultures system have been generated for most commercially important genus. Especially, it has been developed in gymnosperms, such as coniferous and hardwood species. The particle bombardment has proved efficient method for gymnosperms.

3.3 Regeneration of Transformed Cells

Recovery of transgenic plants relies on efficient protocols for the regeneration of transformed cells. The ability of transformed cells to regenerate appears to be affected by stress induced by the transformation process itself. Some of this stress may be imposed during induction of genes for instance, the physiological and biological events associated with bombardment-induced wounding are found to affect regeneration. Also, antibiotics used to kill *Agrobacterium* after transformation may also affect regeneration, as, for example, carbenicilline, which may inhibit morphogenesis.

Regeneration may also be affected by stress imposed during selection of transgenic cells. For example, antibiotics and herbicides, which are common strategies for reducing the survival of nontransformed cells and selection of transformed cells, also reduce the transformation efficiencies of woody plants. This situation might be due to either unstable integration of the transgenes or to the reduced regenerability of transformed cells. The sensitivity to antibiotics differs between plant species and tissues, and for forest tree species without a suitably high regeneration system the effect of selection stress is a key issue in increasing the regeneration ability.

3.4 Repeated Transformation of Forest Tree Species

There are several problems facing the improvement program based on transformation methods of forest tree species. One such problem is that repeated transformation is limited in forest tree species that have long life cycles. To overcome this limitation, Ebinuma et al. (1997a,b) have developed a new vector system (MATVS, multi-auto-transformation vector system), based on the novel principle that morphological changes caused by the *ipt* or *rol* oncogenes of *Agrobacterium* can be used as selection markers. This vector system is characterized by (1) transgenic plants that are autonomously regenerated and visually selected through the endogenous manipulation of plant hormones (positive selection), and (2) marker-free transgenic plants that can be repeatedly transformed without using up the available marker genes and without crossing. Using this system, increased transformation efficiencies have been obtained for both tobacco and hybrid aspen in comparison to that using kanamycin as a selection marker. This MATVS is also expected to improve the transformation efficiencies of recalcitrant plant species, including forest tree species that are sensitive to antibiotics and have a low regeneration ability, and to provide the means to pyramid several valuable genes into perennial plant species that have long germination times. The MATVS is described in detail by Ebinuma et al. (1997a,b).

4. Transformation Research of Populus Species as a Model System for Forest Tree Species

4.1 Transformation Research of Populus Species

Species of the genus *Populus* (poplars and aspens) comprise a genus of great importance in forestry and have been adopted as the experimental model for forest tree species because of their ease of use in clonal propagation, fast growth and short rotation cycle. To date, various genes have been introduced to poplar trees as a model system of forest trees. Transformation research with poplar are summarized in Table 1. In these researches, introduced genes include (1) herbicide tolerance-related gene, such as the mutant *aroA* gene that codes glyphosate tolerance via a 5-enolpyruylshikimate-3-phosphate synthase (Donahue

Table 1. Transgenic research with *Populus* species

Species	Transgenes	Reference
P. trichocarpa × *P. deltoides*	*bar*	De Block (1990)
P. alba × *P. grandidentata*	*aroA*	Fillatti et al. (1987)
P. alba × *P. grandidentata*	*aroA*	Donahue et al.(1994)
P. alba × *P. grandidentata*	*Bt*	Howe et al.(1994)
P. alba × *P. grandidentata*	*PIN2,*	Scott et al.(1997)
P. tremula × *P. alba*	*gor*	Foyer et al.(1995)
P. tremula × *P. alba*	*ggs*	Noctor et al.(1996)
P. tremula × *P. alba*	*gshII*	Foyer et al.(1995)
P. tremula × *P. alba*	*OMT*	Tsai et al. (1998)
P. tremula × *P. alba*	*CAD*	Baucher et al.(1996)
P. tremula × *P. alba*	*4CL*	Hu et al.(1999)
P. tremula × *P. alba*	*ipt*	Schwartenberg et al. (1994)
P. tremula × *P. alba*	*Crs1-1*	Brasileiro et al.(1992)
P. sieboldii × *P. grandidentata*	*POD*	Kawaoka et al.(1998)
P. sieboldii × *P. grandidentata*	*gor*	Endo et al. (1997)
P. sieboldii × *P. grandidentata*	*iaaM*	Ebinuma et al.(1997)
P. tremula × *P. tremuloides*	*OCI*	Leplé et al.(1995)
P. tremula × *P. tremuloides*	*iaaM/H*	Sundberg et al.(1997)
P. tremula × *P. tremuloides*	*LFY*	Weigel and Nilsson (1995)
P. tremula × *P. tremuloides*	*rolC*	Ahuja and Fladung (1996)

AroA, bacterial 5-enolpyruvylshikimate-3-phosphate synthase chimeric gene; *bar*, phosphinotricin acetyltransferase gene; *Bt*, endotocxin gene from *Bacillus thuringiensis*; *crc1-1*, mutant acetolactate synthase gene; *CAD*, cinnamyl alcohol dehydrogenase gene; *ggs*, γ-glutamylcysteine synthetase gene; *gor*, glutathione reductase gene; *gsh*, glutathione synthetase gene; *iaaH*, agorobacterial indoleacetamide hydrolase gene; *iaaM*, agrobacterial tryptophan monooxygenase; *ipt*, agorobacterial isopentenyltransferase gene; *rolC*, one of the genes reponsible for hairy root disease caused by *Agrobacterium*; *LFY*, flower-meristem-identity gene; *OCI*, cysteine proteinase inhibitor gene; *PIN2*, wound-inducible potato proteinase inhibitor II gene; *POD*, peroxidase gene; *OMT*, cafferic acid/5-hydroxylase gene; *4CL*, 4-hydroxycinnamate gene.

et al. 1994, Filatti et al. 1987), a *bar* gene taht encodes the enzyme phosphinotricin acetyltransferase (PAT) which inactivates the herbicide phosphinotricin (glufosinate), and the mutant *crs1-1* gene from a chlorosulfuronherbicideresistant line of Arabidopsis thathliana (De Block 1990, Brasilero et al.1992); (2) an insect and pathogen damage-related gene, such as *OCI* (oryzastatin), a cystein protease inhibitor, and *PIN2* (proteinase inhibitor II), a trypsin/chymotrypsin inhibitor gene for pest resistance (Heuchelin et al.1997; Klopfenstein et al. 1993; Scott et al. 1997; Leplé et al. 1995), insecticidal proterin genes from *Bacillus thuringienis* (Howe et al. 1994); (3) gene which is related with lignification pathway, such as *POD* (Kawaoka et al. 1998), *CAD* (Cinnamyl alchol dehydrogenase) (Baucher et al. 1996), and *4CL* (4-coumarate:coenzyme A ligase) (Hu et al. 1999); (4) a developmental influence genc, such as *iaaM/H*, and *LEAFY* (Sundberg et al. 1997); and (5) a gene related with metabolism, such as *gor* (glutathione reductase), *ggs*(γ glutamylcysteine)or *ghsII* (glutathione synthetase) (Foyer et al. 1995; Noctor et al. 1996). Transgenic poplars resistant to insect and pathogen damage and herbicides and with better wood quality (improvement of lignin biosynthesis) have been obtained (Heuchelin et al. 1997; Klofenstein et al. 1993; Brasileiro et al. 1992; Baucher et al. 1996).

4.2 The Regulation and Expression of Transgenes

A major problem regarding genetic engineering is how to regulate integration and expression of foreign genes in forest trees. Although many promoters of viral, bacterial, and plant origins have been tested in many plant species, little is known about their regulatory control in forest trees. In most researches of transgenic poplars, expression of the transgene is driven by *35S* promoter from cauliflower mosaic virus. A proteinase inhibitor II (*PIN2*) gene was expressed in transgenic hybrid poplar clone (*P. alba* x *P. grandidentata* cv. 'Hansen') (Klopfenstein et al. 1997). The *aroA* gene driven by the *35S* promoter conferred herbicide tolerance in another hybrid poplar clone (*P. alba* x *P. grandidentata* cl. 'NC5339'). In recent investigations, we detected the expression of the *GR* gene or *POD* gene under the *35S* promoter in a hybrid poplar clone (*P. sieboldii* x *P.grandidentata*).

Another problem presented by genetic engineering is stability of the integrated transgenes in forest tree genomes. Forest trees have long life cycles, with an extended vegetative phase raging from 1 year to several decades. Because trees are firmly anchored in one location, they are exposed, over long periods, to changing environments that may influence their physiology and morphogenetic process. Under such conditions, expression of the transgene was variable. Variation in GUS expression was monitored in the 35S-*uidA* transgenic poplar and spruce (Ellis and Raffa 1997). Ellis and Raffa (1997) reported that GUS expression was least variable during *in vitro* culture of regenerated transformants, most variable during field growth of transformants, and most variable during field growth of transgenic plants. These studies on GUS activity suggested that the expression of the introduced gene depends on promoter type, plant genome, and environmental conditions. Research projects are being carried out in greenhouses and in the field with transgenic trees (Raffa et al. 1997).

4.3 Transformation Procedure of Populus Species

A number of populus species have been modified by genetic transformation, mostly through *Agrobacterium*-mediated transformation. Several cases of *A. tumefaciens*-derived transgenic plants, recovered from calli, have been reported (De Block 1990). The hybrid aspen clone (*P. sieboldii* x *P. grandidentata* cl. 'Y63') that is used in our laboratory, is an elite clone of hybrid aspen obtained by crossing fine selected female trees of *Populus siebollidi* with a male elite tree of *Populus grandidentata* (Fig. 2A). For *Agrobacterium*-mediated transformation, stem segments from in vitro cultured plants are cocultured in modified MS medium with LBA 4404 (a disarmed *Agrobacterium* strain) for 3 days, after which stem segments are transferred to a shoot induction medium (containing 0.5 mg/l zeatin, 500 mg/l carbenicillin and 100 mg/l kanamycin) (Ebinuma et al. 1997c). Small calli, emerging from the wounded edges, can be observed about 4 weeks after cocultivation and adventitious buds differentiate from 20%-40% of the calli within 2 months. After a further month of growth under similar conditions, the foliage from each bud grows to a length of 2-3 cm (Fig. 2B), at which time the stems can be aseptically excised and subcultured in root induction medium (a modified two-thirds-strength MS medium containing the artificial auxin, 0.05 mg/l indole-3-butyric acid, IBA). Regenerated plants are transferred to a soil mix and grown in a greenhouse (Fig. 2C).

5. Improving Stress Tolerance in Hybrid Aspen by Genetic Engineering

5.1 Generating Tolerant Plants Through Genetic Engineering

Higher plants respond to environmental stress by triggering various defense mechanisms, one of which responses is the antioxidant defense system. A variety of biochemical responses are commonly observed in the leaves of plants after being exposed to environmental stresses such as air pollutants, low temperature, and herbicide. Such responses include increased amounts of antioxidants or activities of antioxidative enzymes (reviewed by Foyer et al. 1994a) and, in most cases, are accompanied by the formation of active oxygen species.

Recently, transgenic plants expressing various antioxidative enzyme-related genes have been generated and subsequently examined under stress environments. For example, transgenic tobacco, overproducing manganese superoxide (MnSOD)

A. B.

C.

Fig. 2. Hybrid aspen (*Populus sieboldii x P.grandidentata*) **A.** Adult tree of Hybrid aspen. **B.** Adventitious shoots formed from the stem segments cultured on selection medium containing 100 mg/l kanamycin. **C.** Transgenic hybrid aspen

in chloroplasts, or glutathione reductase (GR) in either chloroplasts or mitochondria, has been shown to be less sensitive to environmental stress (see Aono, this volume; Van Camp et al. 1994; Aono et al. 1994; Broadbent et al. 1995). Thus, such strategies of overproducing antioxidative enzymes have considerable potential for improving the stress tolerance of plants.

To evaluate this strategy in forest tree species, we have introduced the *gor* gene, which encodes an *E. coli* designed GR into hybrid aspen. The GR, which normally exists in both the chloroplast and cytosol of plant cells, is involved in the AOS detoxification pathway and catalyzes the connection of oxidized glutathione to reduced glutathione via oxidation of NADPH (Foyer and Halliwell 1976; Rennenberg 1980). Two derivaties of the binary vector, pBI121, in which the *gor* gene is placed under control of the *35S* promoter (pEGR 4, pEGR 6), were used for transformation (Fig. 3). In the pEGR 6 construct, a DNA fragment encoding a chloroplastic transitpeptide was upstream of the *gor* gene to direct the protein to

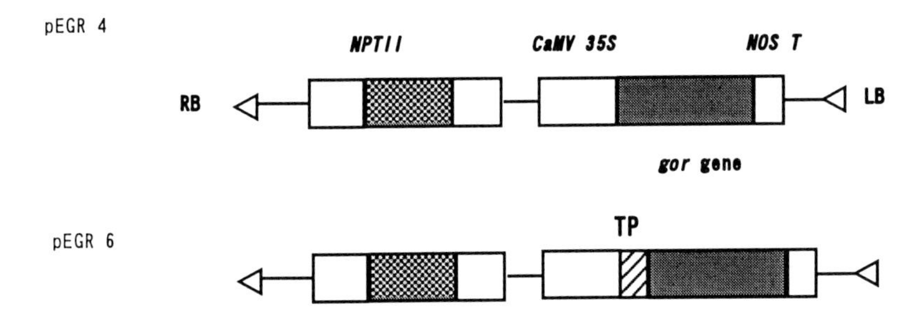

Fig. 3. Construction of the T-DNA region of pEGR 4 and pEGR 6. RB, right border ; LB, left border ; *NPT II*, a chimeric gene for neomycin phosphotransferase ; *CaMV 35S* ; *35S* promoter of cauliflower mosaic virus ; *NOS T*, nopaline synthase terminator ; TP ; chloroplast-specific transit peptide

the chloroplast. Expression and localization of the introduced GR were confirmed by immunochemical analysis. The foliar GR activities of plants transformed with the *gor* gene were substantially higher than those of the untransformed controls (Fig. 4). Despite some variation, the GR activity of the transgenic lines, GR4 and GR6, in which the *gor* gene was expressed in the cytosol and chloroplasts, respectively, were one- to three fold higher than that of the controls.

The physiological consequences of overexpression of the *E.coli* designed *gor* gene have been studied in both tobacco (Foyer et al. 1991; Aono et al.1991, 1994) and poplar (Foyer et al. 1995). The constructs employed in these studies were almost identical to use in our transformations, and so similar increases in GR activity were observed. When the *gor* gene was placed under the *70S* promoter and targeted to chloroplasts of the poplar hybrid, *Populus tremula* X *P. alba*, the level of GR in leaves of transgenic poplar were approximately 100 fold those of the control plants (Foyer et al. 1995).

Increased GR activity has often bean reported under stress situations. Despite constitutive expression of the *gor* gene, the amount of extractable GR activity was found to depend on the conditions under which the plants were grown. Thus, the GR activity of transgenic hybrid aspen grown in the greenhouse was one- to seven foldthat of controls but only one- to three fold greater when grown in vitro. Foyer et al. (1991) reported that the exposure to light increased the GR activity in transgenic tobacco expressing the introduced *gor* gene. The *gor* gene was constitutively expressed and should therefore not have been light inducible, which suggested that the increased GR activity may have resulted from an increased rate of *gor* mRNA translation (Foyer et al. 1991).

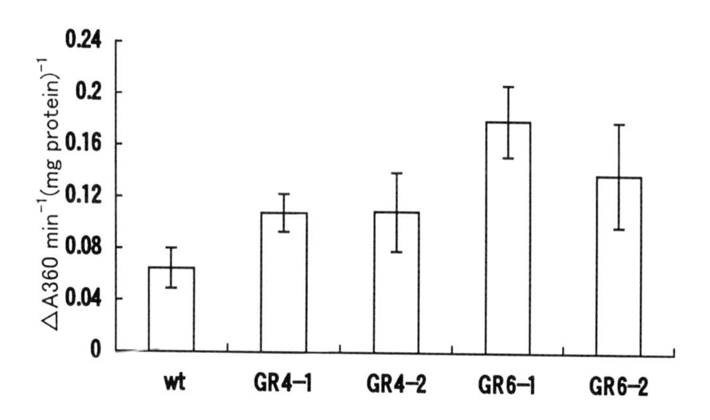

Fig. 4. GR activity of transgenic aspen (GR4-1, GR4-2, GR6-1, GR6-2) and nontransgenic aspen (wt). GR activity of leaf extracts were measured in reaction mixture that contained 0.1 M potassium phosphate (pH7.8), 0.2 mM GSSG and 0.2 mM NADPH in a final volume and monitored by the decrease in absorbance of NADPH at 340nm

5.2 Tolerance of Transgenic Hybrid Aspen to Oxidative Stress

5.2.1 Methyl Viologen

To assess transgenic tolerance to oxidative stress, several studies have exposed leaf disks of both transformed and untransformed lines to the herbicide methyl viologen (MV; 1, 1'-dimethyl-4, 4'-bipyridinum chloride), a redox-active compound that is photoreduced by PSI and subsequently reoxidized by transfer of its electrons to oxygen, forming the superoxide anion (O_2^-). Both O_2^- and other related active oxygen species (AOS), such as hydroxyl radicals(OH^-), are presumably the agents that cause cellular damage (see Poll, this volume). Using leaf disks of our transgenic hybrid aspen, which overproduces GR in the cytosol or chloroplasts, we have found a decrease in the extent of MV-induced leaf injury in comparison with untransformed controls (Fig. 5).

These results concur with those of Aono et al. (1991, 1994, this volume) for transgenic tobacco overproducing GR in the cytosol and chloroplasts. Overexpression of GR in chloroplasts is known to increase the activity of the ascorbate-glutathione cycle, which plays a central role in AOS detoxification, and to result in increased tolerance to MV. However, the mechanism by which overexpression of GR in cytosol leads to increased MV tolerance is unknown.

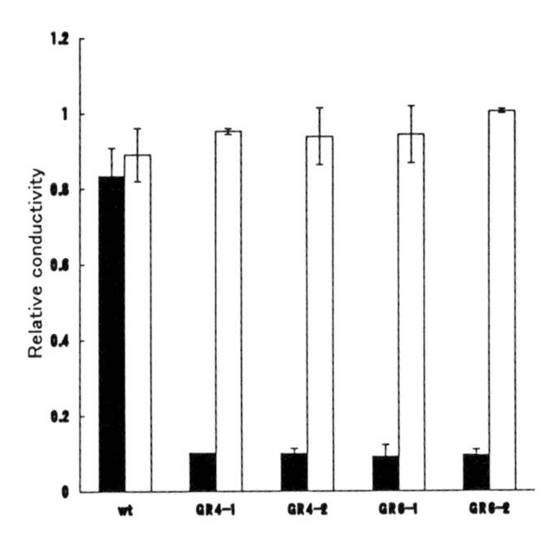

Fig. 5. Electrolyte leakages from paraquat- treatment leaf disks. Electrolyte leakages were assessed with the conductivity of the floating solution. Conductivity was measured after 20 h of illumination and relative values at 1 μ M (filled columns) and 10 μ M (open columns) were determined against conductivity at 100 μ M paraquat

Foyer et al. (1991) reported that high GR expression in the cytosol had no effect on CO_2-dependent O_2 evolution when leaf disks were treated with MV, and suggested that high-GR leaves in the cytosol had played no role in any effect on the detoxification of AOS produced by PSI. However, it is possible that the ascorbate pool might be largely reduced in transgenic plants following MV treatment, but in a largely oxidized form in untransformed control plants.

5.2.2 Sulfur Dioxide (Air Pollutants)

Air pollutants such as ozone (O_3) and sulfur dioxide (SO_2) are thought to be one of the major factors influencing modern forest decline (see Nouchi, this volume). SO_2 injury has been reported in many plant species, especially under intense light or during the daytime. SO_2-induced injury to biological systems is most probably caused by the generation of radicals, including O_2^-, OH^- and SO_3^-, during the oxidation of SO_2 sulfate. In chloroplasts, SO_2 oxidation can be initiated by superoxide generated from the photosythetic electron transport chain and can induce the production of active oxygen species and sulfur trioxide radicals (see Kondo, this volume). To evaluate the response of our transgenic lines to such AOS, we exposed to 4-month-old transgenic hybrid aspen lines (GR4 and GR6) and the wild-type controls to 1 ppm SO_2 fumigation at 25℃ under light (550 μ $E \cdot m^{-2} \cdot s^{-1}$) and 70% humidity in a growth chamber. The transgenic aspen lines, GR4-1 and GR4-2, expressing GR in the cytosol, exhibited two fold higher GR activities than the untransformed controls. The other transgenic lines, GR6-1 and

GR6-2, expressing GR in chloroplasts, showed two- to three fold higher GR activities than the untransformed controls. Three transgenic hybrid aspen plants of each line and a nontransformed control were used in the SO_2 fumigation experiments (Fig. 6). After 7 h of fumigation, all control plants showed severe symptoms of foliar damage, such as necrosis and dehydration. Mature leaves were much more susceptible to SO_2 damage than younger leaves. However, in two independent experiments, leaves of transgenic plants with elevated GR activity (GR 4 and GR 6) exhibited lower damage levels than the untransformed controls (Fig. 7). In addition, transgenic hybrid aspen with GR activity levels similar to those of control plants also showed severe damage symptoms. Both lines of transgenic hybrid aspen lines, GR4 and GR6, showed comparable levels of SO_2 resistance, dependent on the increase in GR activity by overexpression of the *E. coli gor* gene. We believe this result can be best explained by the increased tolerance to AOS by changes in GR activity. SO_2-induced injury to biological systems is most probably caused by the generation of AOS. Augmentation of GR activity may change the balance of the reduced and oxidized glutathione pools in the cell and this may reduce the damage caused by AOS (Foyer et al. 1994b). Now, the balance of the reduced and oxidized glutathione pools will be analyzed in leaves of transgenic plants and untransformed controls.

Glutathione reductase was isolated from cytosol, chloroplasts and mitochondria. However, the effect of cytosolic GR on oxidative stress is not clear. Transgenic tobacco, overexpressing GR in cytosol, showed the increase in GR activity, while it did not show tolerance to air pollutants. Only transgenic plants, overexpressing GR in chloroplasts, showed the increased tolerance to O_3 and SO_2 (Aono et al. 1991, 1993; Foyer et al. 1991). Our results suggested

Fig. 6. Transgenic hybrid aspen plants established in pots containing artificial soil mix

strongly that cytosolic GR affected the tolerance to oxidative stress in hybrid aspen. It is suggested that the detoxification pathway in cytosol may influence the capacity of tolerance to oxidative stress in hybrid aspen. Our next step will be to identify the protective mechanism of overexpressing GR in cytosol.

A

B

Fig. 7. Effect of SO_2 fumigation. Leaves of transgenic and untransformed hybrid aspen after exposure to 1.0 ppm SO_2, 25 ℃, under light, humidity 70%, for 7 hours. A: Leaves of untransformed aspen. B: Leaves of transgenic aspen (GR 6) with elevated GR activity

6. Prospects of Improving Traits of Forest Tree Species Through Genetic Engineering

In the near future, the demand for increased food production by an increasing global population will most certainly reduce the area available for both forest tree growth and the plantation of commercially-important trees. The transgenic forest tree, which has improved ability to grow in highly unfavorable environments through genetic engineering, will have an important role for increasing plantation areas for tree species and for the evolutionary survival of forest area.

Recently, various transgenic crops with new properties have been generated, and more than 40 major genetically modified crop species are now in the process of, or have completed, field trials and commercialization. Although transformation technology has been reached a practical level in crop breeding, some problems exist for forest tree species, that can interfere with the turn of transgenic forest trees to practical use.

A major problem is concerned with the generation of stable transformed plants that express transgenes in a predictable manner. Transformation is believed to result in random integration of transgenes into the genome, causing high variation in quantitative and qualitative expression levels of the transgene in primary transformants or subsequent generations. However, forest tree species are characterized by large size, long life cycle, and long germination time, and so it is difficult to generate the next generation, which stably expresses transgenes. Moreover, forest tree species are exposed to various environmental stresses during their long life cycle, and environmental stresses including mineral deficiency, drought, and low temperatures may significantly influence the expression of the transgene. The forest tree, which shows tolerance to these environmental stresses, may be useful for transformation material. Our groups initiated, in 1999, filed trials with the transgenic aspen introduced *gor* gene. Future research work on transgenic forest trees should focus on the field evaluation of the transgene expressed in forest tree genomes. Our research will focus on understanding the complex environmental stresses in the field.

Another problem is concerned with the repeated transformation system for forest tree species. To improve various traits, stacking of various genes by repeated transformation is needed. For annual plant species, accumulation of the gene is possible by crossing. Because of their long life cycles, these approaches may have limited application in forest tree species. Repeated transformation systems without crossing are needed for forest tree transformation. Future research efforts may focus on the application of the repeated transformation system without crossing, such as the MAT Vector system, to forest tree species.

In future, genetic engineering has the potential possibility to improve various traits of forest tree species that are concerned the tolerance to environmental stress and to improve the ability to grow in highly unfavorable environments. Also the transgenic forest tree will have an important role for increasing plantation areas to areas unfavorable for tree species, such as the desert.

Acknowledgments. The authors thank Dr. N. Kondo, H. Saji, and M. Aono (National Institute for Environmental Studies) for their collaborative studies.

References

Ahuja MR (1986) Gene transfer in forest trees. In: Hanover JW, Keathley DE (eds) Genetic manipulation of woody plants. Plenum Press, New York, pp 25-41

Ahuja MR Fladung M (1996) Stability and expression of chimeric genes in Populus. In: Ahuja MR, Boerjan W, Neale DB (eds) Somatic cells: genetic molecular genetics of trees. Kluwer,Dordrecht, pp 89-96

Aono M, Kubo A, Saji H, Natori T, Tanaka K, Kondo N (1991) Resistance to active oxygen toxicity of transgenic *Nicotiana tabacum* that expresses the gene for glutathione reductase from *E. coli*. Plant Cell Physiol 32:691-697

Aono M, Kubo A, Saji H, Tanaka K, Kondo N (1994) Enhanced tolerance to photooxidative stress of transgenic *Nicotiana tabacum* with high chloroplastic glutathione reductase activity. Plant Cell Physiol34: 129-135

Baucher M, Chabbert B, Pilate G, Van Doorsselaere J, Tollier MT, Petitconil M, Cornu D, Monties B, Van Montagu M, Inzt D, Jouanin L, Boerjan W (1996) Red xylem and higher lignin extractability by downregulating a cinnamyl alcohol dehydrogenase in poplar Plant Physiol 112:1479-1490

Brasileiro ACM, Tourneur C, Leple JC, Combes V, Jouanin L (1992) Expression of mutant *Arabidopsis thaliana* acetolactate synthase gene confers chlorsulfuron resistance to transgenic poplar plants. Transgenic Res 1:398-403

Broadbent P, Creissen GP, Kular B, Wellburn AR, Mullineaux PM (1995) Oxidative stress responses in transgenic tobacco containing altered levels of glutathione reductase activity.Plant J 8(2):247-255

Chilton, MD, Drummond MH, Merlo DJ, Sciaky D, Montoya AL, Gordon MP, Nester EW (1977) Stable incorporation of plasmid DNA into higher plants: the molecular basis of crown gall tumorigenisis Cell 11:263-271

Christou P (1996) Transformation technology. Trends Plant Sci 1(12):423-431

Christey CM (1997) Hairy roots culture. In: Doran M (eds) Hariy roots. Hardwood, Amsterdam, pp 99-111

De Block M (1990) Factors influencing the tissue culture and *Agrobacterium tumefaciens*-mediated transformation of hybrid aspen and poplar clones. Plant Physiol. 93:1110-1116

De Block M, Herrera-Estrella L, van Montagu M, Schell J, Zambryski P (1984) Expression of foreign genes in regenerated plants and their progeny. EMBO J 3: 1681-89

Donahue RA, Davis TD, Michler CH, Riemenschneider DE ,Carter DR, Marquardt PE, Sankhla D, Haissig BE, Isebrands JG (1994) Growth, photosynthesis, and herbicide tolerance of genetically modified hybrid poplar. Can J For Res 24: 2377-2383

Duhoux C, Franche D, Bogung D, et al. (1996) *Casuarina* and *Allocasuarina spesuim* In:Bajaj YPS (eds) Trees. Ⅳ. Biotechnol Agric For. Springer Verlag, Berlin 35: pp76-94

Ebinuma H, Sugita K, Matsunaga, E, Yamakado M (1997a) Selection of marker-free transgenic plants using the isopentenyl transferase gene. Proc. Natl. Acad. Sci USA 94:

2117-2121
Ebinuma H, Sugita K, Matsunaga E, Yamakado M, Komamine A (1997b) Principle of MAT vector system. Plant Biotechnol 14(3):133-139
Ebinuma H, Matsunaga E, Yamada K, Yamakado M (1997c) Transformation of hybrid aspen for resistance to crown gall disease In: Klopfenstein, NB. Young W-C, Kim M-S, Ahuja R (eds) Micropropagation, genetic engineering, and molecular biology of Populus. Rocky Mountain Forest and Rnage Experimental Station, Colorado pp 165-172
Ellis DD, Raffa KF (1997) Expression of transgenic *Bacillus thuringiensis* δ -endotoxin in poplar. In: Klopfenstein NB, Young W-C, Kim M-S, Ahuja R (eds) Micropropagation, genetic engineering, and molecular biology of Populus. Rocky Mountain Forest and Rnage Experimental Station, Colorado pp pp 178-186
Endo S, Matsunaga E, Yamada K, Ebinuma H (1997) Genetic engeineering for air-pollutant resistance in hybrid aspen. In: Ned B. Klopfenstein, Young Woo chun, Mee-Sook Kim, and Rai Ahuja (Eds) Micropropagation, Genetic Engineering, and Molecular Biology of Populus. Rocky Mountain Forest and Rnage Experimental Station, Colorado, pp 187-192
Fillatti JJ, Sellmer J, McCown B, Haissig B, Comai L (1987) Agrobacterium mediated transformation and regeneration of Populus.Mol Gen Genet. 206:192-199
Foyer CH, Halliwell B (1976) The presence of glutathione reductase in chloroplasts: a proposed role in ascorbate acid metabolism. Planta 133:21-25
Foyer CH, Souriau N, Lelandais M, Galap C, Kunert KJ (1991) Effects of elevated cytosolic glutathione reductase activity on cellular gulutathione pool and photosynthesis in leaves under normal and stress conditions. Plant Physiol 97:863-872
Foyer CH, Descourvieres P, Kunert KJ (1994a) Protection against oxygen radicals: an important defence mechanism studied in transgenic plants. Plant Cell Environ17: 507-523
Foyer CH, Descurvieres P, Kunert KL (1994b) Photooxidative stress in plants. Physiol Plant 92: 696-717
Foyer CH, Souriau N, Perret S, Lelandais M, Kunert KJ, Pruvost C, Jouanin L (1995) Overexpression of glutathione reductase but not glutathione synthetase leads to increases in antioxidant capacity and resistance to photoinhibition in poplar trees. Plant Physiol 109:1047-1057
Heuchelin SA, Jouanin L, Klopfenstein NB, McNaBB HS (1997) Potential of proteinase inhibitors for enhanced resistance to populus arthropod and pathogen pests. In: KlopfensteinNB, Young W-C, Kim M-S, Ahuja R (eds) Micropropagation, genetic engineering, and molecular biology of Populus. Rocky Mountain Forest and Rnage Experimental Station, Colorado pp 173-177
Ho C-K, Tsay JT, Vincent L, Chang S-H, et al (1997) Agrobacterium-mediated transformation of eucalyptus camaldulensis and its association with conbentional tree improvement programs. Tappi Proc (1997) 525-528
Howe GT, Goldfarb B, Strauss SH (1994) Agrobacterium mediated transformation of hybrid poplar suspension cultures and regeneration of transformed plants. Plant Cell, Tissue and Organ Culture 36: 59-71
Hu W, Scott AH, Jrhau L, Jacqueline L, Popko T, Vincent L, John R, Douglas DS (1999) Repression of ligin biosythesis in transgenic trees results in high-cellulose and accelerated growth phenotypes. In 10th international symposium on wood and pulping chemistry, pp 516-519

Kawaoka A, Matsunaga E, Yoshida K, Shinmyou A, Ebinuma H (1998) Growth stimulation of hybrid aspen by introduction of peroxidase gene. Abstract 1331, 5th ISPMB Congress, Singapore

Klee H, Horsch R, Rogers S (1987) Annu. Rev, Plant Physiol 38:467-486

Klein TM, Harper EC, Svab Z (1988) Stable genetic transformation of intact *Nicotiana* cells by the particle bombardment process. Proc. Natl. Acad. Sci. USA 85:4305-4309

Klopfenstein NB, McNabb HS, Hart ER et al (1993) Transformation of populus hybrids to study and improve pest resistance. Silvae Genet. 42: 86-90

Leplé JC, Bonadé-Bittino M, Augustin S, Pilate G, Dumanois Lé TânV, Delplanque A, Cornu D, Jouanin L (1995) Toxicity to *Chrysomela tremula* of trangenic poplars expressing a cysteine proteinase inhibitor. Mol Breed 1:319-328

Noctor G, Michael St, Lise J, Karl-josef K, Foyer CH, Rennenberg H (1996) Synthesis of glutathione in leaves of transgenic poplar overexpression γ -glutamylcysteine dynthetase. Plant Physiol 112:1071-1078

Parson TJ, Sinkar, VP, Stettler RF (1986) Transformation of poplar by *Agrobacterium tumefaciens*. Bio/Technology 4:533-536

Potrykus I (1991) Gene transfer to plants; assessment of published approaches and results. Annu. Rev. Plant Physiol. Plnat. Mol. Biol. 42: 205-225

Raffa KF, Kleiner KW, Ellis DD, McCown BH (1997) Environmental risk assessment and deployment strategies for genetic engineering insect-resistant populus. In: KlopfensteinNB, Young W-C, Kim M-S, Ahuja R (eds) Micropropagation, genetic engineering, and molecular biology of Populus. Rocky Mountain Forest and Rnage Experimental Station, Colorado pp 249-263

Rennenberg H (1980) Glutathione metabolism and possible biological roles in higher plant. Phytochemistry 21: 2778-2781

Schwartenberg K, Doumas P, Jouanin L, Pilate G (1994) Enhancement of the endogenous cytokinin concentration in poplar by transformation with *Agrobacterium* T-DNA gene ipt. Tree Physiol. 14:27-35

Scott A, Heuchelin LJ, Klopfenstein NB, Harold S, McNabb Jr (1997) Potential of proteinase inhibitor for enhanced resistance to *Populus Arthropod* and pathogen pests. In: KlopfensteinNB, Young W-C, Kim M-S, Ahuja R (eds) Micropropagation, genetic engineering, and molecular biology of Populus.Rocky Mountain Research Station 173-177

Siemens J, Schieder O (1996) Transgenic plants: genetic transformation recent development and the state of the art. Plant Tissue Culture and Biotechnol 2: 66-75

Sundberg B, Tuominen H, Nilsson O, Moritz T, Little CHA, Sandberg G, Olsson O (1997) Growth and development alternation in transgenic populus:status and potential applications. In: Klopfenstein NB, Young W-C, Kim M-S, Ahuja R (eds) Micropropagation, genetic engineering, and molecular biology of Populus. Rocky Mountain Forest and Rnage Experimental Station, Colorado pp 74-83

Tepfer D (1984) Transformation of several species of higher plants by *Agrobacterium* rhizogenes: sexual transmission of transformed genotype and phenotype. Cell 37:959-967

Teuber M (1996) Genetic modified food and its safety assessment. In: J Tomiuk, K Wohrmann, A Sentker (eds) Transgenic organisms: biological and social implications. Advances in Life Sciences, Birkhauser, Base, lpp 181-189

Tsai C, Popko JL, Mielke MR, Hu W, Podila GK, Chaing VL (1998) Supression of O-methyltransferase gene by homologous sense transgenic in quaking aspen cause red-brown wood phenotypes Plant Physiol 117:101-112

Van Camp W, Willekens H, Bowler C, van Montagu M, Inze D, Reupold-Popp, P, Sandermann Jr H Langebartels C (1994) Elevated levels of superoxide dismutase protect transgenic plants against ozone damage. Bio/Technology 12: 165-168

Weigel D, Nilsson O (1995) A developmental switch sufficient for flower initiation in diverse plants. Nature 377:495-500

Zupan J, Zambryski P (1995) Transfer of T-DNA from *Agrobacterium* to the plant cell. Plant physiol 107: 1041-1047

23
Environmental Risk Assessment of Transgenic Plants: A Case Study of Cucumber Mosaic Virus-Resistant Melon in Japan

Yutaka Tabei

Plant Biotechnology Department, National Institute of Agrobiological Sciences, Kannondai 2-1-2, Tsukuba, Ibaraki 305-8602, Japan

1. Introduction to the Risk Assessment of Genetically Modified Organisms

Genetic engineering is increasingly becoming an essential aspect of plant breeding, and various transgenic crops have been commercialized in the United States, Canada, Australia, and numerous other countries. In 1994, the long-shelf-life transgenic tomato, 'FLAVR SAVR', was launched in the United States market. In 1998, herbicide-tolerant transgenic soybeans comprised about 36% and 55% of the soybean cultivation areas in the United States and Argentina, respectively (James 1998). In the same years, transgenic corn was grown in about 22% of the total cultivation area in United States, while herbicide-tolerant canola in Canada increased to 45% of the total cultivation area (James 1998). The global area used for the cultivation of transgenic crops in 1999 is expected to far exceed that in 1998.

Transgenic crops must be subjected to risk assessment before commercialization. The potential risk of genetically modified organisms (GMOs) on the environment was first pointed out by scientists during the initial stages of the development of transformation technology. Subsequently, at a meeting of scientists from various countries in California, discussions on the regulation of GMOs at the experimental stage led to an agreement that GMOs should be

Air Pollution and Plant Biotechnology
–Prospects for Phytomonitoring and Phytoremediation–
Edited by K. Omasa, H. Saji, S. Youssefian, and N. Kondo
© Springer -Verlag Tokyo 2002

developed and utilized under self-imposed controls (Berg et al. 1974, 1975).

In 1976, the National Institutes of Health (NIH) enacted the first guidelines concerning the experimental use of GMOs (59FR34496-34547, 1994). European countries, including England, France, and Germany, also developed regulatory systems for GMOs. The Organization of Economic Cooperation of Development (OECD) had discussed the need for risk assessment before industrial application of GMOs in the 1980s, and the committee announced their "Recombinant DNA Safety Considerations" to member countries in 1986. In accordance with this recommendation, the United States, Canada, and European countries enacted new items under their existing laws and organized risk assessment systems for the industrial utilization of GMOs. The recommendation of OECD included large-scale industrial applications (tank cultures of microorganisms) and also agricultural/environmental applications (open system applications).

Moreover, OECD has continued to discuss and develop new concepts to evaluate environmental risk and to ensure food safety. In 1993, OECD published a document entitled "Safety Considerations for Biotechnology: Scale-up of Crop Plants" (OECD 1993a) and "Safety Evaluation of Foods Derived by Modern Biotechnology: Concepts and Principles" (OECD 1993b), in which two important principles, familiarity and substantial equivalence, were described. The concept of familiarity can be applied to broad aspects of genetically modified crops from field trials to multiplication of seeds, and that is where environmental risk assessment is conducted, according to experiments on crop breeding, while the environment surrounding the cultivated genetically modified crop plants is also taken into consideration. The other concept, substantial equivalence, used for food safety evaluation, ensures the food safety of transgenic crops by ensuring safety of gene products and comparing the contents of nutrients, toxic substances, and allergens between transgenic crops and original parental lines. These two concepts are now used in many countries as the basis of risk assessment for environmental and food safety.

Subsequently, to promote clear and efficient regulations and to enhance trade of products resulting from biotechnology, the "Harmonization of Regulatory Oversight in Biotechnology" was initiated by OECD in 1997. This working group has continued to publish consensus documents related to the biology of plants, traits introduced into plants, and the biology of microbes to harmonize the baseline of risk assessment in member countries. Moreover, the Internet homepage (http://www.oecd.org/ehs/service.htm) of the working group provides information on the present conditions of risk assessment in member countries and provides links to administrative organizations of these countries.

2. Guidelines for Risk Assessment of Transgenic Crops in Japan

Appropriate risk assessments of the environment, food, and feed are required and conducted depending on the purpose of the transgenic crops. In Japan, four ministries have their own particular set of guidelines. Guidelines of the Ministry of Education, Culture, Sports, Science and Technology Agency (MECSST) (1992) addresses risk assessment of GMOs in laboratory use. On the other hand, the environmental risk of GMOs falls under the guidelines of the Ministry of Agriculture, Forestry and Fisheries (MAFF) (1989), while food safety is evaluated according to the guidelines of the Ministry of Health, Labour and Welfare (MHLW). Feed safety is evaluated by another set of MAFF guidelines.

2.1 Environmental Risk Assessment System for Transgenic Crop Plants

Environmental risk is evaluated on a "step-by-step" and "case-by-case" basis. These steps are classified into four levels, a closed greenhouse, a semiclosed greenhouse, an isolated field, and an open field, each with its own particular requirements. In a closed greenhouse system (1)the windows must be completely closed, air is ventilated through HEPA filters to avoid scattering pollen or dust from genetically modified crop plants to the outside, and the temperature is controlled by an air conditioner. Moreover, the transgenic plants, the soil and pots used to culture these plants, and the waste drainage can be discarded only after sterilization by autoclaving or by other appropriate methods. The semiclosed greenhouse system (2)basically requires the same conditions as the closed greenhouse, except that air is allowed to ventilate between the inside of the greenhouse and outside through open windows that must, however, be covered by mesh (pore size, ~2 mm square) to avoid invasion and scattering of pollinators. In an isolated field system (3)a fence separates the field from ordinary fields. Moreover, an incinerator is required to destroy the transgenic plants after the field trial, and a washing area for the tractors used in the isolated field is necessary. In the final step of cultivation, genetically modified crop plants can be grown in an open field. (4)At this point, there are no restrictive conditions for the cultivation of transgenic plants, although MAFF requests reports of such tests in the first year of cultivation.

By March 2000, 48 applications including 82 breeding lines were confirmed for their biosafety in the environment. Twenty-four of 47 can be cultivated in Japan, and 24 genetically modified crops may be imported from foreign countries. Moreover, official approval for field trial was issued to 76 applications including 128 breeding lines (http://www.s.affrc.go.jp/docs/sentan/index.htm).

2.2 Food Safety Assessment

The MHLW requests food safety assessment data from the developer. The data should include a comparison of the nutritive value of the transgenic and nontransgenic plants, characteristics of the host plant, the sequence and function of the introduced gene, the traits of the vector, and the characteristics of the new protein encoded by the introduced gene. Moreover, the encoded protein must be evaluated for its possible allergen activity and toxicity, such as by intake studies, and comparisons of the sequence to known allergens and toxins. Furthermore, the sensitivity of the protein to heating or artificial gastric juices and artificial intestinal liquids must also be evaluated. Such results would confirm that transgenic and nontransgenic crops are essentially equivalent and that the transgenic plants could be utilized for food. By December 1999, the safety of 29 genetically modified crop plants for food use had been confirmed (http://www.mhlw.go.jp/). Although these 29 genetically modified crop plants evaluated their food safety by guideline, MHLW will conduct food safety assessment under food sanitation low from April 2001.

2.3 Feed Safety Assessment

The safety of transgenic crops as feed must be approved by the Feed Division of the Livestock Industry Department of MAFF. Most evaluated items for feed safety are essentially similar to those for food safety. By December 1999, 25 genetically modified crops had been confirmed for their feed safety.

3. Environmental Risk Assessment of Cucumber Mosaic Virus-Resistant Transgenic Melon

According to the guidelines of MECSST and MAFF, an environmental risk assessment of a transgenic melon plant was conducted from May 1992 to October 1995 at Tsukuba in Japan (Tabei et al. 1994a, 1994b). The environmental impact of the transgenic melon was evaluated mainly for its reproductive traits, weediness, and production of harmful substances in comparison with nontransgenic melon plants. The evaluated items are summarized in Table 1.

3.1 Production of Transgenic Melon Harboring the Coat Protein Gene of Cucumber Mosaic Virus (CMV)

The cDNA of the coat protein gene of strain CMV-Y was cut out of pBR322 and inserted into the binary vector, pBI121 (Clontech, USA), in place of the GUS gene. This reconstructed binary vector was then introduced, by triparental mating, into *Agrobacterium* LBA4404 (Clontech), which was then used for transformation

Table 1. Items of environmental risk evaluation of genetically modified melon

Evaluated items	Closed greenhouse	Semiclosed greenhouse	Isolated field
1. Confirmation and expression of the introduced CMV coat protein gene (CMV-CP) and NPT-II			
1) Expression of NPT-II [a]	●		
2) Confirmation of CMV-CP [a]	●		
3) Expression of CMV-CP [a]	●		
4) Resistance to CMV [a]	●		●
2. Morphological and growth characteristics			
1) Morphological characteristics	●		
2) Fruit maturation period		●	
3. Reproductive characteristics			
1) Pollen form	●		
2) Pollen fertility	●		
3) Pollen longevity	●		
4) Pollen dispersal by wind	●		
5) Pollen dispersal by insect			●
6) Kinds of flower-visiting insects			●
4. Evaluation of harmful effects to other plants			
1) Phenolic acids produced in leaves and stems		●	
2) Phenolic acids released from root	●		
3) Production of volatile compounds	●		
4) Influence of dry powder to other plants		●	
5) Influence on subsequent crop		●	
5. Weediness			
1) Plant body			●
2) Seeds in fruits put on the ground			●
3) Seeds in fruits buried under ground			●
6. Influence on soil microflora		●	●
7. Residual *Agrobacterium* for transformation	●		

a Results reported by Yoshioka et al. 1992, 1993.

of commercial melon (cv. Prince, Sakata Seed Co., Japan; Yoshioka et al. 1992). As a result, nine independent transgenic melon plants were regenerated, and the melon line (designated M5) with the highest resistance to CMV was selected after artificial infection tests with CMV (Yoshioka et al. 1993).

3.2 Environmental Risk Assessment of CMV-Resistant Melon

3.2.1 Morphological Characteristics and Fruit Maturation Periods Between Transgenic and Nontransgenic Melon

Sixteen morphological characteristics of the seed and seedling row (Ministry of Agriculture, Forestry and Fisheries 1977) together with the fruit maturation period, as a growth characteristic, were compared. The morphological characteristics of the transgenic plants did not differ from those of nontransgenic plants. Similarly the fruit maturation period of the transformed plants (44.7 days) did not differ significantly from that of the nontransformed plants (43.0 days).

3.2.2 Pollen Form and Fertility

The size, form, and fertility of pollen from transgenic and nontransgenic melon plants were compared. The size of all pollen measured was about 50～60 μm, while pollen fertility was approximately about 97%. The longevity of viable pollen from transgenic and nontransgenic melon plants was examined on sunny days in May 1992. Pollen was collected from plants at 0930, 1130, 1330, 1530, and 1730 and sown onto pollen germination medium. Most pollen collected at 0930 germinated, but the germination frequency of pollen collected at 1130 was reduced. A few pollen grains collected at 1330 germinated, whereas pollen collected at 1530 did not germinate at all. Therefore, the viability of pollen from both transgenic and nontransgenic melon plants appeared to last until about 1330 in a closed greenhouse on a fine day. These results demonstrate that pollen size, fertility, and longevity did not differ significantly between transgenic and nontransgenic melon plants.

3.2.3 Wind Pollination of Transgenic and Nontransgenic Melon

The wind pollination of transgenic and nontransgenic melon plants was investigated by subjecting the plants to artificial wind generated by an electric fan in a closed greenhouse. Vessels containing pollen germination medium were placed at distances of 0, 5, 10, 15, 50, 100, 200, and 300 cm from the plants at which wind was continuously blown from 1000 to 1530 at a velocity of 0.5～4.0 m/s. Pollen from neither transgenic nor nontransgenic plants could be detected on germination medium at any distance from the plants, confirming that pollen of melon is not dispersed by wind but generally only by insects.

3.2.4 Pollen Dispersal Under Natural Conditions

Recipient melon plants that did not harbor either the kanamycin resistance gene (NPT-II) or the *Fusarium* wilt resistance gene were planted around donor melon plants (transgenic melon [MS] plants and the nontransgenic melon plant cv. 'Ooi' possessing *Fusarium* wilt resistance) (Fig. 1). *Fusarium*-resistant progenies of the recipients were observed at a distance of 15 m from the donor whereas progenies harboring the kanamycin resistance gene were observed at a distance of 10 m from the donor. Because cv. 'Ooi' harbors the homozygous *Fusarium* wilt resistance gene and transgenic melon harbors the heterozygous NPT-II gene, more progenies exhibiting *Fusarium* wilt resistance were observed than those harboring the NPT-II gene at distances of 5, 10, and 15 m. However, progenies resistant to *Fusarium*

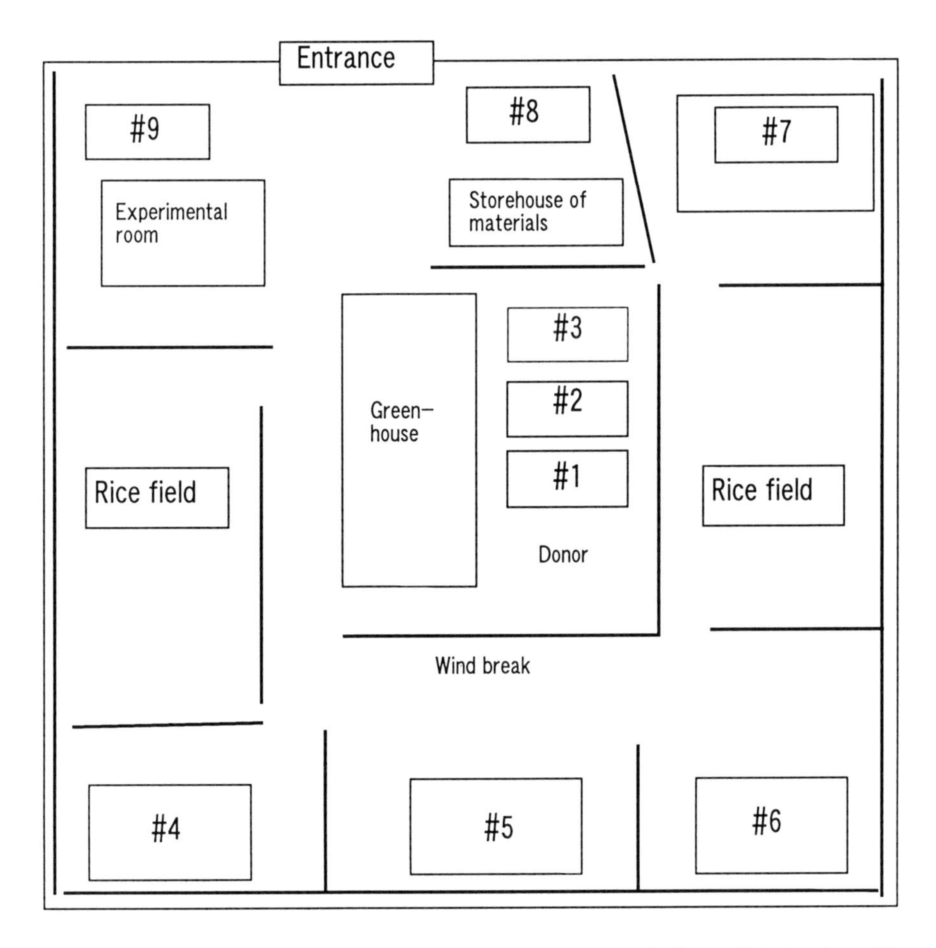

Fig. 1. Isolated field for environmental risk assessment of genetically modified melon. #1-3, 4-6, 8, 9, recipient melon plants of dispersed pollen; #7, experimental field for virus resistance; donor, genetically modified melon plants and non-GMO and *Fusarium*-resistant melon plants

Table 2. Comparison of pollen dispersal between transgenic and nontransgenic melon plants

	Distance from Donor (m)	Number of progenies with NPT-II gene	Number of progenies exhibiting resistance to *Fusarium* wilt
#1	5	5/100 [a]	12/195 [b]
#2	10	3/100	7/192
#3	15	0/100	2/198
#4	25	0/100	0/194
#5	25	0/100	0/194
#6	25	0/100	0/194
#8	30	0/100	0/194
#9	40	0/100	0/194
Cont.#A		0/30	0/197
Cont.#B		12/30	0/196
Cont.#C		0/32	30/30

a Number of progenies with NPT-II gene/total number of progenies examined.
b Number of progenies exhibiting resistance to *Fusarium* wilt/total number of progenies examined.
Cont.#A, progenies of recipient melon plants by self-pollination; Cont.#B, progenies of genetically modified melon plants (M5) by self pollination; Cont.#C, progenies of non-GMO melon plants by self pollination.

wilt as those with the NPT-II gene were not detected at a distance of 25 m from the donor (Table 2). These results indicate that the extent of pollen dispersal in transgenic and non-transgenic melon plants was not significantly different.

3.2.5 Overwintering of Melon

If transgenic plants or their seeds possess an improved/enhanced overwintering trait compared to nontransgenic melon plants, then the transgenic plants would be expected to have more chances of survival and have weed properties in the environment compared to nontransgenic melon plants. Overwintering was examined by the following two experiments from September 1993. In one experiment, both transgenic and nontransgenic plants were cultivated in an unheated greenhouse located on the isolated field. In the other experiment, fruits obtained by artificial pollination were either placed on the ground or buried in the soil of the isolated field. All transgenic and nontransgenic plants were killed as a result of the low-temperature conditions of the greenhouse by the end of December 1993. Germination of seeds was observed from fruits left on the ground following their decomposition; however, these seedlings were also killed by the low-

Table 3. Influence of soil mixed with dried powders of transgenic or nontransgenic melon plants on the germination and growth of cabbage seedlings

Soil additive	Germination (%)	Root length (cm)	Fresh weight of seedling (g)
Transgenic melon	90.0±3.3	7.26±.0.77	10.31±0.21
Nontransgenic melon	95.7±2.3	7.57±0.95	11.28±1.38

Thirty seeds of cabbage were sown in each pot, and three pots were used for each treatment.

temperature conditions before they had set fruit. Although germinated seedlings were not observed from fruits buried in the ground by next early spring, seedlings did emerge from these fruits in the following spring. These observations suggest that melon seeds from fruits buried in the ground could overwinter but that whole plants are not in capable of overwintering. The results for overwintering were not significantly different between transgenic and nontransgenic melon.

3.2.6 Possible Harmful Effects of Transgenic Melon on the Environment

To examine the possibility that the transgenic melon could have harmful effects on the environment, the following compounds were compared between the transgenic melon and nontransgenic melon plants: (1) phenolic acids, generally considered as allelochemical substances produced in the plant body and secreted from the root, measured by high performance liquid chromatography; (2) production of volatile compounds, released from the plant in the atmosphere, analyzed by gas chromatography; and (3) germination ratio, root length, and fresh weight of cabbage grown either in soil used previously for the cultivation of either transgenic or nontransgenic melon plants or in soil mixed with dry powders prepared from these transgenic or nontransgenic melon plants. Specific phenolic acids and volatile compounds were not detected from transgenic melon plants. Furthermore, there were no differences in the germination ratio, root length, or fresh weight of cabbage grown in soil cultivated with either transgenic or nontransgenic melon (Table 3). These results demonstrate that the transgenic melon plants did not produce any specific products that could adversely affect the environment or other plants.

3.2.7 Influence on Soil Microflora

The effect of release of transgenic melon plants on soil microflora was investigated in a greenhouse located in an isolated field. The numbers of bacteria, actinomycetes, and fungi in the soil were monitored at different periods, namely, before transplantation of the melon plants (May 7), at flowering (July 8), and after harvest of the crop (September 3, 1993). Microflora were compared between the soils in which the transgenic and nontransgenic melon plants had been cultivated. Counts of bacteria, actinomycetes, and fungi in the soil cultivated with the transgenic melon plants were slightly higher than in soil cultivated with nontransgenic melon plants by July 8; however, these differences were not statistically significant. It was thus concluded that cultivation of the transgenic melon did not adversely influence the numbers of soil microflora.

3.2.8 Residues of Agrobacterium Tumefaciens Used for Transformation

Residual *Agrobacterium tumefaciens,* strain LBA4404, in or on the plant body was examined in a closed greenhouse. Microorganisms were isolated either by shaking the plants or homogenizing the transgenic or nontransgenic melon plants in sterile distilled water and plating the homogenates on YEB media containing certain selective antibiotics. Strain LBA4404 of *A. tumefaciens* could not be detected either on the surface or in the tissues of the transgenic plants.

3.3 Conclusion of Environmental Risk Assessment of Transgenic Melon

These results provided evidence that there were no significant differences between transgenic melon plants and nontransgenic melon plants as to reproductive traits, weediness, or their impact on other crops and soil microflora. As a result, this CMV-resistant transgenic melon was approved from MAFF in 1966 for cultivation in an ordinary field.

4. Conclusion

Environmental risk assessment of genetically modified organisms (GMO) in practical stage has been considered at OECD since 1983. As a result of the consideration, OECD published two reports, "Safety Considerations for Biotechnology" (OECD 1986) and "Safety Consideration for Biotechnology; Scale-up of Crop Plant" (OECD 1993a). In "Safety Consideration for Biotechnology; Scale-up of Crop Plant", gene flow and weediness are pointed out as the most important evaluation items when influence of transgenic crops on environment is evaluated. According to these reports, evaluation items, such as reproductive trait related to gene flow, morphological traits, weediness, and influence on other living organisms, of virus resistance melon plants should be subjected for risk assessment.

Until December 1999, 47 applications including 82 breeding lines for nonregulated status were confirmed for their biosafety in the environment in Japan. All applications reviewed for environmental risk include results for reproductive and other traits described earlier.

Currently, commercialized transgenic crops are mainly improved herbicide tolerant, insect resistant, virus resistant, and with long-shelf-life. As these traits were controlled by expression of one gene and gene products from the introduced gene are known not to affect metabolism of original plant, it was estimated that traits unpredicted might not appear in transgenic plants without modified traits related to the introduced gene. According to the present environmental risk assessment, if each result of the evaluation items was not different between transgenic and original plants, it could be concluded that the growth (cultivation) of transgenic plants does not have a greater influence on environment as compared to the influence of cultivation of the original plants.

However, recently, many kinds of transgenic plants have been developed for different purposes. The current principle for risk assessment described earlier may not apply to some newly developed transgenic plants. For example, transgenic plants for bioremediation are actually useful for recovering environmental contamination; however, this is an impact on the environment from a different point of view. In this case, new principles for environmental risk assessment or risk management will be necessary.

By 2000, 15 years of first field trial and 7 years of commercialization of genetically modified crops have already passed in the U.S., and other countries including Canada and EU countries. Japan also has accumulated knowledge through many field trials. Although there were several reports to imply adverse effects of transgenic plant on environment (Losey et al. 1999) and food (Ewen and Pusztai 1999), it seems that these reports could not prove actual harmful impacts on the environment and human health from their results of experiments, whereas results of field trials of GMOs suggested that harmful impacts of GMOs on environment had not been reported until now (The International Symposium on Biosafety Results of Field Tests of Genetically Modified Plants and Microorganisms 1994, 1997). It can be concluded that risk assessments systems for environment, food, and feed have been done well.

However, some people feel uncertain about whether foods made from GMOs were safe and the cultivation of transgenic plants had some impact on the environment. Some radical consumer and environmental groups object to cultivation of transgenic plants, consumption of genetically modified foods, and research for developing novel crop plants by genetic engineering. Generally speaking, most consumers who objected may be getting incorrect and biased information about the risk assessment and the safety and merits of GMOs. To promote public acceptance, providing accurate information and communicating with the public will become more important. MAFF had started a project for building public acceptance of GMOs from 1996. In activities of the project, MAFF had organized several kinds of seminars and symposium, published information brochures, and opened a Web site (Tabei 1999a, 1999b).

In any case, it is crucial that reliable risk assessment based on science should be implemented. This research can contribute to promoting public acceptance for utilization of GMOs.

References

Berg P, Baltimore D, Boyer DW, Cohen SN, Davis RW, Hogness DS, Nathans D, Roblin R, Watson JD, Weissman S, Zinder ND (1974) Potential biohazards of recombinant DNA molecules. Science 185:303

Berg P, Baltimore D, Brenner S, Roblin RO, Singer MF, (1975) Summary statement of the Asilomar conference on recombinant DNA molecules. Proc Natl Acad Sci USA 72:1981-1984

Ewen SWB, Pusztai A (1999) Effect of diets containing genetically modified potatoes expressing *Galanthus nivalis* lectin on rat small intestine. Lancet 354:1353-1354

James C. (1998) Global review of commercialized transgenic crops. ISAAA brief no. 8. ISAAA, Ithaca, NY

Ministry of Agriculture, Forestry and Fisheries (MAFF) (1977) The seed and seedling law. MAFF, Tokyo

Ministry of Agriculture, Forestry and Fisheries (MAFF) (1989) Guidelines for Application of Recombinant DNA Organisms in Agriculture, Forestry and Fisheries, the Food Industry and Other Related Industries. MAFF, Tokyo, pp 1-67

Losey L, Rayor L, Carter M (1999) Transgenic pollen harms monarch larvae. Nature 399:214

OECD (1986) Recombinant DNA Safety Considerations. OECD Publication 93 86 02 1

OECD (1993a) Safety considerations for biotechnology; scale-up of crop plants. OECD publication 93 93 08 1

OECD (1993b) Safety evaluation of foods derived by modern biotechnology: concepts and principles. OECD publication 93 93 04 1

Proceeding of the 3rd International Symposium on Biosafety Results of Field Tests of Genetically Modified Plants and Microorganisms (1994), Monterey, CA, USA

Science and Technology Agency (STA) (1992) Guideline for Recombinant DNA (rDNA) Experiments. STA, Tokyo, pp 1-245

Tabei Y (1999a) Environmental risk assessment of transgenic melon in Japan. Plant Biotechnol 16:65-68

Tabei Y (1999b) Addressing public acceptance issues for biotechnology: experiences from Japan In: Cohen JI (ed) Managing agricultural biotechnology: addressing research program needs and policy implications. CABI Publishing, UK, 174-183

Tabei Y, Oosawa K, Nishimura S, Hanada K, Yoshioka K, Fujisawa I, Nakajima K (1994a) Environmental risk evaluation of the transgenic melon with coat protein gene of cucumber mosaic virus in a closed and a semi-closed greenhouses (I). Breed Sci 44:101-105

Tabei Y, Oosawa K, Nishimura S, Watanabe S, Tsuchiya K, Yoshioka K, Fujisawa I, Nakajima K (1994b) Environmental risk evaluation of the transgenic melon with coat protein gene of cucumber mosaic virus in a closed and a semi-closed greenhouses (II). Breed Sci 44:207-211

The 4th International Symposium on Biosafety Results of Field Tests of Genetically Modified Plants and Microorganisms (1997) Tsukuba, Japan

Yoshioka K, Hanada K, Nakazaki Y, Yakuwa T, Minobe Y, Oosawa K (1992) Successful transfer of the cucumber mosaic virus coat protein gene to *Cucumis melo* L. Jpn J Breed 42:227-285

Yoshioka K, Hanada K, Harada T, Minobe Y, Oosawa K (1993) Virus resistance in transgenic melons that express coat protein gene of cucumber mosaic virus and in these progenies. Jpn J Breed 43:629-634

Index

O

P

Q

R

S

T